Progress in Mathematics
Vol. 57

Edited by
J. Coates and
S. Helgason

Birkhäuser
Boston · Basel · Stuttgart

Hans Riesel

Prime Numbers and Computer Methods for Factorization

1985

Birkhäuser
Boston · Basel · Stuttgart

Hans Riesel
Måbärsstigen 2
S–162 39 Vällingby
(Sweden)

This book was produced from a camera-ready manuscript
prepared by the author using the
TEX typesetting system.

Library of Congress Cataloging in Publication Data

Riesel, Hans, 1929–
  Prime numbers and computer methods for factorization.

  (Progress in mathematics ; vol. 57)
  1. Numbers, Prime – – Data processing.   2. Factorization
(Mathematics) – – Data processing.   I. Title.   II. Series:
Progress in mathematics (Boston, Mass.) ; vol. 57.
QA246.R54   1985   512.7   85–5971
ISBN 0-8176-3291-3

CIP-Kurztitelaufnahme der Deutschen Bibliothek

Riesel, Hans:
Prime numbers and computer methods for
factorization / Hans Riesel. — Boston ;
Basel ; Stuttgart : Birkhäuser, 1985.
  (Progress in mathematics ; 57)
  ISBN 3-7643-3291-3 (Stuttgart . . .)
  ISBN 0-8176-3291-3 (Boston)

NE: GT

© 1985 Birkhäuser Boston, Inc.
Printed in Germany
ISBN 0-8176-3291-3
ISBN 3-7643-3291-3

PREFACE

In this book the author treats four fundamental and apparently simple problems. They are: the number of primes below a given limit, the approximate number of primes, the recognition of prime numbers and the factorization of large numbers. A chapter on the details of the distribution of the primes is included as well as a short description of a recent application of prime numbers, the so-called RSA public-key cryptosystem. The author is also giving explicit algorithms and computer programs. Whilst not claiming completeness, the author has tried to give all important results known, including the latest discoveries. The use of computers has in this area promoted a development which has enormously enlarged the wealth of results known and that has made many older works and tables obsolete.

As is often the case in number theory, the problems posed are easy to understand but the solutions are theoretically advanced. Since this text is aimed at the mathematically inclined layman, as well as at the more advanced student, not all of the proofs of the results given in this book are shown. Bibliographical references in these cases serve those readers who wish to probe deeper. References to recent original works are also given for those who wish to pursue some topic further.

Since number theory is seldom taught in basic mathematics courses, the author has appended six sections containing all the algebra and number theory required for the main body of the book. There are also two sections on multiple precision computations and, finally, one section on Stieltjes integrals. This organization of the subject-matter has been chosen in order not to disrupt the reader's line of thought by long digressions into other areas. It is also the author's hope that the text, organized in this way, becomes more readable for specialists in the field. Any reader who gets stuck in the main text should first consult the appropriate appendix, and then try to continue.

The six chapters of the main text are quite independent of each other, and need not be read in order.

For those readers who have a computer (or even a programmable calculator) available, computer programs have been provided for many of the methods described. In order to achieve a wide understanding, these programs are written in the high-level programming language PASCAL. With this choice the author hopes that most readers will be able to translate the programs into the language of the computer they use with reasonable effort.

At the end of the book a large amount of results are collected in the form of tables, originating partly from the author's own work in this field. All tables have been verified and up-dated as far as possible. Also in connection with the tables, bibliographical references are given to recent or to more extensive work in the corresponding area.

The text is an up-dated version of an earlier book by the same author: "En bok om primtal," existing in Swedish only.

The author is very much indebted to Richard Brent, Arne Fransén, Gunnar Hellström, Hans Karlgren, D. H. Lehmer, Thorkil Naur, Andrzej Schinzel, Bob Vaughan and many others for their reading of the manuscript and for suggesting many improvements. Thanks are also due to Beatrice Frock for revising the English, and to the late Ken Clements for reading and correcting one of the last versions of the manuscript.

The author wishes you a pleasant reading!

Stockholm, February 1985

Hans Riesel

CONTENTS

Chapter 3. Subtleties in the Distribution of Primes

Chapter 4. The Recognition of Primes

Chapter 5. Factorization

CONTENTS

Appendix 1. Basic Concepts in Higher Algebra

Appendix 2. Basic Concepts in Higher Arithmetic

Appendix 3. Quadratic Residues

Appendix 4. The Arithmetic of Quadratic Fields

Appendix 5. Continued Fractions

CONTENTS

CONTENTS

Tables

Index

NOTATIONS

Symbol	Meaning

| Symbol | Meaning |

$f(x) \approx g(x)$ $f(x)$ is approximately equal to $g(x)$

$f(x) \sim g(x)$ $f(x)$ is asymptotically equal to $g(x)$, meaning that $\lim f(x)/g(x) = 1$, usually as $x \to \infty$

$[a, b]$ closed interval: all x in $a \le x \le b$

$\overset{c}{=}, \ \overset{c}{\sim}, \ \overset{c}{<}$ conjectured relation

$\bar{a}$ conjugate number: if $a = p + q\sqrt{D}$, then $\bar{a} = p - q\sqrt{D}$

$a \equiv b \pmod{n}$ a is congruent to b modulus n

$a \not\equiv b \pmod{n}$ a is not congruent to b modulus n

$a + \dfrac{b|}{|c} + \dfrac{d|}{|e}$ continued fraction $= a + \dfrac{b}{c + \dfrac{d}{e}}$

$a|b$ a divides b

$a \nmid b$ a does not divide b

$p^\alpha \| n$ highest power of p dividing n, i.e. $p^\alpha | n$, but $p^{\alpha+1} \nmid n$

$\varphi(n)$ Euler's totient function $= \prod p_i^{\alpha_i - 1}(p_i - 1)$, if $n = \prod p_i^{\alpha_i}$

$\Phi_n(x)$ the nth cyclotomic polynomial, having degree $\varphi(n)$

γ Euler's constant $= 0.57721566\,49015328\ldots$

(a, b) greatest common divisor of a and b

$\text{G.C.D.}\,(a, b)$ greatest common divisor of a and b

M_N group of primitive residue classes (mod N)

$[x]$ integer part of x: $[\pi] = 3$, $[-\pi] = -4$

$[a, b]$ the interval $a \le x \le b$

$\left(\dfrac{a}{N} \right)$ Jacobi's symbol, defined if $(a, N) = 1$

$\lambda(n)$ Carmichael's function $= [\lambda(p_i^{\alpha_i})]_i$ if $n = \prod p_i^{\alpha_i}$

$[a, b]$ least common multiple of a and b

$\text{L.C.M.}\,[a, b]$ least common multiple of a and b

Symbol	Meaning
L.H.S.	left hand side
$\left(\dfrac{a}{p}\right)$	Legendres symbol, defined if p is an odd prime
li x	the logarithmic integral of x
$\ln x$	$\log_e x$ $(e = 2.71828182\,84590452\ldots)$
$\log x$	logarithm to unspecified base
$\mu(n)$	Möbius' function
$a \gg b$	a is much larger than b
$a \ll b$	a is much smaller than b
#	"number sign": number of elements in a set
$\omega(N)$	the number of different prime factors of N
$\Omega(N)$	the total number of prime factors of N
$\Omega(f(x))$	"big omega": greater than $Cf(x)$ for some constant $C > 0$ and an infinitude of x, usually as $x \to \infty$
$O(f(x))$	"big ordo": less than $Cf(x)$ for some constant $C > 0$, usually as $x \to \infty$
$o(f(x))$	"little ordo": less than $\epsilon f(x)$, where $\epsilon \to 0$, usually as $x \to \infty$
p	a prime number
$P(N)$	the largest prime factor of N
$P_k(N)$	the kth largest prime factor of N
$\pi(x)$	number of primes $\leq x$
$\prod$	product symbol: $\prod_{i=1}^{n} a_i = a_1 \cdot a_2 \cdot a_3 \cdot \ldots \cdot a_n$
$\Re(z)$	real part x of complex number $z = x + iy$
R.H.S.	right hand side
$\sum$	summation symbol: $\sum_{i=1}^{n} a_i = a_1 + a_2 + a_3 + \cdots + a_n$
$\varsigma(s)$	zeta-function of Riemann, $\varsigma(s) = \sum_{n=1}^{\infty} n^{-s}$
$\chi(A)$	group character
10(5)100	the numbers $10, 15, 20, \ldots, 100$

CHAPTER 1

THE NUMBER OF PRIMES BELOW A GIVEN LIMIT

What Is a Prime Number?

Consider the positive integers $1, 2, 3, 4, \ldots$ Among these there are *composite numbers* and *primes*. A composite number is a product of several factors $\neq 1$, such as $15 = 3 \cdot 5$; or $16 = 2 \cdot 8$. A prime p is characterized by the fact that its only possible factorization apart from the order of the factors is $p = 1 \cdot p$. Every composite number can be written as a product of *primes*, such as $16 = 2 \cdot 2 \cdot 2 \cdot 2$.—Now, what can we say about the integer 1? Is it a prime or a composite? Since 1 has the only possible factorization $1 \cdot 1$ we could agree that it is a prime. We might also consider the product $1 \cdot p$ as a product of two primes; somewhat awkward for a *prime number p*.—The dilemma is solved if we adopt the convention of classifying the number 1 as neither prime nor composite. We shall call the number 1 a *unit*. The positive integers may thus be divided into:

1. The unit 1.

2. The prime numbers $2, 3, 5, 7, 11, 13, 17, 19, 23, \ldots$

3. The composite numbers $4, 6, 8, 9, 10, 12, 14, 15, 16, \ldots$

Frequently, it is of interest to study not only the *positive* integers, but *all* integers:

$$\ldots -4, -3, -2, -1, 0, 1, 2, 3, 4, \ldots$$

The numbers n and $-n$ are called *associated* numbers. The number 0, which is divisible by any integer without remainder (the quotient always being 0), is of a special kind. The integers are thus categorised as:

1. The number 0.

2. The units -1 and 1.

3. The primes $\ldots -7, -5, -3, -2, 2, 3, 5, 7, \ldots$

4. The composite numbers $\ldots -9, -8, -6, -4, 4, 6, 8, 9, \ldots$

Generally, when only *factorization* of numbers is of interest, associated numbers may be considered as equivalent, therefore in this case the two different classifications of the integers given above can be considered equivalent (if we neglect the number 0).—We shall often find that the numbers 0 and 1 have special properties in the theory of numbers which will necessitate some additional explanation, just as in the above case.

The Fundamental Theorem of Arithmetic

When we were taught at school how to find the *least common denominator* of several fractions, we were actually using *the fundamental theorem of arithmetic*. It is a theorem which well illustrates the fundamental role of the prime numbers. The theorem states that every positive integer n can be written as a product of primes, and in one way only:

$$n = p_1^{\alpha_1} p_2^{\alpha_2} \ldots p_s^{\alpha_s} = \prod_{i=1}^{s} p_i^{\alpha_i} \tag{1.1}$$

In order that this decomposition be unique, we have to consider, as identical, decompositions that differ only in the order of the prime factors, and we must also refrain from using associated factors.—Having done much arithmetic, we have become so used to this theorem that we regard it as self-evident, and not necessary to prove. This is, however, not at all the case, and the reader will find a proof on p. 263 in Appendix 2. In Appendix 3 an example is given which shows why the proof is a logical necessity. On p. 298 an arithmetic system is constructed which resembles the ordinary integers in many ways, *but in which the fundamental theorem of arithmetic does not hold.*—As a matter of fact, the logical necessity for a proof of the fundamental theorem was recognized by Euclid, who gave the proof of the almost equivalent Theorem A2.1 on p. 262.

Which Numbers Are Primes? The Sieve of Eratosthenes

All the primes in a given interval can be found by a sieve method invented by Eratosthenes. This method deletes all the composite numbers in the interval, leaving the primes. How can we find all the composites in an interval? To check if a number n is a multiple of p, divide n by p and see whether the division leaves no remainder. This so-called trial division method for finding primes is much too laborious for the construction of

2

prime tables when applied to each number individually, but it turns out that *only one division by each prime p suffices* for the whole interval. To see why this is so, suppose that the first number in the interval is m, and that $m - 1 = p \cdot q + r$, with $0 \leq r < p$. Then, obviously, the numbers

$$m - 1 - r + p, \; m - 1 - r + 2p, \; m - 1 - r + 3p, \; \ldots$$

are precisely those multiples of p which are $\geq m$. Thus all multiples of p can be identified by one single division by p and the number of divisions performed for each number examined will be reduced accordingly. This saves much labour, particularly if the interval is long.

Which values of p need to be tested as factors of any given number n? It obviously suffices to test all values of $p \leq \sqrt{n}$, since a composite number n cannot have two factors, both $> \sqrt{n}$, because the product of these factors would then exceed n. Thus, if all composite numbers in an interval $m \leq x \leq n$ are to be sieved, it will suffice to cross out the multiples of all primes $\leq \sqrt{n}$ (with the possible exception of some primes in the beginning of the interval, a case which may occur if $m \leq \sqrt{n}$). Using these principles, we shall show how to construct a table of primes between 100 and 200. Commence by enumerating all integers from 100 to 200: $100, 101, \ldots, 199, 200$. The multiples to be crossed out are the multiples of all primes $\leq \sqrt{200}$, i.e. of the primes $p = 2, 3, 5, 7, 11$ and 13. First, strike out the multiples of 2, all even numbers, leaving the odd numbers:

$$101, \; 103, \; 105, \ldots, 195, \; 197, \; 199.$$

Next delete all multiples of 3. Since $99 = 3 \cdot 33 + 0$, the smallest multiple of 3 in the interval is $99 + 3 = 102$. By counting 3 steps at a time, starting from 102, we find all multiples of 3 in the interval: 102, 105, 108, 111,... When we are working with paper and pencil, we might strike out the multiples of 5 in the same round, since these are easy to recognize in our decimal number system. After this step the following numbers remain:

101, 103, 107, 109, 113, 119, 121, 127, 131, 133,
137, 139, 143, 149, 151, 157, 161, 163, 167, 169,
173, 179, 181, 187, 191, 193, 197 and 199.

Next, we must locate the multiples of 7, 11 and 13. These are 119, 133, 161 (the remaining multiples of 7), 121, 143, 187 (the remaining multiples of 11) and 169 (the remaining multiple of 13). The 21 numbers now remaining,

101, 103, 107, 109, 113, 127, 131, 137, 139, 149, 151,
157, 163, 167, 173, 179, 181, 191, 193, 197 and 199,

3

are the primes between 100 and 200. This example demonstrates the fact that it requires approximately the same amount of work to construct a prime table as it would take to devise a table for the smallest factor of each number, for a given interval. After having made some simple modifications increasing the efficiency of the sieve of Eratosthenes described above, D. N. Lehmer at the beginning of this century compiled and in 1909 published the largest factor table [1] and the largest prime table [2] ever published as books. These cover the integers up to 10,017,000. It is unlikely that more extensive factor tables or prime tables will ever be published as books since, with a computer, the factorization of a small number is much faster than looking it up in a table.—Before the era of computers a considerable amount of work was invested in the construction of prime and factor tables. Most famous of these is probably Kulik's manuscript, containing a table of the smallest factor of each number up to 10^8. The manuscript, of which parts have been lost was never published. Before computers became readily available, a lot of effort was spent on big prime table projects. For example, in 1956–57 the author of this book compiled a prime table for the interval 10,000,000–70,000,000. The computations were performed on the vacuum-tube computer BESK, which had an addition time of 56 micro-seconds. To sieve an interval containing 48,000 integers took 3 minutes of computing time. The output from the BESK was a teletype paper tape which had to be re-read and printed by special equipment, consisting of a paper tape reader and an electric typewriter.

Finally, in 1959, C. L. Baker and F. J. Gruenberger published a micro-card edition of a table containing the first six million primes [3]. This table comprises all the primes below 104,395,289.—In addition, there exist printed tables for the primes in certain intervals, such as the 1000th million $[999 \cdot 10^6, 10^9]$, see [4], and the intervals $[10^n, 10^n + 150,000]$ for $n = 8$, 9, ..., 15, see [5].

General Remarks Concerning Computer Programs

For all numerical work on primes it is essential to have prime tables accessible in your computer. This may be achieved by the computer generating the primes by means of the sieve of Eratosthenes, and then either immediately performing the computations or statistics required on the primes, or compressing the prime table generated in a way suitable for storing in the computer and then performing the necessary computation or statistics.

4

We shall provide a number of *algorithms* in this book (i.e. schemes for calculating desired quantities). These will be presented in one or more of the following forms:

1. Verbal descriptions.

2. Mathematical formulas.

3. Flow-chart diagrams.

4. Computer programs in PASCAL.

The choice of the high-level programming language PASCAL has been made because this language is simple, reasonably well-known, and close to the mathematical formulas which are to be transformed into computer programs. There are, however, some minor differences between the various versions of PASCAL in use, in particular concerning the input and output routines, hopefully this will not cause any great obstacle. The dialect of PASCAL used in this book is called Hedrick's PASCAL. It is the author's hope that readers trained in computer programming will be able to transform the PASCAL programs presented in this book into the language of their choice without undue difficulty.

A Sieve Program

For a sieve to operate on the interval $[m, n]$, we require all the primes up to p_s, the largest prime $\leq [\sqrt{n}]$, where (as usual) $[x]$ denotes the integer part of x. We assume that these have already been computed and stored in a one-dimensional array, Prime[1:s], with Prime[1]:=2, Prime[2]:=3, Prime[3]:=5,..., Prime[s]:= the sth prime. Now, in order to simplify the computer program, we shall work only with the *odd* integers. Without loss of generality, let both m and n be *odd* (if this is not the case to begin with, we may change m and n during the input part of the program, if we really want the program to work also for *even* values of m and/or n). Next we give each of the $(n - m + 2)/2$ *odd* integers between m and n inclusive a corresponding storage location in the fast access memory of the computer. This may be a computer word, a byte, or even a bit, depending on the level on which you are able to handle data in your computer. We call the array formed by these storage locations Table[1:(n-m+2)/2], where $(n - m + 2)/2$ is the number of elements in the array (its dimension). Suppose that these storage locations are filled with ZEROs to begin with. Each time we find a multiple of some prime $\leq \sqrt{m}$, we shall put the number

5

1 into the corresponding storage location. Here is a PASCAL program to do this for an interval $[m, n]$ below 1000:

```
PROGRAM Eratosthenes
{Prints a prime table between odd m and n < 1000}
(Input,Output);
LABEL 1;
CONST imax=11; jmax=500;
VAR Prime : ARRAY[1..imax] OF INTEGER;
    Table : ARRAY[1..jmax] OF INTEGER;
  m,n,p,p2,q,i,j,start,stop : INTEGER;

BEGIN
Prime[1]:=2;  Prime[2]:=3;  Prime[3]:=5; Prime[4]:=7;
Prime[5]:=11; Prime[6]:=13;  Prime[7]:=17; Prime[8]:=19;
Prime[9]:=23; Prime[10]:=29; Prime[11]:=31;
write('Input m and n: '); read(m,n);
stop:=(n-m+2) DIV 2;
FOR i:=2 TO stop DO
  BEGIN p:=Prime[i]; p2:=p*p;
    IF p2 < m THEN
      BEGIN q:=2*p; start:=(m DIV q)*q+p;
        {start is the odd multiple of p
          which is closest to m}
        IF start < m THEN start:=start+q
      END
        ELSE start:=p2;
    IF p2 > n THEN GOTO 1 ELSE
        BEGIN j:=(start-m) DIV 2 + 1;
          WHILE j <= stop DO
            BEGIN {Here the odd multiples of p are
              marked:} Table[j]:=1; j:=j+p
            END
        END
  END;
1: {When arriving at this point, the elements of Table
    corresponding to the primes are 0, the rest 1}
  FOR i:=1 TO stop DO
    IF Table[i]=0 THEN write(m+2*i-2:4)
    {Here the prime table generated is printed out}
END.
```

The table of small primes required by this program is generated by the preparatory statements at the beginning. When dealing with a larger interval it is practical to produce also the table of small primes required by a sieving program rather than defining them arithmetically. The only necessary augmentation of the program in that case involves starting at the beginning of the generated table, searching for the next entry $= 0$, and converting this to the value of the next prime. Since it is convenient to have computations of general interest, such as sieving with the primes, readily available as computer procedures, we give the modified version of the above sieve program in the form of a PASCAL procedure primes which generates the odd primes in the interval $[3, n]$, n odd. Since it is the first time we show a PASCAL procedure, we provide a complete program Primegenerator containing this procedure, to enable the reader to understand the context in which a PASCAL procedure should be written. The program reads an odd integer $m < 1000$, and generates and prints all odd primes below m:

```
PROGRAM Primegenerator
{Generates and prints the primes up to m < 1000, m odd}
(Input,Output);
CONST n=500;
TYPE vector=ARRAY[0..n] OF INTEGER;
VAR Table : vector; i,j,m : INTEGER;

PROCEDURE primes(n : INTEGER; VAR Prime : vector);
LABEL  1;
VAR i,j,k,p,p2,stop : INTEGER;
BEGIN
  stop:=(n-1) DIV 2; j:=1;
1:  FOR k:=j TO stop DO
      IF Prime[k]=0 THEN {The next prime for sieving
          has been obtained}
        BEGIN p:=2*k+1; j:=k+1; p2:=p*p;
          IF p2 <= n THEN BEGIN
            i:=(p2-1) DIV 2; WHILE i <= stop DO
              BEGIN Prime[i]:=1; i:=i+p END;
            GOTO 1 END END;
      {When arriving here all elements of the array Prime
        corresponding to the primes below m have the value 0,
        the rest have the value 1}
END {primes};
```

```
BEGIN
  write('Input m: '); read(m); primes(m,Table);
  j:=(m-1) DIV 2;
  FOR i:=1 TO j DO IF Table[i]=0 THEN write(2*i+1:4);
END.
```

With computer programs of the type shown above, it is possible to investigate properties of the series of primes, depending on all the individual primes as high as $2 \cdot 10^{10}$, see for instance [6].

Exercise 1.1. Primes in intervals. Write a program for your computer, which performs the following: Read two (odd) integers m and $n \geq m$. Generate an array containing the odd primes below $\sqrt{n}$. Sieve out all composites between m and n by the sieve of Eratosthenes, utilizing the array of primes below $\sqrt{n}$. Print the primes between m and n and their numbers, equalling $\pi(n) - \pi(m-1)$, where the function $\pi(x)$ is defined on p. 11 below. Suitable test values are: $(m, n) =$ (99, 201), (12113, 12553) (check the computer's answer against the last column of Table 1 on p. 371), (9553, 9585) (an interval entirely without primes). More test values can be picked from Table 2 and from Table 3.

Compact Prime Tables

If the prime table generated by the computer needs to be kept for later use or perhaps output, the information can be given in a rather compact form. The simplest method is to store the sequence of ZEROs and ONEs representing the status (composite or prime) of all the odd numbers. Please note that this time we denote the *primes* by ONEs! A table of primes below 107 in this representation will appear as

01110 11011 01001 10010 11010 01001 10010 11001 01001 00010 1101

Here the first five entries in the table, viz. 0, 1, 1, 1, 0, correspond to the integers 1, 3 (=prime), 5 (=prime), 7 (=prime) and 9. The final four entries show that 101, 103 and 107 are primes, while 105 is a composite number.

Since most computers are binary, it is convenient to store this sequence of ZEROs and ONEs (bits) in computer words, where the number of bits per word depends upon the computer's word-size. Thus, in a 36-bit computer we could store a table of the primes among 36 consecutive odd numbers in one computer word. (It might actually be much easier to use only 35 of the 36 bits available, at least when programming in a high-level language.)

8

This method of storing primes is rather simple.—We can, by eliminating also the multiples of 3 and 5 in our example, compress the prime tables even more, but at the cost of simplicity. We are then left with all numbers of the forms

$$30k+1, 30k+7, 30k+11, 30k+13, 30k+17, 30k+19, 30k+23, \text{ and } 30k+29$$

i.e. with only 8 numbers out of 30. Thus, we need to store only one bit for 8 *odd* numbers out of 15, or just about half of them. By removing also multiples of 7, 11 and 13, we can further reduce the storage required by a factor of $\frac{6}{7} \cdot \frac{10}{11} \cdot \frac{12}{13} = 0.72$. If we do so, however, the pattern of the remaining integers is quite complicated and repeats periodically after 5760 steps, leading to some tricky computer programming as well as to a time-consuming table look-up function.—Thus, optimal efficiency requires balancing how large the prime table to be stored against how much computational work required each time a table look-up is requested.—A reasonable compromise is to exclude only multiples of 3 saving 1/3 of the storage otherwise needed, i.e. to store one bit of information for each of the numbers $6k \pm 1$ only. The sequence of primes between 5 and 107 in this representation will be

$$11111\ 11011\ 01111\ 01011\ 01110\ 11010\ 01111$$

which is the sequence previously shown with each third bit removed, the removal starting with the second bit, corresponding to the number 3. The final 5 bits in the string shown represent the character (prime or composite) of the numbers 95, 97, 101, 103 and 107.

In a 36-bit computer, the primes in an interval from $105k$ to $105(k+1)$ can thus be stored in 35 of the 36 bits of a computer word. This string of bits may be reversed and printed out as an integer $< 2^{35}$. A prime table up to $105 \times 96 = 10080$ looks rather strange when printed out in this way (see next page). The reader should compare this with the prime table up to 12553 provided at the end of this book. The table printed there contains slightly more information than the print-out below, as the print-out here comprises all the primes up to 10079 only.—On ordinary computer magnetic tape with a storage capacity of 6250 bits per inch, the primes up to 500,000,000 can be stored on one reel of tape in this way.

Compact prime table up to 10080 (to be read horizontally)

32596917119	19221276355	32294916984	27056746064	13260585324
19153906256	11044217692	10628959443	23930632312	27274595010
12929300524	09758853778	21477751664	18735703058	06820532604
01946775235	27961930040	10687629457	28253630548	10613958227
25803963148	26662686226	21859162944	17449165506	06723734372
27325713986	26053724260	09204998354	06548095832	00563096657
30104951352	26603773969	32489161484	18886509697	30344308812
10336150736	06524317784	08657858241	13223332700	17248824849
02496475956	09263457856	27997067788	17534371331	12956543856
18014274691	30375432704	09021230674	17224718372	01007756482
06518293316	01562192512	10744524916	10695737491	00048255592
27610139283	21483511888	27182047424	17255969348	26401637587
02460821844	00154165328	02423076900	10092832833	25850294296
00746875411	02489932592	01755325507	28234041380	26190874626
21544575008	27592828609	00369648472	01896680659	27988934752
00294462083	21785479944	18472241811	32321401632	01688470608
04297614176	10219687506	11113138744	26578275025	08600449332
09084889235	27959757100	26452510928	15137323024	02040873601
06782734684	18550637056	17558145328	18936318161	10776021528
10336339456	06816596588	09070971010	04330713676	

Hexadecimal Compact Prime Tables

A related method to store the primes is to use one hexadecimal symbol to denote the pattern of primes in each interval of the form $[10k, 10k + 9]$, $k = 1, 2, 3, \ldots$ Thus the primes between 20 and 29 are stored as $0101 = 5$, since 21 is composite (0), 23 is prime (1), 27 is composite (0) and 29 is prime (1). The primes between 10 and 99 are in this way stored as F5AE5AD52. (The hexadecimal symbols have the values A=10, B=11, C=12, D=13, E=14 and F=15.) The storage capacity required is 4 bits out of each 10 numbers.

Difference Between Consecutive Primes

The method demonstrated above is not the only means of storing the primes in compact form. An alternative is to store $(p_{i+1} - p_i)/2$. The table on p. 85 shows how large these values can become. This is convenient if we wish to run sequentially through the list of primes (e.g. for trial division by all small primes).

The Number of Primes Below x

The number of primes $\leq x$, $\pi(x)$, is an important number-theoretic function. Thus $\pi(2) = 1$ because we count 2 as the first prime, $\pi(10) = 4$ and $\pi(\sqrt{1000}) = 11$, since there are 11 primes ≤ 31.6. To calculate arithmetically the exact number of primes below a given limit is an extremely complicated and labour-consuming task, as we shall see, and the early computations of $\pi(x)$ were carried out simply by counting the number of primes in existing prime tables. But since those early prime tables were not free from errors the results of the computations of $\pi(x)$ by this method were somewhat unreliable.

As mentioned above, the existing formulas for $\pi(x)$ are quite complicated. The simplest (but unfortunately also the most labour-consuming) was found by Legendre, and reads

$$1 + \pi(x) = \pi(\sqrt{x}) + [x] - \sum_{p_i \leq \sqrt{x}} \left[\frac{x}{p_i}\right] + \sum_{p_i < p_j \leq \sqrt{x}} \left[\frac{x}{p_i p_j}\right] -$$

$$- \sum_{p_i < p_j < p_k \leq \sqrt{x}} \left[\frac{x}{p_i p_j p_k}\right] + \cdots \tag{1.2}$$

where, as usual, $[z]$ denotes the integer part of z.—Legendre's formula is almost self-evident. It uses the idea that

1 + the number of primes = the number of all integers —
— the number of composites in the interval $[1, x]$.

The term $[x/p]$ enumerates the integers divisible by p in this interval, since $[x/p] = n$ if and only if $np \leq x < (n+1)p$. Since all composite numbers in the interval $[1, x]$ have *some* prime factor $\leq \sqrt{x}$, there are obviously $\sum_{p_i \leq \sqrt{x}} [x/p_i]$ multiples of primes $\leq \sqrt{x}$ in this interval. However, we may not count the multiples $1 \cdot p_i$ as composites; that is why the term $\pi(\sqrt{x})$ has been added. This reasoning accounts for the first three terms in the right-hand-side of the formula. Where do the rest come from? Some of the composites in $[1, x]$ are divisible by *two* of the primes $\leq \sqrt{x}$, p_j as well as p_i. These composites will be counted *twice* when computing the sum $\sum [x/p_i]$; since any integer of the form $a p_i p_j$ will count as a multiple of p_i as well as a multiple of p_j. That is why the total has to be corrected by again adding the number of integers having this particular form, which is carried out by the next term in the formula, $\sum [x/p_i p_j]$. As a result of this new term, all those integers in $[1, x]$ which happen to be divisible by *three*

11

different primes, all $\leq \sqrt{x}$, will not be subtracted at all. This omission has to be corrected, which explains the next term in (1.2), and so forth.

In order to understand the formula better, let us, as an example, compute $\pi(100)$ and $\pi(200)$ and check with the number of primes between 100 and 200, $\pi(200) - \pi(100) = 21$ which we have found on p. 3 by the sieve of Eratosthenes. Legendre's formula gives

$$\pi(100) = \pi(10) + [100] - \left[\frac{100}{2}\right] - \left[\frac{100}{3}\right] - \left[\frac{100}{5}\right] - \left[\frac{100}{7}\right] + \left[\frac{100}{2 \cdot 3}\right] +$$

$$+ \left[\frac{100}{2 \cdot 5}\right] + \left[\frac{100}{2 \cdot 7}\right] + \left[\frac{100}{3 \cdot 5}\right] + \left[\frac{100}{3 \cdot 7}\right] + \left[\frac{100}{5 \cdot 7}\right] - \left[\frac{100}{2 \cdot 3 \cdot 5}\right] -$$

$$- \left[\frac{100}{2 \cdot 3 \cdot 7}\right] - \left[\frac{100}{2 \cdot 5 \cdot 7}\right] - \left[\frac{100}{3 \cdot 5 \cdot 7}\right] + \left[\frac{100}{2 \cdot 3 \cdot 5 \cdot 7}\right] - 1 =$$

$$= 4 + 100 - 50 - 33 - 20 - 14 + 16 + 10 + 7 + 6 + 4 + 2 -$$
$$- 3 - 2 - 1 - 0 + 0 - 1 = 25.$$

When $x = 200$, the primes 11 and 13 also come into play. This gives 66 different terms in all, of which, however, quite a few are equal to 0:

$$\pi(200) = \pi(14) + 200 - 1 - 100 - 66 - 40 - 28 - 18 - 15 + 33 +$$
$$+ 20 + 14 + 9 + 7 + 13 + 9 + 6 + 5 + 5 + 3 + 3 + 2 + 2 +$$
$$+ 1 - 6 - 4 - 3 - 2 - 2 - 1 - 1 - 1 - 1 - 1 - 1 - 1 = 46,$$

which finally gives

$$\pi(200) - \pi(100) = 46 - 25 = 21.$$

Note that this computation agrees with the value found earlier for the number of primes between 100 and 200.

As is obvious from these examples, Legendre's formula as it stands is impracticable for the computation of $\pi(x)$ for large values of x. The large number of terms makes the computation difficult to organize. By various tricks, however, it is possible to collect some of the terms into partial sums and thus arrive at a more feasible calculus. The first efforts in this direction were made already by Legendre himself, who managed to compute $\pi(1, 000, 000)$. The value he found was not completely correct which tells us something about the difficulties involved in these kinds of computations. Thus Legendre [7] found $\pi(10^6)$ to be 78,526 instead of 78,498. The erroneous prime tables that existed at his time displayed 78,492 primes.

The next important progress in the counting of primes was made by Meissel who, with an efficient modification of Legendre's formula, computed $\pi(x)$ for, among other values, $x = 10^7$, 10^8 and 10^9. However, Meissel's computations are not free from errors. For instance, in 1885 Meissel published the value 50,847,478 for $\pi(10^9)$ and this is probably the most often quoted faulty value in the literature on primes. This error passed unnoticed for a long time and was revealed by D. H. Lehmer only in 1958, the correct value of $\pi(10^9)$ being 50,847,534.

Further progress was not made until the era of computers. Even with aid of a computer, Meissel's formula leads to time-consuming computations which have necessitated further reduction of the labour involved. Efforts to achieve this have been made primarily by D. H. Lehmer [8], by David Mapes [9] and, more recently, by J. C. Lagarias, V. Miller and A. M. Odlyzko [10], [11]. We shall now describe their work.

Meissel's Formula

Meissel's formula and its generalizations are based on a detailed analysis of those integers $P_k(x, a)$ in the interval $[1, x]$ which can be written as products of precisely k (not necessarily distinct) prime factors $> p_a$, the ath prime. If k is successively given the values $1, 2, 3, \ldots$, these expressions will account for all integers in $[1, x]$ which have no factor $\leq p_a$. The $[x]$ integers in the interval $[1, x]$ are thus

1. The integer 1.

2. The a first primes, $p_1 = 2$, $p_2 = 3, \ldots, p_a$.

3. $\sum_{1 \leq i \leq a} [x/p_i] - \sum_{1 \leq i < j \leq a} [x/p_i p_j] + \cdots - a$ composites having at least one prime factor $\leq p_a$.

4. $P_1(x, a) = \pi(x) - a$ primes p, where $p_a < p \leq x$.

5. $P_2(x, a)$ integers $n = p_i p_j \leq x$, with $a + 1 \leq i \leq j$.

6. $P_3(x, a)$ integers $n = p_i p_j p_k \leq x$, with $a + 1 \leq i \leq j \leq k$, and so on.

Since this accounts for all the integers in $[1, x]$ we find

$$1 + \sum_{1 \leq i \leq a} \left[\frac{x}{p_i}\right] - \sum_{1 \leq i < j \leq a} \left[\frac{x}{p_i p_j}\right] + \sum_{1 \leq i < j < k \leq a} \left[\frac{x}{p_i p_j p_k}\right] - \cdots +$$
$$+ \pi(x) - a + P_2(x, a) + P_3(x, a) + \cdots = [x]. \tag{1.3}$$

13

How many of the terms $P_2(x, a)$, $P_3(x, a), \ldots$ have to be written down? This depends on the value of a chosen. If a is chosen so large that $p_{a+1} > \sqrt{x}$, then $P_2(x, a)$ will equal 0, since $p_i p_j > \sqrt{x}\sqrt{x} = x$ for all $i, j \geq a + 1$. In this case the original formula of Legendre, (1.2), re-appears. If a is chosen such that $x^{1/3} < p_{a+1} \leq x^{1/2}$, $P_2(x, a)$ will contain some terms, but $P_3(x, a)$ will be an empty sum, since $p_i p_j p_k$ will then be $> x^{1/3}x^{1/3}x^{1/3} = x$. Generally, $P_r(x, a) = 0$ and $P_{r-1}(x, a) \neq 0$, if a is chosen such that $x^{1/r} < p_{a+1} \leq x^{1/(r-1)}$. From this we find it natural to choose a as $\pi(x^{1/r})$ for some suitable value of r, since with such a choice p_{a+1} will just exceed $x^{1/r}$, and the formula will break off at a point with the smallest value of a possible.

Evaluation of $P_k(x, a)$

$P_2(x, a)$ is defined above as the number of integers in the interval $[1, x]$ which are products of two primes p_i and p_j, with $a + 1 \leq i \leq j$. Considering one prime factor at a time, we have

$$
\begin{aligned}
P_2(x, a) = \ & \text{the number of integers } p_{a+1}p_j \leq x \text{ with } a + 1 \leq j \\
& + \text{ the number of integers } p_{a+2}p_j \leq x \text{ with } a + 2 \leq j \\
& + \cdots \\
= \ & \pi\left(\frac{x}{p_{a+1}}\right) - a + \pi\left(\frac{x}{p_{a+2}}\right) - (a+1) + \cdots \\
= \ & \sum \left\{ \pi\left(\frac{x}{p_i}\right) - (i - 1) \right\}, \quad \text{for } p_a < p_i \leq \sqrt{x}.
\end{aligned}
$$

Now, suppose we choose $a < \pi(\sqrt{x})$, the value of which we shall call b for the sake of brevity; then the above sum may be written

$$
\begin{aligned}
P_2(x, a) = \ & \sum_{i=a+1}^{b} \left\{ \pi\left(\frac{x}{p_i}\right) - (i - 1) \right\} = \\
= \ & -\frac{(b - a)(b + a - 1)}{2} + \sum_{i=a+1}^{b} \pi\left(\frac{x}{p_i}\right).
\end{aligned}
\tag{1.4}
$$

Here the sum of $i - 1$ has been calculated with the well-known summation formula for an arithmetic series. If we now choose $a = \pi(x^{1/3}) = c$, in order to make $P_3(x, a)$ and the higher sums $= 0$, we obtain Meissel's

14

formula

$$\pi(x) = [x] - \sum_{i=1}^{c} \left[\frac{x}{p_i}\right] + \sum_{1 \le i < j \le c} \left[\frac{x}{p_i p_j}\right] - \cdots +$$

$$+ \frac{(b+c-2)(b-c+1)}{2} - \sum_{c < i \le b} \pi\left(\frac{x}{p_i}\right). \qquad (1.5)$$

Lehmer's Formula

In the deduction of Meissel's formula we had to analyse the sum $P_2(x, a)$. If we carry the general formula (1.3) one step further, we arrive at Lehmer's formula. To achieve this we need to analyse the term $P_3(x, a)$ which may be described as

$P_3(x, a) = $ the numbers of products $p_{a+1} p_j p_k \le x$ with $a + 1 \le j \le k$

$\quad +$ the number of products $p_{a+2} p_j p_k \le x$ with $a + 2 \le j \le k$

$\quad + \cdots$

$$= P_2\left(\frac{x}{p_{a+1}}, a\right) + P_2\left(\frac{x}{p_{a+2}}, a\right) + \cdots = \sum_{i > a} P_2\left(\frac{x}{p_i}, a\right).$$

Introducing the notation $b_i = \pi(\sqrt{x/p_i})$, formula (1.4) gives

$$P_3(x, a) = \sum_{i > a} P_2\left(\frac{x}{p_i}, a\right) = \sum_{i=a+1}^{c} \sum_{j=i}^{b_i} \left\{\pi\left(\frac{x}{p_i p_j}\right) - (j-1)\right\}. \quad (1.6)$$

Here we have assumed that $a < c = \pi(x^{1/3})$, in order that $P_3(x, a)$ at all contains terms > 0. Now, finally putting $a = \pi(x^{1/4})$ leads to the formula Lehmer has used in [8]:

$$\pi(x) = [x] - \sum_{i=1}^{a} \left[\frac{x}{p_i}\right] + \sum_{1 \le i < j \le a} \left[\frac{x}{p_i p_j}\right] - \cdots +$$

$$+ \frac{(b+a-2)(b-a+1)}{2} -$$

$$- \sum_{a < i \le b} \pi\left(\frac{x}{p_i}\right) - \sum_{i=a+1}^{c} \sum_{j=i}^{b_i} \left\{\pi\left(\frac{x}{p_i p_j}\right) - (j-1)\right\}. \quad (1.7)$$

15

Computations

The most labour-consuming part of the computations done by Meissel and Lehmer is still the "Legendre-sum"

$$\phi(x,a) = [x] - \sum \left[\frac{x}{p_i}\right] + \sum \left[\frac{x}{p_i p_j}\right] - \sum \left[\frac{x}{p_i p_j p_k}\right] + \cdots \qquad (1.8)$$

which counts the positive integers $\leq x$, not divisible by any one of the first a primes $p_1, p_2, \ldots, p_a$. Even if the summations have to be extended only over all primes $\leq x^{1/3}$ in Meissel's formula and over all primes $\leq x^{1/4}$ in Lehmer's formula, as compared with $x^{1/2}$ in Legendre's formula, the labour for large values of x is tedious without using further tricks, which we shall now describe.

Since $\phi(x,a)$ is defined as the number of integers $\leq x$, not divisible by any of the primes $p_1, p_2, \ldots, p_a$, the following recursive formula holds

$$\phi(x,a) = \phi(x, a-1) - \phi\left(\frac{x}{p_a}, a-1\right). \qquad (1.9)$$

This formula expresses the fact that the integers not divisible by any of $p_1, p_2, \ldots, p_a$ are precisely those integers which are not divisible by any of $p_1, p_2, \ldots, p_{a-1}$, with the exception of those which are not divisible by p_a.

Using formula (1.9) repeatedly, we can break down any $\phi(x,a)$ to the computation of $\phi(x,1)$ which is the number of odd integers $\leq x$. However, because the recursion has to be used many times, the evaluation is cumbersome. It is far better to find a way to compute $\phi(x,k)$ for some reasonably large value of k, and then break down the desired value $\phi(x,a)$ just to $\phi(x,k)$ and no further. This can be achieved in the following way: The pattern of multiples of the primes $p_1, p_2, \ldots, p_k$ repeats itself with a period of length of $m_k = p_1 p_2 \cdots p_k$. With Legendre's formula we now easily find for $x = m_k$:

$$\phi(m_k, k) = [m_k] - \sum \left[\frac{m_k}{p_i}\right] + \sum \left[\frac{m_k}{p_i p_j}\right] - \cdots =$$

$$= m_k - \sum \frac{m_k}{p_i} + \sum \frac{m_k}{p_i p_j} - \cdots =$$

$$= m_k \left(1 - \frac{1}{p_1}\right)\left(1 - \frac{1}{p_2}\right) \cdots \left(1 - \frac{1}{p_k}\right) = \prod_{i=1}^{k} (p_i - 1) = \varphi(m_k) \ (1.10)$$

16

It is because all the quotients in the first line are *integers* that we are allowed to remove all integer-part operators and perform the simplification. As a consequence of (1.10) and of the periodicity we find the formula

$$\phi(s \cdot m_k + t, k) = s \cdot \varphi(m_k) + \phi(t, k), \tag{1.11}$$

where t can be chosen between 0 and m_k. Further, if $t > m_k/2$, we can use the periodicity and the symmetry of the multiples about the point 0 to find

$$\phi(t, k) = \varphi(m_k) - \phi(m_k - t - 1, k). \tag{1.12}$$

Thus, a quick way to have access to $\phi(x, k)$ for any x is to create a table of $\phi(x, k)$ up to $m_k/2$. This is readily accomplished by first running a sieve program up to $m_k/2$ for the primes $p_1, p_2, \ldots, p_k$, by means of which all multiples of these primes are marked. Following this a second program is run, which accumulates the number of integers not marked by the sieve program as multiples of any of the primes in question.—For $k = 4$ we have $m_k = 2 \cdot 3 \cdot 5 \cdot 7$, $m_k/2 = 105$, and the table constructed becomes

Critical table of $\phi(x, 4)$											
1	1	19	5	37	9	53	13	71	17	89	21
11	2	23	6	41	10	59	14	73	18	97	22
13	3	29	7	43	11	61	15	79	19	101	23
17	4	31	8	47	12	67	16	83	20	103	24

The table is organized as a so-called critical table, which means that information is stored only *when the function changes its value*. Thus, for instance, we find from the table that $\phi(x, 4) = 12$ for all x, where $47 \leq x < 53$.

A computer program for evaluating $\phi(x, a)$ can now be written, based on the following principles:

1. Place the first a primes, 2, 3, 5, $\ldots, p_a$, in an array, Prime[1:a], just as we did to prepare for the Eratosthenes sieve program.

2. Modify the sieve program previously given, in order to sieve only with the primes $\leq p_a$.

3. Sieve the interval from 1 to $m_a/2 = p_2 p_3 \ldots p_a$. As before, store only the odd integers of this interval.

17

4. Count the zeros below x for all x in the interval $[1, m_a/2]$ in order to find $\phi(x, a) =$ the number of integers $\leq x$, untouched by the sieve.—At this stage you have generated a table of $\phi(x, a)$ for all x in $1 \leq x \leq m_a/2$.

5. Apply the formulas (1.11) and (1.12).

A computer program performing steps 1–5 could resemble

```
PROGRAM phixa {Computes phi(x,a) for a=5}
(input,Output);
LABEL 1;
CONST a=5; ma=2310; {ma=p[1]*...*p[a]a} ma2=1155; {ma2=ma/2}
  stop=578; {Since phi(x,a) never alters its value at even
    x, full information about phi(x,a) is retained if
    phi(x,a) is stored at odd values of x only. stop is
    the number of elements required in tabphia}
  phima=480; {phima=phi(ma,a)=prod(p[i]-1) for i=1,...,a}
VAR Prime : ARRAY[1..a] OF INTEGER;
  tabphia : ARRAY[0..stop] OF INTEGER;
  i,j,s,x : INTEGER;

FUNCTION phia(x : INTEGER) : INTEGER;
{For x<0, phi(x,a) is defined as -phi(-x,a). That is
  why abs(x) and sign(x)=abs(x) DIV x appear!}
VAR r,z : INTEGER;
BEGIN
  r:=abs(x) MOD ma; z:=(abs(x) DIV ma)*phima;
  IF r < ma2
    THEN z:=z+tabphia[(r+1) DIV 2]
    ELSE z:=z+phima-tabphia[(ma-r) DIV 2];
  phia:=abs(x) DIV x * z
END {phia};

BEGIN
Prime[2]:=3; Prime[3]:=5; Prime[4]:=7; Prime[5]:=11;
FOR i:=2 TO a DO
  BEGIN j:=(Prime[i]+1) DIV 2;
    WHILE j <= stop DO
      BEGIN tabphia[j]:=1; j:=j+Prime[i] END
        {Here the sieving is performed}
  END;
s:=0; FOR i:=1 TO stop DO
```

18

```
BEGIN IF tabphia[i]=0 THEN s:=s+1; tabphia[i]:=s END;
   {Here the sum of ZEROs is accumulated, and the table
      phixa, needed in the integer procedure phia(x), is
      ready for use}

1: write('Input x: '); read(x);
IF x <> 0 THEN
   BEGIN
   writeln('phi(',x:5,',',a:1,')=',phia(x):5); GOTO 1
   END
END.
```

For the sake of brevity we have presented a program for the computation of $\phi(x, a)$ for $a = 5$ only. It is evident which changes have to be made in order for the program to work for a different value of a. In a program for computing the number of primes, however, the function $\phi(x, a)$ will be required not only for $a = 5$, but for *all* values of a up to some limit. In such a case the computer program must incorporate all the corresponding PASCAL functions as well as the auxiliary tables needed.

If, for a large value of a, the resulting computation of $\phi(x, a)$ exceeds the storage capacity of the computer, then formula (1.9) may be used. It leads to program statements such as

```
phi6:=phi5(x)-phi5(x/13); phi7:=phi6(x)-phi6(x/17); ...
```

preferably written within functions, so that they can be called in succession. All these functions phia(x) may then be combined to form the two-variable function phi(x,a).—Moreover, for *large* values of a, the basic formula (1.3) can be run *backwards*:

$$\phi(x, a) = 1 + \pi(x) - a + P_2(x, a) + P_3(x, a), \quad \text{for } p_a < x < p_{a+1}^4, \quad (1.13)$$

if we truncate (1.3) after the term $P_3(x, a)$, just as in Lehmer's formula.

Exercise 1.2. A computer program for $\phi(x, a)$. Write a FUNCTION phi(x,a), covering $a \le 10$, following the idea mentioned above. Incorporate this FUNCTION in a computer program which reads x and a and prints $\phi(x, a)$. Since $\phi(x, 10) =$ the number of integers $\le x$, not divisible by any prime ≤ 29,

$$\phi(x, 10) = \pi(x) - 9 + P_2(x, 10) + P_3(x, 10) + \cdots,$$

where, according to (1.4),

$$P_2(x, 10) = \sum_{p_i=31}^{\sqrt{x}} \left\{ \pi\left(\frac{x}{p_i}\right) - (i - 1) \right\}.$$

19

For $x < 31^3$ the terms after $P_2(x, 10)$ have the value zero. Use this fact to check your FUNCTION by computing

$$\phi(10^4, 10) = \pi(10^4) - 9 + \sum_{p_i=31}^{97} \{\pi(10^4/p_i) - (i-1)\},$$

where the 15 values needed for $\pi(10^4/p_i)$ may be taken from Table 1, which is organized to give easy access to $\pi(x)$.

A Computation Using Meissel's Formula

In order to demonstrate the complexity and the computational labour involved we shall compute $\pi(10, 000)$ with Meissel's formula and $\pi(100, 000)$ with Lehmer's formula.

$x = 10, 000$ leads to the following values of the summation limits in Meissel's formula (1.5):

$$b = \pi(10000^{\frac{1}{2}}) = 25, \quad c = \pi(10000^{\frac{1}{3}}) = \pi(21) = 8,$$

which gives

$$\pi(10000) = \phi(10000, 8) + \frac{31 \cdot 18}{2} - \sum_{i=9}^{25} \pi\left(\frac{10000}{p_i}\right).$$

$\phi(10000, 8)$ is "decomposed" in the following way (if, as above, we choose $k = 4$). Observe the frequent use of (1.12) in the calculations.

$$\phi(10000, 8) = \phi(10000, 7) - \phi\left(\frac{10000}{19}, 7\right) = \phi(10000, 7) - \phi(526, 7).$$

$$\phi(10000, 7) = \phi(10000, 6) - \phi\left(\frac{10000}{17}, 6\right) = \phi(10000, 6) - \phi(588, 6).$$

$$\phi(10000, 6) = \phi(10000, 5) - \phi\left(\frac{10000}{13}, 5\right) = \phi(10000, 5) - \phi(769, 5).$$

$$\phi(10000, 5) = \phi(10000, 4) - \phi\left(\frac{10000}{11}, 4\right) = \phi(10000, 4) - \phi(909, 4) =$$

$$= \phi(48 \cdot 210 - 80, 4) - \phi(4 \cdot 210 + 69, 4) =$$

$$= 48 \cdot 48 - \phi(79, 4) - 4 \cdot 48 - \phi(69, 4) =$$

$$= 44 \cdot 48 - 19 - 16 = 2077.$$

20

$$\phi(526,7) = \phi(526,6) - \phi\left(\frac{526}{17},6\right) = \phi(526,6) - \phi(30,6) =$$

$$= \phi(526,5) - \phi\left(\frac{526}{13},5\right) - 5 =$$

$$= \phi(526,4) - \phi\left(\frac{526}{11},4\right) - \phi(40,5) - 5 =$$

$$= \phi(2 \cdot 210 + 106,4) - \phi(47,4) - 8 - 5 =$$

$$= 2 \cdot 48 + \phi(106,4) - 12 - 13 = 95.$$

$$\phi(588,6) = \phi(588,5) - \phi\left(\frac{588}{13},5\right) = \phi(588,4) - \phi\left(\frac{588}{11},4\right) - \phi(45,5) =$$

$$= \phi(3 \cdot 210 - 42,4) - \phi(53,4) - 10 =$$

$$= 3 \cdot 48 - \phi(41,4) - 13 - 10 = 111.$$

$$\phi(769,5) = \phi(769,4) - \phi\left(\frac{769}{11},4\right) = \phi(4 \cdot 210 - 71,4) - \phi(69,4) =$$

$$= 4 \cdot 48 - \phi(70,4) - 16 = 160.$$

From all this we obtain

$$\phi(10000,8) = 2077 - 95 - 111 - 160 = 1711.$$

The sum in Meissel's formula is

$$\sum_{i=9}^{25} \pi\left(\frac{10000}{p_i}\right) = \pi\left(\frac{10000}{23}\right) + \pi\left(\frac{10000}{29}\right) + \cdots + \pi\left(\frac{10000}{97}\right) =$$

$$= \pi(434) + \pi(344) + \pi(322) + \cdots + \pi(103) =$$

$$= 84 + 68 + 66 + \cdots + 27 = 761.$$

The final result is

$$\pi(10000) = 1711 + 9 \cdot 31 - 761 = 1229.$$

21

A Computation Using Lehmer's Formula

$x = 100,000$ leads to the following values of the summation limits in Lehmer's formula (1.7):

$$a = \pi(100000^{\frac{1}{4}}) = \pi(17) = 7$$
$$b = \pi(100000^{\frac{1}{2}}) = \pi(316) = 65$$
$$c = \pi(100000^{\frac{1}{3}}) = \pi(46) = 14$$

and

$$b_i = \pi\left(\sqrt{\frac{100000}{p_i}}\right) \quad \text{for } 8 \le i \le 14,$$

which gives

$$b_8 = \pi\left(\sqrt{\frac{100000}{19}}\right) = \pi(72) = 20, \quad b_9 = \pi\left(\sqrt{\frac{100000}{23}}\right) = \pi(65) = 18,$$

$$b_{10} = \pi\left(\sqrt{\frac{100000}{29}}\right) = \pi(58) = 16, \quad b_{11} = \pi\left(\sqrt{\frac{100000}{31}}\right) = \pi(56) = 16,$$

$$b_{12} = b_{13} = b_{14} = 15.$$

Thus the formula for $\pi(10^5)$ reads

$$\pi(10^5) = \phi(10^5, 7) + \frac{70 \cdot 59}{2} - \sum_{i=8}^{65} \pi\left(\frac{10^5}{p_i}\right) -$$
$$- \sum_{i=8}^{14} \sum_{j=i}^{b_i} \left\{ \pi\left(\frac{10^5}{p_i p_j}\right) - (j-1) \right\}.$$

The first term of the formula is computed as

$$\phi(10^5, 7) = \phi(10^5, 6) - \phi\left(\frac{10^5}{17}, 6\right) = \phi(10^5, 6) - \phi(5882, 6).$$

$$\phi(10^5, 6) = \phi(10^5, 5) - \phi\left(\frac{10^5}{13}, 5\right) = \phi(10^5, 5) - \phi(7692, 5).$$

$$\phi(10^5, 5) = \phi(10^5, 4) - \phi\left(\frac{10^5}{11}, 4\right) = \phi(476 \cdot 210 + 40, 4) - \phi(9090, 4)$$

$$= 476 \cdot 48 + \phi(40, 4) - \phi(43 \cdot 210 + 60, 4) = 20779.$$

$$\phi(5882, 6) = \phi(5882, 5) - \phi\left(\frac{5882}{13}, 5\right) =$$

$$= \phi(5882, 4) - \phi\left(\frac{5882}{11}, 4\right) - \phi(452, 5) =$$

$$= 28 \cdot 48 + \phi(2, 4) - \phi(534, 4) - \phi(452, 4) + \phi\left(\frac{452}{11}, 4\right) = 1128.$$

$$\phi(7692, 5) = \phi(7692, 4) - \phi\left(\frac{7692}{11}, 4\right) = 1598.$$

All this gives the required value of

$$\phi(10^5, 7) = 18053.$$

The first sum in the expression for $\pi(10^5)$ equals

$$\sum_{i=8}^{65} \pi\left(\frac{10^5}{p_i}\right) = \pi(5263) + \pi(4347) + \pi(3448) + \cdots + \pi(321) + \pi(319) =$$
$$= 698 + 593 + 481 + \cdots + 66 + 66 = 9940.$$

The double sum can be split into 7 simple sums:

$$\sum_{j=8}^{20}\left\{\pi\left(\frac{10^5}{19p_j}\right) - (j-1)\right\} + \sum_{j=9}^{18}\left\{\left(\frac{10^5}{23p_j}\right) - (j-1)\right\} + \cdots +$$

$$+ \sum_{j=13}^{15}\left\{\pi\left(\frac{10^5}{41p_j}\right) - (j-1)\right\} + \sum_{j=14}^{15}\left\{\pi\left(\frac{10^5}{43p_j}\right) - (j-1)\right\} =$$

$$= \pi\left(\frac{10^5}{19 \cdot 19}\right) + \pi\left(\frac{10^5}{19 \cdot 23}\right) + \cdots + \pi\left(\frac{10^5}{43 \cdot 47}\right) - 169 -$$

$$- 125 - \cdots - 27 = \pi(277) + \pi(228) + \cdots + \pi(49) - 569 = 586.$$

Finally, the result is

$$\pi(100000) = 18053 + 35 \cdot 59 - 9940 - 586 = 9592.$$

By using (1.7), Lehmer calculated $\pi(x)$ in 1958 with aid of a computer for, among other values, $x = a \cdot 10^6$ with $a = 20, 25, 33, 37, 40, 90, 100, 999, 1000$ and 10000. It was this computation that revealed that Meissel's value of $\pi(10^9)$ was 56 units too low.—In Table 3 at the end of this book, the reader will find the values of $\pi(x)$, x being of the form $s \cdot 10^n$, with s a digit. If the reader would like to attempt some computer programming on $\pi(x)$, these values will be suitable as test values for program debugging and verification.

A Computer Program Using Lehmer's Formula

Having generated in the computer a prime table up to some limit G, we can easily program a function smallpi(x) which gives the values of $\pi(x)$ up to this limit. If we also program the function phi(x,a), giving $\phi(x,a)$ for all a up to A, we can utilize these functions in a program computing $\pi(x)$ with Lehmer's formula. How large are the values of x that such a program is capable of handling? Let us examine which values are needed in Lehmer's formula:

1. $b = \pi(\sqrt{x})$ which is accessible if $x \leq G^2$.
2. $\phi(x,a)$ with $a = x^{1/4}$ which will work if $x < p_{a+1}^4$. Thus, if we limit a to 8, then x must be $< p_9^4 = 23^4 = 279841$.
3. Numerous values of $\pi(x/p_i)$ and $\pi(x/p_ip_j)$. The "worst" of these is $\pi(x/p_{s+1})$. These values will be available as long as $x/p_{s+1} < G$ which will occur if $x < p_{s+1}^{\frac{4}{3}}$, where p_{s+1} is the first prime outside the range of the prime table stored in the computer.—Thus if G is 2000, say, then this part of the program will operate as long as $x < 2003^{\frac{4}{3}} = 25248.8$. — Here it is actually possible to "cheat" a little. Instead of storing an extensive prime table, you might program the computer to ask for a few values of $\pi(x/p_i)$ above the limit G, and then introduce them to the computer after having looked them up in a larger prime table than the one stored in the computer. If you use [3], you will be able to reach as high as $p_{6,000,000} = 104,302,307$ and the limit imposed on x will be $104,302,333^{\frac{4}{3}} = 49,097,422,907.98$. — Another way to raise the limit on x would be

24

to let the PASCAL function Lehmer call itself recursively. However in this case you must be very careful not to exceed the available memory space by overfilling it with stacked information that the computer will automatically store in order to be able to keep track of all the recursive calls involved in the computation.

Now, finally, a PASCAL function for the computation of $\pi(x)$ with Lehmer's formula looks like this:

```
FUNCTION Lehmer(x : INTEGER) : INTEGER;
{Computes pi(x) with Lehmer's formula. Makes use of the
  function phi(x,a) and of an array of integers, Prime,
  containing a "small" prime table}
VAR b,c,a,bi,z,w,sum,i,j : INTEGER;
BEGIN
  z:=trunc(sqrt(x)+0.5); b:=smallpi(z);
  c:=smallpi(trunc(exp(ln(x)/3)+0.5)); {pi(cube root(x)}
  a:=smallpi(trunc(sqrt(z)+0.5)); {pi(fourth root(x)}
  sum:=phi(x,a)+(b+a-2)*(b-a+1) DIV 2;
  FOR i:=a+1 TO b DO
    BEGIN w:=x DIV Prime[i]; sum:=sum-pi(w);
      IF i<=c THEN
        BEGIN
          bi:=smallpi(trunc(sqrt(w)+0.5));
          FOR j:=i TO bi DO sum:=sum-pi(w DIV Prime[j])+j-1
        END
    END;
  Lehmer:=sum
END {Lehmer};
```

Some details in the program look rather complicated. For instance, what role does the constant 0.5 play? Well, this is a safety precaution to ensure that the square roots and the cube root are not too low! Otherwise, it might happen that the built-in standard functions sqrt, exp and log round downwards, which could result in too low values. If this occurs precisely when the integer part of the root is about to change, it may result in a value which is one unit too low. (Thus if $\sqrt{1,000,000}$ emerges as 999.999, the result would be 999 instead of 1000 if the little upward rounding were not added!) Of course it is far better to write your own integer function isqrt(x) for the integer part of the square root of an integer, solving this problem once for all of your programs. However, we cannot overload

every little program with the detail necessary to get everything to work properly—the reader must find these small things out for himself, which actually is part of the fun in computer programming!

Exercise 1.3. Computing $\pi(x)$. Incorporate your FUNCTION phi(x,a) from exercise 1.2 above and the FUNCTION Lehmer into a computer program which reads x and prints $\pi(x)$. Check your program by comparing its results with some entries from Table 3.

Mapes' Method

The methods of Meissel and Lehmer to calculate $\pi(x)$ are both based on Legendre's formula (1.2). As a matter of fact, they are just clever re-arrangements and groupings of the terms of the Legendre sum in order to facilitate its computation. A still more efficient way to compute this sum was found in 1963 [9] by David C. Mapes. The method is rather complicated, and requires the introduction of some notations, but try to keep in mind that it involves merely re-arranging and grouping the terms in Legendre's sum!

Deduction of Formulas

Since Legendre's sum

$$\phi(x,a) = [x] - \sum\left[\frac{x}{p_i}\right] + \sum\left[\frac{x}{p_ip_j}\right] - \sum\left[\frac{x}{p_ip_jp_k}\right] + \cdots \qquad (1.14)$$

where $p_i < p_j < p_k \leq p_a$, and p_a is the ath prime, contains precisely 2^a terms, it is possible to describe each of these 2^a terms as specific cases of the expression

$$T_k(x,a) = (-1)^{\beta_0+\beta_1+\cdots+\beta_{a-1}}\left[\frac{x}{p_1^{\beta_0}p_2^{\beta_1}\ldots p_a^{\beta_{a-1}}}\right]. \qquad (1.15)$$

Here the numbers β_i are the digits (0 or 1) if k is expressed as a binary number

$$k = 2^{a-1}\beta_{a-1} + 2^{a-2}\beta_{a-2} + \cdots + 2^1\beta_1 + 2^0\beta_0.$$

This expression for $T_k(x,a)$ looks difficult, but is just the result of a clever formal trick, based on the following simple idea: Associate the nth prime p_n

26

with the power 2^{n-1}. Thus, 2 is associated with $2^0 = 1$, 3 associated with 2^1, 5 with 2^2, 7 with 2^3, 11 with 2^4, etc. Furthermore, let the product of several primes (all different) be associated with the sum of different powers of 2, corresponding to the individual primes. Thus, $3 \cdot 2$ is associated with $2^1 + 2^0$, $11 \cdot 7 \cdot 3$ with $2^4 + 2^3 + 2^1$ and so on. Finally append a sign to match the sign of the different terms of (1.14), and the result is immediate! Now you can see how clever this description of the terms of (1.14) is, because the denominators of the terms run through all *products of different prime factors*, where all factors are $\leq p_a$, when the "indicator" k extends through all *sums of different powers of 2*, i.e. through all integers $< 2^a$.

As an example, if $a = 3$, the 8 terms from (1.14)

$$[x] - \left[\frac{x}{p_1}\right] - \left[\frac{x}{p_2}\right] - \left[\frac{x}{p_3}\right] + \left[\frac{x}{p_1 p_2}\right] + \left[\frac{x}{p_1 p_3}\right] + \left[\frac{x}{p_2 p_3}\right] - \left[\frac{x}{p_1 p_2 p_3}\right]$$

are written in this notation as $T_0(x,3) = [x]$, and

$$T_1(x,3) = -\left[\frac{x}{p_1}\right], T_2(x,3) = -\left[\frac{x}{p_2}\right], T_3(x,3) = \left[\frac{x}{p_1 p_2}\right], T_4(x,3) = -\left[\frac{x}{p_3}\right],$$

$$T_5(x,3) = \left[\frac{x}{p_1 p_3}\right], T_6(x,3) = \left[\frac{x}{p_2 p_3}\right], T_7(x,3) = -\left[\frac{x}{p_1 p_2 p_3}\right].$$

Thus, Legendre's sum $\phi(x,a)$ may be written

$$\phi(x,a) = \sum_{k=0}^{2^a - 1} T_k(x,a). \tag{1.16}$$

Let M be an integer $< 2^a$ and $2^i \| M$, i.e. let 2^i be the highest power of 2 which divides M. Then, with $\gamma(M,x,a)$ defined as

$$\gamma(M,x,a) = \sum_{k=M}^{M+2^i - 1} T_k(x,a), \tag{1.17}$$

we have

$$\phi(x,a) = T_0(x,a) + \gamma(2^0,x,a) + \gamma(2^1,x,a) + \cdots + \gamma(2^{a-1},x,a). \tag{1.18}$$

Furthermore, defining $T_k(-x,a)$ as $-T_k(x,a)$, we have

$$\text{sign} \, T_k(x,a) = (-1)^{\beta_i + \beta_{i+1} + \cdots + \beta_{n+1}} \quad \text{if} \quad 2^i | k, \tag{1.19}$$

27

since the last i binary digits of k are then 0, and for the same reason

$$|T_k(x,a)| = \left[\frac{x}{p_{i+1}^{\beta_i} p_{i+2}^{\beta_{i+1}} \cdots p_a^{\beta_{a-1}}} \right], \quad \text{if} \quad 2^i | k. \qquad (1.20)$$

By substituting $T_k(x,a)$ for x, and i for a in the definition of $T_{k'}(x,a)$, we obtain

$$T_{k'}\{T_k(x,a),i\} =$$

$$= \text{sign}\{T_k(x,a)\}(-1)^{\beta'_0+\beta'_1+\cdots+\beta'_{i-1}} \times \left\{ \frac{|T_k(x,a)|}{p_1^{\beta'_0} p_2^{\beta'_1} \cdots p_i^{\beta'_{i-1}}} \right\}, \quad (1.21)$$

where the β'_i's are the binary digits of

$$k' = 2^{i-1}\beta'_{i-1} + 2^{i-2}\beta'_{i-2} + \cdots + 2^1\beta'_1 + 2^0\beta'_0,$$

and $0 \le k' < 2^i$. When $2^i | k$, we can substitute the expressions (1.19) and (1.20) defining $T_k(x,a)$ for $T_k(x,a)$ in (1.21), thus arriving at

$$T_{k'}\{T_k(x,a),i\}_{2^i|k} = T_{k+k'}(x,a). \qquad (1.22)$$

This relation can be seen as a consequence of the "logarithmic law" which we expect the "indicator vector" with components β_j to obey—when we *multiply* the denominators of two T-functions (or, when we compose the two T-functions, effectively the same thing) the corresponding indicator vectors are *added*. But only when the binary numbers, representing the indicator vectors, have their ONEs in different positions, can they be added in a manner that conforms with the definition (1.15)—hence the restriction that $k' < 2^i$ and at the same time $2^i | k$—because then the ONEs in the binary representation of k' are all in lower positions than those of k.

Substituting $T_M(x,a)$ for x and i for a in Legendre's sum (1.14) gives

$$\phi\{T_M(x,a),i\} = \sum_{k'=0}^{2^i-1} T_{k'}\{T_M(x,a),i\} =$$

$$= \sum_{k=M}^{M+2^i-1} T_k(x,a) = \gamma(M,x,a). \qquad (1.23)$$

Now, (1.18) becomes

$$\phi(x,a) = T_0(x,a) + \phi\{T_1(x,a),0\} + \phi\{T_2(x,a),1\} + \cdots +$$

$$+ \phi\{T_{2^{a-1}}(x,a),a-1\}. \qquad (1.24)$$

28

By replacing x with $T_M(x,a)$ and a with i in (1.24), we get

$$\phi\{T_M(x,a),i\}_{2^i|M} = T_0\{T_M(x,a),i\} + \phi(T_1\{T_M(x,a),i\},0) +$$
$$+ \phi(T_2\{T_M(x,a),i\},1) + \cdots + \phi(T_{2^i-1}\{T_M(x,a),i\},i-1) =$$
$$= T_M(x,a) + \phi(T_{M+1}(x,a),0) + \phi(T_{M+2}(x,a),1) + \cdots +$$
$$+ \phi(T_{M+2^r}(x,a),r) + \cdots + \phi(T_{M+2^i-1}(x,a),i-1), \tag{1.25}$$

if $2^i|M$, by using (1.22). (Note that $\phi(T_{M+1}(x,a),0) = T_{M+1}(x,a)$, according to (1.16).)

Starting with $\phi(x,a) = \phi(T_0(x,a),a)$, we can now from (1.25) calculate $\phi(x,a)$ by recursive application.—As in the application of Lehmer's method it is favourable to have tables of $\phi(x,a)$ for small values of a, and to use (1.11) and (1.12) to evaluate $\phi(x,a)$. It is also helpful to have an extensive table of primes available, the larger the better.

A Worked Example

As an example, let us re-compute the previously found value of $\phi(10000,8) = 1711$ with Mapes' method:

$$\phi(10000,8) = \phi(T_0(10000,8),8) =$$
$$= T_0(10000,8) + \phi(T_1(10000,8),0) + \phi(T_2(10000,8),1) +$$
$$+ \phi(T_4(10000,8),2) + \phi(T_8(10000,8),3) + \phi(T_{16}(10000,8),4) +$$
$$+ \phi(T_{32}(10000,8),5) + \phi(T_{64}(10000,8),6) + \phi(T_{128}(10000,8),7) =$$
$$= 10000 + T_1(10000,8) - \phi(3333,1) - \phi(2000,2) - \phi(1428,3) -$$
$$- \phi(909,4) + \phi(T_{32}(10000,8),5) + \phi(T_{64}(10000,8),6) +$$
$$+ \phi(T_{128}(10000,8),7) = 10000 - 5000 - 1667 -$$
$$- \left(\frac{2\cdot1998}{6}+1\right) - \left(\frac{8\cdot1440}{30}-3\right) - \left(\frac{48\cdot840}{210}+16\right) +$$
$$+ \phi(T_{32}(10000,8),5) + \phi(T_{64}(10000,8),6) + \phi(T_{128}(10000,8),7) =$$
$$= 2077 + \phi(T_{32}(10000,8),5) + \phi(T_{64}(10000,8),6) + \phi(T_{128}(10000,8),7).$$
$$\phi(T_{32}(10000,8),5) = T_{32}(10000,8) + T_{33}(10000,8) + \phi(T_{34}(10000,8),1) +$$
$$+ \phi(T_{36}(10000,8),2) + \phi(T_{40}(10000,8),3) + \phi(T_{48}(10000,8),4) =$$
$$= -769 + 384 + \phi(256,1) + \phi(153,2) + \phi(109,3) + \phi(69,4) =$$
$$= -769 + 384 + 128 + \left(\frac{2\cdot150}{6}+1\right) + \left(\frac{8\cdot120}{30}-2\right) + 16 = -160.$$

$\phi(T_{64}(10000,8),6) = T_{64}(10000,8) + T_{65}(10000,8) + \phi(T_{66}(10000,8),1) +$

$\qquad + \phi(T_{68}(10000,8),2) + \phi(T_{72}(10000,8),3) + \phi(T_{80}(10000,8),4) +$

$\qquad + \phi(T_{96}(10000,8),5) =$

$\quad = -588 + 294 + \phi(196,1) + \phi(117,2) +$

$\qquad + \phi(84,3) + \phi(53,4) + \phi(T_{96}(10000,8),5) =$

$\quad = -588 + 294 + 98 + \left(\dfrac{2\cdot114}{6}+1\right) + \left(\dfrac{8\cdot90}{30}-1\right) + 13 +$

$\qquad + \phi(T_{96}(10000,8),5) = -121 + \phi(T_{96}(10000,8),5).$

$\phi(T_{128}(10000,8),7) = T_{128}(10000,8) + T_{129}(10000,8) +$

$\qquad + \phi(T_{130}(10000,8),1) + \phi(T_{132}(10000,8),2) + \phi(T_{136}(10000,8),3) +$

$\qquad + \phi(T_{144}(10000,8),4) + \phi(T_{160}(10000,8),5) + \phi(T_{192}(10000,8),6) =$

$\quad = -526 + 263 + \phi(175,1) + \phi(105,2) + \phi(75,3) + \phi(47,4) +$

$\qquad + \phi(T_{160}(10000,8),5) + \phi(T_{192}(10000,8),6) =$

$\quad = -526 + 263 + 88 + \left(\dfrac{2\cdot102}{6}+1\right) + \left(\dfrac{8\cdot60}{30}+4\right) + 12 +$

$\qquad + \phi(T_{160}(10000,8),5) + \phi(T_{192}(10000,8),6) =$

$\quad = -108 + \phi(T_{160}(10000,8),5) + \phi(T_{192}(10000,8),6).$

$\phi(T_{96}(10000,8),5) = T_{96}(10000,8) + T_{97}(10000,8) +$

$\qquad + \phi(T_{98}(10000,8),1) + \phi(T_{100}(10000,8),2) + \phi(T_{104}(10000,8),3) +$

$\qquad + \phi(T_{112}(10000,8),4) =$

$\quad = 45 - 22 - \phi(15,1) - \phi(9,2) - \phi(6,3) - \phi(4,4) =$

$\quad = 45 - 22 - 8 - 3 - 1 - 1 = 10.$

$\phi(T_{160}(10000,8),5) = T_{160}(10000,8) + T_{161}(10000,8) +$

$\qquad + \phi(T_{162}(10000,8),1) + \phi(T_{164}(10000,8),2) + \phi(T_{168}(10000,8),3) +$

$\qquad + \phi(T_{176}(10000,8),4) =$

$\quad = 40 - 20 - \phi(13,1) - \phi(8,2) - \phi(5,3) - \phi(3,4) =$

$\quad = 40 - 20 - 7 - 3 - 1 - 1 = 8.$

$$\phi(T_{192}(10000, 8), 6) = T_{192}(10000, 8) + T_{193}(10000, 8) +$$
$$+ \phi(T_{194}(10000, 8), 1) + \phi(T_{196}(10000, 8), 2) + \phi(T_{200}(10000, 8), 3) +$$
$$+ \phi(T_{208}(10000, 8), 4) + \phi(T_{224}(10000, 8), 5) =$$
$$= 30 - 15 - \phi(10, 1) - \phi(6, 2) - \phi(4, 3) - \phi(2, 4) +$$
$$+ \phi(T_{224}(10000, 8), 5) =$$
$$= 30 - 15 - 5 - 2 - 1 - 1 + \phi(T_{224}(10000, 8), 5) =$$
$$= 6 + \phi(T_{224}(10000, 8), 5).$$

$$\phi(T_{224}(10000, 8), 5) = T_{224}(10000, 8) + T_{225}(10000, 8) +$$
$$+ \phi(T_{226}(10000, 8), 1) + \phi(T_{228}(10000, 8), 2) + \phi(T_{232}(10000, 8), 3) +$$
$$+ \phi(T_{240}(10000, 8), 4) =$$
$$= -2 + 1 + \phi(0, 1) + \phi(0, 2) + \phi(0, 3) + \phi(0, 4) = -1.$$

Summing up all this we obtain

$$\phi(10000, 8) = 2077 - 160 - 121 + 10 - 108 + 8 + 6 - 1 = 1711.$$

Before giving a formal description of Mapes' algorithm, we should like to take advantage of having worked through this tiresome calculation to indicate the *structure* of the whole scheme we have been using to decompose $\phi(10000, 8)$ into simpler terms, together with the binary representations of the subscripts of the different M's occurring during the computations:

$$\phi(10^4, 8) =$$

$$10^4 - \frac{10^4}{2} - \phi\left(\frac{10^4}{3}, 1\right) - \phi\left(\frac{10^4}{5}, 2\right) - \phi\left(\frac{10^4}{7}, 3\right) - \phi\left(\frac{10^4}{11}, 4\right)$$

$$- \frac{10^4}{13} + \frac{10^4}{2\cdot13} + \phi\left(\frac{10^4}{3\cdot13}, 1\right) + \cdots + \phi\left(\frac{10^4}{11\cdot13}, 4\right)$$

$$- \frac{10^4}{17} + \frac{10^4}{2\cdot17} + \phi\left(\frac{10^4}{3\cdot17}, 1\right) + \cdots + \phi\left(\frac{10^4}{11\cdot17}, 4\right)$$

$$+ \frac{10^4}{13\cdot17} - \frac{10^4}{2\cdot13\cdot17} - \phi\left(\frac{10^4}{3\cdot13\cdot17}, 1\right) - \cdots - \phi\left(\frac{10^4}{11\cdot13\cdot17}, 4\right)$$

$$- \frac{10^4}{19} + \frac{10^4}{2\cdot19} + \phi\left(\frac{10^4}{3\cdot19}, 1\right) + \cdots + \phi\left(\frac{10^4}{11\cdot19}, 4\right)$$

$$+ \frac{10^4}{13 \cdot 19} - \frac{10^4}{2 \cdot 13 \cdot 19} - \phi\left(\frac{10^4}{3 \cdot 13 \cdot 19}, 1\right) - \cdots - \phi\left(\frac{10^4}{11 \cdot 13 \cdot 19}, 4\right)$$

$$+ \frac{10^4}{17 \cdot 19} - \frac{10^4}{2 \cdot 17 \cdot 19} - \phi\left(\frac{10^4}{3 \cdot 17 \cdot 19}, 1\right) - \cdots - \phi\left(\frac{10^4}{11 \cdot 17 \cdot 19}, 4\right)$$

$$- \frac{10^4}{13 \cdot 17 \cdot 19} + \frac{10^4}{2 \cdot 13 \cdot 17 \cdot 19} + \phi\left(\frac{10^4}{3 \cdot 13 \cdot 17 \cdot 19}, 1\right) + \cdots + \phi\left(\frac{10^4}{11 \cdot 13 \cdot 17 \cdot 19}, 4\right).$$

The corresponding values of M are:

$$
\begin{array}{ccc}
2^0, & 2^1, \ldots & 2^4 \\
2^5 + 2^0, & 2^5 + 2^1, \ldots & 2^5 + 2^4 \\
2^6 + 2^0, & 2^6 + 2^1, \ldots & 2^6 + 2^4 \\
2^6 + 2^5 + 2^0, & 2^6 + 2^5 + 2^1, \ldots & 2^6 + 2^5 + 2^4 \\
2^7 + 2^0, & 2^7 + 2^1, \ldots & 2^7 + 2^4 \\
2^7 + 2^5 + 2^0, & 2^7 + 2^5 + 2^1, \ldots & 2^7 + 2^5 + 2^4 \\
2^7 + 2^6 + 2^0, & 2^7 + 2^6 + 2^1, \ldots & 2^7 + 2^6 + 2^4 \\
2^7 + 2^6 + 2^5 + 2^0, & 2^7 + 2^6 + 2^5 + 2^1, \ldots, & 2^7 + 2^6 + 2^5 + 2^4
\end{array}
$$

The primes from $p_6 = 13$ and onwards in the denominators are chosen according to the following scheme, where the ZEROs and ONEs are the leading binary digits of the corresponding M-values. The last five binary digits of M are used to indicate which of the five primes ≤ 11 is involved in each term in a row:

19	17	13	M
		1	$1 \cdot 32 + 1, 2, 4, 8, 16$
	1	0	$2 \cdot 32 + 1, 2, 4, 8, 16$
	1	1	$3 \cdot 32 + 1, 2, 4, 8, 16$
1	0	0	$4 \cdot 32 + 1, 2, 4, 8, 16$
1	0	1	$5 \cdot 32 + 1, 2, 4, 8, 16$
1	1	0	$6 \cdot 32 + 1, 2, 4, 8, 16$
1	1	1	$7 \cdot 32 + 1, 2, 4, 8, 16$

Mapes' Algorithm

Comparing this computation with the one previously shown for $\phi(10000, 8)$, it appears more complicated, but for large values of x it is actually much faster. As a matter of fact, for large values of x Mapes' algorithm requires a computing time roughly proportional to $x^{0.7}$, while Lehmer's formula (1.5) is somewhat slower. It is, however, a problem to keep track of all the different uses of formula (1.12) which have to be made. To show how this can be done, we reproduce below the algorithm given in Mapes' paper [9]. It is based on the idea of calculating Legendre's sum by Meissel's or Lehmer's formulas, whenever possible, and to use auxiliary tables, like the one we have used above, whenever *this* is possible. Suppose you possess tables for all small values of a and that you use the tabulated values also to find $\phi(x, a)$ by (1.9) for some larger values of a, as described on p. 19. A reasonable maximal value of a could then be $a = 10$, say.

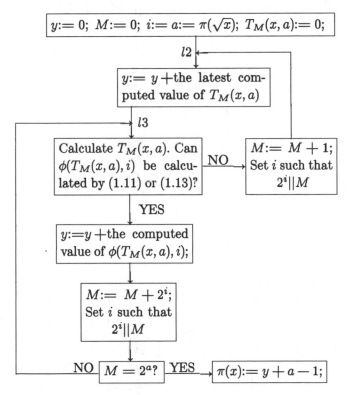

Above is a flow-chart and below a description of the algorithm for computing $\pi(x)$. In the flow-chart, y denotes the sum of the terms of the Legendre sum calculated so far. The flow chart is constructed according to the following rules:

1. For the computation of $\pi(x)$, use the formula

$$\pi(x) = \phi(x,a) + a - 1, \quad \text{with } a = \pi(\sqrt{x}). \qquad (1.26)$$

2. Use tables of $\phi(z,a)$ for all a up to e.g. 6, and all z up to $\frac{1}{2}p_1 p_2 \ldots p_a$ to compute this function for all a up to 10, say. Use also a table of primes, the larger the better.

3. If $p_a < z < p_{a+1}^4$, compute $\phi(z,a)$ as

$$\phi(z,a) = \sum_{s=0}^{3} P_s(z,a) = 1 + \pi(z) - a + P_2(z,a) + P_3(z,a), \qquad (1.27)$$

where $P_2(z,a)$ is the Meissel sum (1.4) and $P_3(z,a)$ is the Lehmer sum (1.6). (If $z \leq p_a$, then $\phi(z,a) = 1$.)

4. Now compute $\phi(z,a)$ by aid of the tables and (1.27), whenever possible, and by re-application of (1.25) if $a >$ the limit chosen for the tables (10 in our suggestion) and if $\pi(z)$ is too large to be found in the table of primes used.

In order to demonstrate how Mapes' method works, let us examine the calculation of $\pi(10^6)$. The largest prime below $\sqrt{x} = 10^3$ is $p_{168} = 997$. Thus, we start by writing

$$\pi(10^6) = \phi(10^6, 168) + 167.$$

Suppose we have a table of the primes < 2000 stored in the computer and also a table of $\pi(x)$ up to $x = 2000$. Suppose, also, that we have a PASCAL function giving $\phi(x,a)$ for all a up to 10 and, whenever possible, that $\phi(x,a)$ is computed backwards from Lehmer's formula (1.7) by aid of a function philehmer(x,a) for $a > 10$. Then the flow-chart above results in the addition of the following entities, a fact which has been recorded by putting a so-called tracer on a computer program representing the flow-chart:

$$x - \phi\left(\frac{x}{2}, 0\right) - \cdots - \phi\left(\frac{x}{31}, 10\right)$$

$$-\frac{x}{37} + \phi\left(\frac{x}{2 \cdot 37}, 0\right) + \cdots + \phi\left(\frac{x}{31 \cdot 37}, 10\right)$$

$$-\frac{x}{41} + \phi\left(\frac{x}{2 \cdot 41}, 0\right) + \cdots + \phi\left(\frac{x}{37 \cdot 41}, 11\right)$$

$$-\frac{x}{43} + \phi\left(\frac{x}{2 \cdot 43}, 0\right) + \cdots + \phi\left(\frac{x}{41 \cdot 43}, 12\right)$$

$$\vdots$$

$$-\frac{x}{499} + \phi\left(\frac{x}{2 \cdot 499}, 0\right) + \cdots + \phi\left(\frac{x}{491 \cdot 499}, 94\right).$$

Here the computation changes pattern, due to the fact that for $p_i > 500$ the value of $10^6/p_i$ is less than 2000, and thus $\pi(10^6/p_i)$ can be found directly in the small prime table stored. In this way the values of $\phi(x/p_i)$ can be computed directly as they stand:

$$-\phi\left(\frac{x}{503}, 95\right) - \phi\left(\frac{x}{509}, 96\right) - \cdots - \phi\left(\frac{x}{997}, 167\right).$$

Programming Mapes' Algorithm

Having prepared a PASCAL function phi(x,a) and a small table of $\pi(x)$, it is now easy to program Mapes' algorithm, following the above flowchart. The only tricky part remaining is how to handle the administrative portion of the program containing the very large numbers M. Since the M's are numbers which in binary representation contain only very few ONE's, a good solution to this problem is to construct a PASCAL procedure addsparse for the addition of "sparse binary integers." A sparse binary integer M may be stored in an array of integers a, containing information about those powers of 2 which occur in M. For instance, $M = 224$, chosen from the computation above, which has the binary representation $224 = 32 + 64 + 128 = 2^5 + 2^6 + 2^7$ could be stored as a[0]=3, a[1]=5, a[2]=6, a[3]=7, where a[0] contains the *number* of ONE's in M.—As a suggestion the procedure could be written as follows:

```
PROCEDURE addsparse(u,v : vector; VAR w : vector);
{Adds two sparse binary integer arrays u and v
```

```
    and places the result in w}
LABEL 1;
VAR  i,j,k,iu,iv,iw,sw,z : INTEGER;

BEGIN iu:=u[0]; iv:=v[0]; iw:=iu+iv; w[0]:=iw;
  i:=1; j:=1; k:=1; WHILE (i <= iu) AND (j <= iv) DO
    BEGIN {Here the arrays u and v are merged into w}
      IF u[i]<v[j]
        THEN BEGIN w[k]:=u[i]; i:=i+1 END
        ELSE BEGIN w[k]:=v[j]; j:=j+1 END;
      k:=k+1
    END;
  IF i=iu+1 THEN FOR i:=j TO iv DO w[k+i-j]:=v[i];
  IF j=iv+1 THEN FOR j:=i TO iu DO w[k+j-i]:=u[j];
  {Here w is reduced to "standard form" with all
   additions of individual ONE's carried out}
1: j:=w[0]; FOR i:=2 TO j DO

  BEGIN
    IF w[i-1]=w[i] THEN
      BEGIN w[i-1]:=-1; w[i]:=w[i]+1 END
  END;
  k:=0; FOR i:=1 TO j DO IF w[i] >= 0 THEN
    BEGIN k:=k+1; w[k]:=w[i] END;
  w[0]:=k; sw:=0; FOR i:=2 TO k DO
    IF w[i-1] > w[i] THEN
      BEGIN sw:=1; z:=w[i]; w[i]:=w[i-1]; w[i-1]:=z END;
  IF sw=1 THEN GOTO 1
  {Here w has been reduced to standard form}
END {addsparse};
```

Recent Developments

J. C. Lagarias, V. S. Miller, and A. M. Odlyzko [10] have recently developed
a new variant of the Meissel-Lehmer method, in which fewer terms of
the form $\phi(y, b)$ are needed in the decomposition of $\phi(x, a)$ according to
(1.9). This leads to an algorithm, asymptotically faster than Mapes', for
computing $\pi(x)$, which its discoverers have used to compute $\pi(x)$ for some
large values of x.

In [11], Lagarias and Odlyzko describe an entirely new analytic method for computing $\pi(x)$, based upon the following formula, which is more general than (2.17):

$$\sum_{m=1}^{\infty} \sum_{p^m \leq x} \frac{1}{m} c(p^m) = \lim_{T \to \infty} \frac{1}{2\pi i} \int_{2-iT}^{2+iT} F(s) \ln \varsigma(s) \, ds \qquad (1.28)$$

in which

$$c(u) = \lim_{T \to \infty} \frac{1}{2\pi i} \int_{2-iT}^{2+iT} F(s) u^{-s} \, ds. \qquad (1.29)$$

The formula is valid whenever the function $F(s)$ is sufficiently well-behaved. (For some background the reader is referred to pp. 47–52.) The use of this formula for efficient computation of $\pi(x)$ requires that a suitable function $F(s)$ is chosen in order that the integral in (1.28) is possible to calculate fast and accurately. (The function F chosen by Lagarias and Odlyzko actually is a function of x as well as of several parameters, but this is not exhibited in (1.28) and (1.29).) After this step the calculation of $\sum_{j=2}^{\infty} \pi(x^{1/j})/j$, and of the improper integral on the right-hand-side of (1.28) to within less than $\pm \frac{1}{2}$ have to be carried out. The sum, which is actually a finite sum since $\pi(x^{1/j}) = 0$ as soon as $x^{1/j}$ becomes < 2, presents no difficulty apart from a prime count up to $x^{\frac{1}{2}}$ (which could be replaced by a recursive computation with the same formula for the argument $x^{\frac{1}{2}}$).

Results

Mapes used his algorithm to compute $\pi(x)$ for every million up to 1000 millions. These computations were further extended in 1972 by Jan Bohman [12], who managed to compute $\pi(x)$ for some isolated values of x as high as $4 \cdot 10^{12}$. He also attempted to compute $\pi(10^{13})$, but, as it has recently been shown, the value he found was slightly in error. Recently Victor Miller has computed $\pi(x)$ for selected values up to $4 \cdot 10^{16}$, this being at present the highest value of x for which $\pi(x)$ is known. Computing this single value took almost 30 of hours running time on a large computer. We give here a very short table of $\pi(x)$:

37

n	$\pi(10^n)$	n	$\pi(10^n)$
3	168	10	455 051 511
4	1 229	11	4 118 054 813
5	9 592	12	37 607 912 018
6	78 498	13	346 065 536 839
7	664 579	14	3 204 941 750 802
8	5 761 455	15	29 844 570 422 669
9	50 847 534	16	279 238 341 033 925

A more complete table is provided at the end of the book.

Comparison Between the Methods Discussed

The six methods to compute $\pi(x)$ discussed in this book, those of Legendre, Meissel, Lehmer, Mapes, Lagarias-Miller-Odlyzko and Lagarias-Odlyzko, differ in various respects. In order to cut down the computational labour, more and more complicated formulas are constructed. However, most of these new formulas also demand that $\pi(x)$ is known higher and higher up, at least for some key values of x. Thus, Bohman's computation of $\pi(4 \cdot 10^{12})$, e.g. made use of various values of $\pi(x)$ as high as $x = 4 \cdot 10^8$.

In the following table the growth of the computing time with x and of the storage space required is given for each of the methods discussed in the text:

Method	Time	Storage
Legendre	$O(x)$	$O(x^{\frac{1}{2}})$
Meissel	$O(x/(\ln x)^3)$	$O(x^{\frac{1}{2}}/\ln x)$
Lehmer	$O(x/(\ln x)^4)$	$O(x^{\frac{1}{3}}/\ln x)$
Mapes	$O(x^{0.7})$	$O(x^{1-\epsilon})$
Lagarias-Miller-Odlyzko	$O(x^{\frac{2}{3}+\epsilon})$	$O(x^{\frac{1}{3}+\epsilon})$
Lagarias-Odlyzko	$O(x^{\frac{3}{5}+\epsilon})$	$O(x^{\epsilon})$

Here the number ϵ occurring in some of the exponents tends to zero through positive values, as $x \to \infty$.

Bibliography

1. D. N. Lehmer, *Factor Table For the First Ten Millions Containing the Smallest Factor of Every Number Not Divisible By 2, 3, 5, Or 7 Between the Limits 0 and 10,017,000,* Hafner, New York, 1956. (Reprint)

2. D. N. Lehmer, *List of Prime Numbers From 1 to 10,006,721,* Hafner, New York, 1956. (Reprint)

3. C. L. Baker and F. J. Gruenberger, *The First Six Million Prime Numbers,* The Rand Corporation, Santa Monica. Published by Microcard Foundation, Madison, Wisconsin, 1959.

4. C. L. Baker and F. J. Gruenberger, *Primes in the Thousandth Million,* The Rand Corporation, Santa Monica, 1958.

5. M. F. Jones, M. Lal and W. J. Blundon, "Statistics on Certain Large Primes," *Math. Comp.* **21** (1967) pp. 103–107.

6. Carter Bays and Richard H. Hudson, "On the Fluctuations of Littlewood for Primes of the Form $4n \pm 1$," *Math. Comp.* **32** (1978) pp. 281–286.

7. A. M. Legendre, *Théorie des Nombres,* Third edition, Paris, 1830. Vol. 2, p. 65.

8. D. H. Lehmer, "On the Exact Number of Primes Less Than a Given Limit," *Ill. Journ. Math.* **3** (1959) pp. 381–388. Contains many references to earlier work.

9. David C. Mapes, "Fast Method for Computing the Number of Primes Less Than a Given Limit," *Math. Comp.* **17** (1963) pp. 179–185.

10. J. C. Lagarias, V. S. Miller and A. M. Odlyzko, "Computing $\pi(x)$: The Meissel-Lehmer Method," *Math. Comp.* (To appear)

11. J. C. Lagarias and A. M. Odlyzko, "Computing $\pi(x)$: An Analytic Method," *Math. Comp.* (To appear)

12. Jan Bohman, "On the Number of Primes Less Than a Given Limit," *Nordisk Tidskr. för informationsbehandling (BIT)* **12** (1972) pp. 576–577.

CHAPTER 2

THE PRIMES VIEWED AT LARGE

Introduction

Not very much is known about the distribution of the primes. On one hand, their distribution in short intervals seems extremely irregular. This is the reason why it appears impossible to find a simple formula describing the distribution of the primes in any detail. On the other hand, the distribution of the primes, viewed at large, can be very well approximated by simple formulas.

As a mathematical theory, the distribution of prime numbers is quite inhomogeneous. Although Euclid has proved that there are infinitely many primes, and Legendre and Gauss, as early as about 1800, conjectured some of the basic theorems, the theory is still a mixture of unsolved problems, more or less reasonable conjectures and a few proved theorems. The proved theorems mostly cover only simple cases, as compared with existing conjectures, and their proofs are often extremely complicated. Many of the proofs are not elementary, relying upon theorems in the theory of functions. This is the reason why in this and the next chapter we sometimes have to refrain from proving even some of the fundamental theorems.—We shall also discuss some of the existing conjectures, together with theoretical or numerical evidence which appears to support or to contradict the conjecture in question.

No Polynomial Can Produce Only Primes

In the search for formulas yielding all primes (and no other numbers) some remarkable polynomials have been found, whose values contain a surprisingly large proportion of primes. One of these is $P(x) = x^2 - x + 17$ which is prime for $x = 0, 1, 2, 3, \ldots, 16$ but obviously is composite for $x = 17$, since $P(17) = 17^2 - 17 + 17$ must be divisible by 17. Still more remarkable is the polynomial $x^2 - x + 41$, found by Euler, yielding primes for $x = 0, 1, 2, \ldots, 40$ but being composite for $x = 41$, since

$41|41^2 - 41 + 41$.—We here take the opportunity to indicate a connection between these remarkable polynomials and those quadratic fields $K(\sqrt{D})$ in which the theorem of unique factorization into prime factors is valid. The two polynomials mentioned above as examples of polynomials rich in primes have precisely the discriminants $D = -67$ and $D = -163$, mentioned in Theorem A4.4 on p. 298.—Another polynomial, being even richer in primes than $x^2 - x + 41$ is Edgar Karst's $2x^2 - 199$, yielding 150 primes (and the number 1) for $x = 0, 1, 2, \ldots, 198$.—The proof that no (non-constant) polynomial can yield only primes is quite simple. The assertion follows from the fact that any polynomial is unbounded when the variables tend to infinity. (It may happen that the values of a polynomial are bounded when the variables tend to infinity in *certain directions*, but not in *all* directions.) Suppose, there were a polynomial in n variables, $P(x, y, z, \ldots)$, yielding only primes for integer values of the variables. First, write $P = Q_n(x, y, z, \ldots) +$ terms of lower degree $+ a$, where Q_n is a *homogenous* polynomial of degree n, representing all terms of the highest degree n of P, and a is the constant term, which we to begin with shall assume is $\neq 0$ and $\neq \pm 1$. If $n > 0$, then Q_n tends to infinity when the variables do so in at least one direction $(\xi, \eta, \varsigma, \ldots)$ in n-space because if Q_n is considered as a polynomial of one of its variables only, it has this property. Next, because Q_n is continuous, Q_n tends to infinity not only in the direction mentioned, but also in some narrow cone, with $(\xi, \eta, \varsigma, \ldots)$ as axis, and the same is true for $P(x, y, z, \ldots)$, being dominated by its highest degree terms, Q_n, as all the variables tend to infinity in the direction considered. Finally, if all the variables are chosen as integer multiples of the constant term a, clearly $a|P(x, y, z, \ldots)$. Since any cone will, only we proceed far enough from the origin, contain points belonging to any point lattice $(al, am, an, \ldots)$, where $l, m, n, \ldots$ all are integers, the above construction leads to *integer* values of the variables $(x, y, z, \ldots)$ for which $a|P(x, y, z, \ldots)$ with $P(x, y, z, \ldots)$ large, i.e. with the quotient $|P(x, y, z, \ldots)/a| > 1$, showing that the value of $P(x, y, z, \ldots)$ is *composite* for the particular set of variables arrived at in this way. (It is only at this very last point of the proof that use is made of the assumption that $a \neq 0$ or $\neq \pm 1$.)

If $a = 0$ or ± 1, then we start by moving the origin to a point $(b, c, d, \ldots)$ with integer coordinates, for which the value of $P(b, c, d, \ldots) = s$ is large. That such a point exists is clear from the proof given above. This transformation $x' = x - b$, $y' = y - c$, $z' = z - d, \ldots$ gives a new polynomial $P'(x', y', z', \ldots)$ with its constant term $P'(0, 0, 0, \ldots) = P(b, c, d, \ldots) = s$, now a *large* integer and thus $\neq 0$ or ± 1. Since $x, y, z, \ldots$ and $x', y', z', \ldots$ take integer values at the same time, the sets of values of P and of P' for integer values of the variables are also the same. Applying our proof for the case when $a \neq 0$ or ± 1 on P' we arrive at the conclusion that neither in one of these cases can P take only prime values. This concludes the proof that no (non-constant) polynomial can give only primes for integer values of the variables.

Formulas Yielding All Primes

There *exist* certain formulas which yield all the primes and no other numbers. However, these are misleading, in that they either presuppose, in some latent way, the knowledge of each individual prime, or rely on the bogus definition of a prime as a non-composite number.—As an example of a formula of the first kind consider the following algorithm: Taking the number $x = 0.203050701101301701902302903\ldots$ as a starting point, extract a suitable total of adjacent digits and the primes will emerge! This is quite obviously cheating, since all the individual primes must be known in advance before the number x can be exploited.

A formula of the second kind mentioned is the polynomial given below, *whose positive values consist of all the primes, when the variables range over all non-negative integers.* The polynomial also yields negative values (as a matter of fact it does so for most values of the variables), but these are not necessarily (negative) primes.

$$(k+2)\Big\{1-[wz+h+j-q]^2-[(gk+2g+k+1)(h+j)+h-z]^2-$$
$$-\left[16(k+1)^3(k+2)(n+1)^2+1-f^2\right]^2-[2n+p+q+z-e]^2-$$
$$-\left[e^3(e+2)(a+1)^2+1-o^2\right]^2-\left[(a^2-1)y^2+1-x^2\right]^2-$$
$$-\left[16r^2y^4(a^2-1)+1-u^2\right]^2-[n+l+v-y]^2-$$
$$-\left[(a^2-1)l^2+1-m^2\right]^2-[ai+k+1-l-i]^2-$$
$$-\left[\{(a+u^2(u^2-a))^2-1\}(n+4dy)^2+1-(x+cu)^2\right]^2-$$
$$-\left[p+l(a-n-1)+b(2an+2a-n^2-2n-2)-m\right]^2-$$
$$-\left[q+y(a-p-1)+s(2ap+2a-p^2-2p-2)-x\right]^2-$$
$$-\left[z+pl(a-p)+t(2ap-p^2-1)-pm\right]^2\Big\}.$$

For the deduction of these types of formula, see [1].—The reader might wonder how this expression, being the product of two factors, $(k+2)$ and the complicated factor within the large curly brackets, can produce any primes at all. Well, this is merely an apparent paradox, since the only *positive* value assumed by the second factor happens to be the value 1. Looking closer at the second factor the reader will notice that it has the form

$$1 - \sum_{i=1}^{14} (\text{expression}_i)^2$$

A factor of this form can obviously take only the values 1, 0, and *negative* values, and thus, once again we have been deceived, the whole thing being the following statement in disguise:

k + 2 is prime if and only if the following Diophantine system of 14 equations in 26 variables has a positive integral solution:

$$
\begin{cases}
wz + h + j - q = 0 \\
(gk + 2g + k + 1)(h + j) + h - z = 0 \\
16(k+1)^3(k+2)(n+1)^2 + 1 - f^2 = 0 \\
2n + p + q + z - e = 0 \\
e^3(e+2)(a+1)^2 + 1 - o^2 = 0 \\
(a^2 - 1)y^2 + 1 - x^2 = 0 \\
16r^2y^4(a^2 - 1) + 1 - u^2 = 0 \\
n + l + v - y = 0 \\
(a^2 - 1)l^2 + 1 - m^2 = 0 \\
ai + k + 1 - l - i = 0 \\
\{(a + u^2(u^2 - a))^2 - 1\}(n + 4dy)^2 + 1 - (x + cu)^2 = 0 \\
p + l(a - n - 1) + b(2an + 2a - n^2 - 2n - 2) - m = 0 \\
q + y(a - p - 1) + s(2ap + 2a - p^2 - 2p - 2) - x = 0 \\
z + pl(a - p) + t(2ap - p^2 - 1) - pm = 0.
\end{cases}
$$

It is the author's hope that the reader has not been greatly disappointed by this revelation of the true nature of the prime-producing polynomial. Remember that *no polynomial can produce only primes*, so that there must be some trick involved in arriving at a polynomial producing all positive primes (and a large number of negative composite integers).

The Distribution of Primes Viewed at Large. Euclid's Theorem

We now give Euclid's extremely elegant proof of the infinitude of primes. (This proof, as a matter of fact, is frequently given as an example of indirect proof.) Suppose there exist only a finite number of primes, $p_1, p_2, \ldots, p_n$. Now, consider the integer $N = p_1 p_2 \cdots p_n + 1$. None of the existing primes divides N, since the division N/p_i will always give the remainder 1. Thus either N is a (new) prime number, or N contains a (new) prime factor, which is different from all the ones given. Therefore we conclude that there must be an infinitude of primes.

Example. The following construction starts with the prime 2 and yields at least one new prime in each step:

$N_2 = 2 + 1 = 3$ (prime)

$N_3 = 2 \cdot 3 + 1 = 7$ (prime)

$N_4 = 2 \cdot 3 \cdot 7 + 1 = 43$ (prime)

$N_5 = 2 \cdot 3 \cdot 7 \cdot 43 + 1 = 1807 = 13 \cdot 139$ (yielding two primes)

$N_6 = 2 \cdot 3 \cdot 7 \cdot 43 \cdot 139 + 1 = 251085 = 5 \cdot 50207$ (yielding two primes)

$N_7 = 2 \cdot 3 \cdot 7 \cdot 43 \cdot 139 \cdot 50207 + 1 = 126\,03664039 =$
$= 23 \cdot 1607 \cdot 340999$ (yielding three primes)

$N_8 = 2 \cdot 3 \cdot 7 \cdot 43 \cdot 139 \cdot 50207 \cdot 340999 + 1 = 4298368\,33293963 =$
$= 23 \cdot 79 \cdot 23653\,47734339$ (yielding three primes)

$N_9 = 2 \cdot 3 \cdot 7 \cdot 43 \cdot 139 \cdot 50207 \cdot 340999 \cdot 23652\,47734339 + 1 =$
$= 10165\,87861619\,05754590\,68761119 =$
$= 17 \cdot 1277\,70091783 \cdot 4\,68022256\,41471129$ (yielding three primes)

The Formulas of Gauss and Legendre for $\pi(x)$

The Prime Number Theorem

Let us start as Legendre and Gauss did and try to estimate the number of primes $\leq x$ by counting the number s of primes in suitable intervals. We choose the intervals $[0.95 \cdot 10^n, 1.05 \cdot 10^n]$ for $n = 3(1)7$ and the intervals $[10^n, 10^n + 150000]$ for $n = 8(1)15$. In each of these intervals of length d we calculate the proportion $(s/d) \times 10,000$ of prime numbers. The result is shown in the following table, where the values given in the last two columns have been rounded:

The number s of primes in different intervals $[x - d/2, x + d/2]$				
x	d	s	$\dfrac{10000s}{d}$	$\dfrac{10000}{\ln x}$
10^3	10^2	15	1500	1448
10^4	10^3	107	1070	1086
10^5	10^4	867	867	869
10^6	10^5	7227	723	724
10^7	10^6	62031	620	620
$10^8 + 75000$	150000	8154	544	543
$10^9 + 75000$	150000	7242	483	483
$10^{10} + 75000$	150000	6511	434	434
$10^{11} + 75000$	150000	5974	398	395
$10^{12} + 75000$	150000	5433	362	362
$10^{13} + 75000$	150000	5065	338	334
$10^{14} + 75000$	150000	4643	310	310
$10^{15} + 75000$	150000	4251	283	290

By studying the figures, we observe that the density of primes in an interval, centred around x, slowly decreases as x grows. Which law does this function obey? Comparing the values found for $x = 10^n$ and $x = 10^{2n}$, we find that the density of primes is approximately halved when x is squared. Mathematically, this is described by the function $1/\ln x$, since $1/\ln(x^2) = 0.5/\ln x$. Let us compare the density of primes with $1/\ln x$ (natural logarithms!), the values of which we have given in the last column of the table above. We see that both columns agree well apart from the smaller values of x. This disagreement is obviously caused by local irregularities in the distribution of primes, which more heavily influence the number of primes in short than in long intervals. This striking agreement between the density of primes in an interval centred around x and the function $1/\ln x$ was discovered independently by Legendre and by Gauss, who formulated the following approximations to $\pi(x)$:

$$\pi(x) \approx \frac{x}{\ln x - B}, \quad B = 1.08366 \quad \text{(Legendre)} \tag{2.1}$$

$$\pi(x) \approx \int_2^x \frac{dx}{\ln x} \quad \text{(Gauss)} \tag{2.2}$$

Nowadays, the latter approximation is usually replaced by the so-called logarithmic integral, defined by

$$\operatorname{li} x = \int\limits_0^x \frac{dx}{\ln x},$$

where this improper integral has to be interpreted as

$$\operatorname{li} x = \lim_{\epsilon \to +0} \left(\int\limits_0^{1-\epsilon} \frac{dx}{\ln x} + \int\limits_{1+\epsilon}^x \frac{dx}{\ln x} \right). \tag{2.3}$$

The approximation by Gauss (2.2) and the logarithmic integral differ only by a constant, $\operatorname{li} 2 = 1.045$.—The approximations by Gauss and Legendre are in fact related, since

$$\int \frac{dx}{\ln x} = \frac{x}{\ln x} + \int \frac{dx}{(\ln x)^2} =$$

$$= \frac{x}{\ln x} + \frac{x}{(\ln x)^2} + 2\int \frac{dx}{(\ln x)^3} =$$

$$= \frac{x}{\ln x - 1} - \frac{x}{(\ln x)^2(\ln x - 1)} + 2\int \frac{dx}{(\ln x)^3}.$$

The last two terms are of smaller order of magnitude than the leading term $x/(\ln x - 1)$ as $x \to \infty$, and hence we have

$$\lim_{x \to \infty} \frac{\operatorname{li} x}{x/\ln x} = \lim_{x \to \infty} \frac{\operatorname{li} x}{x/(\ln x - 1)} = 1.$$

We may thus expect that the approximations given by Legendre and by Gauss should be about equally good, unless there is some particular reason in favour of one or the other.

The first mathematician to *prove* something in the direction of these formulas was Chebyshev, who around 1850 proved that

$$A \operatorname{li} x < \pi(x) < B \operatorname{li} x \tag{2.4}$$

for some suitably chosen values of the constants A and B. This establishes that $\pi(x)$ has the same order of magnitude as $\operatorname{li} x$ (and thus also as $x/\ln x$) as $x \to \infty$. (In honour of this mathematician, $\pi(x) \approx \operatorname{li} x$ is often called Chebyshev's approximation.)

In 1896, Hadamard and de la Vallée-Poussin independently of each other proved *The Prime Number Theorem:*

$$\pi(x) \sim \mathrm{li}\, x, \quad \text{as} \quad x \to \infty. \tag{2.5}$$

(This formula reads: $\pi(x)$ is asymptotically equal to $\mathrm{li}\, x$.) The Prime Number Theorem can be reformulated as

$$\lim_{x \to \infty} \frac{\pi(x)}{\mathrm{li}\, x} = 1. \tag{2.6}$$

The Prime Number Theorem provides information about the error introduced by Gauss' approximation. It says that the relative error of the approximation, $(\mathrm{li}\, x - \pi(x))/\pi(x)$, tends to 0 as x tends to infinity. The *absolute error*, $\mathrm{li}\, x - \pi(x)$, however, may be large, something which will be discussed later.

Unfortunately, the scope of this book does not allow a proof of the Prime Number Theorem. We must refer the reader to [2], which gives an elementary (but very tedious) proof, or to [3], which provides a proof based on the theory of functions.

Exercise 2.1. Computing $\mathrm{li}\, x$. Write a FUNCTION li(x) for $\mathrm{li}\, x$, utilizing the continued fraction expansion

$$\mathrm{li}(e^z) = \frac{e^z}{\lfloor z} - \frac{1}{\lfloor 1} - \frac{1}{\lfloor z} - \frac{2}{\lfloor 1} - \frac{2}{\lfloor z} - \frac{3}{\lfloor 1} - \frac{3}{\lfloor z} - \cdots .$$

Compute the continued fraction backwards, starting with the term $10/z$. Test values can be found in Table 3 (compute $\mathrm{li}\, x$ as $\pi(x) + (\mathrm{li}\, x - \pi(x))$).

The Riemann Zeta-function

The approximations to $\pi(x)$ by Gauss and by Legendre were found by empiric methods. Riemann was the first who with great success systematically deduced relations between the primes and already known mathematical functions. Riemann's starting point was a relation discovered already by Euler,

$$\varsigma(s) = \sum_{n=1}^{\infty} \frac{1}{n^s} = \prod_{p} \frac{1}{1 - p^{-s}} , \tag{2.7}$$

where the infinite product is taken over all primes. The function $\varsigma(s)$ is called the Riemann zeta-function and (2.7) is a highly informative formula

47

from which many properties of the primes can be deduced. It is very important because it relates each individual prime p to the simple sum $\sum n^{-s}$. Thus the properties of the primes are via (2.7) transformed into properties of the sum and this *without the necessity of specifying each individual prime!*—(2.7) can be deduced in the following manner: Write each factor of the infinite product as a (convergent) geometric series

$$\frac{1}{1-p^{-s}} = 1 + p^{-s} + p^{-2s} + p^{-3s} + \cdots$$

Multiply all these series together to obtain the result

$$\sum (2^{\alpha_1} 3^{\alpha_2} \ldots p_n^{\alpha_n})^{-s}$$

where the summation must cover all combinations of non-negative integer exponents α_i and all primes p_i. Now the fundamental theorem of arithmetic tells us that the products so obtained are precisely all the positive integers, raised to the power $-s$, because each integer has a unique representation in the form $2^{\alpha_1} 3^{\alpha_2} \ldots p_n^{\alpha_n}$. But this is exactly what appears in $\sum n^{-s}$.

Riemann's basic idea was to put the so-called theory of analytic functions (differentiable functions of one complex variable) to work. This is effected by extending the variable s, which in (2.7) is restricted to $s > 1$, to a complex variable $s = \sigma + it$. In order for $\sum_{n=1}^{\infty} n^{-s}$ to converge, σ must be > 1. However, with so-called analytic continuation, Riemann was able to extend the function to all real and complex values of s except $s = 1$, which is a singularity, and for which $|\varsigma(s)| = \infty$. The extension of the definition of $\varsigma(s)$ to all s with $\sigma > 0$ is achieved by considering

$$(1 - 2 \cdot 2^{-s})\varsigma(s) = \varsigma(s) - 2 \cdot 2^{-s}\varsigma(s) =$$

$$= 1 + 2^{-s} + 3^{-s} + 4^{-s} + 5^{-s} + 6^{-s} + \cdots -$$

$$- 2 \cdot 2^{-s} \quad - \quad 2 \cdot 4^{-s} \quad - \quad 2 \cdot 6^{-s} - \cdots =$$

$$= 1 - 2^{-s} + 3^{-s} - 4^{-s} + 5^{-s} - \cdots = \sum_{n=1}^{\infty} \frac{(-1)^{n-1}}{n^s}.$$

This series converges for all s with $\sigma > 0$. Thus $\varsigma(s)$ may be written as

$$\varsigma(s) = \frac{1}{1 - 2^{1-s}} \sum_{n=1}^{\infty} \frac{(-1)^{n-1}}{n^s}, \quad \text{when } \sigma > 0, \text{ if } s \neq 1. \tag{2.8}$$

For $\sigma \leq 0$ the so-called functional equation

$$\varsigma(1 - s) = 2^{1-s}\pi^{-s}\cos\frac{\pi s}{2}\,\Gamma(s)\varsigma(s) \qquad (2.9)$$

can be used to obtain the values of $\varsigma(s)$. Here $\Gamma(s)$ is the gamma function, defined for $\sigma > 0$ as

$$\Gamma(s) = \int_0^\infty x^{s-1}e^{-x}\,dx. \qquad (2.10)$$

Exercise 2.2. Computing $\varsigma(s)$. $\varsigma(s)$ may be computed by aid of the Euler-MacLaurin sum formula:

$$\varsigma(s) = \sum_{n=1}^{N-1} n^{-s} + \frac{1}{s-1}N^{1-s} + \frac{1}{2}N^{-s} + \frac{s}{12}N^{-s-1} -$$

$$- \frac{s(s+1)(s+2)}{720}N^{-s-3} + \frac{s(s+1)\cdots(s+4)}{30240}N^{-s-5} - \cdots$$

This is a so-called semiconvergent asymptotic series. The truncation error made in breaking off the summation is always less than the immediately following term. Write a FUNCTION zeta(s) performing the summation of the series. In single precision arithmetic (about 8 decimal digits), a suitable value of N is 10. Break off the series immediately before the last term written out above. Check your values against the known exact values of $\varsigma(2) = \pi^2/6$, $\varsigma(4) = \pi^4/90$, $\varsigma(6) = \pi^6/945$, $\varsigma(8) = \pi^8/9450$ and $\varsigma(10) = \pi^{10}/93555$.

The connection between the Riemann zeta-function and the primes is evident from the infinite product in (2.7). To write $\pi(x)$ with the aid of $\varsigma(s)$ will require theorems and techniques from the theory of functions that we have to omit. A detailed deduction can be found in [4]. We can only hint at some of the highlights of the theory. The first is that Riemann found the function

$$f(x) = \pi(x) + \frac{1}{2}\pi(x^{\frac{1}{2}}) + \frac{1}{3}\pi(x^{\frac{1}{3}}) + \frac{1}{4}\pi(x^{\frac{1}{4}}) + \cdots = \sum_{n=1}^\infty \frac{1}{n}\pi(x^{\frac{1}{n}}) \quad (2.11)$$

to be of more fundamental importance than $\pi(x)$ itself for the study of the desired relation between $\pi(x)$ and $\varsigma(s)$. This sum is only formally infinite, since $\pi(x^{1/n}) = 0$, as soon as $x^{1/n}$ decreases below 2, which will happen as soon as $n > \ln x/\ln 2$. $f(x)$ has jump discontinuities with jumps $1/r$ when x passes a prime power p^r. (When x passes a prime p, this is regarded as the prime power p^1.) Let us now (as is usual in working with trigonometric series) modify the definition (2.11) to read

$$f(p^r) = \lim_{\epsilon \to +0} \frac{f(p^r - \epsilon) + f(p^r + \epsilon)}{2}. \qquad (2.12)$$

This means that all jumps have been split into two equal halves. The graph of the function $f(x)$, modified in this way, is shown on p. 51. Each time x is a prime power p^r, $f(x)$ increases by the amount $1/r$, but $f(x)$ is constant between the jumps.

The simplest relation between the Riemann zeta-function and $f(x)$ is

$$\frac{\ln \varsigma(s)}{s} = \int_1^\infty f(x) x^{-s-1}\, dx, \qquad (2.13)$$

which can be deduced as follows:

$$\ln \varsigma(s) = \ln \prod_p (1 - p^{-s})^{-1} = -\sum_p \ln(1 - p^{-s}) =$$

$$= \sum_p p^{-s} + \frac{1}{2} \sum_p p^{-2s} + \frac{1}{3} \sum_p p^{-3s} + \cdots \qquad (2.14)$$

Using Stieltjes' integrals (see Appendix 9 for this important tool!) and performing integration by parts, we obtain

$$s \int_1^\infty \pi(x) x^{-s-1}\, dx = \left[-x^{-s}\pi(x) \right]_1^\infty + \int_1^\infty x^{-s}\, d\pi(x) = \sum_p p^{-s}, \quad (2.15)$$

since for $s > 0$ the integrated term vanishes both at $x = \infty$ and $x = 1$. In an analogous manner, we find that

$$s \int_1^\infty \pi(x^{\frac{1}{n}}) x^{-s-1}\, dx = \int_1^\infty x^{-s}\pi(x^{\frac{1}{n}})\, dx = \sum_p p^{-ns}. \qquad (2.16)$$

Next, inserting the definition of $f(x)$ from (2.11) in the integral of (2.13), and using (2.15) and (2.16) as well as taking (2.14) into account the integral is reduced to $\ln \varsigma(s)/s$.—Formula (2.13) can, in fact, be transformed in several different ways. One form, in which the so-called Mellin transform has been used, is the following:

50

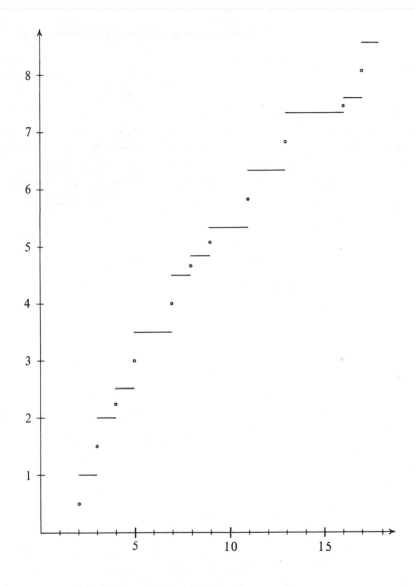

Figure 2.1. The step-function $f(x) = \displaystyle\sum_{n=1}^{\infty} \frac{\pi(x^{1/n})}{n}$

$$\lim_{T \to \infty} \frac{1}{2\pi i} \int_{2-iT}^{2+iT} \frac{x^s}{s} \ln \varsigma(s)\, ds = \begin{cases} \displaystyle\sum_{j=1}^{\infty} \frac{1}{j}\pi(x^{\frac{1}{j}}), & \text{if } x \neq p^m \\[2ex] \displaystyle\sum_{j=1}^{\infty} \frac{1}{j}\pi(x^{\frac{1}{j}}) - \frac{1}{2m}, & \text{if } x = p^m. \end{cases} \qquad (2.17)$$

In this equation $f(x)$ has been expressed with the aid of known functions. However, the integral is very difficult to determine with high accuracy; thus (2.17) is of no immediate value in the computation of $f(x)$ and $\pi(x)$. An efficient formula for numerical computations has recently been devised by Lagarias and Odlysko, and used to calculate $\pi(x)$ for large values of x; see p. 37!

The Zeros of the Zeta-function

There is also a reasonably simple connection between $f(x)$ and the *zeros* of the Riemann zeta-function:

$$f(x) = \operatorname{li} x - \sum_{\rho} \operatorname{li}(x^{\rho}) + \int_{x}^{\infty} \frac{dt}{(t^2 - 1)t \ln t} - \ln 2. \qquad (2.18)$$

This formula was published by Riemann in 1859 and proved by von Mangoldt in 1895. ρ denotes all the complex zeros of the Riemann zeta-function, and $\operatorname{li}(x^{\rho}) = \operatorname{li}(e^{\rho \ln x})$ is the logarithmic integral of a complex variable, defined by

$$\operatorname{li}(e^{u+iv}) = \int_{-\infty+iv}^{u+iv} \frac{e^z}{z}\, dz, \quad v \neq 0 \qquad (2.19)$$

A complication is that $\sum \operatorname{li}(x^{\rho})$ is only a conditionally convergent infinite series; thus the value of its sum is dependent on the order of summation of its terms. The summation has to be carried out in increasing order of magnitude of the complex zeros. It was in connection with these investigations that Riemann formulated his famous conjecture, for which a proof has probably been found by H. Matsumoto in 1984. (Matsumoto's proof has not yet been sufficiently studied by other mathematicians to be considered valid.) Riemann conjectured that the complex zeros of the zeta-function all have their real part $\sigma = 1/2$. By means of extremely laborious but delicate computations, it has been proved by Brent, van de Lune, te Riele and Winter that the first 1,000,000,000 zeros on each side of the σ-axis all lie *exactly* on the line $\sigma = 1/2$ and are simple zeros, see [5] and [6]. This covers the segment $|t| < 372,870,204.6$ of the line $\sigma = 1/2$, since the number of zeros below t is $\approx u \ln u - u - 1/8$, with $u = t/2\pi$. A detailed error analysis of the computations executed in the computer program used proves the correctness of the computer's results.—If the Riemann hypothesis is true, then the order of magnitude of the terms li(x^ρ) in (2.18) will be $O(\sqrt{x})$, and the function $f(x)$ will be approximated by its leading term, li$\,x$, with an error of the order of magnitude $O(\sqrt{x} \ln x)$, as can be proved by a detailed analysis. Thus, *assuming the truth of the Riemann hypothesis*, we have the conjectured error term

$$f(x) \overset{c}{=} \operatorname{li} x + O(\sqrt{x} \ln x). \tag{2.20}$$

Conversion From $f(x)$ Back to $\pi(x)$

If li x is a good approximation to $f(x)$, what can be said about $\pi(x)$? In the definition (2.11), $f(x)$ is a rather complicated function of $\pi(x)$. Fortunately, however, there exists an inversion formula by which $\pi(x)$ can be expressed in terms of $f(x)$:

$$\pi(x) = \sum_{n=1}^{\infty} \frac{\mu(n)}{n} f(x^{\frac{1}{n}}). \tag{2.21}$$

The function $\mu(n)$, which appears here, is called Möbius' function and is defined by

$$\mu(n) = \begin{cases} 1, & \text{if } n = 1 \\ 0, & \text{if } n \text{ contains some multiple prime factor} \\ (-1)^k, & \text{if } n \text{ is the product of } k \text{ distinct primes.} \end{cases} \tag{2.22}$$

The most important property of the Möbius function is

$$\sum_{d|n} \mu(d) = \begin{cases} 1, & \text{if } n = 1 \\ 0, & \text{if } n > 1. \end{cases} \tag{2.23}$$

(Note that also the improper divisors $d = 1$ and $d = n$ of n have to be used in this formula!) To prove (2.23), suppose that $n = \prod_{i=1}^{s} p_i^{\alpha_i}$, with

53

all p_i being different primes. Then $d|n$, and $\mu(d) = (-1)^k$ if d is a product of precisely k different members of the set of s primes p_i. This case will occur for $\binom{s}{k}$ different divisors d of n. All divisors d of n containing one or several of the primes p_i twice or more have $\mu(d) = 0$, according to the definition of $\mu(d)$. Thus

$$\sum_{d|n} \mu(d) = \sum_{k=0}^{s} (-1)^k \binom{s}{k} = (1-1)^s = 0, \quad \text{if} \quad s \geq 1.$$

The relation (2.23) has as one of its consequences that

$$\sum_{n=1}^{\infty} \frac{\mu(n)}{n^s} = \frac{1}{\varsigma(s)}, \tag{2.24}$$

since

$$\varsigma(s) \sum_{n=1}^{\infty} \frac{\mu(n)}{n^s} = \sum_{m=1}^{\infty} \frac{1}{m^s} \sum_{d=1}^{\infty} \frac{\mu(d)}{d^s} = \sum \frac{\sum_{d|md} \mu(d)}{(md)^s} =$$

$$= \sum_{n=1}^{\infty} \frac{\sum_{d|n} \mu(d)}{n^s} = 1 \cdot 1^{-s} = 1.$$

That (2.21) is equivalent to (2.11) is now proved in the following way:

$$\sum_{1}^{\infty} \frac{\mu(n)}{n} f(x^{\frac{1}{n}}) = \sum_{n=1}^{\infty} \left(\frac{\mu(n)}{n} \sum_{m=1}^{\infty} \frac{1}{m} \pi(x^{\frac{1}{mn}}) \right) = \sum_{n=1}^{\infty} \sum_{m=1}^{\infty} \frac{\mu\left(\frac{mn}{m}\right)}{mn} \pi(x^{\frac{1}{mn}})$$

$$= \sum_{u=1}^{\infty} \sum_{m|u} \frac{\mu\left(\frac{u}{m}\right)}{u} \pi(x^{\frac{1}{u}}) = \sum_{u=1}^{\infty} \left(\frac{\pi(x^{\frac{1}{u}})}{u} \sum_{d|u} \mu(d) \right) = \pi(x),$$

according to (2.23).

The Riemann Prime Number Formula

If $f(x)$ in (2.21) is approximated by $\operatorname{li} x$, we obtain Riemann's famous prime number formula:

$$R(x) = \sum_{1}^{\infty} \frac{\mu(n)}{n} \operatorname{li}(x^{\frac{1}{n}}) =$$

$$= \operatorname{li} x - \frac{1}{2} \operatorname{li}(x^{\frac{1}{2}}) - \frac{1}{3} \operatorname{li}(x^{\frac{1}{3}}) - \frac{1}{5} \operatorname{li}(x^{\frac{1}{5}}) + \frac{1}{6} \operatorname{li}(x^{\frac{1}{6}}) - \cdots \tag{2.25}$$

The leading term in (2.25) is the approximation of Gauss, $\operatorname{li} x$, and the maximum error of (2.25) is of the same order of magnitude as it is for the approximation $\operatorname{li} x$. For the numerical computation of $\operatorname{li} x$ and $R(x)$, it is convenient to use their power series expansions in the variable $\ln x$. The deduction for $\operatorname{li} x$ runs

$$\int \frac{dx}{\ln x} = (\text{putting } e^t = x) \int \frac{e^t}{t}\, dt = \int \sum_{n=0}^{\infty} \frac{t^{n-1}\, dt}{n!} =$$

$$= \ln t + \sum_{1}^{\infty} \frac{t^n}{n!\,n} + C_1 = \ln\ln x + \sum_{1}^{\infty} \frac{(\ln x)^n}{n!\,n} + C_1.$$

If the limits of integration are chosen to be 0 and x, it can be shown that the constant of integration assumes the value $\gamma = $ Euler's constant $= 0.5772\ldots$, which gives

$$\operatorname{li} x = \gamma + \ln\ln x + \sum_{n=1}^{\infty} \frac{(\ln x)^n}{n!\,n}. \tag{2.26}$$

The function $R(x)$ can be transformed into the so-called Gram series:

$$R(x) = \sum_{1}^{\infty} \frac{\mu(n)}{n} \operatorname{li}(x^{\frac{1}{n}}) = (\text{putting } x = e^t) \sum_{1}^{\infty} \frac{\mu(n)}{n} \operatorname{li}(e^{\frac{t}{n}}) =$$

$$- \sum_{n=1}^{\infty} \frac{\mu(n)}{n}\left(\gamma + \ln\frac{t}{n} + \sum_{m=1}^{\infty} \frac{\left(\frac{t}{n}\right)^m}{m!\,m}\right) =$$

$$= (\gamma + \ln t) \sum_{1}^{\infty} \frac{\mu(n)}{n} - \sum_{1}^{\infty} \frac{\mu(n)\ln n}{n} + \sum_{n=1}^{\infty}\sum_{m=1}^{\infty} \frac{\mu(n)t^m}{n^{m+1}m!\,m}$$

$$= 1 + \sum_{m=1}^{\infty}\left(\frac{t^m}{m!\,m}\sum_{n=1}^{\infty}\frac{\mu(n)}{n^{m+1}}\right) = 1 + \sum_{m=1}^{\infty}\frac{t^m}{m!\,m\,\varsigma(m+1)}. \tag{2.27}$$

The deduction above is dependent on the following two limits:

$$\sum_{1}^{\infty} \frac{\mu(n)}{n} = \lim_{s\to 1}\sum_{1}^{\infty}\frac{\mu(n)}{n^s} = \lim_{s\to 1}\frac{1}{\varsigma(s)} = 0, \tag{2.28}$$

and

$$\sum_1^\infty \frac{\mu(n)\ln n}{n} = \lim_{s\to 1}\sum_1^\infty \frac{\mu(n)\ln n}{n^s} = \lim_{s\to 1}\sum_1^\infty \mu(n)\frac{d}{ds}\left(-\frac{1}{n^s}\right) =$$

$$= \lim_{s\to 1} -\frac{d}{ds}\sum_1^\infty \frac{\mu(n)}{n^s} = \lim_{s\to 1} -\frac{d}{ds}\varsigma(s) = \lim_{s\to 1}\frac{\varsigma'(s)}{\varsigma^2(s)} = -1. \quad (2.29)$$

The value of the last limit follows from the fact that $\varsigma(s)$ in the vicinity of $s = 1$ has the leading term $1/(s-1)$ and $\varsigma'(s)$ the leading term $-1/(s-1)^2$.

In order to provide the reader with some idea of the accuracy in the approximations $\operatorname{li} x$ and $R(x)$, we give some values in a small table:

Accuracy of the approximations $\operatorname{li} x$ and $R(x)$			
x	$\pi(x)$	$\operatorname{li} x - \pi(x)$	$R(x) - \pi(x)$
10^2	25	5	1
10^3	168	10	0
10^4	1,229	17	−2
10^5	9,592	38	−5
10^6	78,498	130	29
10^7	664,579	339	88
10^8	5,761,455	754	97
10^9	50,847,534	1,701	−79
10^{10}	455,052,511	3,104	−1,828
10^{11}	4,118,054,813	11,588	−2,318
10^{12}	37,607,912,018	38,263	−1,476
10^{13}	346,065,536,839	108,971	−5,773
10^{14}	3,204,941,750,802	314,890	−19,200
10^{15}	29,844,570,422,669	1,052,619	73,218
10^{16}	279,238,341,033,925	3,214,632	327,052

A denser table can be found at the end of the book. We conclude that for large values of x, $\operatorname{li} x$ and in particular $R(x)$ are close to $\pi(x)$. The more detailed table shows that $R(x) - \pi(x)$ has a great number of sign changes, which often characterizes a good approximation.

The Sign of $\operatorname{li} x - \pi(x)$

From the table above we might get the impression that $\operatorname{li} x$ is always $> \pi(x)$. This is, as a matter of fact, an old famous conjecture in the theory of primes. Judging only from the values given in the table, we might even try to estimate the order of magnitude of $\operatorname{li} x - \pi(x)$ and find it to be about $\sqrt{x}/\ln x$. However, *for large values of x, this is completely wrong!* On the contrary, Littlewood has proved that the function $\operatorname{li} x - \pi(x)$ changes sign infinitely often, although nobody so far has found any specific example of $\operatorname{li} x < \pi(x)$. Moreover, Littlewood's bounds for $\operatorname{li} x - \pi(x)$ show that this difference for some large values of x can become much larger than $\operatorname{li} x - R(x)$, showing that $R(x)$ for such values of x is close to $\operatorname{li} x$ rather than to $\pi(x)$. Thus the good approximation of $R(x)$ to $\pi(x)$ is also to a large extent deceptive. If our table could be continued far enough, there would be arguments for which $\operatorname{li} x - \pi(x) = 0$, while $R(x) - \pi(x)$ would be fairly large!

The study of $\operatorname{li} x - \pi(x)$ shows that in some cases reasoning based on numerical evidence can lead to wrong conclusions, even if the evidence is overwhelming!

The Influence of the Complex Zeros of $\varsigma(s)$ on $\pi(x)$

As can be seen from the above table, not even $R(x)$ can reproduce the primes completely. The reason for this is that the contribution to $f(x)$ from the terms $-\operatorname{li}(x^\rho)$ in (2.18) has not been taken into account in $R(x)$. (The other two terms have very little influence on $\pi(x)$, particularly for large values of x.) Since $\operatorname{li} x$ describes well the distribution of primes viewed at large, we may say that the complex zeros of the Riemann zeta-function induce the local variations in the distribution of the primes. It turns out that each pair $\rho_k = \frac{1}{2} \pm i\alpha_k$ of complex zeros of $\varsigma(s)$ gives rise to a correction $C_k(x)$ to $f(x)$, of magnitude

$$C_k(x) = -2\Re(\operatorname{li} x^{\rho_k}) \approx \frac{-2\sqrt{x}\cos(\alpha_k \ln x - \arg \rho_k)}{|\rho_k| \ln x}. \tag{2.30}$$

Each correction $C_k(x)$ to $f(x)$ is an oscillating function whose amplitude $2\sqrt{x}/(|\rho_k| \ln x)$ increases very slowly with the value of x. The curves $y = C_k(x)$, $k = 1, 2, \ldots, 5$, corresponding to the first five pairs of complex zeros of $\varsigma(s)$, have been sketched on the next page.

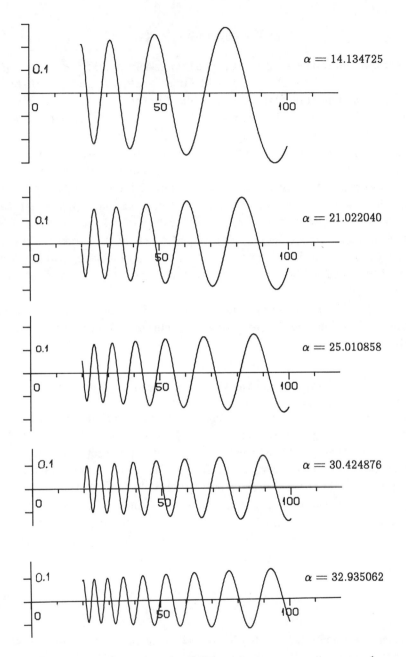

Figure 2.2. The functions $C_k(x) = -\dfrac{2\sqrt{x}}{|\rho_k|\ln x}\cdot\cos(\alpha_k \ln x - \arg\rho_k)$, $\quad \rho_k = \dfrac{1}{2} + i\alpha_k$

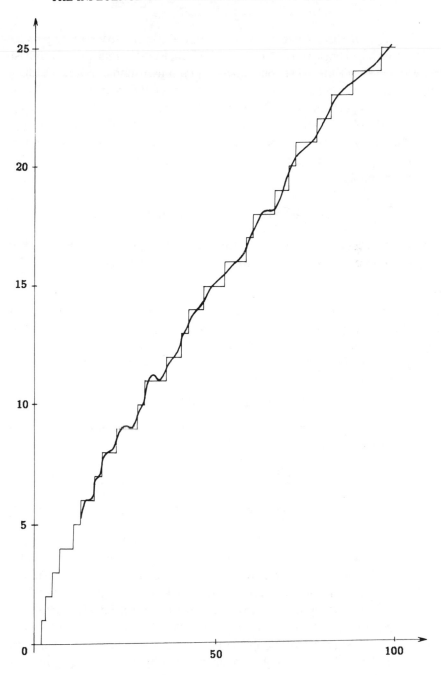

Figure 2.3. $\pi(x)$ vs. $R_{10}(x)$

Finally, all these corrections to $f(x)$ produce, in turn, corrections to $\pi(x)$ which are constructed following the same pattern as that exhibited in formula (2.25) for $R(x)$, but with $C_k(x)$ substituted for li x. Adding up the first ten corrections of the type shown in the graphs on p. 58, we arrive at the approximation $R_{10}(x)$ of $\pi(x)$, depicted on p. 59. Comparing $R_{10}(x)$ with $\pi(x)$, in the same figure, we see that $R_{10}(x)$ reproduces $\pi(x)$ quite faithfully.—The influence of the first 29 pairs of zeros, covering the segment $|t| \leq 100$ of the line $\sigma = 1/2$, has been studied in detail in [7] by the author of this book.

The Remainder Term in the Prime Number Theorem

We have mentioned that assuming the Riemann hypothesis to be true, the Prime Number Theorem with remainder term states that

$$\pi(x) = \text{li}\, x + h(x), \quad \text{with} \quad h(x) \overset{c}{=} O(\sqrt{x}\ln x), \tag{2.31}$$

where "$\overset{c}{=}$" indicates that the result is only conjectured, not proved. Until the Riemann hypothesis is proved, however, only weaker estimates of the remainder term $h(x)$ are known to be valid. If e.g. it could be proved that the upper limit of the real part σ of the zeros of the zeta-function is θ, for some θ between $\frac{1}{2}$ and 1, then it would follow that

$$\pi(x) = \text{li}\, x + O(x^\theta). \tag{2.32}$$

Some actually *proved* estimates of the remainder term are given in the following expressions:

$$\pi(x) = \text{li}\, x + O\left(xe^{-\frac{1}{15}\sqrt{\ln x}}\right) \tag{2.33}$$

$$\pi(x) = \text{li}\, x + O\left(xe^{-0.009(\ln x)^{\frac{3}{5}}/(\ln\ln x)^{\frac{1}{5}}}\right) \tag{2.34}$$

For proofs, see [8] or [9].—In these estimates of the error term, the numerical values of the constants latent in the O-notations are not given. These constants are theoretically computable, but the computations are so complicated that nobody has undertaken them. In spite of this lack of precision, the formulas are very useful in theoretical investigations, but mainly useless when it comes to computing the number of primes less than, say, 10^{100}. In such a situation it is not sufficient to know just only the order of magnitude of $\pi(x) - \text{li}\, x$, when $x \to \infty$, but the numerical values of all the constants in the remainder term must also be known. This remark leads us over to

Effective Inequalities for $\pi(x)$ and p_n

In 1962 some elegant inequalities for $\pi(x)$ have been discovered by J. Barkley Rosser and Lowell Schoenfeld [10]. We quote (without proving) the following:

$$\frac{x}{\ln x}\left(1 + \frac{1}{2 \ln x}\right) < \pi(x) < \frac{x}{\ln x}\left(1 + \frac{3}{2 \ln x}\right) \qquad (2.35)$$
$$x \geq 59 \qquad x \geq 1$$

$$\frac{x}{\ln x - \frac{1}{2}} < \pi(x) < \frac{x}{\ln x - \frac{3}{2}} \qquad (2.36)$$
$$x \geq 67 \qquad x \geq e^{\frac{3}{2}} = 4.48169$$

$$\operatorname{li} x - \operatorname{li} \sqrt{x} \leq \pi(x) \leq \operatorname{li} x \qquad (2.37)$$
$$11 \leq x \leq 10^8 \qquad 2 \leq x \leq 10^8$$

The domain of validity is indicated below the inequality sign for each of the inequalities. We also give some estimates of the nth prime, p_n, the first one given found by Rosser and the second by Robin [11]:

$$n\left(\ln(n \ln n) - \frac{3}{2}\right) < p_n < n\left(\ln(n \ln n) - \frac{1}{2}\right) \qquad (2.38)$$
$$n \geq 2 \quad n \geq 20$$

$$n(\ln(n \ln n) - 1.0072629) \leq p_n < n(\ln(n \ln n) - 0.9385) \qquad (2.39)$$
$$n \geq 2 \quad n \geq 7022$$

The Number of Primes in Arithmetic Progressions

Suppose that a and b are positive integers, and consider all integers forming an arithmetic progression $an + b$, $n = 0, 1, 2, 3, \ldots$ How many of these numbers $\leq x$ are primes? Denote this total by $\pi_{a,b}(x)$. In order for $an+b$ to contain any primes at all, it is apparent that the greatest common divisor (a, b) of a and b must be $= 1$ (except in the obvious case when b is a prime and a is chosen a multiple of b, in which instance $an+b$ will contain just one prime, b). If this condition is fulfilled, a certain proportion of all primes, $1/\varphi(a)$, where $\varphi(a)$ is Euler's function (see Appendix 2, p. 270), belong to the arithmetic series as $x \to \infty$. Utilizing the Prime Number Theorem

61

this gives the following theorem of Dirichlet, the proof of which was not completed until de la Vallée-Poussin gave his proof of the Prime Number Theorem:

$$\lim_{x \to \infty} \frac{\pi_{a,b}(x)}{\text{li}\, x} = \frac{1}{\varphi(a)}. \tag{2.40}$$

Analogous to the remainders of the Prime Number Theorem presented in (2.33)–(2.34), it has been proved that

$$\pi_{a,b}(x) = \frac{\text{li}\, x}{\varphi(a)} + O\!\left(x e^{-\frac{1}{15}\sqrt{\ln x}}\right) \tag{2.41}$$

etc. Proofs are furnished in [8]. Dirichlet's theorem states that the primes are approximately equi-distributed among those arithmetic series of the form $an+b$, for a fixed value of a, which contain several primes. Thus in the limit half of the primes are found in each of the two series $4n-1$ and $4n+1$, or in $6n-1$ and $6n+1$, and 25% of the primes are found in each of the four series $10n\pm1$, $10n\pm3$, etc. Dirichlet's theorem also tells us that every arithmetic series $an+b$ with $(a,b) = 1$ contains infinitely many primes. To give just one example: There are infinitely many primes ending in 33 333, such as 733 333, 1 133 333, 2 633 333, 2 833 333, 3 233 333, 3 433 333, 3 733 333, 4 933 333, 5 633 333, 6 233 333, ..., 1 000 133 333..., because the series $100\,000n + 33\,333$ contains, in the long run, 1/40 000 of all primes.

A readable account in which many of the topics of this chapter are discussed in greater detail is [12], which also contains a large bibliography.

Bibliography

1. James F. Jones, Daihachiro Sato, Hideo Wada and Douglas Wiens, "Diophantine Representation of the Set of Prime Numbers," *Am. Math. Monthly* **83** (1976) pp. 449–464.

2. G. H. Hardy and E. M. Wright, *An Introduction to the Theory of Numbers*, Fifth edition, Oxford, 1979, pp. 359–367.

3. D. J. Newman, "Simple Analytic Proof of the Prime Number Theorem," *Am. Math. Monthly* **87** (1980) pp. 693–696.

4. E. Landau, *Handbuch der Lehre von der Verteilung der Primzahlen*, Chelsea, New York, 1953. (Reprint.)

5. J. van de Lune and H. J. J. te Riele, "On the Zeros of the Riemann Zeta function in the Critical Strip, III," *Math. Comp.* **41** (1983) pp. 759–767.

6. J. van de Lune and H. J. J. te Riele, "Recent Progress on the Numerical Verification of the Riemann Hypothesis," *CWI Newsletter* No. 2 (March 1984) pp. 35–37, and Private communication.

7. Hans Riesel and Gunnar Göhl, "Some Calculations Related to Riemann's Prime Number Formula," *Math. Comp.* **24** (1970) pp. 969–983.

8. A. Walfisz, *Weylsche Exponentialsummen in der neueren Zahlentheorie*, VED Deutscher Verlag der Wissenschaften, Berlin, 1963.

9. W. J. Ellison and M. Mendès France, *Les Nombres Premiers*, Hermann, Paris, 1975.

10. J. Barkley Rosser and L. Schoenfeld, "Approximate Formulas for Some Functions of Prime Numbers," *Ill. Journ. Math.* **6** (1962) pp. 64–94.

11. Guy Robin, "Estimation de la Fonction de Tchebychef Θ sur le k-ième Nombre Premier et Grandes Valeurs de la Fonction $\omega(n)$ Nombre de Diviseurs Premiers de n," *Acta Arith.* **52** (1983) pp. 367–389.

12. Harold G. Diamond, "Elementary Methods in the Study of the Distribution of Prime Numbers," *Bull. Am. Math. Soc., New Series* **7** (1982) pp. 553–589.

63

CHAPTER 3

SUBTLETIES IN THE DISTRIBUTION OF PRIMES

The Distribution of Primes in Short Intervals

There are only very few proved results concerning the distribution of primes in short intervals. The prime number theorem tells us that the *average density* of primes around x is approximately $1/\ln x$. This means that if we consider an interval of length Δx about x and choose any integer t in this interval, then the probability of t being a prime will approach $1/\ln x$ as $x \to \infty$, if Δx is small compared to x. This implies that the primes tend to thin out as x grows larger; an implication that becomes obvious when considering that the condition for a randomly picked integer x to be composite is that it has some prime factor $\leq \sqrt{x}$, and that there are more prime factors $\leq \sqrt{x}$ to choose from when x is larger.

What law governs the thinning out of primes as we go higher up in the number series? *On the average* $p_{n+1} - p_n$ grows slowly with n. However in contrast, certainly as far as current prime tables extend, we repeatedly discover consecutive primes for which $p_{n+1} - p_n = 2$, the so called "twin primes." The reader should at this point examine Table 2 at the end of this book.

Twins and Some Other Constellations of Primes

Twin primes, i.e. pairs of primes of the form x and $x + 2$, occur very high up in the number series. The largest known pair to date is the following incredible number, found in 1983 by Wilfrid Keller:

$$1\,639\,494(2^{4423} - 1) \pm 1.$$

Detailed statistics on the occurrence of primes have been gathered up to 10^9. These statistics indicate that the twins tend to thin out *compared with the primes* as we move higher up in the number series. This is as natural as the thinning out of the primes themselves is because, if p is a prime, it becomes less and less likely that $p+2$ is also a prime, the larger p is. In this context it is of interest to mention one of the few facts that *have* been proved about constellations of primes: Brun's theorem stating that the sum of $1/p$ taken over all twin primes converges:

$$B = \left(\frac{1}{3}+\frac{1}{5}\right) + \left(\frac{1}{5}+\frac{1}{7}\right) + \left(\frac{1}{11}+\frac{1}{13}\right) + \left(\frac{1}{17}+\frac{1}{19}\right) + \cdots \approx 1.9022.$$

The constant B is termed Brun's constant; see [1] for its computation. Brun's theorem tells us that there are not very many twin primes compared with the total number of primes, since $\sum 1/p$ taken over only the twin primes *converges*, while $\sum 1/p$ extended over *all* primes *diverges*.

A careful study of prime tables has also shown that constellations of primes other than twins, known from the beginning of the number series, are repeated over and over again. A detailed study has been made of the occurrence of two different kinds of prime triples, $(p, p+2, p+6)$ and $(p, p+4, p+6)$, and of prime quadruples $(p, p+2, p+6, p+8)$. These constellations are exemplified by (41,43,47), (37,41,43), and (11,13,17,19) respectively.—Nothing has so far been *proved* about the occurrence of such constellations, but Hardy and Littlewood [2] have made a famous conjecture about their numbers. Before a general discussing on constellations, we give below some special instances of this conjecture.

The number of some prime constellations P_x, with their smallest member $\leq x$ is, as x tends to infinity, are given by the following formulas, in which the products are taken over the primes indicated:

$$P_x(p, p+2) \overset{c}{\sim} 2 \prod_{p \geq 3} \frac{p(p-2)}{(p-1)^2} \int_2^x \frac{dx}{(\ln x)^2} = 1.320323632 \int_2^x \frac{dx}{(\ln x)^2} \qquad (3.1)$$

$$P_x(p, p+2, p+6) \overset{c}{\sim} P_x(p, p+4, p+6) \overset{c}{\sim}$$

$$\overset{c}{\sim} \frac{9}{2} \prod_{p \geq 5} \frac{p^2(p-3)}{(p-1)^3} \int_2^x \frac{dx}{(\ln x)^3} = 2.858248596 \int_2^x \frac{dx}{(\ln x)^3} \qquad (3.2)$$

$$P_x(p, p+2, p+6, p+8) \overset{c}{\sim} \frac{1}{2} P_x(p, p+4, p+6, p+10) \overset{c}{\sim}$$

$$\overset{c}{\sim} \frac{27}{2} \prod_{p \geq 5} \frac{p^3(p-4)}{(p-1)^4} \int_2^x \frac{dx}{(\ln x)^4} = 4.151180864 \int_2^x \frac{dx}{(\ln x)^4} \qquad (3.3)$$

65

We have here introduced the symbol $\overset{c}{\sim}$ instead of writing just $\sim$ to indicate that the corresponding relation is only *conjectured*, not proved.

In the table below we give a count of the number of prime constellations, as found by some of the statistics [3] mentioned above, and compare these values with the corresponding numbers calculated by means of the Hardy-Littlewood formulas.

Number of prime constellations $\leq 10^8$		
Constellation	Count	Approx.
$(p, p+2)$	440312	440368
$(p, p+2, p+6)$	55600	55490
$(p, p+4, p+6)$	55556	55490
$(p, p+2, p+6, p+8)$	4768	4734

Admissible Constellations of Primes

Any constellation of integers of the following form, which is not ruled out by divisibility considerations as a candidate for consisting of only prime numbers, is called an *admissible* constellation:

$$t + a_1, \ t + a_2, \ldots, t + a_l.$$

Thus $t + 1, t + 3, t + 5$ is inadmissible. It is eliminated because three consecutive odd (or even) integers always contain a multiple of 3. This particular constellation is ruled inadmissible despite the fact that it contains the sequence of primes 3, 5, 7.

To verify if a given constellation is admissible is easy in principle, but may be a quite laborious task if l is large. The method of verification consists of trying to sieve with the multiples of the primes $2, 3, 5, \ldots$ in such a way that the integers of the constellation are avoided. If this succeeds for all primes $\leq l$ then the constellation is admissible, otherwise it is not. The process is similar to sieving with the sieve of Eratosthenes, but with all possible starting points for the first multiple of the small primes.

As an example let us see if the constellation

$$t + 1, t + 7, t + 11, t + 13, t + 17, t + 19, t + 23, t + 29$$

is admissible. Since all its members have the same parity, we need not consider the factor 2 as long as we choose t even. Represent the constellation by the corresponding integers in the interval $[1, 29]$. Now, there are 3 different ways to sieve with the prime 3:

$$1, 4, 7, 10, 13, 16, 19, 22, 25, 28, \qquad 2, 5, 8, 11, 14, 17, 20, 23, 26, 29,$$

$$\text{and} \quad 3, 6, 9, 12, 15, 18, 21, 24, 27.$$

The last of these 3 sievings does not land upon any number in the constellation. If we thus choose $t \equiv 0 \pmod 3$, everything is satisfactory so far. For the multiples of 5 we also find only one position which does not collide with any member of the constellation:

$$5, 10, 15, 20, 25.$$

The multiples of 7 can be placed in 7 different ways:

$$1, 8, 15, 22, 29, \quad 2, 9, 16, 23, \quad 3, 10, 17, 24, \quad 4, 11, 18, 25,$$

$$5, 12, 19, 26, \quad 6, 13, 20, 27, \quad 7, 14, 21, 28.$$

All these possibilities have at least one number in common with the constellation which, as a result of this fact, is inadmissible. This fact may also be expressed algebraically by saying that the integers of the constellation represent *all seven* residue classes $\pmod 7$:

$$7 \equiv 0, \; 1 \equiv 1, \; 23 \equiv 2, \; 17 \equiv 3, \; 11 \equiv 4, \; 19 \equiv 5, \; 13 \equiv 6 \pmod 7.$$

Actually, this last way of describing the members of the constellation $\pmod 7$ gives a clue to a somewhat simpler procedure to decide whether a constellation is admissible or not. All that is needed is to fix the value of t, to compute $t + a_i \pmod p$ for all a_i's of the constellation and to see if the values found leave *at least one residue class* $\pmod p$ *free or not. If there is at least one free residue class for every prime p, then the constellation is admissible, otherwise not.* Obviously this checking is necessary only for the primes $p \leq l$, where l (as previously) denotes the number of members of the constellation. The reason is that if $p > l$, then there exists at least one free residue class $\pmod p$ (because l elements cannot occupy p different classes, if $l < p$.) So in re-doing the example given above, we successively find (after having fixed the value of $t = 0$):

Occupied residue classes (mod 2) : 1

(mod 3) : 1, 2

(mod 5) : 1, 2, 3, 4

(mod 7) : 0, 1, 2, 3, 4, 5, 6

and we have, as before, arrived at the result that this particular constellation is inadmissible, since all seven residue classes (mod 7) are occupied.

The Hardy-Littlewood Constants

What is the origin of the complicated looking constants in the Hardy-Littlewood formulas (3.1)–(3.3)? Obviously, each residue class (mod p) which is *not represented* by the integers in the constellation gives exactly one possibility for sieving with multiples of p. Enumerating the number of possibilities available for each prime gives the numerical values of these constants. Take as example formula (3.3), concerning the constellation $t, t + 2, t + 6, t + 8$. The multiples of 2 can only be placed in positions 1, 3, 5, 7 and the multiples of 3 can only be inserted in 1, 4, 7 without affecting the constellation. This reduces the number of useful t's to just one residue class (mod 6), *but when this residue class is chosen*, it augments the possibility that $t, t + 2, t + 6$ and $t + 8$ are all primes by a factor $(\frac{2}{1} \times \frac{3}{2})^4 = 81$ when compared with the situation of all four integers being chosen at random. This gives rise to the numerical factor $\frac{1}{6} \cdot 81 = \frac{27}{2}$ in (3.3). Furthermore, for any prime $p \geq 5$ we have $p - 4$ choices to prevent any of $t, t + 2, t + 4$ and $t + 8$ from being multiples of p, as compared with $p - 1$ choices for each of these integers, if they were to be chosen independently. This gives rise to the following factor:

$$\left(1 - \frac{4}{p}\right)\left(1 - \frac{1}{p}\right)^{-4} = \frac{p^3(p - 4)}{(p - 1)^4}. \tag{3.4}$$

This explains the form of the infinite product in (3.3).

As a final example, let us study the constellation representing the pattern of primes between 11 and 67:

$$t + 11, t + 13, t + 17, \ldots, t + 59, t + 61, t + 67. \tag{3.5}$$

We shall prove it admissible and find the Hardy-Littlewood formula for its numbers. Since all its members are primes for $t = 0$, the first primes 2,

3, 5 and 7 obviously do not interfere with this constellation. It has 15 members, so that for large primes p we have at least $p - 15$ residue classes remaining in which to place the multiples of p without hitting any number in the constellation. How about the primes from 11 onwards? Well, a check reveals that (still with $t = 0$) the residue class $\equiv 5 \pmod{11}$ is not used, neither is the residue class $\equiv 12 \pmod{13}$. For $p = 17$ there are 4 free residue classes, for $p = 19$ there are 6, for $p = 23$ there are 9, and finally, for all $p \geq 29$ there are $p - 15$ residue classes available from which to select the multiples of p.

In order to compute the constant in the formula for the number of constellations of this kind, we need to use the information just obtained about the number of residue classes available for each prime. Thus, the exact form of the formula in this case is found to be

$$P_x(t + 11, t + 13, t + 17, \ldots, t + 61, t + 67) \overset{c}{\sim}$$

$$\overset{c}{\sim} \frac{2^{14}}{1^{15}} \cdot \frac{3^{14}}{2^{15}} \cdot \frac{5^{14}}{4^{15}} \cdot \frac{7^{14}}{6^{15}} \cdot \frac{11^{14}}{10^{15}} \cdot \frac{13^{14}}{12^{15}} \cdot \frac{4 \cdot 17^{14}}{16^{15}} \cdot \frac{6 \cdot 19^{14}}{18^{15}} \cdot \frac{9 \cdot 23^{14}}{22^{15}} \times$$

$$\times \prod_{p \geq 29} \frac{p^{14}(p - 15)}{(p - 1)^{15}} \int_2^x \frac{dx}{(\ln x)^{15}} =$$

$$= 187823.7 \int_2^x \frac{dx}{(\ln x)^{15}} \sim 187823.7 \cdot \frac{x}{(\ln x)^{15}}. \tag{3.6}$$

Readers interested in how to compute accurately the numerical values of the infinite products occurring in the constants of these conjectures, are referred to Exercises 3.1 and 3.2 below and to [4]. Despite the fact that the primes are not all individually known, these constants can be computed *to any desired accuracy!* This astonishing result is due to the fact that (2.7) can be transformed (in a rather complicated way) so as to express any of these constants as a convergent series of known functions.

Exercise 3.1. The prime zeta-function. The function $P(s) = \sum p^{-s}$, summed over all primes, is called the prime zeta-function. It is indispensable for the evaluation of expressions containing all primes such as the twin prime constant, occurring in (3.1), or the infinite product in (3.6). Values of $P(s)$ may be computed from values of the Riemann zeta-function $\varsigma(s)$ in the following way: taking logarithms of both sides of (2.7) gives

$$\ln \varsigma(s) = -\sum_{p \geq 2} \ln(1 - p^{-s}) = \sum_{p \geq 2} \sum_{k=1}^{\infty} \frac{p^{-ks}}{k} = \sum_{k=1}^{\infty} \frac{1}{k} \sum_{p \geq 2} p^{-ks} = \sum_{k=1}^{\infty} \frac{P(ks)}{k}.$$

Inversion of this formula gives

$$P(s) = \sum_{k=1}^{\infty} \frac{\mu(k)}{k} \ln \varsigma(ks).$$

Here $\mu(k)$ is Möbius function and the values of the zeta function are to be computed as in exercise 2.2 on p. 49. For large values of ks the following shortcut may be convenient:

$$\ln \varsigma(ks) = \ln(1 + 2^{-ks} + 3^{-ks} + 4^{-ks} + \cdots) =$$

$$= 2^{-ks} + 3^{-ks} + \frac{1}{2} \cdot 4^{-ks} + 5^{-ks} + 7^{-ks} + \cdots$$

Write a FUNCTION P(s) calling the FUNCTION zeta(s) from exercise 2.2 and giving the value of $P(s)$ for $s \geq 2$. Incorporate P(s) in a test program and check your programming of P(s) by re-computing $\ln \varsigma(s)$ from $P(s)$ by the formula above and comparing the result with the result of a direct computation of $\ln \varsigma(s)$.

Exercise 3.2. The twin prime constant. The twin prime constant C in (3.1) can be evaluated by the following computation: Write

$$\ln \frac{C}{2} = \sum_{p \geq 3} \ln \frac{p(p-2)}{(p-1)^2} = \sum_{p \geq 3} \left\{ \ln\left(1 - \frac{2}{p}\right) - 2\ln\left(1 - \frac{1}{p}\right) \right\} =$$

$$= \sum_{p \geq 3} \left(-\frac{2}{p} - \frac{1}{2} \cdot \frac{4}{p^2} - \frac{1}{3} \cdot \frac{8}{p^3} - \cdots + \frac{2}{p} + \frac{1}{2} \cdot \frac{2}{p^2} + \frac{1}{3} \cdot \frac{2}{p^3} + \cdots \right) =$$

$$= \sum_{p \geq 3} \sum_{j=2}^{\infty} \frac{2 - 2^j}{j} \cdot \frac{1}{p^j} = -\sum_{j=2}^{\infty} \frac{2^j - 2}{j} \sum_{p \geq 3} p^{-j}.$$

Now $\sum_{p \geq 3} p^{-s} = P(s) - 2^{-s}$, where $P(s)$ is the prime zeta-function studied in exercise 3.1 above. Use your FUNCTION P(s) in a computer program which evaluates $\ln(C/2)$ after the formula given above. Compare the result with the value calculated from C as given in (3.1)!

A generalization of (3.1)–(3.3) is a very famous conjecture called

The Prime k-Tuples Conjecture

Any admissible constellation occurs infinitely often with all its members primes, and the asymptotic number of its occurrences $\leq x$ is $\Omega(x/(\ln x)^l)$, where l is the number of integers in the constellation.

This has not been proved, not even in its simplest case, for twins! By sieve methods it has been proved, however, that the number of admissible constellations has *upper bounds* of the form $Cx/(\ln x)^2$.—There is, on the other hand, a heuristic line of reasoning which makes the conjecture plausible. It is instructive to examine this line of reasoning, and find out what is lacking for a complete proof.

70

To do so, we start with

Theorem 3.1. Theorem of Mertens.

$$\prod_{2 \leq p \leq x} \left(1 - \frac{1}{p}\right) \sim \frac{e^{-\gamma}}{\ln x} = \frac{0.5615}{\ln x}, \quad \text{as } x \to \infty. \tag{3.7}$$

Here, γ is Euler's constant. For a proof, see $[5]$.

Replacing x by $x^{0.5615}$, we find that

$$\prod_{2 \leq p \leq x^{0.5615}} \left(1 - \frac{1}{p}\right) \sim \frac{1}{\ln x}, \quad \text{for large } x. \tag{3.8}$$

However, $1/\ln x$ is the expectation that a large integer x, chosen at random, is prime. On the other hand, the number x is prime if and only if it is not divisible by any of the primes $\leq \sqrt{x}$. Thus, we might expect

$$\prod_{2 \leq p \leq x^{0.5}} \left(1 - \frac{1}{p}\right) \quad \text{to be } \sim \frac{1}{\ln x} \quad \text{for large } x, \tag{3.9}$$

which clearly is discrepant from Mertens' theorem. This discrepancy must arise from the fact that there are subtleties in the distribution of the primes up to $\sqrt{x}$, which influence the primes near x and which are not accounted for by the simple statistical model utilized, in which divisibility by different primes is considered to be independent. Or, to put it an other way, the sieve of Eratosthenes is special in the sense that it sieves out numbers more efficiently than does a "random" sieve. Thus, when sieving with all primes $p \leq \sqrt{x}$ and utilizing the sieve of Eratosthenes, the Prime Number Theorem ensures that in the vicinity of x the fraction $1/\ln x$ of numbers remain untouched by the sieve, while a random sieve using exactly the same primes leaves, according to (3.7), the larger fraction $1.123/\ln x$ of numbers unaffected by the sieve. Note that a special property of the sieve of Eratosthenes is the existence of *one and only one number*, zero, which is struck by *all* the sieving primes.—The simple model of independent divisibility by the primes $\leq \sqrt{x}$ can, as a matter of fact, be used first when we consider blocks of consecutive integers as huge as the product of all the primes $\leq \sqrt{x}$. The length of such a block is of the order of magnitude $e^{\sqrt{x}}$, which is very much larger than x. We shall make use of the difference in efficiency between the sieve of Eratosthenes and a random sieve later, when discussing superdense constellations.

Theoretical Evidence in Favour of the Prime k-Tuples Conjecture

We now return to the heuristic line of thought as promised in the prime k-tuples conjecture. We confine the deduction to the case of twin primes. Let the notation $E(x, x+2 \in P)$ denote "the expectation that x and $x+2$ both belong to the set of primes P." Then, a similar reasoning to that used previously leads to

$$E(x, x+2 \in P) \overset{c}{\sim} E(x \in P) \prod_{3 \leq p \leq \sqrt{x}} \frac{p-2}{p-1}, \qquad (3.10)$$

unless the above needed transformation from $\sqrt{x}$ to $x^{0.5615}$ (or possibly some other correction) *must be made* in order that the asymptotic number of twin primes emerges correctly. No-one has so far proved anything in this direction. *If* we change $\sqrt{x}$ to $x^{0.5615}$, we easily obtain

$$E(x, x+2 \in P) \overset{c}{\sim} E(x \in P) \prod_{3 \leq p \leq x^{0.5615}} \left(1 - \frac{2}{p}\right)\left(1 - \frac{1}{p}\right)^{-1} \sim$$

$$\sim \frac{2}{(\ln x)^2} \prod_{3 \leq p \leq x^{0.5615}} \left(1 - \frac{2}{p}\right)\left(1 - \frac{1}{p}\right)^{-2} \sim$$

$$\sim \frac{2}{(\ln x)^2} \prod_{p=3}^{\infty} \left(1 - \frac{2}{p}\right)\left(1 - \frac{1}{p}\right)^{-2} = \frac{1.320323632}{(\ln x)^2}. \qquad (3.11)$$

In passing from the first to the second line we have multiplied by a factor

$$\frac{1}{\ln x} \prod_{2 \leq p \leq x^{0.5615}} \left(1 - \frac{1}{p}\right)^{-1} \sim 1,$$

in order that the infinite product in the final line be convergent.

In summary; since the infinite product in (3.10) diverges as $x \to \infty$, the upper limit $x^{0.5615}$ imposed on p *is crucial for the asymptotic formula to be correct!* We complete this theoretical discussion by remarking that the expression $x/(\ln x)^l$ tends to infinity with x, which implies that if the prime k-tuples conjecture is true all kinds of admissible constellations will occur infinitely often with all elements primes. By this remark we actually weaken the conjecture enormously, but it still implies that, although the primes thin out as $x \to \infty$, they tend to form clusters, so that all these constellations are formed over and over again.

Numerical Evidence in Favour of the Prime k-Tuples Conjecture

We have already presented the very impressive comparison between the statistics on the occurrence of four simple constellations and the predictions calculated on the basis of the prime k-tuples conjecture. A similar check has been made for the primes in the intervals $[10^k, 10^k + 150,000]$ for $k \leq 15$, and the agreement between conjecture and reality was strikingly good also in these cases.

In our theoretical discussion of the twin prime conjecture above, we found that a change from $\sqrt{x}$ to $x^{0.5615}$ had to be made in order for the formulas to develop as they should. Such a subtle change may be regarded as having only second-order effects on the distribution of the primes. Thus, if some second-order effects of the empirical model used could be numerically tested, and were found to agree with reality, then this would support other conclusions drawn from the same model, such as the primes k-tuple conjecture. Precisely such a test of second-order effects has been carried out by Richard Brent [6] who, in 1974, published statistics on the occurrence of small gaps between consecutive primes. The distribution of gaps of various sizes can also be predicted from the same model that has led to the twin primes conjecture. Also in these statistics, a very striking agreement between prediction and reality was observed for the whole interval investigated, which ranged from 10^6 to 10^9. To give the reader some idea of Brent's results, we give on the next page the number of prime gaps of length $2r$, as actually counted in a prime table, and as predicted by the model used.

The Prime Number Theorem states that the primes become less dense the higher we reach in the number series. This has led Hardy and Littlewood to formulate another hypothesis, namely

The Second Hardy-Littlewood Conjecture

In this conjecture it is supposed that nowhere can the primes be more dense than in the beginning of the number series, or more precisely: In no interval $[x + 1, x + y]$ of length y are there more primes than in the interval $[1, y]$.—Or

$$\pi(x + y) - \pi(x) \overset{<}{_=} \pi(y) \quad \text{for all } x, y. \tag{3.12}$$

73

At first sight, this conjecture seems very plausible, but in 1974 Douglas Hensley and Ian Richards [7] proved that certain admissible constellations exist which are denser than the series of primes in the beginning of the number series. This result implies that the two conjectures of Hardy and Littlewood are incompatible! Such superdense constellations can be constructed according to the following ideas:

Counted and predicted number of prime gaps of length $2r$ in $\left[10^6, 10^9\right]$					
r	Counted	Predicted	r	Counted	Predicted
1	3 416 337	3 417 060.1	21	953 980	954 689.0
2	3 416 536	3 417 060.1	22	389 432	389 057.1
3	6 076 242	6 077 407.1	23	334 565	335 337.0
4	2 689 540	2 688 560.2	24	577 051	577 898.6
5	3 477 688	3 477 436.8	25	327 960	327 323.5
6	4 460 952	4 460 654.7	26	245 727	245 799.1
7	2 460 332	2 461 360.3	27	410 614	410 578.1
8	1 843 216	1 842 845.7	28	211 409	211 469.0
9	3 346 123	3 347 229.6	29	181 894	182 398.0
10	1 821 641	1 823 424.2	30	371 743	372 007.3
11	1 567 507	1 567 220.8	31	115 542	115 837.8
12	2 364 792	2 362 746.8	32	118 927	118 681.6
13	1 118 410	1 118 419.0	33	216 739	216 467.5
14	1 218 009	1 218 441.9	34	88 383	88 116.0
15	2 176 077	2 176 130.5	35	124 542	125 688.7
16	683 346	682 871.2	36	126 650	126 786.7
17	718 974	718 118.6	37	62 514	62 578.8
18	1 170 757	1 169 307.2	38	55 107	55 325.4
19	548 416	547 688.6	39	105 300	105 390.3
20	648 356	648 539.8	40	53 513	53 578.4

As demonstrated on p. 71, a random sieve leaves untouched about 12% more numbers than does the annihilating sieve of Eratosthenes. The primes below u remain after sieving with the primes up to $\sqrt{u}$ using the algorithm of Eratosthenes. In the construction of an admissible constellation of primes, a random sieve, which is rather more preserving than the sieve of Eratosthenes, might be used, at least in part. Thus, a cleverly constructed admissible constellation of length x could in fact have *more members in it than the $\pi(x)$ numbers unaffected by the sieve of Eratosthenes.*

The Midpoint Sieve

The question is where to look for an admissible constellation with the desired properties? Could the number 1 together with the primes up to x be a suitable candidate? Definitely not, since *the presence of small primes spoil it,* just as in the case of the constellations $t + 2, t + 3$ and $t + 3, t + 5, t + 7$ which are inadmissible. On the other hand, if we omit the small primes then the resulting constellation might be admissible, such as the constellation (3.5) above, but would obviously contain fewer members than $\pi(x)$. At this point the idea of a midpoint sieve enters the discussion. Because the primes, on average, thin out as x grows, the primes below $x/2$ are more than $\frac{1}{2}$ of those below x. Thus, starting with the positive and negative primes between $-x/2$ and $x/2$ (together with the numbers ± 1), there is a possibility that these (apart from two small "holes" near the origin) will constitute an admissible constellation of the desired size. Let us now consider this possibility! Take, for example, the interval $[-73, 73]$, which contains $2\pi(73) = 42$ positive and negative primes. By excluding the primes in $[-23, 23]$ (18 in total) and adding the two numbers ± 1, we are left with 26 numbers. These obviously form an admissible constellation, since all multiples of the primes ≥ 29 can be placed in the positions of numbers already eliminated. However, and this is a crucial point, 26 compares very unfavourably with the desired goal $\pi(146) + 1 = 35$. Can the situation be altered if we choose a very large interval? Unfortunately, no! Let us try to find out why not. Consider now the interval $[-x/2, x/2]$ containing $2\pi(x/2)$ positive and negative primes. Exclude the primes in the interval $[-U, U]$. How large should U be? Well, we expect the final number of members of the constellation to be about $\pi(x)$, so that U ought to be about the same size. Thus, the number of primes about the origin which are excluded is about $2\pi(\pi(x))$, leaving approximately

$$2\pi\left(\frac{x}{2}\right) - 2\pi(\pi(x)) + 2 \qquad (3.13)$$

numbers in the constellation. This expression has now to be evaluated and compared with $\pi(x)$. Later on we shall have to evaluate a slightly

more general expression, viz. (3.14), and thus, since $\pi(x) \sim x/\ln x$, we may evaluate (3.13) by observing that it corresponds to the special case $N = 1$ in the formula immediately following (3.15). Sadly the calculation always results in the value of (3.13) being less than $\pi(x)$, at least for all sufficiently large values of x, and thus it might seem that the midpoint idea was not such a good one after all.

Modification of the Midpoint Sieve

Could the failure of the midpoint idea stem from the fact that the sieve of Eratosthenes annihilates too many of the numbers in the interval $[-x/2,\, x/2]$ as compared to a random sieve? If this is the case, then perhaps the number U should be slightly decreased and chosen smaller than $\pi(x)$. Moreover, in order not to allow multiples of primes between U and $\pi(x)$ to spoil the proposed constellation, they are placed (if possible) on numbers (composite or small primes) in $[-x/2, x/2]$ that have already been employed! This is exactly what Hensley and Richards managed to do, and we now proceed to describe their reasoning in some detail.

Construction of Superdense Admissible Constellations

Consider once again the interval $[-x/2, x/2]$, and place the multiples of all the small primes $\leq U = x/(N \ln x)$ for some constant $N > 2/\ln 2$ in their normal positions, i.e. employ the sieve of Eratosthenes using these primes. Next, try to place the multiples of the primes $> x/(N \ln x)$ on top of the composite numbers already gone or on top of the already annihilated small primes. This is the only tricky step in the entire procedure! If it succeeds, we are left with the positive and negative primes between $x/(N \ln x)$ and $x/2$ and the two numbers $+1$ and -1, in total with

$$2\pi\left(\frac{x}{2}\right) - 2\pi\left(\frac{x}{N \ln x}\right) + 2 \tag{3.14}$$

numbers in the constellation. If this number exceeds $\pi(x)$, which it will do for all sufficently large values of x, then a superdense constellation will result. Here is the calculation which proves that the expression (3.14) will finally take values $> \pi(x)$: as we have already mentioned, it is sufficient to use the Prime Number Theorem with remainder term for the proof. First, from (2.29) it follows that

$$\pi(x) = \text{li}(x) + o\left(\frac{x}{(\ln x)^2}\right) = \frac{x}{\ln x} + \frac{x}{(\ln x)^2} + o\left(\frac{x}{(\ln x)^2}\right) \tag{3.15}$$

Using this, after some calculations we find that

$$2\pi\left(\frac{x}{2}\right) - 2\pi\left(\frac{x}{N\ln x}\right) - \pi(x) = \frac{\left(\ln 2 - \dfrac{2}{N}\right)x}{(\ln x)^2} + o\left(\frac{x}{(\ln x)^2}\right) \sim \frac{Kx}{(\ln x)^2},$$

with a positive constant K, as long as N is chosen $> 2/\ln 2 = 2.89$. Since the expression arrived at tends to infinity with x, there exist, amazingly enough, for all sufficiently large x, superdense admissible constellations of length x, with a number of elements that exceeds $\pi(x)$. The number of elements in the constellation can even be made to exceed $\pi(x)$ *by any fixed amount*, say M. This can be achieved merely by selecting x large enough, and then there will exist an admissible constellation of length x with a number of elements that exceeds $\pi(x)$ by M or more.

By quite complicated and extensive computer runs John Selfridge first showed that any superdense admissible constellation must exceed 500 in length, and Warren Stenberg subsequently showed that there exists such a constellation of length ≤ 20000. Finally, in 1979, Thomas Vehka [8] succeeded in exhibiting a superdense admissible constellation of length 11763 that consists of 1412 numbers, while $\pi(11763)$ is only 1409. If we believe in the prime k-tuples conjecture, which, as we have seen, seems well founded, it would follow that a superdense cluster of primes would exist. It is certainly a challenge to try to exhibit such an entity! Since this cluster (if it exists) probably is very large, the first step to find one is that taken by Vehka, to construct a superdense admissible constellation, i.e. a pattern that these primes could make up. Then, as a second step, different instances of this constellation would have to be verified for primality of all its members, until finally a superdense cluster of primes could be identified.

We have hitherto by-passed the problem of placing the primes between $x/(N\ln x)$ and x on the already eliminated numbers. The possibility to do this depends on a theorem of Westzynthius, Erdös and Rankin which states that, by sieving only with the small primes $< \ln x$ it is actually possible to find here and there in an interval of length x, provided x is sufficiently large, as many as $N\ln x$ integers which have been sieved out by only these primes. N can here be arbitrarily large, provided $x > x_0(N)$. These sequences look much like the following one, taken from a small factor table:

The smallest factor p of 33 consecutive integers n					
n	p	n	p	n	p
60044	2	60055	5	60066	2
45	3	56	2	67	7
46	2	57	3	68	2
47	13	58	2	69	3
48	2	59	19	70	2
49	11	60	2	71	11
50	2	61	17	71	2
51	3	62	2	73	13
52	2	63	3	74	2
53	7	64	2	75	3
54	2	65	5	76	2

In order to make use of the theorem just mentioned Richards and Hensley modified it to apply not only to sequences of *consecutive* integers, but also to integers forming an *arithmetic series with the difference a prime number*, just like the multiples of p in our interval $[-x/2, x/2]$. In this way they managed to prove that their proposed construction of superdense admissible constellations will succeed.

Some Dense Clusters of Primes

Hensley's and Richards' work has some connection with an investigation by the author [9], made in 1969–70, in order to find a dense cluster of large primes. The idea was to seek repetitions of the patterns of primes in the beginning of the number series. It was during the course of this work that it was discovered that the constellation mentioned earlier, consisting of the primes between 11 and 67, is admissible. A search for primes having this pattern was, however, not quite successful. The "closest hit" at that time was the cluster

$$4299\,83158710 + 11, 13, 17, 19, 23, 37, 41, 43, 47, 53, 59,$$

with all its members primes. (Only the primes 29, 31, 61 and 67 of the constellation are not repeated in this cluster.)—In 1982 this search was taken up by Sten Säfholm and Demetre Betsis, who found the prime cluster

$$2\,18172838\,54511250 + 11, 13, 17, 19, 23, 29, 31, 37, 41, 43, 47, 53, 59, 61.$$

The average prime density this high up is only one prime out of about 38 integers, which makes this cluster, with 14 primes out of 51 numbers, rather remarkable.

How high up can we expect to find the cluster searched for, repeating the 15 primes between 11 and 67? Well, using the prime k-tuples conjecture we have above found the expected numbers of this cluster below x to be approximately $187823.7x/(\ln x)^{15}$. This expression takes the value 1 for $x \approx 3.3 \cdot 10^{19}$, which is the order of magnitude where the first repetition of this cluster is to be expected. It is thus quite a bit higher than the author's search in 1969–70 had attained, and this explains why no repetition of this cluster of primes was found.

The Distribution of Primes Between the Two Series $4n+1$ and $4n+3$

The Prime Number Theorem for arithmetic series (2.40) tells us that the number $\pi_{4,1}(x)$ of primes below x of the form $4n+1$ is about the same as the number $\pi_{4,3}(x)$ of primes below x of the form $4n+3$, in the sense that

$$\lim_{x \to \infty} \frac{\pi_{4,1}(x)}{\pi_{4,3}(x)} = 1.$$

Not much is known about the finer details in the distribution of primes between these two series, but prime number counts can give the general impression that primes of the form $4n+3$ are more abundant than primes of the form $4n+1$, so that

$$\pi_{4,3}(x) - \pi_{4,1}(x)$$

is > 0 for most values of x. It is, however, known [10] that there are infinitely many integers x with this difference negative, indeed the difference may be as large as

$$\pi_{4,1}(x) - \pi_{4,3}(x) = \Omega_{\pm}\left(\frac{\sqrt{x}\ln\ln\ln x}{\ln x}\right).$$

In 1978, Carter Bays and Richard Hudson [11] have made a numerical investigation as high up as to $2 \cdot 10^{10}$. We shall now describe their results.

79

Graph of the Function $\pi_{4,3}(x) - \pi_{4,1}(x)$

In order to provide a rough idea of how regular the distribution of the odd primes may be between the two series $4n+3$ and $4n+1$, we produce below a computer drawn graph of the function $\Delta(x) = \pi_{4,3}(x) - \pi_{4,1}(x)$, and since the values of x and of the function will finally be large, a logarithmic scale on both axes in the diagram was used. This is a little bit awkward, since the function value may occasionally be zero or negative and therefore impossible to represent in a logarithmic diagram, so we actually drew

$$\text{sign}(\Delta(x))\log_{10}(|\Delta(x)| + 1). \tag{3.16}$$

On the Δ-axis we mark, of course, the values of Δ and not the values of (3.16).

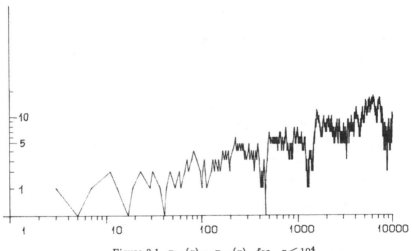

Figure 3.1. $\pi_{4,3}(x) - \pi_{4,1}(x)$ for $x \leq 10^4$

Above we have shown the curve for x up to 10^4 and on the next page for x between 10^4 and 10^6. We observe that the x-axis is first crossed at $x = 26,861$. The graphs show the "general behaviour" of the function, which is approximately linear in the log-log-scale with local oscillations superimposed like a band around the general curve.

80

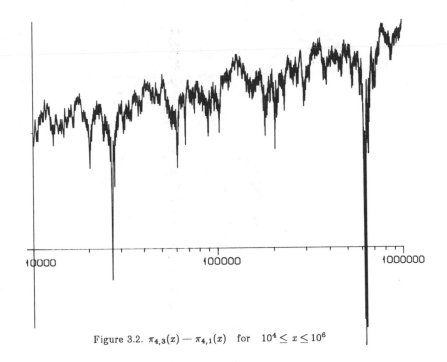

Figure 3.2. $\pi_{4,3}(x) - \pi_{4,1}(x)$ for $10^4 \leq x \leq 10^6$

The Negative Regions

The study of the function $\Delta(x)$ has revealed six regions of x below $2 \cdot 10^{10}$, in which $\Delta(x)$ is < 0. The first two of these regions are visible in the above graph of $\Delta(x)$. The first region consists of only one point, the value $x = 26,861$ with $\Delta(x) = -1$. In the second region of negative values of $\Delta(x)$, starting at $x = 616,841$, $\Delta(x)$ makes several short dips into the negative. The deepest of these lies between 623,437 and 623,803 where $\Delta(x)$ runs as low as -8 for $x = 623,681$. From $x = 633,799$, the last value in the region with $\Delta(x) \leq 0$, $\Delta(x)$ rapidly increases again to the value 40 for $x = 641,639$ and then continues on the same high level that it had reached before it entered into this negative region. The third region with negative values of $\Delta(x)$ is much higher up, about $x = 12,366,589$, for which the lowest value of $\Delta(x)$ in this region, -24, occurs. We reproduce on the following page the graph of $\Delta(x)$ in this region as well as, for comparison, the graph of a curve depicting the results of coin tossing:

$$f(x) = \sum_{p \leq x} \epsilon_p, \quad p \text{ prime.}$$

81

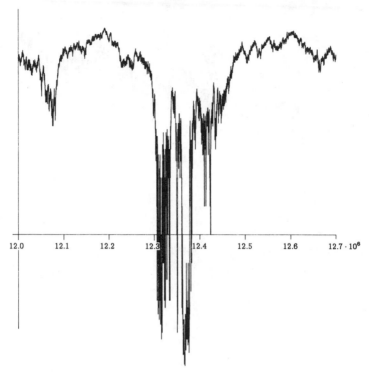

Figure 3.3. $\pi_{4,3}(x) - \pi_{4,1}(x)$ for $12.0 \cdot 10^6 \leq x \leq 12.7 \cdot 10^6$

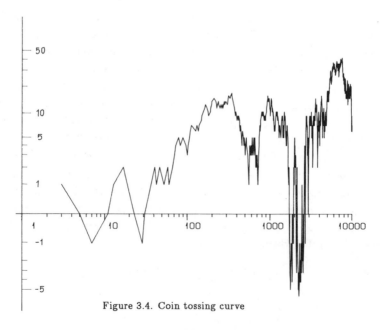

Figure 3.4. Coin tossing curve

In the coin-tossing curve $\epsilon_p = +1$ or -1 with equal probability. Observing the general resemblance between the coin-tossing curve and the curve for $\Delta(x)$, we have yet another confirmation of the reasonability of using probabilistic models in studying the distribution of primes, at least locally. The coin-tossing curve does not, however, explain why $\pi_{4,3}(x) - \pi_{4,1}(x)$ resumes positive values after the visits into the negative. Quite the contrary—on average in half the cases the coin-tossing curve continues in the negative after an axis-crossing region. See [12].

The third negative region is somewhat different from the first two. In this region there is an area in which $\Delta(x)$ stays negative for a long while. This is the interval $[12\,361933,\ 12\,377279]$ *in which* $\Delta(x) < 0$ *continuously.* This means that there are over 15,000 consecutive integers here for which $\Delta(x)$ is negative!

This property of remaining negative for a period will be even more pronounced in the following negative regions.

The Negative Blocks

In the three regions described so far, the "band" around the "main curve" descends only now and then below the x-axis, while the general tendency is still positive, and only occasionally are there short intervals when $\Delta < 0$. In each of the subsequent three negative regions, however, the curve plunges into and stays in the negative very long, resulting in long blocks of negative values of $\Delta(x)$. We give some key values for these three long negative blocks:

Region	Negative block	min $\Delta(x)$	max $\Delta(x)$
4	$951,850,000$–$951,880,000$	-48	-12
5	$6.34 \cdot 10^9$–$6.37 \cdot 10^9$	-1374	-19
6	$18.54 \cdot 10^9$–$18.95 \cdot 10^9$	-2719	-54

The negative blocks are actually larger than shown in the table but their extent is at present not precisely known. This is because the computer program with which Bays and Hudson detected them only provided output at regular intervals, and so the exact beginning and end of these blocks was lost.—The existence of these very long negative blocks came as a big surprise. In the largest block there are more than 410 million (consecutive)

83

integers with $\Delta(x)<0$. This is more than 2% of the numbers in the range investigated (up to $2 \cdot 10^{10}$).

Bays and Hudson have, in addition, made a corresponding study of the distribution of the primes between the two series $6n \pm 1$, with a similar result. Their report on this is published in [13].

Large Gaps Between Consecutive Primes

Another local property of the series of primes that has been studied is the difference $\Delta_n = p_{n+1} - p_n$ between consecutive primes. The following assertion is easy to prove: Δ_n can be made as large as we wish.

Proof. Let $N = 2 \cdot 3 \cdot 5 \ldots p_n$. Then the consecutive integers

$$N+2,\ N+3,\ N+4,\ldots,\ N+p_{n+1}-2,\ N+p_{n+1}-1 \qquad (3.17)$$

are all composite. This is easy to understand, if we make a sieve of Eratosthenes for the given interval. Since N is divisible by all primes $\leq p_n$, the multiples of these primes will strike upon integers forming exactly the same pattern as the multiples of these primes form in the interval $[2, p_{n+1} - 1]$. But every one of these integers is certainly divisible by some prime $\leq p_n$, which will therefore also be the case for the corresponding integer of (3.17). Thus, every one of the integers (3.17) is composite. Since p_{n+1} can be made as large as we like, the assertion follows. Let us attempt to estimate how high up in the number series we have to advance in order to find a gap of width Δ_k. The mean value $\overline{\Delta}_k$ of Δ_k, $k = 0, 1, 2, \ldots, n-1$, is, if p_0 is defined as 0,

$$\overline{\Delta}_k = \frac{1}{n}\sum_{k=0}^{n-1}\Delta_k = \frac{p_n}{n}.$$

According to (2.39),

$$\frac{p_n}{n} > \ln(n \ln n) - 1.0073 = \ln n + \ln \ln n - 1.0073 > \ln n,$$

if $n > e^{e^{1.0073}} = 15.5$. Since the largest value of Δ_k certainly is $\geq$ its mean value $\overline{\Delta}_k$

$$\max_{k \leq n-1} \Delta_k \geq \frac{p_n}{n} > \ln n \quad \text{for } n \geq 16. \qquad (3.18)$$

84

From this it follows that a difference $p_{k+1} - p_k$ of size $\geq \Delta_k$ certainly will occur for some

$$x < G = e^{\Delta_k}. \tag{3.19}$$

This value is, unfortunately, a grotesque over-estimate of the smallest value of x possible, but this is what is easily proved.

		Maximal gaps between consecutive primes $\leq 4.444 \cdot 10^{12}$			
$\max \Delta_k$	p_{k+1}	$\dfrac{\ln p_{k+1}}{\sqrt{\Delta_k}}$	$\max \Delta_k$	p_{k+1}	$\dfrac{\ln p_{k+1}}{\sqrt{\Delta_k}}$
1	3	1.10	248	1919 13031	1.21
2	5	1.14	250	3870 96383	1.25
4	11	1.20	282	4362 73291	1.19
6	29	1.37	288	12942 68779	1.24
8	97	1.62	292	14531 68433	1.23
14	127	1.29	320	23009 42869	1.21
18	541	1.48	336	38426 11109	1.20
20	907	1.52	354	43024 07713	1.18
22	1151	1.50	382	1 07269 05041	1.18
34	1361	1.24	384	2 06780 48681	1.21
36	9587	1.53	394	2 23670 85353	1.20
44	15727	1.46	456	2 50560 82543	1.12
52	19661	1.37	464	4 26526 18807	1.14
72	31469	1.22	468	12 76763 35139	1.18
86	1 56007	1.29	474	18 22268 96713	1.19
96	3 60749	1.31	486	24 11606 24629	1.19
112	3 70373	1.21	490	29 75010 76289	1.19
114	4 92227	1.23	500	30 33714 55741	1.18
118	13 49651	1.30	514	30 45995 09051	1.17
132	13 57333	1.23	516	41 66086 96337	1.18
148	20 10881	1.19	532	46 16905 10543	1.16
154	46 52507	1.24	534	61 44874 54057	1.17
180	170 51887	1.24	540	73 88329 28467	1.18
210	208 31533	1.16	582	134 62943 11331	1.16
220	473 26913	1.19	588	140 86954 94197	1.15
222	1221 64969	1.25	602	196 81885 57063	1.15
234	1896 95893	1.25	652	261 49417 11251	1.12

When Δ_k takes a larger value for some k than it has for all smaller values of k, this is called a *maximal gap*. The maximal gaps have been found up to $4.444 \cdot 10^{12}$ by Daniel Shanks [14], by L. J. Lander and T. R. Parkin [15] and by Richard Brent [16], [17], and are given in the table above.

The (over-)estimate (3.19) proves that a gap of length 100 must occur before $e^{100} = 10^{43.43}$, while the table shows that such a gap occurs already for $10^{5.57}$, a significantly smaller value of G. From the values of the table Shanks has conjectured that the first gap of width Δ_k will appear at approximately $G = e^{\sqrt{\Delta_k}}$, which may also be expressed as

$$\max_{k \leq G} \sqrt{\Delta_k} \sim \ln G, \quad \text{as } G \to \infty. \tag{3.20}$$

The Cramér Conjecture

A related conjecture has been made much earlier by Harald Cramér [18], who conjectured that

$$\varlimsup_{n \to \infty} \frac{\Delta_n}{(\ln p_n)^2} \stackrel{c}{=} 1. \tag{3.21}$$

Since Cramérs argument actually is quite simple and is a nice example of statistical reasoning applied to the distribution of primes, we shall present it. The starting point is the Prime Number Theorem $\pi(x) \sim x/\ln x$ which is utilized to create sequences of integers which in some respect resemble the sequence of primes. Cramér constructs a model in which black and white balls are drawn with the chance of drawing a white ball in the nth trial being $1/\ln n$ for $n > 2$ and being arbitrarily chosen for $n = 1$ and $n = 2$. If the nth white ball is produced in the $P_n{}^{\text{th}}$ trial, then the sequence $P_1, P_2, \ldots$ will form an increasing sequence of integers. The class C of all possible sequences $\{P_n\}$ is considered. Obviously the sequence S of ordinary primes $\{p_n\}$ belongs to this class.

Next, Cramér considers the order of magnitude of $P_{n+1} - P_n$ by aid of the following construction. Let $c > 0$ be a given constant and let E_m denote the event that black balls are obtained in all the trials $m + \nu$ for $1 \leq \nu \leq c(\ln m)^2$. Then the following two events have the same probability

1. The inequality

$$P_{n+1} - P_n > c(\ln P_n)^2 \tag{3.22}$$

   is satisfied for an infinity of values of n.

2. An infinite number of the events E_m are realized.

If ϵ_m denotes the probability of the event E_m, then

$$\epsilon_m = \prod_{\nu=1}^{c(\ln m)^2}\left(1 - \frac{1}{\ln(m+\nu)}\right).$$

Thus

$$-\ln \epsilon_m = -\sum_{\nu=1}^{c(\ln m)^2}\ln\left(1 - \frac{1}{\ln(m+\nu)}\right) \sim$$

$$\sim \sum_{\nu=1}^{c(\ln m)^2}\left(\frac{1}{\ln(m+\nu)} + O\left(\frac{1}{(\ln m)^2}\right)\right) \sim \int_1^{c(\ln m)^2}\frac{dx}{\ln(m+x)} + O(1) \sim$$

$$\sim c\ln m + O(1),$$

and thus we can find two positive constants A and B such that for all sufficiently large values of m

$$\frac{A}{m^c} < \epsilon_m < \frac{B}{m^c}. \tag{3.23}$$

Thus if $c > 1$ the series $\sum_{m=1}^{\infty}\epsilon_m$ is convergent, and consequently, due to a lemma by Cantelli (see [12], pp. 188–189), the probability of the realization of an infinite number of events E_m is equal to zero.

If, on the other hand, $c < 1$, consider the events $E_{m_1}, E_{m_2}, \ldots$, where $m_1 = 2$ and

$$m_{r+1} = m_r + [c(\ln m_r)^2] + 1.$$

Then for some constant K and all sufficiently large r

$$m_r \le Kr(\ln r)^2$$

(since the function $Kx(\ln x)^2$, having the derivative $K(\ln x)^2 + 2K\ln x$ grows faster than the function m_x having a difference $\le c(\ln x)^2 + 1$). Thus, according to (3.23), in this case series $\sum \epsilon_{m_r}$ is divergent, because $c \le 1$. Since the events E_{m_r} are mutually independent, we conclude, again using a lemma by Cantelli, that with a probability $= 1$ an infinite number of these events will be realized.

To sum up: The probability of an infinite number of solutions of (3.22) is equal to zero if $c > 1$ and equal to one if $c < 1$. Combining these two results, Cramér obtains the following

Theorem 3.2. With a probability $= 1$

$$\overline{\lim_{n \to \infty}} \frac{P_{n+1} - P_n}{(\ln P_n)^2} = 1.$$

This suggests (3.21).

Cramér's conjecture is another example of the danger of relying on heuristic arguments which seem very persuasive and are supported by numerical evidence. H. Maier has recently shown that the number of primes in intervals about x of length $(\ln x)^c$, where c is any positive constant, varies much more than the Cramér model predicts.

In order to compare the gaps predicted by Shanks' conjecture (3.20) with reality, the quotient between the left-hand-side and the right-hand-side of (3.20) has also been given in the table above. The quotient may well tend to 1 as $x \to \infty$, and its largest value in the table is 1.62. Thus, if no dramatic changes occur higher up, we may conjecture that always

$$1.62 \max_{k \leq G} \sqrt{\Delta_k} \overset{c}{>} \ln p_k. \tag{3.22}$$

The inequality (3.22) would imply that a gap Δ_k occurs before

$$G \overset{c}{=} e^{1.62\sqrt{\Delta_k}}. \tag{3.23}$$

If this holds, a prime free interval of length 1 million ought to be found below $e^{1620} < 10^{704}$, an enormous number, but nevertheless smaller than many known large primes.

Bibliography

1. Daniel Shanks and John W. Wrench, Jr., "Brun's Constant," *Math. Comp.* **28** (1974) pp. 293–299.

2. G. H. Hardy and J. E. Littlewood, "Some Problems of 'Partitio Numerorum' III: On the Expression of a Number as a Sum of Primes," *Acta Math.* **44** (1922) pp. 1–70 = G. H. Hardy, *Coll. Papers*, vol. 1, pp. 561–630.

3. F. J. Gruenberger and G. Armerding, *Statistics on the First Six Million Prime Numbers*, Reviewed in *Math. Comp.* **19** (1965) pp. 503–505.

4. Hans Riesel and Robert. C. Vaughan, "On Sums of Primes," *Arkiv för Mat.* **21** (1983) pp. 45–74.

5. G. H. Hardy and E. M. Wright, *An Introduction to the Theory of Numbers,* Fifth edition, Oxford, 1979, p. 351.

6. Richard P. Brent, "The Distribution of Small Gaps Between Successive Primes," *Math. Comp.* **28** (1974) pp. 315–324.

7. Ian Richards, "On the Incompatibility of Two Conjectures Concerning Primes," *Bull. Amer. Math. Soc.* **80** (1974) pp. 419–438.

8. Thomas Vehka, "Explicit Construction of an Admissible Set for the Conjecture that Sometimes $\pi(x+y) > \pi(y)+\pi(x)$," *Notices Am. Math. Soc.* **26** (1979) p. A-453.

9. Hans Riesel, "Primes Forming Arithmetic Series and Clusters of Large Primes," *Nordisk Tidskr. för Informationsbehandling (BIT)* **10** (1970) pp. 333–342.

10. J. E. Littlewood, "Sur la Distribution des Nombres Premiers," *Comptes Rendus* **158** (1914) pp. 1869–1872.

11. Carter Bays and Richard H. Hudson, "On the Fluctuations of Littlewood for Primes of the Form $4n \pm 1$," *Math. Comp.* **32** (1978) pp. 281–286.

12. William Feller, *An Introduction to Probability Theory and its Applications,* vol. I, Second edition, Wiley, New York, 1957, pp. 73–87.

13. Carter Bays and Richard H. Hudson, "Details of the First Region of Integers x with $\pi_{3,2}(x) < \pi_{3,1}(x)$," *Math. Comp.* **32** (1978) pp. 571–576.

14. Daniel Shanks, "On Maximal Gaps Between Successive Primes," *Math. Comp.* **18** (1964) pp. 646–651.

15. L. J. Lander and T. R. Parkin, "On the First Appearance of Prime Differences," *Math. Comp.* **21** (1967) pp. 483–488.

16. Richard P. Brent, "The First Occurrence of Large Gaps Between Successive Primes," *Math. Comp.* **27** (1973) pp. 959–963.

17. Richard P. Brent, "The First Occurrence of Certain Large Prime Gaps," *Math. Comp.* **35** (1980) pp. 1435–1436.

18. Harald Cramér, "On the Order of Magnitude of the Difference Between Consecutive Prime Numbers," *Acta Arith.* **2** (1936) pp. 23–46.

CHAPTER 4

THE RECOGNITION OF PRIMES

Introduction

One very important concern in number theory is to establish whether a given number N is prime or composite. At first sight it might seem that in order to decide the question an attempt must be made to factorize N and if it fails, then N is a prime. Fortunately there exist primality tests which do not rely upon factorization. This is very lucky indeed, since all factorization methods developed so far are rather laborious. Such an approach would admit only numbers of moderate size to be examined and the situation for deciding on primality would be rather bad. It is interesting to note that methods to determine primality, other than attempting to factorize, do not give any indication of the factors of N in the case where N turns out to be composite.—Since the prime 2 possesses certain particular properties, we shall, in this and the next chapter, assume for most of the time that N is an *odd* integer.

Tests of Primality and of Compositeness

Every logically stringent primality test has the following form: If a certain condition on N is fulfilled then N is a prime, otherwise N is composite. Very many primality tests are of this form. However, they are often either rather complicated, or applicable only to numbers N of a certain special form, such as $N = 5 \cdot 2^k - 1$ or $N = 9 \cdot 10^k + 1$. Conversely, there exist numerous tests which are mathematically simple and computationally fast, but which on certain, very rare occasions, fail. These failures always result in a composite number being indicated as a prime, and never vice versa. We shall call such tests *compositeness tests*, while those that never fail on deciding primality will be termed *primality tests*. We have thus introduced the following terminology: A successful compositeness test on N proves that N is composite and a successful primality test proves that N is prime. If a primality test is performed on N and the condition for primality is not fulfilled, then this constitutes a proof that N is composite. However, if a compositeness test is performed on N and the condition for compositeness is *not* fulfilled, then primality of N is *not necessarily proved*.

Factorization Methods as Tests of Compositeness

In the next chapter we shall discuss the most important factorization methods in current use. Of course, a successful factorization of N resolves the question of compositeness/primality for N and it follows that each of the factorization methods presented in Chapter 5 can be used as a compositeness test for N. However, as has already been mentioned, factorization is quite expensive and is therefore used only as a compositeness test for factors that are easy to find, for instance very small factors. In the short table on p. 148 we see that 76% of all odd integers have a prime factor < 100 and thus, by simply performing trial divisions by all odd primes below 100, we have a good chance of proving N to be composite!—Normally, however, factorization methods play a rôle as compositeness tests only, as mentioned above, when used as a fast preliminary means of settling the question for the majority of N, before some more elaborate test needs to be applied for the remainder.

Fermat's Theorem as Compositeness Test

The foundation-stone of almost all efficient primality tests and compositeness tests is Fermat's Theorem A2.8 (see p. 270): If p is a prime and $(a, p) = 1$, then

$$a^{p-1} \equiv 1 \pmod{p}. \tag{4.1}$$

The logical converse of Fermat's Theorem immediately yields

Theorem 4.1. Fermat's Theorem used as compositeness test. If $(a, N) = 1$ and

$$a^{N-1} \not\equiv 1 \pmod{N}, \tag{4.2}$$

then N is composite.

Fermat's Theorem as Primality Test

Could Fermat's Theorem also be used as a primality test? Unfortunately not! Despite the fact that the overwhelming majority of composite integers

91

N can be shown to be composite by applying Theorem 4.1, there do exist certain combinations of a and *composite* N for which $a^{N-1} \equiv 1 \pmod{N}$, and these values of N are thus not revealed as composite by this criterion. One of the simplest examples is $N = 341 = 11 \cdot 31$, which gives $2^{340} \equiv 1 \pmod{341}$. In this case, however, a change of base from 2 to 3 helps: $3^{340} \equiv 56 \pmod{341}$, which proves N to be composite.

Pseudoprimes and Probable Primes

A composite number which, like $N = 341$ above, behaves like a prime in Fermat's Theorem, is called a *pseudoprime* (meaning false prime). To render the definition of pseudoprime a little more precise, we give

Definition 4.1. An odd *composite* number N, for which

$$a^{N-1} \equiv 1 \pmod{N} \tag{4.3}$$

is called a (Fermat) *pseudoprime for the base a*.

Remark. The term pseudoprime has been used in older literature for any N satisfying (4.3), *primes as well as composites*, thus including both primes and false primes! In more modern texts the term "probable prime" is often used for prime number candidates until their primality (or compositeness) has been established.

Now, if all the pseudoprimes for some particular base were known, then a Fermat test performed for this base in combination with a list of pseudoprimes would actually constitute a fast primality test. As a matter of fact this works very well for numbers of limited size! The technique has been made use of by D. H. Lehmer, see [1] and [2], who prepared a table of all Fermat pseudoprimes below $2 \cdot 10^8$ for the base 2 with no factor < 317. Lehmer's table of pseudoprimes starts at 10^7 which was the limit of the existing prime tables at the time. In order to test a number N between 10^7 and $2 \cdot 10^8$ for primality with aid of Lehmer's table this scheme was followed:

1. Perform trial divisions by the 65 primes ≤ 313. If a divisor is found N is composite.
2. Compute $2^{N-1} \pmod{N}$. If the residue is $\not\equiv 1$, then N is composite.
3. For the residue 1, check with Lehmer's table of pseudoprimes. If N is in the table, it is composite and its factors are given in the table. If not in the table, N is a prime.

Recently this technique of settling the question of primality for numbers of limited size has been further investigated by Carl Pomerance, John Selfridge and Samuel Wagstaff and is reported in [3]. It appeared that below $25 \cdot 10^9$ there are 21853 pseudoprimes for the base 2, of which 4709 remain if the base 3 is also tested. Of these, base 5 leaves 2552 and, finally, base 7 eliminates all but 1770. Thus if a list is available of these 1770 numbers below $25 \cdot 10^9$, the pseudoprimes for all four bases 2, 3, 5 and 7, we can decide upon the primality of any number up to this limit by performing at most four Fermat tests. We will show below that Fermat tests are computationally fast. One Fermat test can, as a matter of fact, be carried out in at most $2 \log_2 N$ steps, each step consisting of the multiplication and reduction (mod N) of two integers (mod N). In the next subsection we shall demonstrate in detail how this can be implemented.

A Computer Program for Fermat's Test

The problem is: How can we compute a^d (mod N) without too much labour? The algorithm used relies upon the *binary representation* of the exponent d:

$$d = \beta_0 + \beta_1 \cdot 2 + \beta_2 \cdot 2^2 + \cdots + \beta_s \cdot 2^s, \qquad (4.4)$$

where the binary digits β_i are 0 or 1. The number of digits, $s + 1 = [\log_2 d] + 1$, since for $d = 2^k$ we have the binary representation $d = 1 \cdot 2^k$, and so $\beta_k = 1$ in this case, while all the other digits are 0. Now we have

$$a^d = a^{\sum_{i=0}^{s} \beta_i 2^i} = \prod_i a^{\beta_i 2^i} = \prod_{\beta_j = 1} a^{2^j}. \qquad (4.5)$$

To clarify this with an example: In order to compute a^{13}, first write 13 as $1 + 4 + 8$, then compute a^2, a^4, a^8 by successively squaring starting from a, since $a^{2^{i+1}} = (a^{2^i})^2$ and, finally, compute $a^{13} = a^1 \cdot a^4 \cdot a^8$. This procedure is very efficient, especially in a binary computer, where d is already stored in binary form. However, we shall not make use of this shortcut here, since our intention is to express all computer programs in the high-level language PASCAL. In this case (or when d is stored in decimal form, as in a programmable calculator), we must find the binary digits of d before we can choose which factors a^{2^j} to multiply together in (4.5). There are several good algorithms for computing the binary digits of an integer and in this particular situation it would be nice to start the multiplications of the expressions a^{2^j} with the lower powers of 2 first, so that we must use an algorithm which gives the least significant digit β_0 first. This is easy if you include, in your PASCAL program, the code

```
IF odd(d) THEN beta0:=1 ELSE beta0:=0; d:=(d-beta0) DIV 2;
```

These statements compute β_0 and remove it from d, so that β_1 can then be found by a similar operation. This fragment of code can, however, be simplified considerably, due to the fact that the expression (d-beta0) DIV 2 will assume the value d DIV 2 *regardless of whether* β_0 *is* = 0 *or* = 1. (Since if $\beta_0 = 1$, then d is odd and thus $(d - \beta_0)/2$ is the integer part of $d/2$, precisely what the integer division d DIV 2 in PASCAL furnishes in this case.) But, since the value of β_0 is of no consequence, the variable beta0 is not needed at all in the program and the section of code can be reduced to

```
IF odd(d) THEN statement; d:=d DIV 2;
```

Here statement stands in place of the computation that is to be carried out only in case d is odd. In order to compute $a^d \pmod{N}$, we may now construct a loop in the program in which a squaring and reduction (mod N) of a^{2^j} is performed, and also a multiplication by a^{2^j} followed by a reduction (mod N) of the product $\prod a^{2^j}$ so far accumulated, in case $\beta_j = 1$. If $\beta_j = 0$ then this part of the computation is omitted. The program loop becomes

```
WHILE d > 0 DO
  BEGIN IF odd(d) THEN prod:=prod*a2j mod N;
    d:=d DIV 2; a2j:=sqr(a2j) mod N
  END;
```

Now the loop has only to be "initialized" by suitable values: a2j:=a; and prod:=1; Incorporating this in a complete program, we arrive at the following code for computing $a^d \pmod{N}$:

```
PROGRAM Fermat
{Computes a↑d mod N}
(Input,Output);
VAR a,d,N,beta,a2j,prod : INTEGER;

BEGIN
write('Input a, d and N: '); read(a,d,N);
prod:=1; a2j:=a;
```

```
WHILE d>0 DO
  BEGIN IF odd(d) THEN prod:=prod*a2j mod N;
    d:=d DIV 2; a2j:=sqr(a2j) mod N
  END;
{When arriving here, a↑d mod N has been computed}
write('a↑d mod N=',prod:5)
END.
```

This short program will operate only if N^2 is less than the largest integer that can be stored in a computer word. If N is larger, multiple precision arithmetic must be used for the computations.—As in several similar situations throughout this book the reader must consider the code above as a model of how the computation should be organized rather than as a general program which covers all situations.

The Labour Involved in a Fermat Test

From the computer program just shown, it is obvious that the number of multiplications and reductions (mod N) when computing a^d (mod N) lies between $[\log_2 d]$ and $2[\log_2 d]$, depending on the number of binary ONEs of d. Since about half of the binary digits of a number chosen at random have the value 1, the average value will be $1.5[\log_2 d]$ operations of the described kind for evaluating a^d (mod N). Performing a Fermat test (with $d = N - 1$) therefore takes at most $2\log_2 N$ and on average $1.5\log_2 N$ multiplications and reductions (mod N). Thus it is an algorithm of *polynomial order* (for this concept, see p. 241 in Appendix 1).

Carmichael Numbers

For a composite number N it is usually fairly easy to find, by trial and error, an a, such that $a^{N-1} \not\equiv 1$ (mod N). Hence, in most cases a small number of Fermat compositeness tests will reveal N as being composite. *However, there exist composite numbers N such that $a^{N-1} \equiv 1$ (mod N) for all a satisfying $(a, N) = 1$.* These exceptional numbers are called *Carmichael numbers,* see [4], of which the smallest is $561 = 3 \cdot 11 \cdot 17$. A Carmichael number never reveals its compositeness under a Fermat test, unless a happens to be a divisor of N, a situation which can be avoided by first testing whether (a, N) is > 1.—Although the Carmichael numbers are rather scarce (see (4.12) below for an upper bound of their numbers), there do exist enough of them to create frustration when a number of Fermat tests all yield the result $a^{N-1} \equiv 1 \pmod{N}$. (There are 2163 Carmichael numbers below $25 \cdot 10^9$.—The reader might find it strange that there are *more* Carmichael numbers than the 1770 composites, mentioned on p. 93, which "survived" pseudoprime tests for the bases $a = 2, 3, 5$ and 7, since a Carmichael number should pass a pseudoprime test to *any* base. However, it should be borne in mind that a pseudoprime test with a base a, where $(a, N) > 1$, will certainly reveal N to be composite. Thus, the discrepancy between these two figures is due to the fact that there are $2163 - 1770 = 393$ Carmichael numbers below $2.5 \cdot 10^9$ which possess one or more factors among 2, 3, 5 and 7.)—As a matter of fact, infinitely many formulas for the structure of possible Carmichael numbers can be constructed, such as

$$N = (6t + 1)(12t + 1)(18t + 1). \tag{4.6}$$

Here all three factors $6t + 1$, $12t + 1$ and $18t + 1$ must be primes for the same value of t. To verify that this formula actually gives Carmichael numbers, we compute $\lambda(N) = [6t, 12t, 18t] = 36t$, and thus $N - 1 = 36t(36t^2 + 11t + 1) = s \cdot \lambda(N)$, s integer, and moreover from Carmichael's Theorem A2.11 on p. 276 we find that $a^{N-1} = (a^{\lambda(N)})^s \equiv 1 \pmod{N}$ for all a such that $(a, N) = 1$. Examples of Carmichael numbers of the particular structure given in (4.6) are

$$7 \cdot 13 \cdot 19 = 1729, \quad 37 \cdot 73 \cdot 109 = 294409, \quad 211 \cdot 421 \cdot 631 = 56052361, \ldots$$

$$9091 \cdot 18181 \cdot 27271 = 4507445537641, \ldots$$

The general structure of Carmichael numbers is covered by the following rule: N is a Carmichael number if and only if $p - 1 | N - 1$ for every prime factor p of N and N is composite and squarefree.—Therefore, due to the existence of Carmichael numbers, it is essential to improve the Fermat test in order that the result "character undetermined" occurs less often. One such improvement is described in the next subsection.

Euler Pseudoprimes

According to Euler's criterion for quadratic residues, Theorem A3.3 on p. 280, we have

$$a^{(N-1)/2} \equiv \pm 1 \pmod{N} \quad \text{for} \quad (a, N) = 1, \tag{4.7}$$

if N is an odd prime. Hence, (4.7) can be used as a test of compositeness on N as well as Fermat's theorem eq. (4.1):

Theorem 4.2. Euler's criterion used as compositeness test. If N is odd, $(a, N) = 1$, and

$$a^{(N-1)/2} \not\equiv \pm 1 \ (\text{mod } N), \tag{4.8}$$

then N is composite. If $a^{(N-1)/2} \equiv \pm 1 \ (\text{mod } N)$, then this value should be compared with the value of Jacobi's symbol (a/N) (see p. 284), and if

$$a^{(N-1)/2} \not\equiv \left(\frac{a}{N}\right) (\text{mod } N), \tag{4.9}$$

then N is composite.—If, however,

$$a^{(N-1)/2} \equiv \left(\frac{a}{N}\right) (\text{mod } N), \tag{4.10}$$

then the test is not conclusive.

As in Fermat's compositeness test there are "false primes." So there are odd composites which for *certain bases* a satisfy (4.10), and so we have use for a new

Definition 4.2. An odd *composite* number N such that

$$a^{(N-1)/2} \equiv \left(\frac{a}{N}\right) (\text{mod } N), \quad \text{with} \quad (a, N) = 1,$$

is called an *Euler pseudoprime for the base* a. Many Carmichael numbers are revealed as composites by Euler's criterion, so that it distinguishes primes from composites more efficiently than does Fermat's theorem. The above remark applies also to many composite numbers which are *not* Carmichael numbers, but which are still not shown to be composite by Fermat's theorem for a *specific* base a. We find for instance that

$$2^{170} \equiv 1 \ (\text{mod } 341), \ 5^{280} \equiv 67 \ (\text{mod } 561), \ 11^{864} \equiv 1 \ (\text{mod } 1729).$$

Since $67 \not\equiv \pm 1 \ (\text{mod } 561)$, there is no need to compute Jacobi's symbol in this case. In the remaining two cases we obtain

$$\left(\frac{2}{341}\right) = -1, \quad \text{and} \quad \left(\frac{11}{1729}\right) = -1,$$

so that all these integers will be revealed as composite by Euler's criterion. Of these three numbers, 561 and 1729 are Carmichael numbers, while 341, having $2^{340} \equiv 1 \ (\text{mod } 341)$, is not.

97

Strong Pseudoprimes and a Primality Test

The idea of using Euler's criterion instead of Fermat's theorem to distinguish between primes and composites can be carried a little further. In doing so we arrive at the concept *strong pseudoprime*, defined in

Definition 4.3. An odd *composite* number N with $N - 1 = d \cdot 2^s$, d odd, is called a *strong pseudoprime for the base a* if either

$$a^d \equiv 1 \pmod{N} \text{ or } a^{d \cdot 2^r} \equiv -1 \pmod{N}, \text{ for some } r = 0, 1, 2, \ldots, s - 1.$$

Neither Euler pseudoprimes nor strong pseudoprimes admit the possibility of analogues to Carmichael numbers. Therefore, using Euler's criterion or testing for strong primality with enough bases will finally reveal any composite number or, alternatively, prove the primality of any prime N. As an example, there exist only 13 numbers below $25 \cdot 10^9$ which are strong pseudoprimes for all the bases 2, 3 and 5 simultaneously and these are listed below.

List of strong pseudoprimes for all the bases 2, 3 and 5	
Number	Factorization
25326001	2251 · 11251
1 61304001	7333 · 21997
9 60946321	11717 · 82013
11 57839381	24061 · 48121
32 15031751	151 · 751 · 28351
36 97278427	30403 · 121609
57 64643587	37963 · 151849
67 70862367	41143 · 164569
143 86156093	397 · 4357 · 8317
155 79919981	88261 · 176521
184 59366157	67933 · 271729
198 87974881	81421 · 244261
212 76028621	103141 · 206281

This fact can be utilized to identify any number below $25 \cdot 10^9$ as being either composite or prime by means of the following very simple scheme, given in [3]:

1. Check whether N is a strong pseudoprime for the base 2. If not, then N is composite.
2. Check whether N is a strong pseudoprime for the base 3. If not, then N is composite.
3. Check whether N is a strong pseudoprime for the base 5. If not, then N is composite.
4. If N is one of the 13 numbers in the table below, then N is composite, otherwise N is prime.

Note that if N is prime three tests will be necessary (in addition to looking in the table) to determine this fact. On the other hand, if N is composite, then this is most frequently revealed in step 1, so that steps 2 and 3 are not often required.—If one is willing to do pseudoprime tests also for the bases 7 and 11, then no table look-up at all is needed below $25 \cdot 10^9$, see [3], p. 1023.

Remark 1. A strong primality test is more stringent than an Euler pseudoprime test. In [3], p. 1009 the following is proved: every strong pseudoprime for the base a is also an Euler pseudoprime for the base a.

Remark 2. Assuming the truth of the so-called Extended Riemann Hypothesis it has been proved by Gary Miller [5] that consecutive, strong pseudoprime tests can constitute a primality test working in polynomial time. The Extended Riemann Hypothesis is needed to prove the existence of a "small" non-residue of order q (mod p) for every prime p.—This hypothesis is a generalization of the ordinary Riemann hypothesis (see p. 52) to other functions than $\varsigma(s)$, such as the Dirichlet L-functions mentioned on p. 259. In the Extended Riemann Hypothesis it is assumed that all of the L-functions have all their non-trivial zeros exactly on the line $s = 1/2$.

Since the various types of pseudoprimes behave like disguised primes in the tests performed, it is practical to introduce a term for all integers which satisfy a certain (false) primality test. This is formalized in the following

Definition 4.4. An odd number (prime or composite), which is not revealed as composite by

1. a Fermat test with the base a is a *probable prime* for the base a
2. Euler's criterion with the base a is an *Euler probable prime* for the base a
3. a strong pseudoprime test with the base a is a *strong probable prime* for the base a.

A Computer Program for Strong Pseudoprime Tests

Using the program for a^d (mod N) above and the scheme just given, we can devise the following PASCAL-program for testing the primality of any odd number below $25 \cdot 10^9$:

```
PROGRAM Strong Pseudoprime Test
{Identifies any odd N below 25*10↑9 as prime or composite}
(Input,Output);
LABEL 1,2,3;
VAR a,d,d1,N,s,beta,a2j,prod,i : INTEGER;

FUNCTION abmodn(a,b,N : INTEGER) : INTEGER;
{Computes a*b mod N. Here the reader must program his own
  function in double precision, depending on the word length
  of his computer. If assembly language is accessible from
  the high-level language the reader is using then an
  assembly code is preferable. The function shown here is
  only a model which operates as long as N↑2 < the largest
  integer which can be stored}
BEGIN
  abmodn:=a*b MOD N
END {abmodn};

BEGIN
write('Input N: '); read(N);
d:=N-1; s:=0; WHILE not odd(d) DO
  BEGIN d:=d DIV 2; s:=s+1 END;
d1:=d;
FOR a:=2 TO 5 DO IF a<>4 THEN
  BEGIN
    prod:=1; a2j:=a; d:=d1; WHILE d>0 DO
      BEGIN IF odd(d) THEN prod:=abmodn(prod,a2j,N);
          {Here the last binary digit of d was used}
        d:=d DIV 2; a2j:=abmodn(a2j,a2j,N)
      END;
  {When arriving here, prod=a↑d mod N has been computed}

  IF prod=1 THEN GOTO 1
    ELSE FOR i:=1 TO s DO
      BEGIN prod:=abmodn(prod,prod,n);
        IF prod=N-1 THEN
```

```
1:          BEGIN writeln('N is a spsp(',a:1,')');
               GOTO 2
             END
           END;
      writeln('N is composite'); GOTO 3;
2: END;

{Here the checking of the list begins}
IF (N=25326001) OR (N=161304001) OR (N=960946321)
   OR (N=1157839381) OR (N=3215031751) OR (N=3697278427)
   OR (N=5764643587) OR (N=6770862367) OR (N=14386156093)
   OR (N=15579919981) OR (N=18459366157) OR (N=19887974881)
   OR (N=21276028621) THEN
writeln('N is composite') ELSE writeln('N is prime');

3: END.
```

The remark we made about the program FERMAT above applies equally well to this coding: The reader should regard it only as a structural model of the particular version of the program which would operate in his computer.

Exercise 4.1. Strong pseudoprime test. Write the FUNCTION abmodn(a,b,N) hinted at in the program above and incorporate it in a complete program for your computer. In order to save computing time, precede the pseudoprime test by trial divisions by the small primes $\leq G$, thereby discarding many composites before the more time-consuming pseudoprime test is applied. Make computer experiments with various values of G in order to find its optimal value. Use the integers between 10^8 and $10^8 + 100$ as test numbers in these experiments.

Pseudoprime Counts

The density of pseudoprimes determines how often probable primes are composite. It is thus of interest to count the number of pseudoprimes of various types $\leq x$ and to find asymptotic formulas for their numbers as $x \to \infty$. Let $P_2(x)$ be the number of pseudoprimes $\leq x$ for the base 2 and $C(x)$ the number of Carmichael numbers $\leq x$. Not too much has been proved about the behavior of these counting functions, but some fairly precise conjectures have recently been made (see [6]). The following inequalities have been proved so far:

$$e^{(\ln x)^{5/14}} \leq P_2(x) \leq x e^{-\frac{1}{2}\ln x \ln_3 x / \ln_2 x} \qquad (4.11)$$

101

$$C(x) \leq xe^{-\frac{\ln x}{\ln_2 x}\left(\ln_3 x + \ln_4 x + \frac{\ln_4 x - 1}{\ln_3 x} + O\left(\frac{\ln_4 x}{\ln_3 x}\right)^2\right)}.$$ (4.12)

Here $\ln_2 x$ stands for $\ln \ln x$ and so on. Comparing these expressions with $\pi(x) \sim x/\ln x$, we can confirm our earlier assertion that the pseudoprimes for the base 2 and the Carmichael numbers are scarce. To give just a few examples: for $x = 10^{20}$, $P_2(x)/\pi(x) \leq 1.43 \cdot 10^{-2}$ and $C(x)/\pi(x) \leq 7.2 \cdot 10^{-5}$ and for $x = 10^{50}$ these quotients are $\leq 7.2 \cdot 10^{-7}$ and $\leq 5.7 \cdot 10^{-16}$, respectively. Thus the fraction of probable primes $\leq x$ which actually are composite turns out to be very tiny indeed for large values of x.

The following two conjectures are stated in [6]:

$$P_2(x) \overset{c}{\leq} xe^{-\frac{\ln x}{\ln_2 x}(\ln_3 x + \ln_4 x + o(x))}, \quad \text{as} \quad x \to \infty$$

$$C(x) \overset{c}{=} xe^{-\frac{\ln x}{\ln_2 x}\left(\ln_3 x + \ln_4 x + \frac{\ln_4 x - 1}{\ln_3 x} + O\left(\frac{\ln_4 x}{\ln_3 x}\right)^2\right)}, \quad \text{as} \quad x \to \infty.$$

Rigorous Primality Proofs

For large primes N it is not easy actually to *prove* their primality. The reason for this is that any rigorous proof must exclude the possibility of N being only a pseudoprime. The more compositeness tests we perform on N the greater the likelihood that N is prime, but this in itself is no mathematical proof. Only fairly recently have *general* primality tests been devised which do not exhibit the flaw of admitting pseudoprimes. These new tests are, however, very complicated in theory (and even to program) but have (almost) polynomial run-time, which means that they are computationally fast. Eventually a method will probably be discovered to replace these complicated primality tests by some combination of simpler tests in such a way that the combination of tests chosen can never admit a *particular* composite number N as a pseudoprime in *all* the tests.

We now proceed by demonstrating how the possibility of N being a pseudoprime may be excluded by demanding supplementary conditions in the pseudoprime test. The "old" rigorous primality proofs for N all depend in some degree upon the factorization of $N-1$, or $N+1$, or $N^2 \pm N + 1$ or the like. Unfortunately, this dependence upon factorization makes these techniques *inefficient in the general case*, since there is always the risk of chancing upon a number N such that *none* of these factorizations can be achieved. On the other hand, in cases when these techniques work—and they do in many practical examples—they work very well indeed, so in spite of any impending, more general, approach to the problem of proving primality, the author finds it well worthwhile dwelling on the subject.— Moreover, besides the theoretical interest of these proofs, they constitute the fastest known primality proofs for large classes of interesting numbers, such as $N = 15 \cdot 2^n - 1$, $N = 7 \cdot 2^n + 1$ and so on. We commence our account with

Lehmer's Converse of Fermat's Theorem

One way of actually *proving* the primality of N is to employ a theorem by Édouard Lucas, for which a general proof has been given by D. H. Lehmer:

Theorem 4.3. Suppose $N - 1 = \prod_{j=1}^{n} q_j^{\beta_j}$, with all q_j's distinct primes. If an integer a can be found, such that

$$a^{(N-1)/q_j} \not\equiv 1 \pmod{N} \quad \text{for all } j = 1, 2, \ldots, n \qquad (4.13)$$

and such that

$$a^{N-1} \equiv 1 \pmod{N}, \qquad (4.14)$$

then N is a prime.

We may regard the conditions (4.13) as extending Fermat's condition (4.14), and securing that N is really a prime and not merely a pseudo-prime.—Before giving a formal proof of Lehmer's theorem, we shall discuss the idea behind it. (The reader who is unfamiliar with the concepts in the following paragraph should now consult Appendix 1 and Appendix 2 before continuing.)

One difference between a prime and a composite number N is that the number $\varphi(N)$ of primitive residue classes (mod N) $= N - 1$ if N is a prime, and is $< N - 1$ if N is composite. The same is thus true of the order of the group M_N of primitive residue classes (mod N) since $M_N = \varphi(N)$. Furthermore, if N is prime, the group M_N is always a cyclic group and thus contains a generator of order $N - 1$. If N is composite, on the other hand, the order of any group element is, at most, $\varphi(N) < N - 1$. If the reader carefully considers the conditions (4.13) and (4.14) he should realize that they establish a as an element of order $N - 1$ of the group M_N of primitive residue classes (mod N); precisely what is needed to ensure the primality of N. (Such an element a of M_N is in number theory termed a *primitive root* of N.)

Formal Proof of Theorem 4.3

Consider all exponents e such that $a^e \equiv 1$ (mod N). These exponents constitute a module M consisting of only integers. See Appendix 1, p. 237. According to (4.14), $N-1$ is an element of this module M. The module is thus generated by some integer $d \leq N-1$, which divides $N-1$. But every divisor d of $N-1$, with the exception of $N-1$ itself, divides at least one of the numbers $(N-1)/q_j$, $j = 1, 2, \ldots, n$. Now, if d were $< N-1$, then at least one of the numbers $(N-1)/q_j$ would belong to the module M, and thus for this particular value of j, the corresponding $a^{(N-1)/q_j}$ would be $\equiv 1$ (mod N), contrary to the assumptions (4.13). Thus the combination of (4.13) and (4.14) implies that the generator d of the module M is $= N-1$. Euler's theorem A2.9 on p. 272, however, grants that always $a^{\varphi(N)} \equiv 1$ (mod N), if $(a, N) = 1$ and that $\varphi(N) < N-1$ for all *composite* numbers N. But these facts imply that also $\varphi(N)$ belongs to the module M, which is impossible if $\varphi(N) < N-1$, because the generator d is the *least* positive element of the module. This proves that N is in this case a prime.

Remark. We had to introduce the condition $(a, N) = 1$ in order to be able to use Euler's theorem. This requirement, however, need not be explicitly stated in the wording of Theorem 4.3, since if $(a, N) > 1$, then it can be proved that $a^{N-1} \not\equiv 1$ (mod N). Thus the condition $(a, N) = 1$ which is implied by (4.14), may be omitted in the wording of the theorem.

Example. $N = 3^{59} - 2^{59} = 14130\,38609116\,22737524\,61387579$. Factorizing $N-1$ we find

$$N-1 = 2 \cdot 3 \cdot 7 \cdot 59 \cdot 1151 \cdot 58171 \cdot 123930193 \cdot 687216767.$$

The two larger factors were found by applying Shanks' factoring method (see p. 191) to the number $8\,51669065\,67146031$, which remained after all the small factors had been removed from $N-1$. Those two factors were found after the program computed only 573 partial denominators of the continued fraction expansion of $\sqrt{N-1}$.—It turns out that $a = 7$ is a primitive root of N, and so N is a prime. The computations which verify this are shown below:

$$
\begin{array}{rl}
7^{(N-1)/2} & \equiv \qquad\qquad\qquad\qquad\qquad\qquad\quad -1 \\
7^{(N-1)/3} & \equiv 14039\,52407176\,60958441\,81052225 \\
7^{(N-1)/7} & \equiv 11292\,25508857\,85053499\,63835029 \\
7^{(N-1)/59} & \equiv 2802\,49189705\,71158036\,52416186
\end{array}
$$

$$7^{(N-1)/1151} \quad \equiv \quad 2706\,58963050\,54768415\,59851274$$
$$7^{(N-1)/58171} \quad \equiv \quad 10951\,68902411\,53649984\,82043016$$
$$7^{(N-1)/123930193} \equiv 14036\,31931644\,46024491\,67220257$$
$$7^{(N-1)/687216767} \equiv 10974\,48314837\,68234524\,28433200 \pmod{N}.$$

Ad Hoc Search for a Primitive Root

The problem of searching for a primitive root is that there is no efficient *deterministic* method known to produce a primitive root nor even to find a quadratic non-residue. In using Lehmer's theorem 4.3 we frequently need to make numerous trials before an a is found which fulfills *all* the conditions in (4.13). The general situation may be described as follows: We are seeking a primitive root a of the prime N. There exist $\varphi(N-1)$ such a's, but by (A2.21)

$$\varphi(N-1) = \varphi\Big(\prod q_j^{\beta_j}\Big) = (N-1)\prod_j \Big(1 - \frac{1}{q_j}\Big).$$

Thus, the more small prime factors $N-1$ has, the smaller is the proportion of a's which are primitive roots of N and the longer we have to search to find one by trial and error. However, since $\varphi(N-1) > CN/\ln\ln N$, almost certainly a good a can be found after, say, $(10/C)\ln\ln N$ trials—not too many.—When applying Theorem 4.3 we also have to factor $N-1$, sometimes a formidable task when N is large. (We shall subsequently explain how this difficulty can, at least partially, be overcome.) Let us now go back to the example given above. The search for a primitive root of N proceeds as follows: When $a^{(N-1)/q_j} \equiv 1 \pmod{N}$, a is called a *power residue of order* q_j with respect to N. About one a out of q_j is such a $q_j{}^{\text{th}}$-power residue of N and is therefore not a primitive root. Thus the factor 2 of $N-1$ eliminates *half of all* a's as possible primitive roots of N. In order to avoid computing in vain all the $a^{(N-1)/2} \pmod{N}$ having the value 1, we calculate instead the value of the Jacobi symbol (a/N) (see Appendix 3), and discard those values of a for which $(a/N) = 1$. This shortcut enables us to escape the most common failure of the trial and error approach and thus speeds up the process considerably. The shortcut is demonstrated in the computation below.

$$a = 2 \quad \text{gives} \quad 2^{(N-1)/2} \equiv -1, \quad 2^{(N-1)/3} \equiv 1 \pmod{N},$$

so $a = 2$ is a cubic residue of N, ending this trial!

$$a = 3 \quad \text{yields} \quad 3^{(N-1)/2} \equiv -1, \quad 3^{(N-1)/3} \equiv 1 \pmod{N},$$

and thus $a = 3$ is also a cubic residue, so we have a new failure in our search for a primitive root of N. We continue in this way until the smallest absolute primitive root $a = -5$ is found or, if we prefer to look only for positive a's, until $a = 7$ is found. In order to reduce the amount of unnecessary work we have, for each new value of a, started by evaluating the value of $a^{(N-1)/q_j}$ with the smallest value of q_j first, since this is most likely to violate (4.13). This will obviously lead to a smaller average computing time than first evaluating all the other $a^{(N-1)/q_j}$'s (which would probably not violate (4.13) in any case!) and then not until at the end of the entire computation perhaps discovering that the attempted value of a was in fact unsuitable.—If N is not too large then a primitive root of N is usually found quite fast in the manner described above. However, if the values of a are chosen at random, the test may cycle forever without proving N prime by an unlikely, but theoretically possible sequence of bad luck. It is this behavior that classes the algorithm as non-deterministic.

The Use of Several Bases

As we have just seen, a great deal of computing may have to be done before a primitive root a is discovered. Fortunately, John Selfridge has in [7] succeeded in relaxing the condition (4.13) of Theorem 4.3 in such a way that a primitive root a need not be determined in order to prove the primality of N. The relaxed version of the theorem reads:

Theorem 4.4. Suppose $N-1 = \prod_{j=1}^{n} q_j^{\beta_j}$, with the q_j's all distinct primes. If for every q_j there exists an a_j such that

$$a_j^{(N-1)/q_j} \not\equiv 1 \,(\text{mod } N) \quad \text{while} \quad a_j^{N-1} \equiv 1 \,(\text{mod } N), \qquad (4.15)$$

then N is a prime.

Proof. Let e_j be the order of a_j in the group M_N of primitive residue classes (mod N). Since $e_j | N - 1$ while $e_j \nmid (N-1)/q_j$, we have $q_j^{\beta_j} | e_j$. But for each j, $e_j | \varphi(N)$ (regardless of whether N is prime or composite) and thus also $q_j^{\beta_j} | \varphi(N)$, which implies that $\prod q_j^{\beta_j} = N - 1$ divides $\varphi(N)$. However, if $N - 1 | \varphi(N)$, then $\varphi(N)$ cannot be $< N-1$, and consequently N is prime.

Example. We consider the above example once again, $N = 3^{59} - 2^{59}$. The primality of N can now be established by means of the following congruences. We have also indicated below the two "failures" in the trial

106

and error method for determining a primitive root of N: $2^{(N-1)/3} \equiv 1$ (mod N) and $3^{(N-1)/3} \equiv 1$ (mod N):

$2^{(N-1)/2}$	$\equiv$	-1
$2^{(N-1)/3}$	$\equiv$	1
$3^{(N-1)/3}$	$\equiv$	1
$5^{(N-1)/3}$	$\equiv$	$14039\,52407176\,60958441\,81052225$
$5^{(N-1)/7}$	$\equiv$	$782\,66109729\,95267547\,70837537$
$5^{(N-1)/59}$	$\equiv$	$10636\,29203818\,09458018\,79749999$
$5^{(N-1)/1151}$	$\equiv$	$3216\,43070546\,34805980\,22736901$
$5^{(N-1)/58171}$	$\equiv$	$13450\,33889565\,61733879\,77763600$
$5^{(N-1)/123930193}$	$\equiv$	$3732\,50753518\,56196918\,18435804$
$5^{(N-1)/687216767}$	$\equiv$	$9167\,67553110\,06092700\,57486746$ (mod N).

(Note that the value of $5^{(N-1)/3}$ (mod N) happens to be identical to $7^{(N-1)/3}$ (mod N), found on p. 104 above! This is not at all the marvellous coincidence that it seems, since the congruence $x^3 \equiv 1$ (mod N) has (at most) two solutions $\not\equiv 1$ (mod N) if N is prime, and every number $a^{(N-1)/3}$ is obviously a solution! Thus, if neither $5^{(N-1)/3}$ nor $7^{(N-1)/3}$ is $\equiv 1$ (mod N), then the chances of these two values coinciding will in fact be one in two.)

The use of several bases may save much computing, especially in those cases where the least primitive root of N is rather large.

Fermat Numbers and Pepin's Theorem

Lehmer's theorem is particularly simple to apply whenever $N - 1$ has only a small number of distinct prime factors. The simplest case of all is when $N - 1$ is a power of 2, in which case $N = 2^s + 1$. However, these numbers are known to be composite unless the exponent s is a power of 2 (cf. Appendix 6). The numbers

$$F_n = 2^{2^n} + 1 \tag{4.16}$$

are called Fermat numbers, and they can be primes. Next, we investigate what is required by Lehmer's theorem in order that the number F_n is a prime. We must find an a such that

$$\begin{cases} a^{(F_n-1)/2} = a^{2^{2^n-1}} \not\equiv 1 \ (\text{mod } F_n) \quad \text{and} \\ a^{F_n-1} \ = a^{2^{2^n}} \ \equiv 1 \ (\text{mod } F_n). \end{cases} \tag{4.17}$$

If we put $a^{2^{2^n-1}} = x$, then this gives

$$x^2 \equiv 1 \pmod{F_n} \quad \text{and} \quad x \not\equiv 1 \pmod{F_n}. \tag{4.18}$$

Now, if F_n is prime, then x is an element of a ring with no zero divisors (see p. 253). Hence, the congruence $x^2 - 1 \equiv (x+1)(x-1) \equiv 0 \pmod{F_n}$ will have two, and only two, solutions due to Theorem A1.5 on p. 253, $x \equiv 1 \pmod{F_n}$ and $x \equiv -1 \pmod{F_n}$. Since the solution $x \equiv 1 \pmod{F_n}$ violates the first condition of (4.17), the only remaining possibility is $x \equiv -1 \pmod{F_n}$. Thus, we seek an integer a satisfying

$$x \equiv a^{(F_n-1)/2} \equiv -1 \pmod{F_n} \tag{4.19}$$

when F_n is a prime. Here the theory of quadratic residues (Appendix 3) comes to our aid. Euler's criterion on p. 280 tells us in this case that a must be a *quadratic non-residue* of F_n. But it is known that the integer 3 happens to be a quadratic non-residue of all primes of the form $12n \pm 5$, see Table 32. Is F_n of this form? Let us check: $2^{2^1} = 4$, $2^{2^2} = 16 \equiv 4 \pmod{12}, \ldots$, so that $F_n \equiv 4 + 1 \equiv 5 \pmod{12}$. Therefore, if F_n is a prime then $a = 3$ certainly will be a quadratic non-residue of F_n and we arrive at

Theorem 4.5. Pepin's Theorem. A necessary and sufficient condition for the Fermat number $F_n = 2^{2^n} + 1$, $n \geq 1$, to be prime is that

$$3^{2^{2^n-1}} \equiv -1 \pmod{F_n}. \tag{4.20}$$

Example. We demonstrate the primality of the largest known Fermat prime, $F_4 = 2^{16} + 1 = 65537$. The proof runs as follows:

$$3^8 = 6561, \quad 3^{16} \equiv 54449, \quad 3^{32} \equiv 61869, \quad 3^{64} \equiv 19139,$$

$$3^{128} \equiv 15028, \quad 3^{256} \equiv 282, \quad 3^{512} \equiv 13987,$$

$$3^{1024} \equiv 8224 = 8192 + 32 = 2^{13} + 2^5,$$

$$3^{2048} \equiv 2^{26} + 2^{19} + 2^{10} \equiv -2^{10} - 2^3 + 2^{10} \equiv -8,$$

$$3^{4096} \equiv 2^6, \quad 3^{8192} \equiv 2^{12}, \quad 3^{16384} \equiv 2^{24} \equiv -2^8.$$

$$3^{32768} \equiv 2^{16} \equiv -1 \pmod{F_4}.$$

This concludes the proof.

With the aid of computers, Pepin's Theorem has been used in practice to test the primality of some higher Fermat numbers. Pepin's test is particularly attractive to use on a binary computer (almost all computers are binary), since the most complicated part of the programming, the multiple precision division by F_n, needed in each loop to reduce the number $t_s \equiv 3^{2^s} \pmod{F_n}$, is extremely simple to perform due to the very simple binary representation of F_n.—Even before the era of computers all the Fermat numbers up to F_8 had been investigated for primality. F_8 has 78 digits, so the proof of its compositeness, using only a desk-calculator was quite an achievement! Today the first Fermat number of unknown character is F_{20}, containing the incredible figure of 315653 digits! The largest Fermat number so far exposed to Pepin's test is F_{14}, having 4933 digits. This number was proved to be composite by John Selfridge and Alexander Hurwitz in 1963. See Table 4 at the end of this book for more details on Fermat numbers.

A Relaxed Converse of Fermat's Theorem

We have already encountered the major difficulty in using Lehmer's converse of Fermat's theorem as an algorithm for general primality proofs. We come to a halt when unable to factor $N - 1$ completely! There are several ways of escaping this situation. It turns out that it is possible to relax the conditions in Lehmer's theorem so that a *partial factorization only* of $N - 1$ suffices for the primality proof. As a matter of fact, we need not assume that $N - 1$ is *completely* split into prime factors, since it is possible to prove a theorem, analogous to Lehmer's theorem, where $N - 1 = R \cdot F$, $R < F$, and only F is split into prime factors. In this way the frequently very laborious task of factorizing a large cofactor $R = (N - 1)/F$ is avoided.—In this notation, which we shall frequently use, F stands for the factored part of $N - 1$ and R for the remaining part.

Theorem 4.6. Suppose $N - 1 = R \cdot F = R \prod_{j=1}^{n} q_j^{\beta_j}$, with all q_j's distinct primes, with $(R, F) = 1$ and $R < F$. If an integer a can be found, such that

$$\left(a^{(N-1)/q_j} - 1, N \right) = 1 \quad \text{for all} \quad j = 1, 2, \ldots, n \qquad (4.21)$$

and satisfying

$$a^{N-1} \equiv 1 \pmod{N}, \qquad (4.22)$$

then N is a prime.

109

Proof. Consider any possible prime factor p of N. (4.21) implies that

$$\left(a^{(N-1)/q_j} - 1, p\right) = 1 \quad \text{for all} \quad j \tag{4.23}$$

and (4.22) implies that

$$a^{N-1} \equiv 1 \pmod{p}. \tag{4.24}$$

Further, suppose that the element a of the group M_p has order d. Then $d|N-1$, while $d \nmid (N-1)/q_j$ for any j, i.e.

$$d \Big| R \prod_{\nu} q_{\nu}^{\beta_{\nu}}, \quad \text{while} \quad d \nmid R q_j^{\beta_j - 1} \prod_{\nu \neq j} q_{\nu}^{\beta_{\nu}}.$$

But this is possible only if $q_j^{\beta_j} | d$ for every j, which implies that $F|d$. Now, since $d|p-1$, it follows that also $F|p-1$. But then the smallest value which p could possibly take would be $p = F + 1 > \sqrt{N}$, since the completely factored part, F, is assumed to be larger than $\sqrt{N}$. However if N were composite then this would be a contradiction if applied to the smallest prime factor p of N, which in this case would be $\leq \sqrt{N}$. Thus, the only remaining possibility is that N is prime.—There is a related theorem, on factorization, which has been much used also for primality proofs. This is Theorem 4.13 on p. 129.

Proth's Theorem

As an immediate consequence of Theorems 4.5 and 4.6 we obtain the following analogue of Pepin's theorem concerning the Fermat numbers:

Theorem 4.7. Proth's Theorem. Suppose N has the form $N = h \cdot 2^n + 1$, with $2^n > h$ and h an odd integer. If there exists an integer a such that

$$a^{(N-1)/2} \equiv -1 \pmod{N}, \tag{4.25}$$

then N is a prime.

It should be emphazised that the theorems mentioned so far are very efficient when a computer is put to work. As examples of large primes which have been identified in this way, we give

$$180(2^{127} - 1)^2 + 1, \quad 5 \cdot 2^{13165} + 1, \quad 29 \cdot 2^{7927} + 1.$$

For more detailed information, the reader is referred to Table 5 at the end of this book.

Tests of Compositeness for Numbers of the Form $N = h \cdot 2^n \pm k$

Using Euler's criterion for (a/p) (see Appendix 3, p. 280), it is easy to construct a necessary but not always sufficient condition for $N = h \cdot 2^n \pm k$ to be prime, i.e. a compositeness test for N, provided that h and k are odd and k is not too large. Choose an integer a such that $(a/N) = -1$. Then $a^{(N-1)/2} \equiv -1 \pmod{N}$ if N is prime, which gives

$$a^{h \cdot 2^{n-1} + (k-1)/2} \equiv -1 \pmod{N}, \quad \text{if } N = h \cdot 2^n + k$$

and

$$a^{h \cdot 2^{n-1} - (k+1)/2} \equiv -1 \pmod{N}, \quad \text{if } N = h \cdot 2^n - k.$$

The computation is carried out by first calculating $a^h \pmod{N}$ and then, by means of $n-1$ successive squarings and reductions $\pmod{N}$, obtaining $a^{h \cdot 2^{n-1}} \pmod{N}$. Finally, we must verify whether

$$a^{h \cdot 2^{n-1}} \cdot a^{(k-1)/2} \equiv -1 \pmod{N}, \quad \text{if } N = h \cdot 2^n + k \qquad (4.26)$$

or

$$a^{h \cdot 2^{n-1}} \equiv -a^{(k+1)/2} \pmod{N}, \quad \text{if } N = h \cdot 2^n - k. \qquad (4.27)$$

If the left-hand-side agrees with the right-hand-side, then N is an Euler probable prime for the base a, otherwise N is composite.

An Alternative Approach

Another way out of the problem to factor $N - 1$ would be to look for a theorem, analogous to Lehmer's, but based on another probable prime test rather than Fermat's theorem. In order to explain this we must examine the reason why Fermat's Theorem is applicable in proofs of primality. This is related to the divisibility properties of the numbers in the sequence

$$T_1 = a - 1, \ T_2 = a^2 - 1, \ldots, T_s = a^s - 1, \ldots \qquad (4.28)$$

If the prime p divides a term T_s of this sequence, it will also divide all terms T_{ks} for $k = 2, 3, 4, \ldots$, since $T_{ks} = a^{ks} - 1$ is divisible by $T_s = a^s - 1$. This fact is the principal element of the proof of Theorems 4.3 and 4.4. The sequence (4.28) can now be described in a recursive form, a form which will help us to search for similar sequences having different divisibility properties in the case where $N - 1$ is too difficult to factor. Using the definition of T_n we find immediately that

$$T_{n+1} = (T_n + 1)a - 1 = aT_n + (a - 1). \tag{4.29}$$

This shows that a first-order linear recurrence relation holds for the terms T_n.—Now, there exist certain slightly different sequences, termed Lucas sequences, which exhibit divisibility properties different from those of the sequence T_n and which can, in certain cases, be used to construct probable prime tests analogous to Fermat's test *but with $N + 1$ substituted for $N - 1$*. Thus, we have at our disposal an alternative means of proving the primality of N if we are able to factor $N + 1$ (at least in part) rather than $N - 1$.—In addition, there exist algorithms devised for cases where one of $N^2 + 1$, $N^2 + N + 1$ or $N^2 - N + 1$ has been partially factored. This can be useful if both $N - 1$ and $N + 1$ are too difficult to factor (see [7]). All of these partial factorizations can be used in one single combined primality test and factor bounds can also be worked into this test. See Ferrier's example on p. 129.—These types of primality tests often admit a cascade of possibilities in their application: $N - 1 = 2 \cdot 3 \cdot H_1$, say, where H_1 is probably prime. Maybe we cannot factor $H_1 - 1$, but $H_1 + 1$ factors etc. However, we shall mention only one of these more complicated techniques, namely Theorem 4.18 on p. 136, since general primality tests are now available, which actually avoid factorization altogether. Nevertheless, it might still be of interest to mention an instance when *all* five of $N \pm 1$, $N^2 + 1$ and $N^2 \pm N + 1$ were found too hard to factor. Such a case is the following one, reported in [8]. For the integer $N = (10^{84} + 17)/9$ the following (insufficient) partial factorizations are easily obtained:

$$N - 1 = 2^3 \cdot 1531 \cdot H_1, \quad N + 1 = 2 \cdot 3 \cdot H_2,$$

$$N^2 + 1 = 2 \cdot 5 \cdot 2069 \cdot 2157989 \cdot H_3,$$

$$N^2 + N + 1 = 7 \cdot 14869 \cdot H_4, \quad N^2 - N - 1 = 3 \cdot 271 \cdot H_5.$$

All of these *five* composite integers H_i lack factors below the search limit $5,988,337,680$ and as a result the methods described here all failed to prove N prime, a rather unusual situation.—However, by finally applying a version of the more general method of Adleman, Pomerance and Rumely mentioned on p. 138, N was easily proved to be prime by Lenstra and Cohen.

112

Primality Tests of Lucasian Type

As we have discussed above, Lehmer's converse of Fermat's theorem is not applicable unless we succeed in factoring $N - 1$ at least until the "completely factored part" exceeds the remaining cofactor. By using results from the arithmetic theory of quadratic fields (see Appendix 4), it is possible to find a converse of the analogue of Fermat's theorem in these number fields. It turns out that in certain cases the factorization of $N + 1$ comes into play in exactly the same way as does the factorization of $N - 1$ in the ordinary field of rationals. Thus we have a choice between $N - 1$ and $N + 1$ and can choose which of these two numbers is the easiest to factor. We shall give a fairly detailed account of this technique, which has in fact led to many useful primality tests, mainly due to Lucas.

Lucas Sequences

Suppose a and $b = \bar{a}$ are two numbers satisfying a quadratic equation with integer coefficients, the so-called characteristic equation

$$\lambda^2 - P\lambda + Q = 0. \tag{4.30}$$

Now, for $n \geq 0$ define the numbers U_n and V_n by

$$U_n = \frac{a^n - b^n}{a - b}, \qquad V_n = a^n + b^n. \tag{4.31}$$

We can then demonstrate that these numbers are rational integers in the following way: Firstly, we have $U_0 = 0$, $U_1 = 1$, $V_0 = 2$ and $V_1 = a + b = P$. Next, because we have assumed P and Q to be integers, all the numbers U_m and V_m must be integers since the "initial values" U_0, U_1 and V_0, V_1 in the recurrence relations (4.34) and (4.35) below are integers. The numbers U_n and V_n constitute what are termed *Lucas sequences*. It can easily be shown that these satisfy certain interesting formulas:

$$U_{m+n} = \frac{a^{m+n} - b^{m+n}}{a - b} =$$

$$= \frac{(a^m - b^m)(a^n + b^n)}{a - b} - \frac{a^n b^n (a^{m-n} - b^{m-n})}{a - b} =$$

$$= U_m V_n - a^n b^n U_{m-n} \tag{4.32}$$

113

and, similarly

$$V_{m+n} = a^{m+n} + b^{m+n} =$$
$$= (a^m + b^m)(a^n + b^n) - a^n b^n(a^{m-n} + b^{m-n}) =$$
$$= V_m V_n - a^n b^n V_{m-n}. \tag{4.33}$$

For the special case $n = 1$ we obtain the second-order recurrence relations

$$U_{m+1} = (a + b)U_m - abU_{m-1} = PU_m - QU_{m-1} \tag{4.34}$$

$$V_{m+1} = (a + b)V_m - abV_{m-1} = PV_m - QV_{m-1}, \tag{4.35}$$

already utilized above when proving that U_m and V_m are integers. The deduction of all these formulas requires the use of elementary algebra including the well-known relations between the roots and the coefficients of a quadratic equation, i.e. $a + b = P$ and $ab = Q$.

The Fibonacci Numbers

Taking $U_0 = 0$, $U_1 = 1$ yields the Fibonacci sequence 0, 1, 1, 2, 3, 5, 8, 13, 21, 34, 55, 89, 144,... satisfying $U_{m+1} = U_m + U_{m-1}$. This recurrence relation corresponds to the characteristic equation $\lambda^2 = \lambda + 1$, with roots $a, b = (1 \pm \sqrt{5})/2$. The numbers U_m are given by

$$U_m = \frac{1}{\sqrt{5}}\left\{\left(\frac{1 + \sqrt{5}}{2}\right)^m - \left(\frac{1 - \sqrt{5}}{2}\right)^m\right\}, \tag{4.36}$$

and the V_m's are

$$V_m = \left(\frac{1 + \sqrt{5}}{2}\right)^m + \left(\frac{1 - \sqrt{5}}{2}\right)^m. \tag{4.37}$$

The V_m's have the values 2, 1, 3, 4, 7, 11, 18, 29, 47, 76, 123,... The simplest way of calculating the numerical values of all the V_m's is to start with $V_0 = 2$, $V_1 = 1$ and then apply the recursion $V_{m+1} = V_m + V_{m-1}$.

Large Subscripts

Let us give another example demonstrating how (4.32) and (4.33) can be used to compute U_n and V_n easily for an isolated, large subscript n. We now choose $n = 100$. In order to calculate U_{100} and V_{100}, we let $m = n$ in (4.32) and (4.33) and arrive at the special case

$$U_{2n} = U_n V_n \tag{4.38}$$

$$V_{2n} = V_n^2 - 2(ab)^n = V_n^2 - 2Q^n. \tag{4.39}$$

If the subscript is odd $= 2n + 1$, we instead put $m = n + 1$ and obtain

$$U_{2n+1} = U_{n+1} V_n - Q^n \tag{4.40}$$

$$V_{2n+1} = V_{n+1} V_n - PQ^n. \tag{4.41}$$

Alternatively, we could have substituted n for m and $n + 1$ for n in (4.32) and would then have arrived at

$$U_{2n+1} = U_n V_{n+1} - Q^{n+1} U_{-1} = U_n V_{n+1} + Q^n, \tag{4.42}$$

where the value of U_{-1} is calculated as $(a^{-1} - b^{-1})/(a - b) = -1/(ab) = -1/Q$.

The computation of U_{100} and V_{100} can now be performed as follows (letting $P = 1$ and $Q = -1$ for the Fibonacci numbers):

$$U_{100} = U_{50} V_{50}, \qquad V_{100} = V_{50}^2 - 2,$$

$$U_{50} = U_{25} V_{25}, \qquad V_{50} = V_{25}^2 + 2.$$

Since all subscripts involved were even hitherto U_{100} and V_{100} have been generated in a straight-forward manner. Now consider the next step:

$$U_{25} = U_{13} V_{12} - 1, \quad V_{25} = V_{13} V_{12} - 1.$$

From this point on a difficulty presents itself. We are forced to keep track of *four* new quantities U_{12}, U_{13}, V_{12} and V_{13} and their descendants as we continue to decompose. Does this mean a doubling of the number of U's and V's to keep in mind each time we arrive at an *odd* subscript? Fortunately not! No further doubling occurs when we apply (4.38)–(4.41) again. Observe how the calculation proceeds:

$$U_{12} = U_6 V_6, \quad V_{12} = V_6^2 - 2, \quad U_{13} = U_7 V_6 - 1, \quad V_{13} = V_7 V_6 - 1.$$

Here we still have only *four* new quantities U_6, V_6, U_7 and V_7 to keep track of! The process extends in this manner:

$$U_7 = U_4 V_3 + 1, \quad V_7 = V_4 V_3 + 1, \quad U_6 = U_3 V_3, \qquad V_6 = V_3^2 + 2,$$

115

$$U_4 = U_2 V_2, \qquad V_4 = V_2^2 - 2, \qquad U_3 = U_2 V_1 + 1, \quad V_3 = V_2 V_1 + 1$$

until, finally,

$$U_2 = U_1 V_1, \qquad V_2 = V_1^2 + 2.$$

U_{100} and V_{100} have thus been decomposed in a chain of calculations which can be run backwards, starting with $U_1 = V_1 = 1$:

$$U_2 = U_1 V_1 = 1, \qquad\qquad V_2 = V_1^2 + 2 = 3,$$
$$U_3 = U_2 V_1 + 1 = 2, \qquad\qquad V_3 = V_2 V_1 + 1 = 4,$$
$$U_4 = U_2 V_2 = 3, \qquad\qquad V_4 = V_2^2 - 2 = 7,$$
$$U_6 = U_3 V_3 = 8, \qquad\qquad V_6 = V_3^2 + 2 = 18,$$
$$U_7 = U_4 V_3 + 1 = 13, \qquad\qquad V_7 = V_4 V_3 + 1 = 29,$$
$$U_{12} = U_6 V_6 = 144, \qquad\qquad V_{12} = V_6^2 - 2 = 322,$$
$$U_{13} = U_7 V_6 - 1 = 233, \qquad\qquad V_{13} = V_7 V_6 - 1 = 521,$$
$$U_{25} = U_{13} V_{12} - 1 = 75025, \qquad V_{25} = V_{13} V_{12} - 1 = 167761,$$
$$U_{50} = U_{25} V_{25} = 125\,86269025, \quad V_{50} = V_{25}^2 + 2 = 281\,43753123$$

and, finally,

$$U_{100} = U_{50} V_{50} \;\; = 35422\,48481792\,61915075,$$
$$V_{100} = V_{50}^2 - 2 = 79207\,08398483\,72253127.$$

The calculation looks rather unwieldy at first sight! Suppose we want to compute $U_n \pmod{N}$ or $V_n \pmod{N}$ for a very large subscript n, say about 10^{100}. How many steps will the scheme include? Well, at each stage the subscripts are roughly halved, so that approximately $\log_2 10^{100} \approx 332$ steps will be required before the subscripts have been reduced to 1. In each step four values U_s, V_s, U_{s+1} and $V_{s+1} \pmod{N}$ must be calculated, except possibly in some of the initial steps in the deduction, a case which occurs when n is divisible by a power of 2. Moreover, of course, if *only* V_n is required, then the entire chain of calculations makes no use whatsoever of the U's, so in such a situation the labour demanded is halved. However, in general this is an *algorithm of polynomial growth* as are some of the algorithms encountered earlier such as Euclid's algorithm (Appendix 1,

p. 238) or the algorithm for computing $a^d \pmod{N}$ using the binary representation of d given on p. 93 above.—Hence, the times taken in a computer for the calculation of $a^s \pmod{N}$ and for $U_s \pmod{N}$ will be of the same order of magnitude for large values of s and N. The computation of U_s in actual fact is slower, but only by some relatively small numerical factor.

Finally, we remark that if the subscript n possesses many factors 2, say $n = h \cdot 2^s$, h odd, then the chain of calculations takes a particularly simple form:

$$U_n = U_{h \cdot 2^s} = V_{h \cdot 2^s - 1} U_{h \cdot 2^s - 1} =$$

$$\vdots \tag{4.43}$$

$$= V_{h \cdot 2^s - 1} V_{h \cdot 2^s - 2} V_{h \cdot 2^s - 3} \cdots V_h U_h$$

and

$$V_n = V_{h \cdot 2^s} = V_{h \cdot 2^s - 1}^2 - 2Q^{h \cdot 2^{s-1}}$$

$$\vdots \tag{4.44}$$

$$V_{2h} = V_h^2 - 2Q^h.$$

If Q is chosen as 1 or -1 then the last set of formulas (4.44) turns out to be particularly simple. As an example, $U_{96} = V_{48} V_{24} V_{12} V_6 V_3 U_3$ and $V_6 = V_3^2 + 2$, $V_{12} = V_6^2 - 2$, $V_{24} = V_{12}^2 - 2$, $V_{48} = V_{24}^2 - 2$. Starting from $U_3 = 2$ and $V_3 = 4$, we obtain successively $V_6 = 4^2 + 2 = 18$, $V_{12} = 18^2 - 2 = 322$, $V_{24} = 322^2 - 2 = 103682$, $V_{48} = 103682^2 - 2 = 107\,49957122$, giving

$$U_{96} = 2 \cdot 4 \cdot 18 \cdot 322 \cdot 103682 \cdot 107\,49957122.$$

An Alternative Deduction

There is an alternative approach which also leads to the algorithm sketched above for the fast computation of U_n and V_n in $O(\log n)$ steps. This makes use of 2 by 2 matrices in order to condense the two scalar recursion formulas (4.34) and (4.35) into one matrix equation:

$$A_m = \begin{pmatrix} U_{m+1} & V_{m+1} \\ U_m & V_m \end{pmatrix} = \begin{pmatrix} P & -Q \\ 1 & 0 \end{pmatrix} \begin{pmatrix} U_m & V_m \\ U_{m-1} & V_{m-1} \end{pmatrix} \tag{4.45}$$

Applying (4.45) m times, we arrive at the formula

$$\begin{pmatrix} U_{m+1} & V_{m+1} \\ U_m & V_m \end{pmatrix} = \begin{pmatrix} P & -Q \\ 1 & 0 \end{pmatrix}^m \begin{pmatrix} U_1 & V_1 \\ U_0 & V_0 \end{pmatrix} \tag{4.46}$$

Thus, we can compute A_m by the usual "square and multiply" algorithm for obtaining m^{th} powers in $O(\log m)$ steps, but note that here the algorithm is applied to 2 by 2 matrices rather than the usual scalars.

Exercise 4.2. Fibonacci numbers. Use (4.45)–(4.46) and the binary representation of m to deduce a set of formulas for the computation of U_m and V_m. Write a computer **PROGRAM Fibonacci** which reads P, Q, m and N, and gives $U_m \pmod{N}$ and $V_m \pmod{N}$. Use the same technique of utilizing the binary representation of m as in the **PROGRAM Fermat** on p. 94!

With the notation introduced in (4.45) and (4.46) our previous calculation of U_{100} and V_{100} runs as follows:

$$A_{100} = \begin{pmatrix} U_{101} & V_{101} \\ U_{100} & V_{100} \end{pmatrix} = \begin{pmatrix} 1 & 1 \\ 1 & 0 \end{pmatrix}^{100} \begin{pmatrix} 1 & 1 \\ 0 & 2 \end{pmatrix} \tag{4.47}$$

Now, successively

$$\begin{pmatrix} 1 & 1 \\ 1 & 0 \end{pmatrix}^2 = \begin{pmatrix} 2 & 1 \\ 1 & 1 \end{pmatrix}$$

$$\begin{pmatrix} 1 & 1 \\ 1 & 0 \end{pmatrix}^3 = \begin{pmatrix} 1 & 1 \\ 1 & 0 \end{pmatrix}\begin{pmatrix} 2 & 1 \\ 1 & 1 \end{pmatrix} = \begin{pmatrix} 3 & 2 \\ 2 & 1 \end{pmatrix}$$

$$\begin{pmatrix} 1 & 1 \\ 1 & 0 \end{pmatrix}^6 = \begin{pmatrix} 3 & 2 \\ 2 & 1 \end{pmatrix}^2 = \begin{pmatrix} 13 & 8 \\ 8 & 5 \end{pmatrix}$$

$$\begin{pmatrix} 1 & 1 \\ 1 & 0 \end{pmatrix}^{12} = \begin{pmatrix} 13 & 8 \\ 8 & 5 \end{pmatrix}^2 = \begin{pmatrix} 233 & 144 \\ 144 & 89 \end{pmatrix}$$

$$\begin{pmatrix} 1 & 1 \\ 1 & 0 \end{pmatrix}^{24} = \begin{pmatrix} 233 & 144 \\ 144 & 89 \end{pmatrix}^2 = \begin{pmatrix} 75025 & 46368 \\ 46368 & 28657 \end{pmatrix}$$

$$\begin{pmatrix} 1 & 1 \\ 1 & 0 \end{pmatrix}^{25} = \begin{pmatrix} 1 & 1 \\ 1 & 0 \end{pmatrix}\begin{pmatrix} 75025 & 46368 \\ 46368 & 28657 \end{pmatrix} = \begin{pmatrix} 121393 & 75025 \\ 75025 & 46368 \end{pmatrix}$$

$$\begin{pmatrix} 1 & 1 \\ 1 & 0 \end{pmatrix}^{50} = \begin{pmatrix} 121393 & 75025 \\ 75025 & 46368 \end{pmatrix}^2 = \begin{pmatrix} 203\,65011074 & 125\,86269025 \\ 125\,86269025 & 77\,78742049 \end{pmatrix}$$

$$\begin{pmatrix} 1 & 1 \\ 1 & 0 \end{pmatrix}^{100} = \begin{pmatrix} 203\,65011074 & 125\,86269025 \\ 125\,86269025 & 77\,78742049 \end{pmatrix}^2 =$$

$$= \begin{pmatrix} 57314\,78440138\,17084101 & 35422\,48481792\,61915075 \\ 35422\,48481792\,61915075 & 21892\,29958345\,55169026 \end{pmatrix} = \begin{pmatrix} a_{11} & a_{12} \\ a_{21} & a_{22} \end{pmatrix},$$

say. Thus,

$$\begin{pmatrix} U_{101} & V_{101} \\ U_{100} & V_{100} \end{pmatrix} = \begin{pmatrix} a_{11} & a_{12} \\ a_{21} & a_{22} \end{pmatrix}\begin{pmatrix} 1 & 1 \\ 0 & 2 \end{pmatrix},$$

yielding $U_{100} = a_{21}$, $V_{100} = a_{21} + 2a_{22}$ and finally giving the values for U_{100} and V_{100} previously found on p. 116.

Divisibility Properties of the Numbers U_n

A necessary and sufficient condition for the roots of the characteristic equation (4.30) $\lambda^2 - P\lambda + Q = 0$ to be irrational is that the discriminant $P^2 - 4Q$ is not a square number. If we assume that this equation has irrational roots a and b, then these are conjugate integers in the number field $K(\sqrt{D})$ if $P^2 - 4Q = c^2 D$ holds and D is a square-free (rational) integer. According to Fermat's theorem for the quadratic field $K(\sqrt{D})$ (see Appendix 4, p. 296) we then have, if p is an odd prime such that $p \nmid D$,

$$a^p \equiv a \ (\text{mod } p), \quad \text{if} \quad \left(\frac{D}{p}\right) = +1 \tag{4.48}$$

or

$$a^p \equiv \bar{a} \ (\text{mod } p), \quad \text{if} \quad \left(\frac{D}{p}\right) = -1. \tag{4.49}$$

If $(a, p) = 1$ in $K(\sqrt{D})$, i.e. if $(Q, p) = 1$, then (4.48) implies

$$a^{p-1} \equiv 1 \ (\text{mod } p), \quad \text{if} \quad \left(\frac{D}{p}\right) = 1. \tag{4.50}$$

If $(D/p) = -1$ then always

$$a^{p+1} \equiv a\bar{a} \ (\text{mod } p). \tag{4.51}$$

Thus, if $(Q, p) = 1$, we find

$$U_{p-1} = \frac{a^{p-1} - \bar{a}^{p-1}}{a - \bar{a}} \equiv \frac{1 - 1}{a - \bar{a}} \equiv 0 \ (\text{mod } p) \quad \text{if} \quad \left(\frac{D}{p}\right) = +1 \tag{4.52}$$

and

$$U_{p+1} = \frac{a^{p+1} - \bar{a}^{p+1}}{a - \bar{a}} \equiv \frac{a\bar{a} - \bar{a}a}{a - \bar{a}} \equiv 0 \ (\text{mod } p) \quad \text{if} \quad \left(\frac{D}{p}\right) = -1. \tag{4.53}$$

The congruences (4.52) and (4.53) are fundamental for the factorization of the Lucas numbers U_n. If p is an odd prime not dividing Q, then obviously either U_{p-1} or U_{p+1} will contain the prime p as a factor. Note that the condition $p \nmid Q$ is necessary. To see why this is so, we can study the case when

$$U_n = \frac{1}{2\sqrt{-15}} \left\{ (2 + \sqrt{-15})^n - (2 - \sqrt{-15})^n \right\}.$$

The characteristic equation is $\lambda^2 - 4\lambda + 19 = 0$ and the corresponding recursion formula becomes $U_n = 4U_{n-1} - 19U_{n-2}$. The integers U_n successively assume the values 0, 1, 4, -3, -88, -295, ... Reducing U_n (mod 19) gives $U_n \equiv 4U_{n-1}$ (mod 19) because of the recursion formula, and hence $U_n \equiv$ 0, 1, 4, 16, 7, 9, 17, 11, 6, 5, 1, 4, 16, 7, ... (mod 19), a sequence which is periodic apart from its first element $U_0 \equiv 0$. Thus, in this particular case neither U_{18} nor U_{20} contains the factor 19. The reason for this "failure" is that we *cannot cancel* the factor $2 + \sqrt{-15}$ in both sides of the congruence, although Fermat's theorem in $K(\sqrt{-15})$ guarantees that

$$(2 + \sqrt{-15})^{19} \equiv 2 + \sqrt{-15} \ (\text{mod } 19),$$

since -15 is a quadratic residue (mod 19). This is because $19 = (2 + \sqrt{-15})(2 - \sqrt{-15})$ and, therefore, according to the cancellation rule for congruences, we do *not* arrive at $(2 + \sqrt{-15})^{18} \equiv 1$ (mod 19), but only at the weaker result

$$(2 + \sqrt{-15})^{18} \equiv 1 \ (\text{mod } 2 - \sqrt{-15}).$$

The reader who is unaccustomed to computations in quadratic fields might, at this stage, wish to see a verification of the fact that the congruence $(2 + \sqrt{-15})^{18} \equiv 1$ (mod 19) actually is false. One way of demonstrating this is as follows:

$$(2 + \sqrt{-15})^2 = -11 + 4\sqrt{-15} \equiv 8 + 4\sqrt{-15} \equiv 4(2 + \sqrt{-15}) \ (\text{mod } 19).$$

Putting $2 + \sqrt{-15} = a$, this is $a^2 \equiv 4a$ (mod 19). Thus we successively find $a^4 \equiv 16a^2 \equiv 2^6 a$, $a^8 \equiv 2^{12}a^2 \equiv 2^{14}a$, $a^9 \equiv 2^{16}a$ and, finally,

$$a^{18} \equiv 2^{34}a \equiv 2^{16}a \equiv 5a \equiv 10 + 5\sqrt{-15} \not\equiv 1 \ (\text{mod } 19).$$

Please note that this is *not* a typical example of arithmetic in a quadratic field. The extraordinarily simple relation $a^2 \equiv 4a$ (mod 19) makes the calculations much easier than ought to be expected.

It is also easy to verify that

$$5(2 + \sqrt{-15}) \quad is \quad \equiv 1 \ (\text{mod } 2 - \sqrt{-15}).$$

This is demonstrated in the following way:

$$\frac{5(2 + \sqrt{-15}) - 1}{2 - \sqrt{-15}} = \frac{9 + 5\sqrt{-15}}{2 - \sqrt{-15}} = \frac{(9 + 5\sqrt{-15})(2 + \sqrt{-15})}{19} =$$

$$= \frac{-57 + 19\sqrt{-15}}{19} = -3 + \sqrt{-15} = \text{an integer in } K(\sqrt{-15}).$$

Now, let us return to our main line of thought. What can be said about the divisibility of the U_i's by prime powers? Using the more general congruences (A4.18) and (A4.19) rather than Fermat's theorem in $K(\sqrt{D})$,

$$a^{p^{n-1}(p-1)} \equiv 1 \pmod{p^n} \text{ if } \left(\frac{D}{p}\right) = 1 \text{ and } (a,p) = 1 \text{ in } K(\sqrt{D}) \quad (4.54)$$

and

$$a^{p^{n-1}(p+1)} \equiv (a\bar{a})^{p^{n-1}} \pmod{p^n} \quad \text{if} \quad \left(\frac{D}{p}\right) = -1, \quad (4.55)$$

we obtain the following analogues to (4.52) and (4.53):

$$U_{p^{n-1}(p-1)} \equiv 0 \pmod{p^n}, \text{ if } \left(\frac{D}{p}\right) = 1 \text{ and } (a,p) = 1 \text{ in } K(\sqrt{D}) \quad (4.56)$$

and

$$U_{p^{n-1}(p+1)} \equiv 0 \pmod{p^n} \quad \text{if} \quad \left(\frac{D}{p}\right) = -1. \quad (4.57)$$

These two congruences together with the conditions imposed on a, p and D may be condensed into the following formula:

$$U_{p^{n-1}\{p-(D/p)\}} \equiv 0 \pmod{p^n}, \quad \text{if} \quad (2QD,p) = 1. \quad (4.58)$$

Primality Proofs by Aid of Lucas Sequences

The divisibility properties of the numbers U_n, discussed above, suffice to prove a theorem analogous to Lehmer's converse of Fermat's theorem.

Theorem 4.8. Suppose $N + 1 = \prod_{j=1}^{n} q_j^{\beta_j}$, with all q_j's distinct primes. If a Lucas sequence U_ν satisfying $(2QD, N) = 1$ can be found such that

$$(U_{(N+1)/q_j}, N) = 1 \quad \text{for all} \quad j = 1, 2, \ldots, n \quad (4.59)$$

and

$$U_{N+1} \equiv 0 \pmod{N}, \quad (4.60)$$

then N is a prime.

121

Proof. Consider any prime factor p of N. The conditions in the Theorem imply that

$$(U_{(N+1)/q_j}, N) \equiv 1 \ (\text{mod } p) \quad \text{for} \quad j = 1, 2, \dots, n \tag{4.61}$$

and that

$$U_{N+1} \equiv 0 \ (\text{mod } p). \tag{4.62}$$

In exactly the same way as in Theorem 4.6 we arrive at the conslusion that

$$d = N + 1. \tag{4.63}$$

We shall now prove how this induces primality of N. The idea on which the proof is based is to calculate d for a composite number N and show that $d < N + 1$ always holds for *composite* numbers, just as the fact $\varphi(N) < N - 1$ for composite numbers is used in the proof of Theorem 4.3.

To achieve this, suppose $N = \prod p_i^{\alpha_i}$ and $(2QD, N) = 1$. Then, by (4.58), we have

$$U_{p_i^{\alpha_i - 1}\{p_i - (D/p_i)\}} \equiv 0 \ (\text{mod } p_i^{\alpha_i}). \tag{4.64}$$

Using the fact that all the subscripts satisfying precisely *one* of the congruences

$$U_n \equiv 0 \ (\text{mod } p_i^{\alpha_i})$$

constitute (the non-negative part of) a module, we now find that the subscripts satisfying *all* the congruences simultaneously also form (the non-negative part of) a module with a generator which is the least common multiple of the generators of all the individual modules. Thus, it is clear that

$$U_m \equiv 0 \ (\text{mod } N) \tag{4.65}$$

with

$$m = \text{L.C.M.}\left\{ \text{all } p_i^{\alpha_i - 1}\left(p_i - \left(\frac{D}{p_i}\right)\right)\right\}. \tag{4.66}$$

Now, since all p_i are odd (we have assumed $(2QD, N) = 1$!), all $p_i - (D/p_i)$ are even and therefore

$$m = 2 \times \text{L.C.M.}\left\{ \frac{p_i - (D/p_i)}{2} p_i^{\alpha_i - 1}\right\} \leq 2 \prod_i \frac{1 - \frac{1}{p_i}\left(\frac{D}{p_i}\right)}{2} p_i^{\alpha_i} \leq$$

$$\leq 2N \prod_i \frac{1}{2}\left(1 + \frac{1}{p_i}\right) = T, \text{ say.}$$

The simplest case of a composite N is a prime power, $N = p_1^{\alpha_1}$, $\alpha_1 \geq 2$. In this instance we calculate the exact value of m and obtain

$$m = N\left(1 \pm \frac{1}{p_1}\right) = p_1^{\alpha_1} \pm p_1^{\alpha_1 - 1} \neq N + 1.$$

If N contains at least two distinct prime factors, we have

$$T = N\left(1 + \frac{1}{p_1}\right)\left(1 + \frac{1}{p_2}\right) \cdot \frac{1}{2} \cdot \prod_{i>2} \frac{1}{2}\left(1 + \frac{1}{p_i}\right) \leq$$

$$\leq N\left(1 + \frac{1}{3}\right)\left(1 + \frac{1}{5}\right) \cdot \frac{1}{2} = 0.8N < N + 1.$$

This proves that $d \neq N + 1$ for a composite number N, and hence the value obtained previously, $d = N + 1$, implies primality of N.

A very simple case is $N = M_n = 2^n - 1$. In this case $N + 1$, being a power of 2, has only one prime factor and the conditions of theorem 4.8 simplify to the following:

If a Lucas sequence U_n can be found such that

$$U_{N+1} \equiv 0 \pmod{N} \quad \text{while} \quad U_{(N+1)/2} \not\equiv 0 \pmod{N}, \tag{4.67}$$

then $N = 2^n - 1$ is prime.

Also in the case when $N + 1$ is not a power of 2 the value of U_{N+1} (mod N) is, as previously shown, relatively easy to compute, leading to fairly simple primality tests. Among these, the tests for the numbers M_n, which are called *Mersenne numbers*, are outstanding in their simplicity. This makes these numbers very attractive to investigate, especially on binary computers due to the simple binary representation of N. Thanks to this fact, the Mersenne numbers are the largest numbers which have been exposed to primality tests and the largest known primes are also found among these numbers. In the following, we shall describe this interesting pursuit of large primes in more detail.

Lucas Tests for Mersenne Numbers

First, we combine the two conditions (4.67) for the primality of M_n into one single condition. Since $U_{N+1} = U_{(N+1)/2}V_{(N+1)/2}$, the integer $V_{(N+1)/2}$

must be the one that introduces the factor N, finally leading to $U_{N+1} \equiv 0$ (mod N). Thus, the replacement of the two conditions by: $V_{(N+1)/2} \equiv 0$ (mod N) is necessary and sufficient for the primality of M_n. Next, we change notation in order to write down the chain of computations leading to $V_{(N+1)/2} = V_{2^{n-1}}$ in a more convenient form. Put $V_{2^s} = v_s$ so that $V_{2^s} = V_{2^{s-1}}^2 - 2Q^{2^{s-1}}$ transforms into $v_s = v_{s-1}^2 - 2Q^{2^{s-1}}$. Starting the computation with $v_0 = V_1 = P$, we see that the test $V_{(N+1)/2} \equiv 0$ (mod N) is equivalent to the following: Put $v_0 = P$. Let $v_s \equiv v_{s-1}^2 - 2Q^{2^{s-1}}$ (mod N). Then $v_{n-1} \equiv 0$ is necessary and sufficient for $M_n = 2^n - 1$ to be a prime.

So far, we have not yet approached the problem of how to find a Lucas sequence which is appropriate to the Mersenne numbers. The simplest way of doing this is to construct a sequence fulfilling the requirement $(D/N) = -1$. This condition is not *explicitly* mentioned in Theorem 4.8, but if N is a prime and if $U_{N+1} \equiv 0$ (mod N), then (4.53) demands that $(D/N) = -1$. The other requirement, i.e. $(2QD, N) = 1$, causes no problem, since before a large number N is put to a Lucas test, N has normally been searched for small factors in any event and, moreover, since Q and D are of moderate size this will guarantee that N has no divisors in common with $2QD$. There are always many different Lucas sequences which are suitable (about half of all D-values have $(D/N) = -1$). The following is one possible sequence: the choice $a = 1 + \sqrt{3}$, $b = \bar{a} = 1 - \sqrt{3}$, which gives

$$P = a + \bar{a} = 2, \quad Q = a\bar{a} = -2, \quad P^2 - 4Q = 12, \quad D = 3.$$

Now, $D = 3$ happens to be a quadratic non-residue of all primes of the form $12n \pm 5$ (Table 32). Furthermore, for the powers of 2 (mod 12) we find

$$2^1 \equiv 2, \ 2^2 \equiv 4, \ 2^3 \equiv 8, \ 2^4 \equiv 4, \ 2^5 \equiv 8, \ 2^6 \equiv 4, \dots$$

Here only the *odd* values of the exponents are interesting, since if $M_n = 2^n - 1$ is to be prime at all then n must be odd (except in the trivial case $2^2 - 1 = 3$). Hence, for $n \geq 3$, $M_n = 2^n - 1 \equiv 7$ (mod 12) and the number 3 is thus a quadratic non-residue of M_n.—Combining these results we arrive at the following primality test for the Mersenne numbers:

If n is odd, then $M_n = 2^n - 1$ is prime if and only if $v_{n-1} \equiv 0$ (mod M_n), where

$$v_s = v_{s-1}^2 - 2 \cdot 2^{2^{s-1}} \quad \text{with } v_1 = v_0^2 + 4 = 8.$$

Example. $n = 7$. $M_7 = 2^7 - 1 = 127$, and we successively find

$$v_2 = 64 - 8 = 56, \quad v_3 = 3136 - 32 = 3104 \equiv 56, \quad v_4 \equiv 3136 - 512 \equiv 84,$$

$$v_5 \equiv 7056 - 2^{17} \equiv 63, \quad v_6 \equiv 3969 - 2^{33} \equiv 3969 - 2^5 \equiv 0 \pmod{127}.$$

Can the Lucas test for Mersenne numbers be simplified? Well, the powers of 2 in the recursion formula for v_s would certainly vanish if we could choose $Q = 1$ or $Q = -1$. However, this would severely limit the possibilities of finding a suitable Lucas sequence. $Q = 1$ together with $(D/p) = -1$ causes already $U_{(p+1)/2} \equiv 0 \pmod{p}$, creating the problem that Theorem 4.8 cannot be used without some modification. Nevertheless, there exist somewhat more general Lucas sequences than those we have so far studied which are appropriate to our problem. The principle is to *insert new elements* between the terms of the original Lucas sequence. This will double the value of all the subscripts in the old sequence and, in this way, the troublesome $U_{(p+1)/2} \equiv 0 \pmod{p}$ will be transformed into the desired $U'_{p+1} \equiv 0 \pmod{p}$.—Next, in order to achieve this, suppose we have a Lucas sequence U_n such that

$$U_{(N+1)/(2q_j)} \not\equiv 0 \pmod{N} \quad \text{and} \quad U_{(N+1)/2} \equiv 0 \pmod{N}, \quad (4.68)$$

with a and b satisfying $\lambda^2 - P\lambda + Q = 0$. Now we shall try to make this Lucas sequence U_n denser by doubling the number of its elements to obtain the new sequence U'_n. Because the original U_n's are given by the formula

$$U_n = \frac{a^n - \bar{a}^n}{a - \bar{a}} = \frac{\sqrt{a^{2n}} - \sqrt{\bar{a}^{2n}}}{a - \bar{a}}, \tag{4.69}$$

the sequence U'_n defined by

$$U'_n = \frac{\sqrt{a^n} - \sqrt{\bar{a}^n}}{(\sqrt{a} - \sqrt{\bar{a}})(\sqrt{a} + \sqrt{\bar{a}})} \tag{4.70}$$

obviously satisfies the desired subscript-doubling relation

$$U'_{2n} = U_n. \tag{4.71}$$

It turns out, however, that this new sequence U'_n is slightly more general than those we have encountered so far, and that the divisibility properties of its elements cannot immediately be inferred from what we have already proved for quadratic fields. We have to omit the details of this deduction, but mention only that the main conclusion is still valid, namely that N is prime if

$$U'_{(N+1)/q_j} = U_{(N+1)/(2q_j)} \not\equiv 0 \pmod{N} \quad \text{for all} \quad j = 1, 2, \ldots, n \tag{4.72}$$

and

$$U'_{N+1} = U_{(N+1)/2} \equiv 0 \pmod{N}. \tag{4.73}$$

Thus we finally arrive at the Lucas test for Mersenne numbers: Choose $a = 2 + \sqrt{3}$, $b = 2 - \sqrt{3}$, which gives

$$P = 4, \ Q = 1, \ \sqrt{a} = \frac{\sqrt{3}+1}{\sqrt{2}}, \ \sqrt{b} = \frac{\sqrt{3}-1}{\sqrt{2}}, \ D = 3, \ \left(\frac{D}{M_n}\right) = -1$$

$$P' = \sqrt{a} + \sqrt{b} = \sqrt{6}, \quad Q' = \sqrt{a}\sqrt{b} = 1,$$

resulting in the Lucas sequence

$$V'_n = \left(\frac{\sqrt{3}+1}{\sqrt{2}}\right)^n + \left(\frac{\sqrt{3}-1}{\sqrt{2}}\right)^n. \tag{4.74}$$

This sequence obeys $V'_{2n} = V'^2_n - 2$ and starts with $V'_1 = \sqrt{6}$ and $V'_2 = 4$. If we change the notation (as done previously) by letting $V'_{2^s+1} = v_s$, then we have

Theorem 4.9. Lucas' test for Mersenne numbers. If n is odd, then $M_n = 2^n - 1$ is a prime if and only if $v_{n-2} \equiv 0 \pmod{M_n}$, where

$$v_s = v^2_{s-1} - 2 \quad \text{with } v_0 = 4. \tag{4.75}$$

Further on we shall prove a result (Theorem 4.17) which is a generalization of Theorem 4.9.

Example. For $N = 2^{19} - 1 = 524287$ the computation runs as follows:

$$v_0 = 4, \qquad v_1 = 14, \qquad v_2 = 194, \qquad v_3 = 37634, \quad v_4 \equiv 218767$$

$$v_5 \equiv 510066, \ v_6 \equiv 386344, \ v_7 \equiv 323156, \ v_8 \equiv 218526, \ v_9 \equiv 504140$$

$$v_{10} \equiv 103469, \ v_{11} \equiv 417706, \ v_{12} \equiv 307417, \ v_{13} \equiv 382989, \ v_{14} \equiv 275842$$

$$v_{15} \equiv 85226, \ v_{16} \equiv -2^{10},$$

and, finally,

$$v_{17} \equiv 2^{20} - 2 \equiv 0 \pmod{2^{19} - 1}.$$

This proves the primality of M_{19}.

Using Theorem 4.9 and another very similar primality test, Lucas in 1876 proved M_{127} (39 digits!) to be prime. This was the record for the largest prime discovered before the advent of computers. (In fact, larger numbers have been investigated without the use of a computer. The author believes that the largest *prime* found without a computer is A. Ferrier's $(2^{148}+1)/17$, but this result was published in 1952 when there were already several larger primes known which had been found by aid of computer.) All Mersenne numbers with exponents up to 100000 and some higher ones have been tested to date, among which the following 29 exponents p have been found to lead to Mersenne primes M_p:

$$2, \ 3, \ 5, \ 7, \ 13, \ 17, \ 19, \ 31, \ 61, \ 89, \ 107, \ 127, \ 521, \ 607,$$

$$1279, \ 2203, \ 2281, \ 3217, \ 4253, \ 4423, \ 9689, \ 9941, \ 11213,$$

$$19937, \ 21701, \ 23209, \ 44497, \ 86243 \ \text{and} \ 132049.$$

M_{132049} (39751 digits!) is the largest prime known at present. It was found with the aid of a CRAY-computer at the Lawrence Radiation Laboratory in California in 1983 by David Slowinski.

A Relaxation of Theorem 4.8

As the reader may remember, we have given two converses to Fermat's theorem, namely Lehmer's Theorem 4.3 and a version with relaxed conditions on the factorization of $N - 1$, Theorem 4.6. The relaxed version is very helpful indeed when $N - 1$ has many small prime factors and a large cofactor $< \sqrt{N}$. In such a case we can avoid the labour of factorizing this large cofactor. The situation is identical when Lucas sequences are used rather than Fermat's theorem. The condition of Theorem 4.8 that $N + 1$ is *completely* factored may be relaxed so that a cofactor $< \sqrt{N}$ can be left unfactored. This is stated in

Theorem 4.10. Suppose that $N + 1 = R \cdot F = R\prod_{j=1}^{n} q_j^{\beta_j}$, with all q_j's distinct primes, $R < F$ and $(R, F) = 1$. If a Lucas sequence U_ν with $(2QD, N) = 1$ exists satisfying

$$(U_{(N+1)/q_j}, N) = 1 \quad \text{for all} \quad j = 1, 2, \ldots, n \tag{4.76}$$

and such that

$$U_{N+1} \equiv 0 \,(\text{mod } N), \tag{4.77}$$

then N is a prime.

Proof. As in the proof of Theorem 4.8, we must consider a possible prime factor p of N. In exactly the same way as in Theorems 4.6 and 4.8 we arrive at the conclusion that $F|d$. By (4.52) and (4.53), either $d|p-1$ or $d|p+1$. Thus $F|p\pm1$. Hence, the smallest possible value p can assume is one of $F\pm1$. If the sign is positive, then $p=F+1>\sqrt{N}$, and so N is certainly prime. If, however, the sign is negative, then the magnitude of p could be just below $\sqrt{N}$, so we need to provide some other reason why this case also leads to a contradiction. Let us then consider $N=RF-1=R(p+1)-1=Rp+R-1\equiv0\,(\text{mod }p)$. This congruence requires R to be $\equiv1\,(\text{mod }p)$. But since $0<R<F=p+1$, the only solution to $R\equiv1$ $(\text{mod }p)$ is $R=1$, for which $N=p$, and thus, again, N is prime.

In the same way as we used Theorem 4.6 earlier to find elegant primality tests for numbers of the form $h\cdot2^n+1$, we can now apply Theorem 4.10 to devise primality tests for numbers of the form $h\cdot2^n-1$. Using the notation, introduced on p. 124, we obtain

Theorem 4.11. If h is odd and $2^n>h$, then $N=h\cdot2^n-1$ is prime if there exists a Lucas sequence V_ν with $(2QD,N)=1$ such that

$$V_{(N+1)/4}\equiv0\,(\text{mod }N).\qquad(4.78)$$

Pocklington's Theorem

Theorem 4.12. Pocklington's Theorem. Let $N-1=Rq^n$, where q is prime and $q\nmid R$. If there exists an integer a, satisfying

$$\left(a^{(N-1)/q}-1,N\right)=1\quad\text{and such that}\quad a^{N-1}\equiv1\,(\text{mod }N),$$
$$(4.79)$$

then each prime factor p of N has the form $p=q^n m+1$.

Proof. Exactly as in the proof of Theorem 4.6 on p. 109, each (possible) prime factor p of N must satisfy $q^n|p-1$, which means that $p=q^n m+1$ for some positive integer m.

Remark. Pocklington's theorem can in some cases help to find a factor, but its main application is in primality proofs, as illustrated in the example below, where it is used to establish that any prime factor of N must be $>\sqrt{N}$, thereby implying the primality of N. The theorem is only a slight generalization of the reasoning used to prove the primality of a number by proving the existence of a primitive root $g\,(\text{mod }p)$.

Lehmer-Pocklington's Theorem

If the factorization of $N-1$ is, instead, given as $N-1 = R \cdot F = R \prod q_j^{\beta_j}$, as in Theorem 4.6, then the modular conditions arising from the different prime powers $q_j^{\beta_j}$ can be combined to yield:

Theorem 4.13. Lehmer-Pocklington's Theorem. Let $N-1 = R \cdot F = R \prod_{j=1}^{n} q_j^{\beta_j}$, with the q_j's distinct primes and $(R, F) = 1$. If there is an integer a, satisfying

$$\left(a^{(N-1)/q_j} - 1, \, N \right) = 1 \quad \text{for all} \quad j = 1, 2, \ldots, n, \qquad (4.80)$$

and such that

$$a^{N-1} \equiv 1 \,(\text{mod } N),$$

then each prime factor p of N has the form $p = Fm + 1$.—Note that it is not necessary to have $R < F$ here.

Example. The primality of $N = (2^{148} + 1)/17$, the largest prime found using a desk calculator, was established by A. Ferrier in 1951 by means of the following arguments:

Firstly, it was verified that $3^{N-1} \equiv 1 \,(\text{mod } N)$.

Secondly, $N - 1 = 2^4(2^{72} - 1)(2^{72} + 1)/17$ contains a large prime factor, namely $Q = 4878\,24887233$. Applying Pocklington's theorem with $q^n = Q$, Ferrier showed that $3^{17(N-1)/Q} - 1$ is prime to N, which implies that $3^{(N-1)/Q} - 1$ is also prime to N. Thus, every prime factor of N must be of the form $Qm + 1$.

Next, $17N$ being of the form $a^{148} + b^{148}$, every prime factor of N (excepting possible factors 17) are of the form $296k + 1$, due to Legendre's Theorem on p. 184. Combining the two forms, Ferrier found that every prime factor of N is of the form

$$p = p(y) = 1443961\,66620968y + 1 = qy + 1.$$

Now, for $y = 1, 2, 3, \ldots, 11$, p is not prime, having a small divisor. Hence, every possible prime factor p of N exceeds $p(12) - 1 = 17327539\,99451616$.

Further, applying a type of argument which often is useful in connexion with Fermat's factorization method, Ferrier wrote

$$N = A^2 - B^2 = (A - B)(A + B) = p \cdot \frac{N}{p}.$$

Here

$$2A = (A - B) + (A + B) = p + \frac{N}{p} < 1.3 \cdot 10^{28}$$

since $\sqrt{N} > p > 17327539\,99451616$. Now, on the other hand, since *both* factors p and N/p are of the form $qy + 1$, we have

$$N + 1 = (qy + 1)(qz + 1) + 1 = q^2 yz + (qy + 1) + (qz + 1) =$$
$$= q^2 yz + 2A \equiv 2A \,(\text{mod } q^2). \tag{4.81}$$

But $N + 1 \equiv 18859\,78087126\,35496631\,61494450 \,(\text{mod } q^2)$ so that $2A > 1.8 \cdot 10^{28}$. Since this inequality contradicts the previous one, A does not exist and therefore N is prime.

Pocklington-Type Theorems For Lucas Sequences

If $N + 1$ is easier to factor than $N - 1$, then the following analogues of theorems (4.12) and (4.13), utilizing Lucas sequences $\{U_n\}$ instead of powers a^n, may be of use in the search for factors of N:

Theorem 4.14. Lehmer's analogue to Pocklington's Theorem. Let $N + 1 = R \cdot q^n$, where q is prime and $q \nmid R$. IF $\{U_n\}$ is a Lucas sequence with $(D/N) = -1$ and such that

$$(U_{(N+1)/q}, N) = 1 \quad \text{with} \quad U_{N+1} \equiv 0 \,(\text{mod } N), \tag{4.82}$$

then each prime factor p of N has the form $p = q^n m \pm 1$.

Proof. In Pocklington's (original) theorem 4.12 the result follows from the fact that a prime p can divide N only if $q^n | p - 1$. For a Lucas sequence, q^n will instead have to be a factor of $p - (D/p)$ (compare (4.66)), i.e. $q^n | p \pm 1$, yielding the present theorem first proved by D. H. Lehmer.

If the factorization of $N + 1$ is instead given as $N + 1 = R \cdot F = R \prod q_j^{\beta_j}$, then again combining the modular conditions for the different prime powers $q_j^{\beta_j}$ surprisingly enough yields

Theorem 4.15. Let $N + 1 = R \cdot F = R \prod_{j=1}^n q_j^{\beta_j}$, with q_j distinct primes and $(R, F) = 1$. If $\{U_n\}$ is a Lucas sequence with $(D/N) = -1$, for which

$$(U_{(N+1)/q_j}, N) = 1 \quad \text{for all} \quad j = 1, 2, \ldots, n \tag{4.83}$$

and such that

$$U_{N+1} \equiv 0 \,(\text{mod } N),$$

then each prime factor of N has the form $p = Fm \pm 1$.

The surprise is that only *two* residue classes instead of 2^n are possible for the module $F = \prod q_j^{\beta_j}$, being a product of n modules, each admitting two residue classes for p.

Proof. The proof of this Theorem is similar to the proof of theorem 4.10 on p. 127, that F must divide $p \pm 1$ if p is to be a prime factor of N. So p must be $\equiv \pm 1 \pmod{F}$—This theorem was discovered by the author of this book in 1967 and re-discovered by Michael Morrison and published in 1974.

Primality Tests for Integers of the Form $N = h \cdot 2^n - 1$, when $3 \nmid h$

In order to satisfy (4.76) we need as usual to require that $(D/N) = -1$. Furthermore, as in the case of Mersenne numbers, the computation will be particularly simple if we can choose $Q = \pm 1$. Now, if $3 \nmid h$, then it is possible to cover all interesting cases $(3 \nmid N)$ by the same choice of a and b as on p. 124, i.e. by once again considering the Lucas sequence (4.74). $V_{(N+1)/4} = V'_{(N+1)/2} = v'_{h \cdot 2^{n-1}}$ is most easily calculated by adopting the initial value

$$v_0 = \left(\frac{\sqrt{3}+1}{\sqrt{2}} \right)^{2h} + \left(\frac{\sqrt{3}-1}{\sqrt{2}} \right)^{2h} = (2+\sqrt{3})^h + (2-\sqrt{3})^h \quad (4.84)$$

and then performing the usual recursion $v_s = v_{s-1}^2 - 2$. The condition for primality of N, (4.78), will then take the form $v_{n-2} \equiv 0 \pmod{N}$. We thus arrive at the following

Theorem 4.16. Lucas' primality test. Suppose that h is an odd integer, that $2^n > h$ and that neither $N = h \cdot 2^n - 1$ nor h is divisible by 3. Then N is prime if and only if $v_{n-2} \equiv 0 \pmod{N}$, where $v_s = v_{s-1}^2 - 2$ and $v_0 = (2+\sqrt{3})^h + (2-\sqrt{3})^h$.

Note that the initial value v_0 can easily be computed using (4.39) and (4.41). This is because we can consider v_0 itself to be an element of a suitably chosen Lucas sequence!

Example. $N = 5 \cdot 2^{14} - 1 = 81919$. First we have to compute $v_0 = (2+\sqrt{3})^5 + (2-\sqrt{3})^5$. The Lucas sequence with $V_m = (2+\sqrt{3})^m + (2-\sqrt{3})^m$ satisfies $V_{m+1} = 4V_m - V_{m-1}$ with $V_0 = 2$ and $V_1 = 4$. Thus $V_2 = 4 \cdot 4 - 2 = 14$, $V_3 = 4 \cdot 14 - 4 = 52$, $V_4 = 4 \cdot 52 - 14 = 194$ and $V_5 = 4 \cdot 194 - 52 = 724$. (This initial value v_0 is identical for the

Lucas test on *all* numbers of the form $5 \cdot 2^n - 1$ not divisible by 3.) Next, compute

$$v_1 \equiv v_0^2 - 2 \equiv 32660, \qquad v_2 \equiv 8299, \qquad v_3 \equiv 61439, \qquad v_4 \equiv 5118,$$

$$v_5 \equiv 61761, \quad v_6 \equiv 26722, \quad v_7 \equiv 59278, \quad v_8 \equiv 47696,$$

$$v_9 \equiv 17784, \quad v_{10} \equiv 63314, \quad v_{11} \equiv 38248, \quad v_{12} \equiv 0 \ (\text{mod } 81919),$$

which proves 81919 to be prime.—By applying Theorem 4.16 and using a computer, the author of this book has identified many primes, for instance

$$5 \cdot 2^{248} - 1, \quad 7 \cdot 2^{177} - 1, \quad 11 \cdot 2^{246} - 1, \quad 17 \cdot 2^{150} - 1.$$

More detailed information can be found in Table 6 at the end of the book.

Primality Tests for $N = h \cdot 2^n - 1$, when $3|h$

What now remains in order to complete the picture is to construct Lucas sequences which are applicable in the case when $N = h \cdot 2^n - 1$ and 3 divides h. This problem is actually more difficult than when $3 \nmid h$. The difference between the two situations lies in the choice of a suitable initial value v_0 for the recursion $v_s = v_{s-1}^2 - 2$. In the case treated above, where $3 \nmid h$, v_0 is dependent on h only and not on n (as long as $3 \nmid N$), whereas in the case when $3|h$, v_0 is dependent on n as well as on h. Moreover, still worse, also D depends on h and n, making the search for a suitable Lucas sequence theoretically complicated and computationally tedious. The reason for this is the following. Suppose that $N = 3A \cdot 2^n - 1$ and that $3A$ has an odd prime factor p. Then (p/N) is always $= 1$, since

$$\left(\frac{p}{N}\right) = \left(\frac{N}{p}\right)(-1)^{\frac{1}{2}(p-1)\frac{1}{2}(N-1)} = \left(\frac{-1}{p}\right)(-1)^{\frac{1}{2}(p-1)\frac{1}{2}(N-1)}$$

by the theorem of quadratic reciprocity, see p. 281. Now, working through the two cases $p = 4k - 1$ and $p = 4k + 1$, we find

$$\left(\frac{p}{N}\right) = (-1)(-1)^{(2k-1)(3A \cdot 2^{n-1}-1)} = +1$$

and

$$\left(\frac{p}{N}\right) = (+1)(-1)^{2k(3A \cdot 2^{n-1}-1)} = +1$$

in these two cases, respectively. Finally, if $n > 2$ then

$$\left(\frac{2}{N}\right) = (-1)^{(N-1)(N+1)/8} = (-1)^{3A \cdot 2^{n-2}(N-1)/2} = +1,$$

so that it is impossible to find a value of D that is a quadratic non-residue of all values of A simultaneously. Therefore, according to the value of A selected, we need to work with different values of D, leading to varying mathematical formulas for v_0, depending on the chosen value of A. Furthermore, the requirement $(D/N) = -1$ cannot be satisfied for all (interesting) values of N by simply choosing a fixed value of D, when D is large. This fact leads to different values of D (and therefore of v_0), even if we fix the value of A. A rather surprising situation! The reader is invited to examine the table on the p. 135 to see just how confusing the situation does appear. Look e.g. at the choice $A = 5$. For the numbers $N = 15 \cdot 2^n - 1$ we first have a "main case", $D = 21$, covering $n \equiv 1$ (mod 3). If $n \equiv 0$ (mod 3), then it turns out that $7|N$ and so we can dispose of this case. How about the remaining case, $n \equiv 2$ (mod 3)? We will need several different values of D for this, and even so, it will not be completely covered. As can be seen from the table, $D = 11$ covers $n \equiv 0, 2, 3, 7$ and 9 (mod 10), $D = 13$ covers $n \equiv 2, 3, 4, 5, 7$ and 10 (mod 12) while $D = 17$ covers $n \equiv 0, 1, 5$ and 6 (mod 8). But these congruences together *do not exhaust all cases*, as can be quite easily seen by re-writing them (mod 120). The cases $n \equiv 44, 116$ (mod 120) are still missing! This is indicated by the text "not covered" in the table. Finding, by trial and error, a suitable value of D covering the remaining cases is an almost hopeless task.

Fortunately, the search for useful values of D can be carried out in a systematic way, by using a quite general theorem discovered by the author of this book in 1969, see [10]. This is

Theorem 4.17. Suppose that h is an odd integer and that $2^n > h$. Then $N = h \cdot 2^n - 1$ is prime if and only if $v_{n-2} \equiv 0$ (mod N), where $v_s = v_{s-1}^2 - 2$ with $v_0 = a^h + a^{-h}$. In this expression a is a unit of $K(\sqrt{D})$ of the form

$$a = \frac{(k + l\sqrt{D})^2}{r} \quad \text{where} \quad \left(\frac{D}{N}\right) = -1 \quad \text{and} \quad \frac{k^2 - l^2 D}{r}\left(\frac{r}{N}\right) = -1.$$

Proof. The integers $V_s = a^s + a^{-s}$, according to (4.31), constitute a Lucas sequence since a is a unit of $K(\sqrt{D})$ and this implies that the characteristic equation of the sequence, $\lambda^2 - (a + \bar{a})\lambda + a\bar{a} = 0$, has integer coefficients $P = a + \bar{a}$ and $Q = a\bar{a} = \pm 1$. In order to demonstrate that the conditions in the Theorem are necessary, suppose first that N is prime. Then

$$v_{n-2} = V_{(N+1)/4} = a^{(N+1)/4} + a^{-(N+1)/4} =$$

$$= a^{-(N+1)/4}\left(1 + a^{(N+1)/2}\right) =$$

$$= a^{-(N+1)/4}\left(1 + (k + l\sqrt{D})^{N+1}r^{-(N+1)/2}\right) \equiv$$

(by Fermat's theorem in $K(\sqrt{D})$, since $\left(\dfrac{D}{N}\right) = -1$)

$$\equiv a^{-(N+1)/4}\left(1 + \frac{k^2 - l^2 D}{r}\left(\frac{r}{N}\right)\right) \equiv 0 \ (\mathrm{mod} \ N).$$

Here we have been using the condition on (r/N) assumed in the Theorem. For the sufficiency of the conditions we can refer to Theorem 4.11 since $v_{n-2} = V_{(N+1)/4} \equiv 0 \ (\mathrm{mod} \ N)$ implies the primality of N.

The Lucas tests proved earlier are special cases of Theorem 4.17. If h and n are such that $3 \nmid h$ and $3 \nmid N$, then the unit

$$a = 2 + \sqrt{3} = \frac{(1 + \sqrt{3})^2}{2} \quad \text{of} \quad K(\sqrt{3})$$

can be used in the Theorem since

$$\left(\frac{D}{N}\right) = \left(\frac{3}{N}\right) = -1 \quad \text{and} \quad \frac{k^2 - l^2 D}{r}\left(\frac{r}{N}\right) = \frac{-2}{2} \cdot \left(\frac{2}{N}\right) = -1$$

in this instance.—By a further restriction to $h = 1$, we find

$$v_0 = a^h + a^{-h} = (2 + \sqrt{3})^1 + (2 - \sqrt{3})^{-1} = 4,$$

which reverts us to Theorem 4.9.

At this point, a natural question is how best to apply Theorem 4.17 in practice. One method is to investigate which combinations of D, h and n satisfy all the requirements and note these down for future use. This has been done in the Table below, where useful Lucas sequences have been compiled for all $h \leq 29$. Since Legendre's symbol occurs in the conditions of Theorem 4.17, the possible values of n for a given value of D belong to certain arithmetic series; frequently the case in problems which can be solved by aid of Legendre's symbol.—As pointed out earlier, this technique of extracting valid combinations of h, n and D is quite complicated and is thus principally used if only a few numbers need to be checked for primality.

| \multicolumn{5}{c}{Lucas tests for numbers of the form $N = h \cdot 2^n - 1$} |
h	D	v_0	$n \pmod{d}$	d
1	3	$6 \cdot 1^2 - 2 = 4$	1	2
3	21	$7 \cdot 4^2 - 2 = 110$	0, 2	3
3	39	$13 \cdot 98^2 - 2 = 124852$	0, 1, 2, 4, 7, 10, 11	12
5	3	$6 \cdot 11^2 - 2 = 724$	0	2
7	3	$6 \cdot 41^2 - 2 = 10084$	1	2
9	5	$5 \cdot 34^2 - 2 = 5778$	0, 1	4
9	17	$287298^3 - 3 \cdot 287298$	3, 4, 6, 7	8
11	3	$6 \cdot 571^2 - 2 = 1956244$	0	2
13	3	$6 \cdot 2131^2 - 2 = 27246964$	1	2
15	21	$2525^3 - 3 \cdot 2525 = 16098445550$	1	3
15	11	$22 \cdot 1197677521^2 - 2$	0, 2, 3, 7, 9 $(n > 2)$	10
15	13	$13 \cdot 16835050^2 - 2$	2, 3, 4, 5, 7, 10	12
15	17	$17 \cdot 10730180955650^2 - 2$	0, 1, 5, 6	8
15	—	not covered by the above	44, 116	120
17	3	$6 \cdot 29681^2 - 2$	0	2
19	3	$6 \cdot 110771^2 - 2$	1	2
21	5	$5 \cdot 10946^2 - 2 = 599074578$	2, 3	4
21	13	$13 \cdot 21851881930^2 - 2$	0, 1, 2, 3, 5, 8	12
23	3	$6 \cdot 1542841^2 - 2$	0	2
25	3	$6 \cdot 5757961^2 - 2$	1	2
27	5	$5778^3 - 3 \cdot 5778 = 192900153618$	1, 2	4
27	21	$1330670^3 - 3 \cdot 1330670$	1, 2	3
29	3	$6 \cdot 80198051^2 - 2$	0	2

In a situation where *all* values of h and n in a certain domain are to be tested, there exists a better approach. This is to systematically allow a computer to check through all values of D and list those combinations of h and n which fit the conditions of Theorem 4.17. In order to obtain the smallest possible value of v_0, it is practical to organize this search with increasing values of V_1. The author has carried this out for all n and all odd h in the domain $2 \leq n \leq 1000$ and $1 \leq h \leq 105$. To limit the amount of labour demanded as much as possible, all $N = h \cdot 2^n - 1$ having some factor < 10000 were first sieved out. After this preliminary reduction a suitable V_1 was sought for each of the remaining numbers. The largest V_1 necessary in this investigation was $V_1 = 57$, required for $N = 63 \cdot 2^{354} - 1$. For this V_1 the value of D was found to be 3245 and

$$v_0 = \left(\frac{57 + \sqrt{3245}}{2} \right)^{63} + \left(\frac{57 - \sqrt{3245}}{2} \right)^{63}.$$

Finally, for each of the numbers under investigation, the starting value v_0 of the recursion was computed and the number tested. The results are given in Table 6 at the end of the book, which has been brought up-to-date by including also later results, found by others.

Compositeness Tests with Lucas Sequences for $N = h \cdot 2^n \pm k$

Lucas sequences as an alternative to Fermat's theorem can be used to construct compositeness tests for numbers of the form $N = h \cdot 2^n \pm k$, provided k is not too large. We shall not give details here, but mention only that this has been accomplished by K. Inkeri and J. Sirkesalo in [11] and [12].

The Combined $N - 1$ and $N + 1$ Test

From Theorems 4.6 and 4.10 we infer that the primality of N can easily be proved if *one* of $N - 1$ or $N + 1$ can be factored up to its square root. What if this is not possible? Then we have to use the following *combined primality test*, relying upon the *simultaneous* partial factorizations of $N-1$ and $N + 1$. The test has also been relaxed by using the "several bases" idea from Theorem 4.4. Search limits for the factors of $N - 1$ and $N + 1$ have also been included.

Theorem 4.18. The combined $N - 1$ and $N + 1$ test. Assume that

1. $N - 1 = F_1 R_1$, with R_1 odd > 1 and $(F_1, R_1) = 1$.
2. $N + 1 = F_2 R_2$, with R_2 odd > 1 and $(F_2, R_2) = 1$. Here F_1 and F_2 are completely factored.

136

3. For each prime q_i dividing F_1 there exists an integer a_i such that $a_i^{N-1} \equiv 1 \pmod{N}$ and $(a_i^{(N-1)/q_i} - 1, N) = 1$.

4. There exists an integer a such that $a^{N-1} \equiv 1 \pmod{N}$ and $(a^{(N-1)/R_1} - 1, 1) = 1$.

5. For each prime q_i dividing F_2 there exists a Lucas sequence $\{U_\nu\}$ with $(D/N) = -1$ such that $N|U_{N+1}$ and $(U_{(N+1)/q_i}, N) = 1$.

6. There exists a Lucas sequence $\{U_\nu\}$ with $(D/N) = -1$ such that $N|U_{N+1}$ and $(U_{(N+1)/R_2}, N) = 1$.

Suppose that the prime factors of R_1 and R_2 are respectively $\geq B_1$ and $\geq B_2$. Define r and s by $R_1 = (F_2/2)s + r$, $0 \leq r < F_2/2$, and let m be the smallest non-negative integer for which $mF_1F_2/2 + rF_1 + 1|N$ (or the smallest *positive* integer in case r happens to be 0). Moreover, let

$$G = \max(B_1F_1 + 1, B_2F_2 - 1, mF_1F_2/2 + rF_1 + 1), \quad m \geq 1.$$

Then

$$\text{if} \quad G(B_1B_2F_1F_2/2 + 1) > N, \quad N \text{ is prime.}$$

A proof of this theorem is rather straight-forward, following the lines of proof given in Theorems 4.4, 4.6, 4.8, 4.10, 4.12–4.15. We omit the proof and refer the reader to p. 635 of [7].—This theorem is one of the more important forerunners to the modern primality tests.

Lucas Pseudoprimes

One essential step in proving the primality of a number N by aid of Theorem 4.10 is to prove that $U_{N+1} \equiv 0 \pmod{N}$ for the Lucas sequence chosen. Unfortunately, this condition is not always sufficient for the primality of N, i.e. there do exist certain *composite* numbers N which satisfy this congruence. They are termed Lucas pseudoprimes and are covered by

Definition 4.5. An odd *composite* number N with $N \nmid Q$, $(D/N) = -1$ and $U_{N+1} \equiv 0 \pmod{N}$ is a Lucas pseudoprime with parameters D, P and Q, where the Lucas sequence $\{U_\nu\}$ is as usual defined by (4.30)–(4.31) and D is the square-free part of $P^2 - 4Q$.

Any odd N satisfying all the conditions in this definition is called a *Lucas probable prime.*—The significance of the concept "Lucas probable prime" lies in the fact that it could well be possible that some combination of Lucas probable prime tests and strong probable prime tests is sufficient to actually *prove* primality. This conjecture has, however, not yet been proved. A discussion of this topic can be found in [3] and a related conjecture is discussed in [13]. The proof would constitute a demonstration that there exists no number which is a pseudoprime with respect to *all* these tests simultaneously. Such a combination of (a limited number of) Fermat-type primality tests would require only polynomial running time on a computer, since each of the tests involved has this property. In anticipation of such a proof we must at present content ourselves with some results achieved recently, which we shall now very briefly describe.

Recent Progress in General Primality Proofs

Leonard Adleman, Carl Pomerance and Robert Rumely (see [16]–[17]) have recently devised an algorithm for primality testing which is of nearly polynomial growth. The algorithm in itself is quite complicated, both in theory, since the so-called cyclotomic fields are used, and in practice, as many different cases have to be covered in computer program implementations. We shall not enter into all the details here because the importance of this breakthrough lies not so much in the specific algorithm designed, but rather in the fact that it has now been proved that a fast primality testing algorithm for the number N does exist which is completely independent of any factorization of some number dependent on N. Although the original algorithm discovered is complicated, it has been programmed for computers and works very well. Its running time can be described as "nearly" polynomial, as the labour involved is

$$O(\ln N)^{c \ln \ln \ln N}, \qquad (4.85)$$

where c is a positive constant. Since the exponent $\ln \ln \ln N$ increases extremely slowly, it is almost as good as a constant exponent, provided N is of limited size, say has less than a million digits. (For $N = 10^{1000000}$, $\ln \ln \ln N$ is only 2.68, and thus the growth of the exponent as N increases is very moderate indeed.)—Recently a simplified version of this algorithm based on results by H. Cohen and H. W. Lenstra, Jr. [18]–[19] has been published by John Dixon [20]. This version of the algorithm has been

138

implemented on a CDC Cyber 170-750 computer and is able to deal with 100 digit numbers in about 30 seconds and 200 digit numbers in 8 minutes. We shall in the next subsection describe Dixon's proof.—It will probably not be long now before a computationally simple and reasonably fast algorithm is constructed, taking care of general primality proofs.

A General Primality Testing Algorithm

In this and the next two subsections we shall describe a simplified version of the generalized pseudoprime test to which we have referred above. We shall closely follow the proof given by Dixon in [20], pp. 347–350. Readers not familiar with the theory of group characters should now read pp. 258–261 in Appendix 1 before proceeding.

To prepare for Theorem 4.19 below and its proof we begin by discussing some elementary properties of so-called Gaussian sums.

Let N be the number whose primality ultimately has to be proved and let p and q be primes not dividing N such that $p|q - 1$. Furthermore, let ς_p and ς_q be primitive p^{th} and q^{th} roots of unity. Let g be a generator of the cyclic group M_q consisting of all primitive residue classes (mod q) and let C be the cyclic group generated by ς_p. Since p is a divisor of $q - 1$ there exists a homomorphism which maps M_q onto C. Utilizing the technique demonstrated in the example on p. 261, we associate the function-value ς_p^j with the element g^j of M_q, thereby defining a certain group character χ_{pq}, or χ for short, on M_q. Finally, define the Gaussian sums $\tau(\chi)$ and $\tau(\chi^{-1})$ by

$$\tau(\chi) = \sum_a \chi(a)\varsigma_q^a \quad \text{and} \quad \tau(\chi^{-1}) = \sum_a \{\chi(a)\}^{-1}\varsigma_q^a, \tag{4.86}$$

respectively, where the summations are extended over all the $p-1$ primitive residue classes (mod q). Since $\varsigma_q^q = 1$, the exponent of ς_q in (4.86) runs (mod q), and ς_q^a takes the same value, regardless of the representative chosen of the residue class a (mod q).

Three Lemmas

We shall now proceed by proving three lemmas.

Lemma 1. $$\tau(\chi)\tau(\chi^{-1}) = \chi(-1)q. \tag{4.87}$$

Proof. Let the variables a and b range over all elements of the group M_q. Then

$$\tau(\chi)\tau(\chi^{-1}) = \sum_a \chi(a)\varsigma_q^a \sum_b \{\chi(b)\}^{-1}\varsigma_q^b = \sum_a \sum_b \chi(a)\chi(b^{-1})\varsigma_a^{a+b} =$$

(since ab ranges over M_q at the same time as a does)

$$= \sum_a \sum_b \chi(ab)\chi(b^{-1})\varsigma_q^{ab+b} = \sum_a \sum_b \chi(a)\varsigma_q^{b(a+1)} = \sum_a \chi(a) \sum_b \varsigma_q^{b(a+1)}.$$
(4.88)

Using the fact that $\varsigma_q^q = 1$ we can evaluate the sum $\sum_b \varsigma_q^{b(a+1)} = \varsigma_q^{a+1} + \varsigma_q^{2(a+1)} + \cdots + \varsigma_q^{(q-1)(a+1)} = (\varsigma_q^{q(a+1)} - 1)/(\varsigma_q^{a+1} - 1) - 1$. As long as $a \not\equiv -1 \pmod q$ this expression takes the value -1. If $a \equiv -1 \pmod q$, on the other hand, each term of the sum is $= 1$ and the sum is $= q - 1$. Plugging this into (4.88), the double sum reduces to

$$\chi(-1)(q-1) + \sum_{a \not\equiv -1} \chi(a)(-1) = q\chi(-1) - \sum_a \chi(a).$$

By Theorem A1.6 the above sum reduces to zero, unless χ is the principal character of M_q, which is obviously not the case. This proves Lemma 1.

Lemma 2. If N is prime, then

$$\tau(\chi)^{N^{p-1}-1} \equiv \chi(N) \pmod N \tag{4.89}$$

in the ring R_{pq} generated by $\varsigma_{pq} = e^{2\pi i/pq}$.

Proof. In the ring R_{pq} of cyclotomic integers, many arithmetic properties of ordinary integers still hold. Thus, if α_i are cyclotomic integers and N is prime, then

$$(\alpha_1 + \alpha_2 + \cdots + \alpha_k)^N \equiv \alpha_1^N + \alpha_2^N + \cdots + \alpha_k^N \pmod N. \tag{4.90}$$

This congruence is easily proved by starting with $(\alpha_1 + \alpha_2)^N$ which is $\equiv \alpha_1^N + \alpha_2^N \pmod N$ according to the binomial theorem (remember that all binomial coefficients $\binom{N}{k}$, $1 \le k \le N - 1$, are divisible by N, if N is prime). Having proved (4.90) for *two* integers, the full expression (4.90) is deduced by adding one variable at a time. Repeating the step taken in (4.90) $m - 1$ times we arrive at

$$(\alpha_1 + \alpha_2 + \cdots + \alpha_k)^{N^m} \equiv \alpha_1^{N^m} + \alpha_2^{N^m} + \cdots + \alpha_k^{N^m} \pmod N. \tag{4.91}$$

Thus

$$\tau(\chi)^{N^{p-1}} = \sum_a \left(\chi(a)\varsigma_q^a\right)^{N^{p-1}} (\text{mod } N). \qquad (4.92)$$

Now $\chi(a)^{N^{p-1}} = \chi(a)$ since, by Fermat's Theorem, $p|N^{p-1}-1$ and $\chi(a)$ is a pth root of unity. Moreover, there exists an element b of M_q such that $bN^{p-1} \equiv 1 \pmod q$ since $q \nmid N^{p-1}$. Thus

$$\sum_a \chi(a)^{N^{p-1}} \varsigma_q^{aN^{p-1}} = \sum_a \chi(ab)\varsigma_q^{abN^{p-1}} = \sum_a \chi(ab)\varsigma_q^a =$$

$$= \chi(b)\sum_a \chi(a)\varsigma_q^a = \chi(b)\tau(\chi) = \chi(N)^{1-p}\tau(\chi),$$

again because $\chi(N)$ is a pth root of unity. We have now arrived at the congruence

$$\tau(\chi)^{N^{p-1}} \equiv \chi(N)\tau(\chi) \pmod N \qquad (4.93)$$

and should only have to cancel one factor $\tau(\chi)$ in order to arrive at Lemma 2. This cancellation, however, might not lead to a valid result in the ring R_{pq} and has to be avoided by instead performing another operation: multiplication by $\tau(\chi^{-1})$. Since $(\chi(-1)q, N) = (\pm q, N) = 1$, there exists an integer a such that $\chi(-1)qa \equiv 1 \pmod N$. Now multiply (4.93) by $a\tau(\chi^{-1})$ and utilize (4.87):

$$a\tau(\chi)^{N^{p-1}}\tau(\chi^{-1}) \equiv \chi(N)\tau(\chi)\tau(\chi^{-1}) = a\chi(N)\chi(-1)q \equiv \chi(N)(\text{mod } N).$$

Making use of (4.87) once more, this time in the left-most part of the expression above, we finally find (4.89).

Lemma 3. If for some prime $r \neq p$ we have

$$\varsigma_p^i \equiv \varsigma_p^j \pmod r, \quad \text{then} \quad \varsigma_p^i = \varsigma_p^j. \qquad (4.94)$$

Proof. Let $\Phi_p(z)$ be the cyclotomic polynomial $(z^p-1)/(z-1) = z^{p-1} + z^{p-2} + \cdots + 1$. The assumption in the lemma is equivalent to $\varsigma_p^{i-j} \equiv 1 \pmod r$, and we have

$$\Phi_p(\varsigma_p^{i-j}) \equiv \Phi_p(1) \equiv p \pmod r. \qquad (4.95)$$

Since $r \neq p$, $\Phi_p(\varsigma_p^{i-j}) \neq 0$ and so ς_p^{i-j} is *not* a primitive p^{th} root of unity. Hence $\varsigma_p^{i-j} = 1$, implying $\varsigma_p^i = \varsigma_p^j$, which proves Lemma 3.

By this proof we are finished with the preparatory work for Lenstra's Theorem.

THE RECOGNITION OF PRIMES

Lenstra's Theorem

The idea behind the modern, uniform algorithms for primality testing is to obtain information from generalized pseudoprime tests on N, working in cyclotomic rings. The information obtained is used to construct a sieve which restricts the possible prime divisors of N so severely that in the end N is proved to be prime.

The primality test utilizes two finite sets of primes *not dividing* N, one set P and one set Q, consisting of primes q, all having $q-1$ as products of *distinct* primes from P (i.e. all $q-1$'s are *squarefree*). Put $z = \prod_{p\in P} p$ and $w = \prod_{q\in Q} q$. This construction guarantees that $q-1|z$ for each $q\in Q$, and so the order of each element in M_w must divide z, because of Lagrange's Theorem. Thus $a^z \equiv 1 \pmod{w}$ for each integer a with $(a,w)=1$.

Theorem 4.19. Lenstra's Primality Test. Suppose that N is an odd integer satisfying the following conditions:

1. For all primes p and q with $q \in Q$ and $p|q-1$ we have a group character $\chi_{pq} = \chi$, not the principal character, and $\tau(\chi)^{N^{p-1}-1} \equiv \chi(N) \pmod{N}$.

2. For each prime $p \in P$ and each prime $r|N$, $p^{e_p}|r^{p-1}-1$, where p^{e_p} is the largest power of p dividing $N^{p-1}-1$.

Then each prime factor r of N is $\equiv N^i \pmod{w}$ for some $0\le i\le z-1$.

Proof. In the quotient $(r^{p-1}-1)/(N^{p-1}-1)$, condition 2 above guarantees that we can cancel a common factor p^{e_p}, simplifying the fraction to $b_p(r)/a_p \equiv l_p(r) \pmod{p}$, say, since $p\nmid a_p$. Now, Lemma 2 shows that for each pair of primes p and $q \in Q$ with $p|q-1$,

$$\chi(r) \equiv \tau(\chi)^{r^{p-1}-1} \pmod{r}.$$

Raising this to the power a_p we find

$$\chi(r)^{a_p} \equiv \tau(\chi)^{a_p(r^{p-1}-1)} = \tau(\chi)^{b_p(r)(N^{p-1}-1)} \equiv \chi(N)^{b_p(r)} \pmod{N}. \quad (4.96)$$

Here condition 1 in the assumptions of the Theorem has been used. Taking (4.96) (mod r) we conclude that

$$\chi(r)^{a_p} \equiv \chi(N)^{b_p(r)} \pmod{r}$$

142

and so $\chi(r)^{a_p} = \chi(N)^{b_p(r)}$ by Lemma 3. Since both sides of this equation are p^{th} roots of unity and $p \nmid a_p$ we conclude that

$$\chi(r) = \chi(N)^{b_p(r)/a_p(\text{mod } p)} = \chi(N)^{l_p(r)(\text{mod } p)}. \qquad (4.97)$$

This holds for each prime factor r of N and all $p \in P$ and $q \in Q$ with $p|q-1$.

Finally, by the Chinese Remainder Theorem we can find a congruence class $l(r) \pmod{z}$ such that $l(r) \equiv -l_p(r) \pmod{p}$ for each $p \in P$ (remember that $z = \prod_{p \in P} p$). Then, since $\chi(N)$ is a pth root of unity, from (4.97) we find that

$$\chi(r)N^{l(r)} = \chi(r)\chi(N)^{l(r)(\text{mod } p)} = \chi(r)\chi(N)^{-l_p(r)(\text{mod } p)} = 1. \qquad (4.98)$$

This equation tells us that the group element $rN^{l(r)}$ belongs to the kernel of the mapping from M_q to the subgroup C, mentioned above. Moreover, (4.98) holds for each prime $p \in P$. Thus we can apply Theorem A2.10 on p. 274 and conclude that $rN^{l(r)}$ is the neutral element of M_q, i.e. $rN^{l(r)} \equiv 1 \pmod{q}$. Furthermore, this congruence holds for each prime $q \in Q$ and hence it holds $\pmod{\prod q_i}$, yielding $rN^{l(r)} \equiv 1 \pmod{w}$. Finally, since $l(r)$ is a residue class $\pmod{z}$, $z - l(r)$ can be chosen between 0 and $z-1$, and hence $r \equiv N^{z-l(r)} \equiv N^i \pmod{w}$ for some i between 0 and $z-1$, which proves the theorem.

The only thing which remains in order to prove the primality of N is to choose the set Q so large that $w > \sqrt{N}$, checking that $(N, zw) = 1$ and that the least positive residue of $N^i \pmod{w}$ for $1 \le i \le z-1$ is never a proper divisor of N, all this implying that a possible prime factor r of N would be $> \sqrt{N}$ and thus N prime.

Exercise 4.1. Theorem 4.19 gives sufficient conditions for N to be prime. Show that the conditions are also necessary, i.e. that they are satisfied if N is prime (for any choice of the sets P and Q compatible with the conditions stated above).

The Sets P and Q

It is in [17] shown that the set $P = \{2, 3, 5, 7, 11, 13, 17\}$ and a set Q consisting of 32 primes q covers the primality testing of all numbers $N \le 6.5191 \cdot 10^{166}$. In order to be able to check the assumptions of Theorem 4.19, a prime q must be found, for which the character χ_{pq} is not the principal character. Occasionally it happens that $\chi_{pq}(N) \ne 1$ fails for some q in the set Q. In such a case another squarefree q with $p|q-1$

has to be found. Usually this represents no problem, since asymptotically the fraction $1 - 1/p$ of all primes $q \equiv 1 \pmod{p}$ also satisfy $\chi_{pq}(N) \neq 1$. However, since there is at present no *proof* that a small q with the required properties exists, this little detail places Lenstra's test among the probabilistic tests.—An alternative approach which avoids this problem is given in [19].

The Running Time

The version of the test given above is theoretically simple but not practical to implement on computers. Although the asymptotic growth of the running time is the same for a whole class of uniform primality testing algorithms, in practice, their efficiency depends much on the constant hidden in the O-notation, i.e. on the number of individual tests which have to be made in order to check all the assumptions of Theorem 4.19.—In [19] Cohen and Lenstra have given another version of the test which is more computer practical.

The running time of the deterministic versions of the algorithm, the expression (4.85), can be confirmed heuristically in the following way. Suppose P consists of the t first primes. From these 2^t numbers of the form $1 + \prod p_i$ can be built up. By the Prime Number Theorem the average size of the logarithms of these numbers should be $\frac{1}{2} t \ln t$, and about $2^t / (\frac{1}{2} t \ln t)$ of these numbers should be prime. Thus we might expect the size of $\ln w$ to be

$$\frac{2^t}{\frac{1}{2} t \ln t} \cdot \frac{1}{2} t \ln t = 2^t.$$

Putting $w = \sqrt{N}$ and solving for t gives $t = \ln \ln \sqrt{N} / \ln 2$. Now, the number of residues $N^i \pmod{w}$, which have to be computed is z, and the Prime Number Theorem yields

$$\ln z \sim t \ln t \sim \frac{\ln(\frac{1}{2} \ln N)}{\ln 2} \ln \ln \ln N.$$

Thus we expect z to grow asymptotically as

$$\left(\frac{1}{2} \ln N \right)^{\ln \ln \ln N / \ln 2},$$

which is (4.85).

Bibliography

1. D. H. Lehmer, "On the Converse of Fermat's Theorem," *Amer. Math. Monthly* **43** (1936) pp. 347–354.

2. D. H. Lehmer, "On the Converse of Fermat's Theorem II," *Amer. Math. Monthly* **56** (1949) pp. 300–309.

3. Carl Pomerance, John L. Selfridge and Samuel S. Wagstaff, Jr., "The Pseudoprimes to $25 \cdot 10^9$," *Math. Comp.* **35** (1980) pp. 1003–1026.

4. Oystein Ore, *Number Theory and Its History*, McGraw-Hill, New York, 1948, pp. 331–339.

5. Gary Miller, "Riemann's Hypothesis and Tests for Primality," *Journ. of Comp. and Syst. Sc.* **13** (1976) pp. 300–317.

6. Carl Pomerance, "On the Distribution of Pseudoprimes," *Math. Comp.* **37** (1981) pp. 587–593.

7. John Brillhart, D. H. Lehmer and John Selfridge, "New Primality Criteria and Factorizations of $2^m \pm 1$," *Math. Comp.* **29** (1975) pp. 620–647.

8. Daniel Shanks, "Corrigendum," *Math. Comp.* **39** (1982) p. 759.

9. H. C. Williams and J. S. Judd, "Some Algorithms for Prime Testing Using Generalized Lehmer Functions," *Math. Comp.* **30** (1976) pp. 867–886.

10. Hans Riesel, "Lucasian Criteria for the Primality of $N = h \cdot 2^n - 1$," *Math. Comp.* **23** (1969), pp. 869–875.

11. K. Inkeri and J. Sirkesalo, "Factorization of Certain Numbers of the Form $h \cdot 2^n + k$," *Ann. Univ. Turkuensis, Series A* **No. 38** (1959)

12. K. Inkeri, "Tests for Primality," *Ann. Acad. Sc. Fenn., Series A* **No. 279** (1960)

13. William Adams and Daniel Shanks, "Strong Primality Tests That Are Not Sufficient," *Math. Comp.* **39** (1982) pp. 255–300.

14. Leonard Adleman and Frank T. Leighton, "An $O(n^{1/10.89})$ Primality Testing Algorithm," *Math. Comp.* **36** (1981) pp. 261–266.

15. Carl Pomerance, "Recent Developments in Primality Testing," *The Mathematical Intelligencer* **3** (1981) pp. 97–105.

16. Carl Pomerance, "The Search for Prime Numbers," *Sc. Amer.* **247** (Dec. 1982) pp. 122–130.

17. Leonard M. Adleman, Carl Pomerance and Robert S. Rumely, "On Distinguishing Prime Numbers from Composite Numbers," *Ann. of Math.* **117** (1983) pp. 173–206.

18. H. W. Lenstra, Jr., "Primality Testing Algorithms," *Séminaire Bourbaki* **33** (1980–81) No. 576, pp. 243–257.

19. H. Cohen and H. W. Lenstra, Jr., "Primality Testing and Jacobi Sums," *Math. Comp.* **42** (1984) pp. 297–330.

20. John D. Dixon, "Factorization and Primality Tests," *Am. Math. Monthly* **91** (1984) pp. 333-352. Contains a large bibliography.

CHAPTER 5

FACTORIZATION

Introduction

The art of decomposing large integers into prime factors has advanced considerably during the last 15 years. It is the advent of high-speed computers that has rekindled interest in this field. This development has followed several lines. In one of these, already existing theoretical methods and known algorithms have been carefully analyzed and perfected. As an example of this work we mention Michael Morrison and John Brillhart's analysis of an old factorization method, the continued fraction algorithm, going back to ideas introduced already by Legendre and developed further by Maurice Kraïtchik, D. H. Lehmer and R. E. Powers.

The progress achieved in this area is due also to another line of development, namely the introduction of new ideas. We shall discuss two of these in connection with the continued fraction method, namely the factor base by Morrison and Brillhart and the quadratic sieve by Carl Pomerance. The clever use of all of these developments has resulted in the most efficient *practically applicable* general method so far implemented for factoring large numbers. The practical upper limit for this method with today's computers is the factorization of integers having around 75 digits.

We shall also describe two entirely new methods, Daniel Shanks' "square forms factorization" and J. M. Pollard's ρ-method.

Due to the present lively interest in this field, it is not possible at the moment to provide a complete overview of all partial results found, nor to give a good classification of all the methods used. Nevertheless, despite the risk of this text becoming obsolete quite soon, the author has attempted to classify the main methods in at least some sort of logical scheme.

When Do We Attempt Factorization and Which Method Should We Use?

In Chapter 4 we saw that the identification of prime numbers is generally quite a fast operation. Since factorization methods so far invented are not that fast, we should attempt factorization only of numbers *which we know in advance to be composite.* Otherwise, we will certainly, sooner or later, hit upon the "worst possible case" for the method we are using, and perhaps finally end up with a very laborious proof of the primality of the number being investigated. There are, however, occasional exceptions to this rule but it is always advisable to verify that the numbers involved are composite.

The choice between the various methods available is not always easy. Firstly, there are certain *special factorization methods* (or adaptations of general methods), which are advantageous if the number to be factored has a particular mathematical form. If it has not, or if no such method is available on your computer, then one of the *general factorization methods* must be tried. These are of two kinds, namely those which find the factors of N in approximate order of magnitude, and those for which the computing time *does not depend on the size of the factors.* It is obvious that much computing time may be used up achieving little, if a method of the latter kind is applied to a large number having a small prime factor, when it could easily have been detected by one of the former methods.— The more intricate details of how to choose between different methods will be discussed after we have described the most important factorization methods.

Trial Division

Trial division consists of making trial divisions of the number N by the small primes. *Either* you store a table of primes up to some limit in your computer, after which you may have a fast running program at the expense of using a good deal of storage space, *or alternatively* you generate the small primes, which leads to a program running a little slower, but demanding less storage capacity. In the latter case, it is actually advantageous *not to use only the primes* as trial divisors, as it is quite time-consuming to run a sieve program in order to generate the primes. Instead, it is better to use the integers 2 and 3 and then *all* positive integers of the form $6k \pm 1$ as trial divisors. This covers all the primes and includes making divisions by some composite numbers, namely 25, 35, 49, ..., but the loop in the program will be very fast, so that we can afford some waste in performing these unnecessary divisions! (At the expense of some extra programming, however, the number of unnecessary divisions can be reduced and the efficiency of the computer program increased.) If you search for small divisors up to the search limit G, then you will carry out roughly $G/3$ trial divisions with the procedure indicated rather than the $\pi(G) \approx G/\ln G$ divisions required if dividing by the primes only. With $G = 10^6$, the proportion of "useful" divisions will be only $3/\ln 10^6 \approx 22\%$, which looks a little discouraging! Whether this pays or not depends on your ability to write a computer program for division by the primes only, which is *less than about* $1/0.22 = 4.6$ *times slower per single loop* to execute! Remember, even if you store a prime table up to $G = 10^6$, you probably will have to write it in compact form as shown in Chapter 1, in which case a time-consuming table look-up function might be involved. The methods mentioned on p. 10, however, are generally fast on most computers if programmed in assembly code, and need little storage space. Thus $(p_{i+1} - p_i)/2$ can be stored in one 6-bit byte for the primes below 1.35 million and in one 8-bit byte for the primes below $3 \cdot 10^{11}$. Moreover, in judging the speed by which a computer program runs, there are actually *two* figures to consider. One is the *maximal running time for the worst possible case*; for example, if we find no factor $\leq G$ in the case just discussed. The other important figure is the *mean running time* which determines the total running cost. Since small prime factors are quite common, it seldom occurs that no factor $\leq 10^6$, say, is found, so that the program will probably in most cases terminate long before the search limit G is reached.

Proportion α of odd numbers having no factor $\leq G$					
G	α	G	α	G	α
10^2	0.2406	10^5	0.0975	10^8	0.0610
10^3	0.1619	10^6	0.0813	10^9	0.0542
10^4	0.1218	10^7	0.0697	10^{10}	0.0488

To demonstrate the consequences of this fact, we show above in a small table, the proportion of *odd* numbers which "survive" trial divisions by all primes below 10^n for n up to 10. This proportion, if the search limit 10^n is again called G, will be

$$\prod_{p \geq 3}^{G} \left(1 - \frac{1}{p} \right) \approx 2\frac{e^{-\gamma}}{\ln G} = \frac{0.4877}{\log_{10} G}, \tag{5.1}$$

according to Mertens' Theorem (see p. 71). Hence if there is no special reason to act otherwise, trial division by 2, 3 and all integers $6k \pm 1$ is sufficiently efficient to be preferred, because of its simplicity, to other ways of implementing the trial division method.

A Computer Implementation of Trial Division

Although quite simple, we think it worth while discussing in some detail how to implement trial division. One question is: how can we generate all integers of the form $6k \pm 1$? The answer is by starting at 5 and then alternately adding 2 and 4: $5 + 2 = 7, 7 + 4 = 11, 11 + 2 = 13, 13 + 4 = 17, \ldots$ These alternate additions of 2 and 4 can be carried out as follows: Define a variable d, let d:=2 initially and then perform d:=6-d in the loop! This will cause d to take the values 2 and 4 alternately, which is exactly what needs to be added to proceed from one trial divisor to the next. Another way to achieve the divisions by $6k \pm 1$ is to perform *two* divisions in the same loop, one by p and another by $p + 2$, and then augment p by 6 within the loop. With this approach the administrative operations involved will be less burdensome and the program faster. We have chosen the second alternative in the procedure divide below.—In order to terminate the computation correctly, you must verify whether the trial divisor p has reached the search limit G or $\sqrt{N}$, whichever is the smaller. If $p > \sqrt{N}$, then p is also $> N/p$, so this latter test can be performed without calculating any square-root. However, the calculation of a square-root at the beginning of the computation and each time a factor has been removed will not influence the computing time very much, since most of the time is expended in performing trial divisions without finding any factors. Moreover, it is essential to reduce N immediately by any factor discovered because this will speed the termination, since in each case the limit will be lowered to $\min(G, \sqrt{N/p})$. Also, do not forget to test repeatedly, as divisors of N, all the prime factors found, continuing until they are no longer divisors of what remains of N. Otherwise, multiple prime factors of N will be overlooked!

Now, finally, below is a PASCAL procedure divide which stores the factors $\leq G$ of N in the integer array Factor, and the number of factors found in Factor [0]. If N is not completely factored by the procedure, then the last co-factor (of unknown character; it may be prime or composite) is returned in m for further processing by other factorization procedures.

Further, in order to make the logical structure of the computation easily recognizable, a subprogram `reduce` is introduced. This subprogram is used in the search for factors 2 and 3 and in the pursuit of multiple prime factors. In addition to this, it is used for dividing N by any prime found by the main procedure.—The procedure `divide` is shown here in the context of a small test program by which an integer N can be input and factored.

```
PROGRAM Trial division {Factors an integer N}
(Input,Output);
LABEL 1;
CONST k=22;
TYPE vector=ARRAY[0..k] OF INTEGER;
VAR Factor : vector; N,G,i,m : INTEGER;

PROCEDURE divide(N,G : INTEGER; VAR Factor : vector;
                 VAR m : INTEGER);
{Places all prime factors <G in the integer array Factor. The
number of factors found is placed in Factor[0]. The remaining
co-factor is returned in m, unless the divisions show this
co-factor to be prime, in which case it is included among the
factors in the array Factor and the value of m returned is 1}
VAR p,plimit : INTEGER;

PROCEDURE reduce(r : INTEGER);
 {Divides N by r, if possible}
BEGIN IF r>1 THEN WHILE m MOD r = 0 DO
  BEGIN m:=m DIV r; Factor[0]:=Factor[0]+1;
    Factor[Factor[0]]:=r
  END;
  plimit:=isqrt(m);
  IF plimit>G THEN plimit:=G
END {reduce};

BEGIN m:=N; Factor[0]:=0;
  reduce(2); reduce(3); p:=5;
  WHILE p<=plimit DO
    BEGIN IF m MOD p  = 0 THEN reduce(p);
          IF m MOD(p+2)=0 THEN reduce(p+2);
          p:=p+6
    END;
  IF p>isqrt(m) THEN reduce(m)
END {divide};
```

```
BEGIN
1: write('Input an integer N>0 for factorization: '); read(N);
   IF N>0 THEN
   BEGIN
     write('Input the search limit G: '); read(G);
     divide(N,G,Factor,m);
     FOR i:=1 TO Factor[0] DO write(Factor[i]:8);
     writeln; IF m=1 THEN writeln('Factorization complete')
                ELSE
     writeln('Factorization incomplete. Co-factor=',m:8);
     GOTO 1
   END
END.
```

Note that in order to compute the correct search limit G or $[\sqrt{N/p_i}]$, we have assumed the availability of a procedure isqrt for calculating the integral part of the square-root of an integer. This procedure is not shown here. However, for test purposes, provided that N is chosen *less than the smallest integer that cannot be stored as a* REAL *variable without rounding off*, the construction

$$\texttt{isqrt:=trunc(sqrt(m)+0.5);}$$

will do. If N is larger, but still a single-precision integer, then it is advisable to verify that the above construction has not returned a value z one unit too low, which could occasionally happen. This verification and correction is achieved by writing

$$\texttt{IF sqr(z+1)<=m THEN z:=z+1;}$$

Exercise 5.1. Trial division. Explore the practical upper limit of N for the PROGRAM Trial division above when used on your computer. (It is for primes N that the longest running times occur.) Plot the maximal running time vs. the size of N in a logarithmic diagram. Decide on a value of G in order to obtain reasonable running times even in the case when N happens to be a product of two nearly equal prime factors.

Euclid's Algorithm as an Aid to Factorization

Euclid's algorithm can be (and has been!) used to search for factors. The method is as follows: In order to search for prime factors of N between g

and G, multiply together all primes between the two limits. Then, apply Euclid's algorithm (which is fast) on the product formed and N. Any prime factor of N within the given search limits will then show up. In the days before the invention of computers, it was advantageous to use pre-computed values of

$$P_0 = \prod_2^{97} p = \qquad 23055\ 67963945\ 51842475\ 31021473\ 31756070$$

$$P_1 = \prod_{101}^{199} p = 338308\ 05092969\ 17481189\ 79876079\ 64806707\ 71162183$$

$$P_2 = \prod_{211}^{293} p = \qquad 2620256\ 64754470\ 33438281\ 30718839\ 84477441$$

$$P_3 = \prod_{307}^{397} p = \qquad 4\ 97665233\ 93936228\ 75013859\ 80827529\ 80119549$$

$$P_4 = \prod_{401}^{499} p = 122745\ 35402370\ 14997887\ 61565110\ 91819579\ 03188941$$

$$P_5 = \prod_{503}^{599} p = \qquad 2530563\ 06993037\ 84009224\ 45621969\ 81381959$$

$$P_6 = \prod_{601}^{691} p = 87903\ 18928189\ 78804933\ 06530627\ 46911120\ 09314693$$

$$P_7 = \prod_{701}^{797} p = \qquad 1\ 68664678\ 15653776\ 12724390\ 71676293\ 19108817$$

$$P_8 = \prod_{809}^{887} p = \qquad 8362\ 34357346\ 06723958\ 85255200\ 21529016\ 29917681$$

$$P_9 = \prod_{907}^{997} p = \qquad 50\ 01399907\ 16305530\ 69693302\ 73892941\ 39003181$$

to find all prime factors of N below 1000.

    With the aid of a desk-calculator Euclid's algorithm was utilized on N and all the P_i's, one at a time. As long as N did not exceed the register capacity of the calculator used this was quite simple to carry out; only the initial division in Euclid's algorithm (the long number divided by N) required some care, but was not too difficult to perform in a step-wise fashion.

Even nowadays this method is occasionally applied on a computer, particularly when N is too large to be easily divided by a small prime using the computer's built-in arithmetic. Instead of carrying out the search for factors by performing numerous slow divisions with each of the small primes using a multiple-precision arithmetic package, it is faster to divide only one huge number P_i by N using the package, and then employ Euclid's algorithm on the remainder and N, i.e. on two numbers of magnitude N.

Another application of Euclid's algorithm is when only small divisors are sought, as in the continued fraction method, described on pp. 199–206 below. In order to save computing time, rather than store the small primes of the factor base and dividing by each of these, we store products of such primes, where each product is smaller than a single or a double word, depending on the kind of machine arithmetic available to us. Then N is divided by each of these products and, finally, Euclid's algorithm performed on the remainder found by the division and the corresponding product of small primes, an operation which can be efficiently programmed in assembly language.

Fermat's Factoring Method

With the exception of trial division Fermat's factoring method is the oldest systematic method of factoring integers. Although, in general, the method is not very efficient, it is of theoretical as well as of some practical interest. We therefore find it instructive to discuss the method in some detail and to study the ideas behind it.

Fermat's idea was to try to write an odd composite number $N = a \cdot b$ as a difference between two square numbers. If we succeed in writing N as $x^2 - y^2 = (x-y)(x+y)$, we immediately deduce a factorization of N. How can such a representation of N be found? Well, obviously x must be $> \sqrt{N}$, so let us start by computing $m = [\sqrt{N}] + 1$, which is the smallest possible value of x (unless, of course, N happens to be a square number x^2, in which case we have the representation $N = x^2 - 0^2$). Now, we have to consider $z = m^2 - N$ and check whether this number is a square. If it is, then we have found $N = x^2 - y^2$, and are finished. Otherwise, we try the next possible x, i.e. $m+1$, and compute $(m+1)^2 - N = m^2 + 2m + 1 - N = z + 2m + 1$, test whether *this* is a square, and so on. The procedure to follow is most easily understood by looking at an

153

Example. $N = 13199$. $\sqrt{N} = 114.88\ldots$, so that N is not a square. Thus we proceed:

$$m = 115, \quad z = 115^2 - 13199 = 26,$$

which is not a square. After this start, the calculation runs as follows:

m	$2m+1$	z		m	$2m+1$	z
115	231	26		124	249	2177
116	233	257		125	251	2426
117	235	490		126	253	2677
118	237	725		127	255	2930
119	239	962		128	257	3185
120	241	1201		129	259	3442
121	243	1442		130	261	3701
122	245	1685		131	263	3962
123	247	1930		132	265	4225

The numbers z in the third column are calculated by adding the numbers $2m+1$ and the numbers z in the preceding line: $257 = 231+26$, $490 = 233 + 257$, and so on.—In the last line we have arrived at a square, $z = 4225 = 65^2$, and from this we immediately find the factorization of N: $N = 132^2 - 65^2 = (132 - 65)(132 + 65) = 67 \cdot 197$.

We are now going to answer the following question: Using Fermat's algorithm, how much work is required before N is factored? If $N = ab$, with $a < b$, is a product of only *two* primes (the hard case), then the factorization will be achieved when m reaches $x = (a + b)/2$. Since the starting value of m is $\approx \sqrt{N}$, and $b = N/a$, this will take

$$\approx \frac{1}{2}\left(a + \frac{N}{a}\right) - \sqrt{N} = \frac{(\sqrt{N} - a)^2}{2a} \tag{5.2}$$

lines in the table above. From this expression it can be seen directly that if N is the product of two almost equal factors, then the amount of work needed to find a factorization is small, because in this case the factor a is just below $\sqrt{N}$. Consider e.g. the 19-digit number $N = (10^{22}+1)/(89 \cdot 101)$, which happens to have two almost equal prime factors. (It takes only 1803 cycles with Fermat's method to arrive at the factorization. The reader can confirm this!)—Suppose that $a = k\sqrt{N}$, $0 < k < 1$ in (5.2). Then the number of cycles necessary to obtain the factorization is

$$\frac{(1-k)^2}{2k}\sqrt{N}.$$

This has the order of magnitude $O(\sqrt{N})$, and the reader might be inclined to believe that the labour involved in using Fermat's method is also of that order of magnitude. Note, however, that the constant latent in the O-notation may be very large! For instance if $k = 0.001$, then $(1-k)^2/(2k) \approx 499$, and thus it is really only when the two factors are very close to $\sqrt{N}$ that Fermat's method is of practical use. Let us consider just one more case, a sort of "ordinary case", where the two factors are $a \approx N^{\frac{1}{3}}$ and $b \approx N^{\frac{2}{3}}$. In this situation the number of cycles necessary will be

$$\frac{(\sqrt{N} - \sqrt[3]{N})^2}{2\sqrt[3]{N}} = \frac{(\sqrt[3]{N})^2(\sqrt[6]{N} - 1)^2}{2\sqrt[3]{N}} \approx \frac{1}{2}N^{\frac{2}{3}},$$

an order of magnitude considerably higher than $O(N^{\frac{1}{2}})$, and therefore totally impractical!

This analysis of Fermat's method is very discouraging indeed, but in spite of this, the method has been used with some success. This is due to the simplicity of the algorithm, which can be modified by using suitable shortcuts. One of these shortcuts is the following. As the reader can observe, the end-figures of squares cannot form any number, but only certain two-digit numbers, namely the following 22 combinations:

$$00, 01, 04, 09, 16, 21, 24, 25, 29, 36, 41, 44, 49, 56, 61, 64, 69, 76, 81, 84, 89, 96.$$

These may be written as $00, e1, e4, 25, o6, e9$, where e is an even digit and o an odd digit. Using this fact greatly simplifies the search for squares in the column for z above. Only when a number with a possible square ending occurs need we check whether a square has appeared. This simple trick can also be used in a binary computer, since the square of any number is $\equiv 0, 1, 4,$ or $9 \pmod{16}$, conditions that can be very easily tested in a binary computer. Looking closer, it turns out that the values of x must also belong to certain arithmetic series in order for the end-figures to be possible square-endings. Calculating square-endings for other moduli, such as 25, 27 etc., a sieve on x can be constructed, thereby leading, if we wish, to a reduction by a large numerical factor in the number of x-values to be tried. However, this gain in speed will be at the expense of increased complexity in the programming.

Many attempts have been made to remedy the main difficulty of Fermat's method, which is that the number of steps required in the "worst possible case" is unreasonable. One way, if $N = pq$, with $p < q$, would be to multiply N by a factor f, such that $N \cdot f$ admits a composition into two very close factors. If we choose $f \approx q/p$, then $N \cdot f$ will have two factors $p \cdot f \approx q$ and q, which are close. However, how can we find f when we do not know the size of q/p? We could *either* successively try all factors $f = 1, 2, 3, \ldots$ up to $N^{1/3}$, say, which means that we run one or more steps at a time in the algorithm above, applied to the number $N \cdot f$, for each value of f, *or* we could multiply N by a suitably chosen large factor f containing many factors of different sizes and, in this way, apply Fermat's method to a number $N \cdot f$ which certainly does have two close factors. Suitable choices of f could be factorial numbers $1 \cdot 2 \cdot 3 \cdots k$ of appropriate size or, better still, so-called highly composite numbers. *Finally,* a combination of these two ideas might be used, i.e. to test successively, as candidates for the factor f, all multiples of a set of factors, each having many divisors. An analysis of how this can be done in a most efficient manner has been carried out by R. Sherman Lehman and is reported in [1]. The report contains also a computer program. The order of Lehman's factoring method is $O(N^{1/3})$, if N has no factors below $N^{1/3}$ so if combined with trial division up to $N^{1/3}$, Lehman's method is $O(N^{1/3})$ in all cases.

Legendre's Congruence

The methods described in Chapter 4 for recognizing primes use some characteristic which distinguishes composites from primes. Can such characteristics be also used to factor numbers? To some extent, yes! One of the simplest examples of this is Legendre's factoring method which we shall describe in some detail a little later. The method utilizes the following distinction between primes and composites: For every N, prime or composite, the congruence

$$x^2 \equiv y^2 \pmod{N} \tag{5.3}$$

has the solutions $x \equiv \pm y \pmod{N}$, which we shall call the *trivial* solutions. For a *composite N*, however, the congruence also has *other solutions*, which can be used to factor N. To understand this, let us start with the congruence

$$u^2 \equiv y^2 \pmod{p}, \quad p \text{ an odd prime.}$$

If we consider y as a fixed integer $\not\equiv 0 \pmod{p}$, then there are precisely two solutions $u \equiv \pm y \pmod{p}$ to this congruence. This is because Theorem

A1.5 on p. 253 indicates that there are *at most* two solutions of a quadratic congruence in a ring without zero divisors and, since we can observe two solutions here, these must be the only ones. We have only to check that u and $-u$ belong to different residue classes (mod p), but $u \equiv -u$ (mod p) gives $2u \equiv 0$ (mod p), which implies $u \equiv 0$ (mod p), if $(p, 2) = 1$. Thus if $u \not\equiv 0$ (mod p) then everything will be right. That is why we have to exclude the case $y^2 \equiv 0$ (mod p), leading to $u \equiv 0$, and, because of $(p, 2) = 1$, also the case $p = 2$.

Now, since $u^2 \equiv y^2$ (mod p) has two solutions $u \equiv \pm y$ (mod p), so has $v^2 \equiv y^2$ (mod q), namely $v \equiv \pm y$ (mod q), if q is an odd prime $\neq p$. Thus the congruence $x^2 \equiv y^2$ (mod pq) has *four* solutions, which we may find by combining in four ways the two solutions (mod p) and the two solutions (mod q):

$$\begin{cases} u \equiv y \pmod{p} \\ v \equiv y \pmod{q} \end{cases} \qquad \text{giving } x \equiv y \pmod{pq},$$

$$\begin{cases} u \equiv -y \pmod{p} \\ v \equiv -y \pmod{q} \end{cases} \qquad \text{giving } x \equiv -y \pmod{pq},$$

$$\begin{cases} u \equiv y \pmod{p} \\ v \equiv -y \pmod{q} \end{cases} \qquad \text{giving } x \equiv z \pmod{pq}, \quad \text{say,}$$

$$\text{and} \qquad \begin{cases} u \equiv -y \pmod{p} \\ v \equiv y \pmod{q} \end{cases} \qquad \text{giving } x \equiv -z \pmod{pq}.$$

Thus, we see that if $N = pq$, the congruence $x^2 \equiv y^2$ (mod N) has one more pair of solutions $x \equiv \pm z$ (mod N) than the trivial pair $x \equiv \pm y$ (mod N). This fact can now be used to find the factorization of $N = pq$. First, we need to find a non-trivial solution to $x^2 \equiv y^2$ (mod N). Since $x^2 - y^2 = (x+y)(x-y) \equiv 0$ (mod pq), and $x+y$ or $x-y$ is not divisible by *both* p and q (remember, the combinations where one of $x+y$ or $x-y$ is divisible by both p and q lead to the trivial solutions!), then one of $x+y$ and $x-y$ must be divisible by p and the other by q. The factor p (or q) can be extracted by using Euclid's algorithm on $x+y$ and N.

If N has more than two prime factors, the method still works in a similar way since the above reasoning can be applied to one of the prime factors p and the corresponding co-factor $q = N/p$, which in this case will be composite.

To summarize the use of Legendre's congruence for factorization: Find a non-trivial solution to the congruence $x^2 \equiv y^2$ (mod N). Compute a factor of N as the G.C.D. $(x - y, N)$ by means of Euclid's algorithm.

Several very important factorization methods make use of Legendre's congruence. They differ only in the way in which the solution to $x^2 \equiv y^2$ (mod N) is found. As examples, we can mention the method of Fermat already discussed (a solution to $x^2 - y^2 \equiv 0$ (mod N) is obtained if we have a representation of $k \cdot N$ as $x^2 - y^2$), Shanks' method and the continued fraction method. Although these methods use very different factorization algorithms, fundamentally they are all based on Legendre's idea. Also Euler's method, discussed in the next sub-section, will be shown to rely upon Legendre's idea.

Euler's Factoring Method

This method is special because it is applicable only to integers which can be written in the form $N = a^2 + Db^2$ in two different ways (for the same value of D). It depends on a famous identity given by Lagrange:

$$(x^2 + Dy^2)(u^2 + Dv^2) = \begin{cases} (xu + Dyv)^2 + D(yu - xv)^2 \\ (xu - Dyv)^2 + D(yu + xv)^2. \end{cases} \quad (5.4)$$

This identity shows that a product of two integers, both of the form $a^2 + Db^2$, is itself an integer of the same form, and that this product *has two different representations* as $r^2 + Ds^2$. Euler proved the converse: If N has two different representations as $r^2 + Ds^2$, $N = a^2 + Db^2$ and $N = c^2 + Dd^2$, with $(bd, N) = 1$, then N can be written as a product of two numbers of this same form. The factors of N can be found in the following manner. Start by

$$a^2 d^2 \equiv -Db^2 d^2 \ (\text{mod } N) \equiv b^2 c^2 \ (\text{mod } N). \quad (5.5)$$

Thus we have a case for Legendre's congruence and the factors of N are found as

$$(N, ad - bc) \quad \text{and} \quad (N, ad + bc). \quad (5.6)$$

The algorithm for finding two different representations of N as $r^2 + Ds^2$ is completely analogous to the algorithm used in Fermat's method. However, since not all integers have even *one* representation as $r^2 + Ds^2$ (a necessary condition for this is that $-D$ is a quadratic residue of *all* prime factors of N appearing with an odd exponent, but often even this is not sufficient), the search for a second representation is normally undertaken only for integers already possessing one known representation in the form $r^2 + Ds^2$. If $D > 0$ this will take, at most, $\sqrt{N/D}$ steps, as in this case Ds^2 must be $< N$ and thus there are, at most, $\sqrt{N/D}$ possible values for s. We demonstrate the algorithm in an

Example. $N = 34889 = 143^2 + 10 \cdot 38^2$. The search for a second representation of N as $r^2 + 10s^2$ starts as follows: Put $z = r^2 = N - 10s^2$ with $s_0 = [\sqrt{N/10}] = 59$, and $z_0 = N - 10s_0^2 = 34889 - 34810 = 79$. The computation runs as below:

s	$10(2s-1)$	z	s	$10(2s-1)$	z
59	1170	79	45	890	14639
58	1150	1249	44	870	15529
57	1130	2399	43	850	16399
56	1110	3529	42	830	17249
55	1090	4639	41	810	18079
54	1070	5729	40	790	18889
53	1050	6799	39	770	19679
52	1030	7849	38	750	143^2
51	1010	8879	37	730	21199
50	990	9889	36	710	21929
49	970	10879	35	690	22639
48	950	11849	34	670	23329
47	930	12799	33	650	23999
46	910	13729	32	630	157^2

Thus, N is also $= 157^2 + 10 \cdot 32^2$, and (5.6) gives the two factors

$$(N, ad - bc) = (34889, 143 \cdot 32 - 38 \cdot 157) = (34889, -1390) = 139$$

and

$$(N, ad + bc) = (34889, 10542) = 251. \tag{5.7}$$

Gauss' Factoring Method

Though quite complicated, Gauss' factoring method [2] is very important because it is the basis of many sieving methods for factorization. Such methods have been in use for over 150 years of computing by hand and by desk-calculator and have yielded many factorizations. Gauss' method is a sort of exclusion method which, by finding more and more quadratic residues (mod N), excludes more and more primes from being possible factors of N, and so finally reduces the number of possible factors so markedly that it is possible to undertake trial division by those remaining. The reader who is not familiar with the theory of quadratic residues should now study Appendix 3 before proceeding further with this chapter.

The idea behind Gauss' method of factorization is the following: Since an integer a is a quadratic residue of about half of the primes, as described in Appendix 3 and in Table 32 at the end of this book, the fact that a is a quadratic residue (mod N) will exclude about half of all primes as divisors of N. This is because $a \equiv x^2$ (mod N) implies $a \equiv x^2$ (mod p) for any divisor p of N. Thus, by statistical laws, k independent quadratic residues (mod N) will leave about the fraction 2^{-k} of all primes as possible divisors of N. Thus, 20 known quadratic residues will reduce the number of trial divisions necessary by a factor of about $2^{20} \approx 1,000,000$.

How can we find many quadratic residues? Principally this is very easy. We simply have to square many integers and then reduce the squares (mod N). However, since it is a complicated matter to find out which primes are excluded as divisors of N based on the existence of a *large* quadratic residue (compare Table 32), it would be useful to have a number of *small* quadratic residues such as the ones ready for examination in Table 32. This is achieved by the following trick: If $a \equiv x^2$ (mod N), then $x^2 - a = kN$ for some integer k. Letting x be close to $[\sqrt{kN}]$ will give comparatively small values of a; and, since the product of two quadratic residues is again a quadratic residue, it is of interest to use only those quadratic residues which can be completely factorized by using only *small* prime factors. These quadratic residues can then easily be combined by multiplication and exclusion of square factors to yield new quadratic residues. Let us consider an

Example. $N = 12\,007\,001$. If we confine our search for small quadratic residues to representations of $kN = x^2 - a$ for which $|a| < 50\,000$ and having no prime factor of a larger than 100, then we have the following expressions

$$N = 3459^2 + 42320 = x^2 + 2^4 \cdot 5 \cdot 23^2$$
$$N = 3461^2 + 28480 = x^2 + 2^6 \cdot 5 \cdot 89$$
$$N = 3463^2 + 14632 = x^2 + 2^3 \cdot 31 \cdot 59$$
$$N = 3464^2 + 7705 = x^2 + 5 \cdot 23 \cdot 67$$
$$N = 3465^2 + 776 = x^2 + 2^3 \cdot 97$$

$$2N = 4898^2 + 23598 = x^2 + 2 \cdot 3^3 \cdot 19 \cdot 23$$
$$2N = 4900^2 + 4002 = x^2 + 2 \cdot 3 \cdot 23 \cdot 29$$
$$3N = 6003^2 - 15006 = x^2 - 2 \cdot 3 \cdot 41 \cdot 61$$
$$5N = 7745^2 + 49980 = x^2 + 2^2 \cdot 3 \cdot 5 \cdot 7^2 \cdot 17$$
$$8N = 9799^2 + 35607 = x^2 + 3 \cdot 11 \cdot 13 \cdot 83$$
$$10N = 10957^2 + 14161 = x^2 + 7^2 \cdot 17^2$$
$$11N = 11491^2 + 33930 = x^2 + 2 \cdot 3^2 \cdot 5 \cdot 13 \cdot 29$$
$$11N = 11492^2 + 10947 = x^2 + 3 \cdot 41 \cdot 89$$
$$14N = 12964^2 + 32718 = x^2 + 2 \cdot 3 \cdot 7 \cdot 19 \cdot 41$$
$$14N = 12965^2 + 6789 = x^2 + 3 \cdot 31 \cdot 73$$
$$17N = 14287^2 + 648 = x^2 + 2^3 \cdot 3^4$$
$$19N = 15105^2 - 28006 = x^2 - 2 \cdot 11 \cdot 19 \cdot 67$$
$$21N = 15879^2 + 4380 = x^2 + 2^2 \cdot 3 \cdot 5 \cdot 73$$

Now we can systematically combine the quadratic residues found to obtain new ones:

$$10N = x^2 + 7^2 \cdot 17^2 \qquad \text{gives} \quad a = -1,$$
$$N = x^2 + 2^4 \cdot 5 \cdot 23^2 \qquad \text{gives} \quad a = 5,$$
$$N = x^2 + 2^6 \cdot 5 \cdot 89 \qquad \text{gives} \quad a = 89,$$
$$17N = x^2 + 2^3 \cdot 3^4 \qquad \text{gives} \quad a = 2,$$
$$N = x^2 + 2^3 \cdot 97 \qquad \text{gives} \quad a = 97,$$

$$14N = x^2 + 3 \cdot 31 \cdot 73 \text{ and } 21N = x^2 + 2^2 \cdot 3 \cdot 5 \cdot 73 \text{ give } a = 31,$$

$$N = x^2 + 2^3 \cdot 31 \cdot 59 \text{ gives } a = 59,$$

$$3N = x^2 - 2 \cdot 3 \cdot 41 \cdot 61 \text{ and } 11N = x^2 + 3 \cdot 41 \cdot 89 \text{ give } a = 61.$$

In summary, we have so far established that the following integers are quadratic residues of N: $a = -1, 2, 5, 31, 59, 61, 89$ and 97. (Primarily we have found $a = -5, -89, -2, \ldots$, but since -1 is a quadratic residue, we also find the corresponding *positive* numbers to be residues.)

Now follows the complicated part of Gauss' method, namely to use this information in a systematic manner to find a factor of N. Since -1 is a quadratic residue of N, only primes of the form $p = 4k + 1$ can divide

N. In addition, because 2 is a residue, every p that divides N must be of the form $p = 8k \pm 1$. Combining these two conditions, we find that only $p = 8k + 1$ is possible as a divisor of N. Using also the quadratic residue 5, which restricts possible divisors to one of the four forms $p = 20k \pm 1, \pm 9$, we conclude that p is of the form $p = 40k + 1$ or $40k + 9$ (since p must be $8k+1$). Now it is no longer practical to proceed by deducing the arithmetic forms of the possible divisors of N, because the number of feasible residue classes for the larger moduli used will be too big. Instead we determine, by computing the value of Legendre's symbol (a/p), which of the primes of the two forms $p = 40k + 1$ and $p = 40k + 9$ below $[\sqrt{N}] = 3465$ have, as quadratic residues, all of the previously obtained a's. As soon as one of the values of a is found to be a quadratic non-residue of one of the primes in question, then this prime is eliminated as a possible divisor of N and need no longer be considered. This procedure excludes about half of the remaining primes for each new value of a used. (Note that the values of a must be *independent*, i.e. if r and s have already been used as quadratic residues, then the quadratic residue rs will *not* rule out any further primes!) The computation in our example proceeds as below. The following primes below 3465 of the form $p = 40k + 1$ or $p = 40k + 9$ have $(31/p) = +1$:

$$p = 41, 281, 521, 769, 1249, 1289, 1321, 1361, 1409, 1489, 1601, 1609,$$

$$1721, 2081, 2281, 2521, 2609, 2729, 3001, 3089, 3169, 3209, 3449.$$

Among these, $(59/p) = 1$ for

$$p = 41, 281, 521, 1361, 1609, 2081, 2729, 3001, 3089 \text{ and } 3449.$$

Of these $(61/p) = 1$ for

$$p = 41, 1361, 2729, 3001 \text{ and } 3089.$$

Finally, the quadratic residue 89 leaves as only possible divisor

$$p = 3001.$$

We also verify that Legendre's symbol $(97/3001) = 1$. Therefore, if N is not a prime, it must be divisible by 3001: $N = 3001 \cdot 4001$.

The procedure described above is too tedious in practice to carry through in a case of, say, 100 known quadratic residues, and has therefore been replaced by another device in the Morrison-Brillhart method, which likewise starts by determining lots of small quadratic residues of N and breaking these up into prime factors. We shall describe this method in detail later on.

Legendre's Factoring Method

Legendre's factoring method is very similar to that of Gauss discussed above. It differs only in the procedure for finding small quadratic residues of N. For this purpose Legendre introduced the continued fraction expansion of $\sqrt{N}$ and used the relations (A5.31) and (A5.34) on pp. 315–316. (In fact, the proof of (A5.31) given in Appendix 5 is due to Legendre.) After a sufficient number of small quadratic residues have been found, a sieve is constructed, just as in Gauss' method, in which each residue annihilates about half of the primes as possible divisors of N.—Because continued fractions are used for finding quadratic residues, this and certain related methods are sometimes called *continued fraction methods* for factorization.

The Number of Prime Factors of Large Numbers

Before we proceed with our description of the factoring methods, we require information about what we can expect when we attempt to factorize a large unknown number. This information has to be drawn from the theoretical parts of number theory. We begin by introducing the function $\omega(n)$, the number of different prime factors of n. This function has the so-called average order $\ln \ln n$, meaning that

$$\sum_{n \leq x} \omega(n) = x \ln \ln x + B_1 x + o(x), \qquad (5.8)$$

where the constant B_1 has the value

$$B_1 = \gamma + \sum_p \left(\ln \left(1 - \frac{1}{p}\right) + \frac{1}{p} \right) = 0.2614972128\ldots \qquad (5.9)$$

Proof.

$$\sum_{n \leq x} \omega(n) = \sum_{n \leq x} \sum_{p|n} 1 = \sum_{p \leq x} \left[\frac{x}{p}\right] = \sum_{p \leq x} \frac{x}{p} + O(\pi(x)). \qquad (5.10)$$

Now, $\sum_{p \leq x} 1/p$ can be evaluated by taking logarithms of both sides in Mertens' formula on p. 71, yielding

$$\sum_{p \leq x} \ln \left(1 - \frac{1}{p}\right) = -\sum_{p \leq x} \frac{1}{p} - \sum_{p \leq x} \frac{1}{2p^2} - \cdots = -\gamma - \ln \ln x + o(1). \qquad (5.11)$$

Since the sums $\sum_{p=2}^{\infty} p^{-2}$, $\sum_{p=2}^{\infty} p^{-3}, \ldots$, all converge, this can be written as

$$\sum_{p \leq x} \frac{1}{p} = \ln \ln x + B_1 + o(1), \qquad (5.12)$$

where

$$B_1 = \gamma + \sum_{p=2}^{\infty} \left(\ln \left(1 - \frac{1}{p} \right) + \frac{1}{p} \right). \qquad (5.13)$$

After these preliminaries we shall give the following useful theorem, which generalizes the Prime Number Theorem, eq. (2.5) on p. 47:

Theorem 5.1. The number $\pi_k(x)$ of integers $\leq x$ which have exactly k prime factors, all different, is

$$\pi_k(x) \sim \frac{x}{\ln x} \cdot \frac{(\ln \ln x)^{k-1}}{(k-1)!} = \pi_k^*(x), \text{ say}, \qquad (5.14)$$

when $x \to \infty$. We do not prove this theorem, but refer the reader to [3].

We note that with $k = 1$ we recover the old, familiar Prime Number Theorem.—It also follows from the Theorem that integers having many prime factors (so-called round numbers) are very rare. A computation shows that in the vicinity of $N = 10^{100}$ numbers having 15 prime factors or more are very rare indeed, only 0.15% fall in this category.

Remark. The approximations $\pi_k^*(x)$ to $\pi(x)$, implied by (5.14), are not *uniform* for all k, x. This occurs only for small $k = o(\ln \ln x)$ and for k in the neighbourhood of the value $\ln \ln x$, $k = (1 + o(1)) \ln \ln x$.

Another useful theorem, whose proof can also be found in [3], is

Theorem 5.2. The "normal" number of different prime factors of N is about $\ln \ln N$.

This is due to the fact that the function $\pi_k^*(N)$, for a fixed value of N, has its maximal value for $k \approx \ln \ln N$. The theorem is based upon the fact that the average order of $\omega(n)$ is $\ln \ln N$.—We must here define the significance of the word "normal" in the theorem. It means that the number of prime factors of *almost all integers* about N is between

$$(1 - \epsilon) \ln \ln N \quad \text{and} \quad (1 + \epsilon) \ln \ln N,$$

for every $\epsilon > 0$. And *almost all integers* means a fraction of all integers *as close to 1 as we wish*.—Of course, this set of integers will depend on ϵ and on how close to 1 we decide to approach.

How Does a Typical Factorization Look?

From Theorem 5.2, we can draw conclusions about "typical" factorizations of integers. The reader must be warned here that the reasoning is not at all rigorous, but rather is a heuristic argument.—Since an integer of size N has about $\ln \ln N$ prime factors, suppose these are arranged in increasing order of magnitude:

$$N = P_s(N)P_{s-1}(N)\ldots P_2(N)P(N) \text{ with } P_s(N) \le P_{s-1} \le \cdots \le P(N).$$

How large ought $P(N)$ typically to be? How large ought $N/P(N)$ typically be? As $N/P(N)$ has only $s-1$ prime factors, we have

$$s - 1 \approx \ln \ln \frac{N}{P(N)} = \ln(\ln N - \ln P(N)) =$$

$$= \ln \ln N + \ln\left(1 - \frac{\ln P(N)}{\ln N}\right) = s + \ln\left(1 - \frac{\ln P(N)}{\ln N}\right).$$

This leads to the estimation

$$\ln\left(1 - \frac{\ln P(N)}{\ln N}\right) \approx -1 \quad \text{or} \quad 1 - \frac{\ln P(N)}{\ln N} \approx \frac{1}{e},$$

typically giving

$$\ln P(N) \approx \left(1 - \frac{1}{e}\right)\ln N = 0.632 \ln N. \tag{5.15}$$

In the analysis of factorization algorithms (5.15) is frequently re-written in the still less stringent form

$$P(N) \approx N^{0.632}. \tag{5.16}$$

Since $\ln N$ is proportional to the number of digits of N, the following result seems plausible:

"Typically" the largest prime factor $P(N)$ of an integer N has a length in digits being about 63% of the length of N, i.e. $\log P(N) \approx 0.63 \log N$. The second largest prime factor is expected to have a length of 63% of the remainder of N, i.e. $\log P_2(N) \approx 0.23 \log N$, and so on.

Remark. We have to be very careful when applying this result. As a matter of fact, there are hardly any integers to which it applies to the full extent. *But it shows the general tendency.* Take a good look at factor tables of large numbers, and remarkably you will find many instances where the factorizations do indeed look a little as this rule of thumb predicts, e.g.

$$2^{96} + 1 = 641 \cdot 67\,00417 \cdot 18446\,74406\,94145\,84321.$$

The principal value of the rule is, of course, not to predict the general appearance of factor tables, but rather to aid in finding good strategies for factoring large integers. Since in many factoring methods the *size of the second largest prime factor* determines the amount of computing work needed before the integer is completely factored, it is of some help to know that we typically have to expect this factor to be about $N^{0.23}$.—It is also of value to know that integers consisting of many prime factors are very rare, contrary to what is frequently believed.

It is because of the consequences of this rule that, in analyzing Fermat's factoring method on p. 155, we dared to introduce the term "ordinary case" for a number N with its largest factor of about the size $N^{\frac{2}{3}}$, which is close to $N^{0.63}$.

The Erdös-Kac Theorem

A beautiful theorem, from which further information on the distribution of prime factors of large numbers can be drawn, is

Theorem 5.3. The Erdös-Kac Theorem. Denote by $N(x, a, b)$ the number of integers in the interval $[3, x]$ for which the inequalities

$$a \leq \frac{\omega(n) - \ln\ln n}{\sqrt{\ln\ln n}} \leq b \tag{5.17}$$

hold, where $\omega(n)$ is the number of different prime factors of n. Then, as $x \to \infty$,

$$N(x, a, b) = (x + o(x)) \frac{1}{\sqrt{2\pi}} \int_a^b e^{-t^2/2} dt. \tag{5.18}$$

A stringent proof is quite complicated and we have to refer the reader to [4]. A heuristic argument for (5.18) can be found in Kac's own little book [5]. The Erdös-Kac Theorem reflects the fact that the function $\omega(n)$, the number of different prime divisors of n, may be written as a sum of statistically independent functions $\rho_p(n)$, defined by

$$\rho_p(n) = \begin{cases} 1, & \text{if } p \mid n \\ 0, & \text{if } p \nmid n. \end{cases} \tag{5.19}$$

This suggests that the distribution of values of $w(n)$ may be given by the normal law in statistics, which is indeed the case.—Another way of describing this is to say that $\Omega(N)-1$ can be approximated by a normally distributed random variable with mean $\ln \ln N$ and standard deviation $\sqrt{\ln \ln N}$, when N is large. See [6], p. 345–346, where a proof is given that the variable $\Omega(N)-1.03465\ldots$ has the statistical properties just mentioned.

Remark. In theorems 5.1–5.3 it does not really matter if the *total* number $\Omega(N)$ of prime factors of N is considered rather than the number $w(N)$ of *different* prime factors. That the distinction between $\Omega(N)$ and $w(N)$ makes no importance in these theorems follows from the fact that the number of integers $m \leq x$ for which $\Omega(n) - w(n) > (\ln \ln x)^{1/2}$ is $o(x)$, as proved in [3].

A brief account of a more detailed theory, leading to more precise results will be given on pp. 169–171.

The Distribution of Prime Factors of Various Sizes

Finally, we shall prove some theorems on the distribution of various sizes of prime factors of large numbers. We commence by

Theorem 5.4. The prime factors p of numbers chosen at random in a short interval about a sufficiently large number N are such that the variable $\ln \ln p$ is approximately equally distributed. More precisely: the number of prime factors p of integers in the interval $[N - x, N + x]$ such that

$$a < \ln \ln p < b$$

is proportional to $b - a$ if $b - a$ as well as x are sufficiently large as $N \to \infty$.

Proof. Since the presence of multiple prime factors complicates the proof, we shall first assume that every prime factor p of a number M is counted only once, *regardless of its multiplicity* in M. Subsequently, we shall indicate how to modify the proof if, instead, the multiplicity of p is counted.

Now in the interval $[N - x, N + x]$ containing $2x + 1$ integers, there are between

$$\frac{2x + 2}{p} - 1 \quad \text{and} \quad \frac{2x}{p} + 1$$

multiples of p, dependent on where these multiples happen to occur. This is $2x/p + O(1)$. Thus, the number of events of a prime p between u and v occurring as a factor of one of the numbers in $[N - x, N + x]$ becomes

$$f(u, v, x) = \sum_{u \leq p \leq v} \left(\frac{2x}{p} + O(1) \right). \tag{5.20}$$

167

Using (5.8), we immediately find

$$f(u, v, x) \sim 2x(\ln \ln v - \ln \ln u), \tag{5.21}$$

if x is supposed to be sufficiently large, compared with u and v. Substituting $\ln \ln u = a$ and $\ln \ln v = b$, we have $a < \ln \ln p < b$ and $f(u, v, x)$ becomes $\sim 2x(b - a)$, thus proving the theorem.

We have promised the reader to indicate how this proof could be adapted to the case where each prime factor is counted by its multiplicity. In this situation the number of prime factors p of numbers in $[N - x, N + x]$ is no longer $2x/p + O(1)$, but rather

$$\frac{2x}{p} + \frac{2x}{p^2} + \frac{2x}{p^3} + \cdots + O\left(\frac{\ln N}{\ln p}\right). \tag{5.22}$$

Since $2x \sum_p p^{-s} < 2x \sum_n n^{-s}$ is convergent for $s > 1$ while $\sum_p 2x/p$ is divergent—as can be seen by taking $x \gg v$ in the asymptotic formula for $f(u, v, x)$ just found—the new terms in the series (5.22) cannot replace $2x \ln \ln v$ as the dominating term of the result. The term $O(\ln N / \ln p)$, however, gives rise to an error term of the order of magnitude of

$$O(\ln N) \int_{u-0}^{v+0} \frac{d\pi(t)}{\ln t} = \frac{O(\ln N)}{\ln \xi} \int_{u-0}^{v+0} d\pi(t) = \frac{\ln N}{\ln \xi} O(v - u). \tag{5.23}$$

Here $u < \xi < v$ and we have used the mean value theorem for Stieltjes integrals (Theorem A9.3 on p. 365). The expression (5.23) surpasses the previous error term by a factor of $\ln N / \ln \xi$ in magnitude. Therefore in this case x must be just a little larger than previously in order for this new error term to be small compared with the dominant term.

Dickman's Version of Theorem 5.4

A related result, due to Dickman, see [7], is the following:

Theorem 5.5. Dickman's Theorem. The probability of a number N chosen at random having a prime factor between N^α and $N^{\alpha(1+\delta)}$ is approximately $= \delta$, independent of the magnitude of α, if δ is small.

To see that Dickman's theorem is equivalent to Theorem 5.4 above, consider the number of prime factors of a number N chosen at random in the interval $\left[N^\alpha, N^{\alpha(1+\delta)}\right]$ which, according to Theorem 5.4 is

$$\ln\ln N^{\alpha(1+\delta)} - \ln\ln N^\alpha = \ln(1+\delta) \approx \delta,$$

i.e. *independent of* α. Thus, the length of the interval $\left[N^\alpha, N^{\alpha(1+\delta)}\right]$ is such that the average number of prime factors of numbers about N, falling in this interval, will be precisely δ, showing that Theorem 5.4 is only a different formulation of Dickman's result.

For future reference we also give the following slight variation of Theorem 5.2:

Theorem 5.6. The "typical" number of prime factors $\leq x$ in sufficiently large numbers N is about $\ln\ln x$.

Proof. Just as theorem 5.2 follows from the fact that the average order of $\omega(n)$ for the interval $[3, N]$ is $\ln\ln N$, the present Theorem is equivalent to the fact that the average order of $\omega(n)$ in $[3, x]$ is $\ln\ln x$.

Exercise 5.2. Statistics on prime factors. Collect some statistics on the size of the prime factors p of large numbers N by counting the number of cases for which $\ln\ln p$ falls in various intervals of length 0.2, e.g. $0.2k < \ln\ln p < 0.2(k+1)$, $k = 1, 2, 3, \ldots, 10$. In producing these statistics, utilize the computer program for factorization by trial division, mentioned on pp. 149–150, to factor 1000 consecutive integers of the size about 10^6, 10^7 and 10^8. Compare your results with those predicted by Theorem 5.4.

A More Detailed Theory

The theory sketched above for the distribution of the prime factors of large numbers is very simple and yields sufficiently accurate results to be useful in estimating running times of various factorization algorithms. There exists, however, a more detailed and slightly more precise theory, given in [6], by which the average size of the logarithm of the largest, second largest, third largest etc. prime factors of N have been calculated rigourously instead of intuitively, yielding the mean value of $\ln P(N)/\ln N$ and $\ln P_2(N)/\ln N$ as 0.624 and 0.210 for the largest and second largest prime factor, respectively, as compared to our values 0.632 and 0.233. This more accurate theory is, however, much more elaborate and complicated, and we can only briefly touch upon it here and must refer the reader to [6] for a more detailed account.

The Size of the kth Largest Prime Factor of N

The problem of determining the distribution of the kth largest prime factor of large numbers N is, as is often the case with mathematical problems, very much a question of posing the "correct" problem, i.e. of formulating a problem which admits a reasonably simple solution. In this case, the correct question to ask is: What is the proportion $\rho_k(\alpha)$ of numbers whose k^{th} largest prime factor $P_k(N)$ is smaller than $N^{1/\alpha}$?

It turns out that the functions $\rho_k(\alpha)$ can be expressed in terms of the so-called polylogarithm functions $L_k(\alpha)$, defined recursively by

$$L_0(\alpha) = \begin{cases} 0, & \text{for} \quad \alpha \leq 0 \\ 1, & \text{for} \quad \alpha > 0 \end{cases}$$

$$L_k(\alpha) = \begin{cases} 0, & \text{for} \quad \alpha \leq k \\ \displaystyle\int_k^{\alpha} L_{k-1}(t-1)\frac{dt}{t}, & \text{for} \quad \alpha \geq k. \end{cases} \tag{5.24}$$

It is easy to see that the first two of the functions $L_k(\alpha)$ are

$$L_1(\alpha) = \begin{cases} 0, & \text{for} \quad \alpha \leq 1 \\ \ln \alpha, & \text{for} \quad \alpha \geq 1 \end{cases} \tag{5.25}$$

and

$$L_2(\alpha) = \int_2^{\alpha} \frac{\ln(t-1)\,dt}{t}, \quad \text{for} \quad \alpha \geq 2. \tag{5.26}$$

Now, the functions $\rho_k(\alpha)$ can be expressed by means of the relations

$$\begin{aligned} 1 - \rho_1(\alpha) &= L_1(\alpha) - L_2(\alpha) + L_3(\alpha) - L_4(\alpha) + L_5(\alpha) - \cdots \\ 1 - \rho_2(\alpha) &= \qquad\quad L_2(\alpha) - 2L_3(\alpha) + 3L_4(\alpha) - 4L_5(\alpha) + \cdots \\ 1 - \rho_3(\alpha) &= \qquad\qquad\qquad\quad\; L_3(\alpha) - 3L_4(\alpha) + 6L_5(\alpha) - \cdots \\ &\;\;\vdots \end{aligned} \tag{5.27}$$

$$1 - \rho_k(\alpha) = \sum_{n=0}^{\infty} \binom{-k}{n} L_{n+k}(\alpha).$$

The function $\rho_2(\alpha)$ is of particular interest in factorization. This is because of the existence of a whole class of factorization algorithms which yield the prime factors of a number N in approximate order of magnitude.

If factors discovered are always removed, and the co-factors tested for primality, then such algorithms will terminate *as soon as the second largest factor has been found.* Thus, the distribution of the size of the second largest prime factor of large numbers N is essential when the average running time of these algorithms is to be estimated.—We conclude this sub-section by providing a small table of the functions $\rho_1(\alpha)$ and $\rho_2(\alpha)$ as well as a graph of the distribution functions $F_1(x)$, $F_2(x)$ and $F_3(x)$ for the largest, the second largest and the third largest prime factors, respectively. Both the table and the graph are reproduced from [6].

Distribution of largest and second largest prime factors		
α	$\rho_1(\alpha)$	$\rho_2(\alpha)$
1.5	0.59453 48919	1.00000 00000
2.0	0.30685 28194	1.00000 00000
2.5	0.13031 95618	0.95338 97063
3.0	0.04860 83883	0.85277 93230
3.5	0.01622 95932	0.73348 11652
4.0	0.00491 09256	0.62368 10600
4.5	0.00137 01177	0.53365 25720
5.0	0.00035 47247	0.46322 21870
6.0	0.00001 96497	0.36521 77517
7.0	0.00000 08746	0.30178 60103
8.0	0.00000 00323	0.25743 57108
9.0	0.00000 00010	0.22459 21627

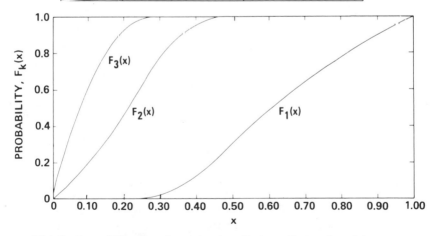

Distribution of the three largest prime factors of a random integer.
For example, the probability that the largest factor $P(N) < N^{0.44}$ is 20%.

Pollard's $(p-1)$-Method

J. M. Pollard's $(p-1)$-method, published in [8], formalizes several rules, which have been known for some time. The principle here is to use information concerning the order of some element a of the group M_N of primitive residue classes (mod N) to deduce properties of the factors of N. (See Appendix 1 for details of the structure of M_N!) In this method the idea employed is that if $p-1|Q$ then $p|a^Q-1$, if $(Q,p)=1$. The assertion follows immediately from Fermat's Theorem A2.8 on p. 270. Since $a^{p-1} \equiv 1$ (mod p), also $a^Q \equiv 1$ (mod p), and thus p can be found by applying Euclid's algorithm to a^Q-1 (mod N) and N.

Well, how can this observation be of any value in factoring N? Would it not work equally well to divide N by all primes up to p, as to check if $(a^Q-1, N) > 1$ for all Q up to $p-1$? Yes, it would certainly be much faster to perform the trial divisions, but the point is that *if we can use only a few probable values* of Q, then we may gain advantage by carrying out the other computation. It is here that Pollard's idea comes into play: suppose that $p-1$, for *one of the factors p* of N, happens to have a factorization *containing only small prime factors*; then if we compute (a^Q-1, N) on these comparatively rare occasions, i.e. for those integers Q which have many prime factors, we might be able to determine $p-1$ (or a multiple thereof) rather soon, and thus find a relatively large factor p of N with a limited amount of work.

Now, the original problem has been replaced by the problem of generating (multiples of) all integers containing only small divisors. This is actually quite easy. You could use the numbers $Q_k = k!$, which contain all primes $\leq k$, but then the small primes would be over-represented. It is better to utilize $Q_k = p_1 p_2 \ldots p_k$ with some additional factors of the small primes appended, in order not to miss such $p-1$ which are divisible by 4, 8, 9 or other such small prime powers. This can be achieved by multiplying Q by an extra factor p each time a power of p is passed, when multiplying by the primes p_k.

To arrive at a practical algorithm, proceed as follows: First, generate a list of all primes and prime-powers up to some search limit G, say 100,000. For each prime square, prime cube, etc., write the corresponding prime instead of the prime power in question. Divide this list into blocks of suitable length for the fast memory of your computer, for instance let a block contain all entries in an interval of length 1000.

Next, choose a value of a, e.g. $a = 13$, and compute

$$b_{i+1} \equiv b_i^{p_i} \pmod{N}, \tag{5.28}$$

where p_i is the ith integer in your list of prime-powers. Start the sequence (5.28) with $b_1 = a$. Finally, compute also the accumulated product

$$Q_n \equiv \prod_{i=1}^{n} (b_i - 1) \pmod{N}, \tag{5.29}$$

and check (Q_n, N) periodically, to see whether a factor p of N has emerged, e.g. at regular intervals of 100 cycles.

How soon will Pollard's $(p-1)$-algorithm determine a factor of N? Suppose as usual that

$$N = \prod_i p_i^{\alpha_i}, \quad \text{and that} \quad p_i - 1 = \prod_j q_{ij}^{\beta_{ij}}. \tag{5.30}$$

Let the largest prime-power in the factorization of $p_i - 1$ be q^β. Then the factor p_i will be obtained as soon as the computation has passed the value q^β in the list of prime-powers used. This means that the factor p_i of N, for which the value of q^β is the smallest for all the factors p of N, appears first, and occasionally very large prime factors may be rapidly detected by this method. As examples of this we refer to [9], in which the factors

$$p = 1214505\ 06296081 \ \text{ of } 10^{95} + 1$$

and

$$q = 267\ 00917351\ 08484737 \ \text{ of } 3^{136} + 1$$

are reported to have been found by the $(p-1)$-algorithm. The factorizations of $p - 1$ and $q - 1$ are

$$p - 1 = 2^4 \cdot 5 \cdot 13 \cdot 19^2 \cdot 15773 \cdot 20509$$

$$q - 1 = 2^7 \cdot 3^2 \cdot 7^2 \cdot 17^2 \cdot 19 \cdot 569 \cdot 631 \cdot 23993.$$

Since all prime factors of $p-1$ and $q-1$ are small, this explains the success of Pollard's $(p-1)$-method in these cases.

Remark. Even if $p_i - 1$ has only small *prime* factors it might still happen that p_i is not found because it contains an extremely large power, exceeding the search limit, of a small prime.—Also, N might have *several* prime factors p_i, all detectable by this method, and it may so happen that two or more of these factors are extracted between two consecutive computations of (Q_n, N). In such a case these factors will appear multiplied together. They can be separated, in most situations, if the last values of Q_{100k} and b_{100k} have been saved. Then one can *recompute from this point on, only with more frequent use of Euclid's algorithm.* Only in the rare case when N has such special structure that several factors are found at the same value of i, is the computation a failure.

The $(p+1)$-Method.

A variant of this method, which uses Lucas sequences instead of powers and achieves a rapid factorization if some factor p of N has a decomposition of $p+1$ in only small prime factors, has been implemented by Williams and is described in [9]. A problem associated with this method is that of finding, in an efficient manner, a quadratic non-residue such that all the conditions for a Lucas-sequence that will work, can be met. The far from obvious solution is due to Brent and Kahan, for which we must refer the reader to Williams' paper [9]. Using this algorithm the factor

$$p = 22597\ 40655\ 03889 \text{ of } 10^{102}+1$$

was computed. Here

$$p+1 = 2 \cdot 5 \cdot 11 \cdot 79 \cdot 401 \cdot 7867 \cdot 8243.$$

Pollard's ρ-Method

This method is also called Pollard's second factoring method. It is based on a "statistical" idea [10] and has been refined by Richard Brent [11]. The ideas involved for finding the factor p of the number N are described below:

1. Construct a sequence of integers $\{x_i\}$ which is periodically recurrent (mod p).
2. Search for the period, i.e. find i and j such that $x_i \equiv x_j$ (mod p).
3. Identify the factor p of N.

The first requirement, that of finding a periodic sequence (mod m), where m is an arbitrary integer, is quite easy to fulfil. Consider any recursively defined sequence of the following type (s is assumed to be a constant, i.e. independent of i, and F is a polynomial):

$$x_i \equiv F(x_{i-1}, x_{i-2}, \ldots, x_{i-s}) \pmod{m} \tag{5.31}$$

with given initial values for $x_1, x_2, \ldots, x_s$. Then $x_{s+1}, x_{s+2}, \ldots$ can be computed successively by the formula given. However, since all the x_k's are given (mod m), there are only m different values that each x_k can take and thus there are, at most, m^s distinct sequences $x_{i-1}, x_{i-2}, \ldots, x_{i-s}$

of s consecutive numbers x_k. Thus, after, at most, $m^s + 1$ steps in the recursion, two identical sequences of s consecutive numbers must have occurred. Let us call these $x_{i-1}, x_{i-2}, \ldots, x_{i-s}$ and $x_{j-1}, x_{j-2}, \ldots, x_{j-s}$. Now, since the definition of the next x_k uses only the preceding s values $x_{k-1}, x_{k-2}, \ldots, x_{k-s}$, it is clear that, if these sequences of values are identical for two different values of k, then the values x_i and x_j, computed from these in a similar manner will also be the same. Thus, we have two new sequences of s identical values, viz. $x_i, x_{i-1}, \ldots, x_{i+1-s}$ and $x_j, x_{j-1}, \ldots, x_{j+1-s}$ which lead to identical x_{i+1} and x_{j+1}, and so on. But this means that the sequence $\{x_i\}$ is periodically repeated, except possibly for a part at the beginning called the aperiodic part.

As a sequel to this somewhat abstract reasoning, let us look at a simple example, the Fibonacci sequence (mod 11). This sequence is defined by

$$x_i \equiv x_{i-1} + x_{i-2} \ (\text{mod } 11), \quad \text{with } x_1 \equiv x_2 \equiv 1.$$

We successively obtain the following elements of the sequence

$$1, 1, 2, 3, 5, 8, 2, 10, 1, 0, 1, 1, 2, 3, \ldots \ (\text{mod } 11).$$

After 10 elements the sequence is repeated. Since this particular sequence is repeated from the very beginning, it has no aperiodic part.

We now proceed to the second step of the algorithm, the search for the period. To determine it in the most general case would require finding where a sequence of consecutive elements is repeated if the period is long. This is quite a formidable task and is ruled out by the large amount of labour involved. In the simplest case, however, where x_i is defined by means of x_{i-1} only and by no other x_k's, the sequence is periodically repeated as soon as any single element x_j is the same as a previous one. Therefore, this case requires only a comparison of each new x_j with the previous x_i's to find the period. If the period is very long (several million elements), however, it is not feasible to save all the elements and to compare them pairwise. Instead, the following technique can be used:

Suppose the periodic sequence $\{x_i\}$(mod m) has an aperiodic part of length a and a period of length l. The period will then ultimately be revealed by the test: Is $x_{2i} \equiv x_i \ (\text{mod } m)$? This is called *Floyd's cycle-finding algorithm*.

Proof. First, if $x_{2i} \equiv x_i \pmod{m}$ then the sequence is obviously periodic from x_{2i} onwards, possibly even earlier. Conversely, for any periodic sequence with period-length l, $x_j \equiv x_i \pmod{m}$ for $j = i + kl$, $k = 1, 2, 3, \ldots$ and all $i > a$ (i.e. for all elements following the aperiodic part with subscripts differing by a multiple of l), there will eventually be an i with $x_{2i} \equiv x_i \pmod{m}$. The first such value of i is $i = (l+1)[a/l]$. If $a > l$, then this search will reveal the period only after several complete periods have passed, but nevertheless the periodicity of the sequence will finally be detected.

Now, how can x_{2i} be compared with x_i without saving all x_i's? Simply by recomputing the x_i's in parallel with the x_{2i}'s. Suppose that $x_{i+1} = f(x_i)$. Then this period-finding algorithm may be described by the following computer program code:

```
x:=x1; y:=f(x1);
WHILE x<>y DO
  BEGIN
    x:=f(x); y:=f(y); y:=f(y)
  END;
{When arriving here x=y and the period has been run through!}
```

Finally, consider the third and last requirement of Pollard's ρ-method. If we have a sequence $\{x_i\}$ that is periodic $\pmod{p}$, by which means can we find p, the unknown factor of N? In the same manner as we did in the $(p-1)$-method, simply by using Euclid's algorithm for finding the greatest common divisor d of $x_{2i} - x_i \pmod{N}$ and N. Normally d will turn out to be 1, but as soon as $x_{2i} \equiv x_i \pmod{p}$, d will be divisible by p.

Following these preliminaries, we are now in a position to discuss what an efficient factor searching algorithm based on these ideas ought to be like. Firstly, the sequence $\{x_i\}$ should be very easy to calculate (especially because it must be calculated twice!). Secondly, the period length (or rather the number of steps in Floyd's cycle-finding algorithm) should be small. Thirdly, the use of Euclid's algorithm should be organized in an efficient manner, so as not to use up too much computing time in merely calculating the G.C.D. $(N, x_{2i} - x_i \pmod{N}) = 1$.

Pollard found that in a sequence $\{x_i\}$ of random integers $\pmod{p}$ an element is usually recurring after only about $C\sqrt{p}$ steps. This is easy to understand if you consider the solution of the so-called *birthday problem*:

How many persons need to be selected at random, in order that the probability of at least two of them having the same birthday, exceeds $\frac{1}{2}$?

Solution. The probability that q persons all have different birthdays is

$$\left(1 - \frac{1}{365}\right)\left(1 - \frac{2}{365}\right)\left(1 - \frac{3}{365}\right) \cdots \left(1 - \frac{q-1}{365}\right)$$

This expression is < 0.5 when $q \geq 23$.

Generalization. How large must q be, in order that at least two randomly chosen integers out of q will be congruent (mod p) with probability $> \frac{1}{2}$?

Obviously this will be the case if

$$\left(1 - \frac{1}{p}\right)\left(1 - \frac{2}{p}\right)\left(1 - \frac{3}{p}\right) \cdots \left(1 - \frac{q-1}{p}\right) < \frac{1}{2}$$

The left-hand-side is

$$\approx \left(1 - \frac{q}{2p}\right)^{q-1} \approx e^{-q(q-1)/2p}.$$

This expression is $= 0.5$, if

$$\frac{q(q-1)}{2p} = \ln 2, \quad \text{i.e. if} \quad q \approx \sqrt{2p \ln 2} + 0.5 \approx 1.18\sqrt{p}.$$

At this point we can describe Pollard's factor-searching algorithm. Instead of random integers $\{x_i\}$, we must recursively compute what is called a sequence of pseudo-random integers. The simplest choice would be to select $x_{i+1} \equiv a x_i \pmod{p}$ for a fixed value of a. It turns out, however, that this choice does *not* produce numbers that are sufficiently random to give a short period of only $C\sqrt{p}$ steps for $\{x_i\}$. The next simplest choice is to take a quadratic expression, say

$$x_{i+1} \equiv x_i^2 + a \pmod{p}. \tag{5.32}$$

It is an empirical observation that (5.32) possesses the required properties (at least if a is neither 0 nor -2) but this has not been proved so far.

How shall we effect the search for p with Euclid's algorithm on $x_{2i} - x_i \pmod{N}$ and N in each cycle? Once again, the same trick we used in the $(p-1)$-method will do: accumulate the product

$$Q_i \equiv \prod_{j=1}^{i} (x_{2j} - x_j) \pmod{N}, \tag{5.33}$$

and apply Euclid's algorithm only occasionally, e.g. when i is a multiple of 100. In this way, the burden of using Euclid's algorithm is, in practice, reduced to one multiplication and one reduction $\pmod{N}$ per cycle.

A Computer Program for Pollard's ρ-Method

The following is an outline of a computer program for what we have so far discussed

```
PROGRAM Pollard
(Input,Output);
LABEL 1,2;
VAR a,x1,x,y,Q,i,p,N : INTEGER;

FUNCTION Euclid(a,b : INTEGER) : INTEGER;
{Computes (a,b) with Euclid's algorithm}
VAR m,n,r : INTEGER;
BEGIN m:=a; n:=b;
   WHILE n <> 0 DO BEGIN r:=m MOD n; m:=n; n:=r END;
   Euclid:=m
END {Euclid};

BEGIN
1: write('Input a<>0,2 and x1: '); read(a);
   IF a=0 THEN GOTO 2;
   read(x1); x:=x1; y:=x1; Q:=1;
   write('Input N for factorization: '); read(N);
   FOR i:=1 TO 10000 DO
      BEGIN
         x:=(x*x-a) MOD N; y:=(y*y-a) MOD N; y:=(y*y-a) MOD N;
         Q:=Q*(y-x) MOD N;
         IF i MOD 20 = 0 THEN
            BEGIN p:=Euclid(Q,N);
               IF p>1 THEN WHILE N MOD p = 0 DO
```

```
        BEGIN
          writeln('p=',p:8,' found for i=',i:4);
          N:=N DIV p;
          {Here a factor of N is found and divided out}
          IF N=1 THEN GOTO 1
        END
      END
    END;
  writeln('No factor found in 10,000 cycles');
  GOTO 1;
2: END.
```

Note that the algorithm can fail! If N has several factors, it could happen that some of these are extracted between two consecutive computations using Euclid's algorithm precisely as in the $(p-1)$-method. Exactly as in that case we have to save the latest values of Q_{100k}, x_{100k} and x_{200k} and rerun the computation from this point on with a more frequent use of Euclid's algorithm. If also this fails the whole algorithm has to be rerun with a different value of a.

Note also that the above PASCAL program is only a model of what a computer code implementing Pollard's method could be! It certainly works, but only for small values of N, that is values for which N^2 is less than the largest integer that can be stored in a computer word. In order to transform this model to a program of real life, you must use at least double precision arithmetic or, better still, a package for multiple precision arithmetic so that the multiplications do not cause arithmetic overflow. Moreover it is advantageous not to perform the computations of the G.C.D. using Euclid's algorithm at equidistant points, but rather using a smaller interval at the beginning, and then successively letting this interval grow to 100 or more. This is in order not to obtain all the small factors of N multiplied together at some stage.

Since the labour involved in obtaining a factor p is proportional to $\sqrt{p}$, it is essential to use Pollard's algorithm only on numbers which are known to be composite.

Example. Factor 91643 with $x_{i+1} = x_i^2 - 1$, $x_0 = 3$.

$$x_1 = 8, \qquad x_2 = 63, \qquad y_1 = 63 - 8 = 55, \qquad (55, N) = 1$$
$$x_3 = 3968, \qquad x_4 \equiv 74070, \qquad y_2 \equiv x_4 - x_2 \equiv 74007, \qquad (74007, N) = 1$$

179

$$x_5 \equiv 65061, \quad x_6 \equiv 35193, \quad y_3 \equiv x_6 - x_3 \equiv 31225, \quad (31225, N) = 1$$
$$x_7 \equiv 83746, \quad x_8 \equiv 45368, \quad y_4 \equiv x_8 - x_4 \equiv 62941, \quad (62941, N) = 113.$$

Factorization of N: $N = 113 \cdot 811$.

Using this method, the author has found many factorizations, such as

$$11^{23} - 6^{23} = 5 \cdot 7345513129 \cdot 8198620554281$$
$$3^{53} + 2^{53} = 5 \cdot 107 \cdot 24574686173 \cdot 1474296171913$$
$$6^{29} - 5^{29} = 8212573801 \cdot 4463812400371$$
$$11^{23} + 4^{23} = 3 \cdot 5 \cdot 47 \cdot 12172517977 \cdot 104342735227$$
$$11^{25} - 4^{25} = 7 \cdot 22861 \cdot 2603070851 \cdot 260198475451.$$

Exercise 5.3. Pollard's ρ-method. Write the necessary double precision arithmetic in order to be able to implement the PROGRAM Pollard above on your computer. Explore the possibilities of the method on some single precision integers N. Compare the running times with those of the trial division program on p. 150.

An Algebraic Description of Pollard's ρ-Method

There are fairly simple algebraic arguments to show why a factor p of N is found after about $O(\sqrt{p})$ cycles in Pollard's ρ-method. The argument runs as follows:

$$y_i = x_{2i} - x_i = x_{2i-1}^2 + a - (x_{i-1}^2 + a) = x_{2i-1}^2 - x_{i-1}^2 =$$
$$= (x_{2i-1} + x_{i-1})(x_{2i-1} - x_{i-1}) =$$
$$= (x_{2i-1} + x_{i-1})(x_{2i-2} + x_{i-2})(x_{2i-2} - x_{i-2}) = \qquad (5.34)$$
$$\vdots$$
$$= (x_{2i-1} + x_{i-1})(x_{2i-2} + x_{i-2}) \ldots (x_i + x_0)(x_i - x_0).$$

Thus, the factor y_i, which is included in the product Q_i for the computation of (Q_i, N), contains at least $i + 1$ algebraic factors. How many different prime factors $\leq p$ does a typical factor, $x_k - x_l$, contain? From Theorem 5.6 on p. 169 we expect the number of prime factors $\leq G$ of a number N to be about $\ln \ln G$, if N is sufficiently large. Now, what can be said about the size of $x_k - x_l$? The numbers x_k grow extremely fast—their *number of digits* is doubled from one k to the next because a squaring is performed in the recursion formula $x_{k+1} = x_k^2 + a$. Hence x_k will be of the order of magnitude $x_0^{2^k}$, and thus will exceed the critical limit for being "sufficiently large" in the application of Theorem 5.6 very rapidly. So we find the expected number of prime factors $\leq G$ of y_i, which is a product of $i + 1$ very large numbers, to be $\approx (i + 1) \ln \ln G$. Running Pollard's ρ-algorithm for n cycles, we accumulate in Q_n the primes of all the factors $y_1, y_2, \ldots, y_n$ which together can be expected to include

$$\ln \ln G \sum_{i=1}^{n} (i + 1) \approx 0.5 n^2 \ln \ln G$$

prime factors $\leq G$. How far must we proceed in order to ensure that *all* primes below some factor p of N have been accumulated in Q_n? (This limit will secure success in the search for factors, regardless of the particular value of p.) Below p there are $\pi(p) \approx p/\ln p$ primes, and therefore n ought to be so large that, by choosing some suitable factor C, $C \cdot 0.5 n^2 \ln \ln p$ attains the magnitude of $\pi(p)$ (the factor C allowing for some of the primes appearing in several of the factors of Q_n as well as taking into account possible bad luck in the pseudo-random process we use for generating the small primes):

$$C \cdot 0.5 n^2 \ln \ln p \approx \frac{p}{\ln p},$$

i.e.

$$n = \text{const.} \times \left(\frac{p}{\ln p \; \ln \ln p} \right)^{\frac{1}{2}}. \tag{5.35}$$

Thus the order of magnitude of the number of cycles necessary to discover a factor p with Pollard's method actually emerges slightly better than $C\sqrt{p}$. Now, since experience seems to indicate that Pollard's method *is* $C\sqrt{p}$, and not slightly better, we therefore conclude that the factor C we introduced, is in fact not a constant, but changes very slowly with p.

This algebraic model of Pollard's method is somewhat crude, but nevertheless we shall use it later on when we study some modifications of the method, and also in discussing the very important question: *How fast can a factorization algorithm be?*

Brent's Modification of Pollard's Method

The ρ-method of Pollard has been made about 25% faster by a modification due to Brent [11]. Pollard searched for the period of the sequence x_i (mod p) by considering $x_{2i} - x_i$ (mod p) (Floyd's cycle-finding algorithm).

Instead, Brent halts x_i when $i = 2^k$ is a power of 2 and subsequently considers $x_j - x_{2^k} \pmod{p}$ for $3 \cdot 2^{k-1} < j \le 2^{k+1}$. The values computed in Brent's method are:

$$
\begin{array}{ll}
x_1, x_2, & x_2 - x_1, \\
x_3, x_4, & x_4 - x_2, \\
x_5, x_6, x_7, & x_7 - x_4, \\
x_8, & x_8 - x_4, \\
x_9, x_{10}, x_{11}, x_{12}, x_{13}, & x_{13} - x_8, \\
x_{14}, & x_{14} - x_8, \\
x_{15}, & x_{15} - x_8, \\
x_{16}, & x_{16} - x_8, \\
x_{17}, x_{18}, \ldots, x_{25}, & x_{25} - x_{16}, \\
x_{26}, & x_{26} - x_{16}, \\
x_{27}, & x_{27} - x_{16}, \\
x_{28}, & x_{28} - x_{16}, \\
\text{etc.} &
\end{array}
\tag{5.36}
$$

In this way a period of length l is discovered after fewer arithmetic operations than demanded by the original algorithm of Pollard. The saving in Brent's modification stems from not needing to compute the lower x_i's twice as in Floyd's algorithm.

Alternatively, we could explain the economization by applying the algebraic model introduced above. Analogously to (5.34) we have

$$
x_j - x_i = (x_{j-1} + x_{i-1})(x_{j-2} + x_{i-2}) \ldots (x_{j-i} + x_0)(x_{j-i} - x_0). \tag{5.37}
$$

The number of algebraic factors of $x_j - x_i$ is $i+1$, precisely as in the analysis of Pollard's original algorithm for $x_{2i} - x_i$. Since the numbers involved are very large, we know that the number of prime factors $\le p$ is proportional to the number of algebraic factors in Q_i. If in the course of Pollard's and Brent's algorithm we compare the number of algebraic factors arrived at versus the number of "heavy" operations performed, viz. multiplication of two numbers of size N, followed by a reduction $\pmod{N}$, we obtain the values given in the table below.

Pollard		Brent	
# op:s	# factors	# op:s	# factors
4	2	3	2
8	5	6	5
12	9	10	10
16	14	12	15
20	20	18	24
24	27	20	33
28	35	22	42
32	44	24	51
36	54	34	68
40	65	36	85
$\vdots$	$\vdots$	$\vdots$	$\vdots$
$\vdots$	$\vdots$	$\sim \frac{4}{3}i$ (worst)	$\sim \frac{2}{27}i^2$
$4i$	$\sim \frac{1}{2}i^2$	$\frac{3}{2}i$ (best)	$\sim \frac{1}{6}i^2$

In Brent's version the efficiency is dependent upon the value of i at which p is discovered. The worst possible case is when $i = 3 \cdot 2^{k-1} + 1$, immediately following a long run of computing new x_i's only and not testing for factors. The best case is when $i = 2^{k+1}$, when many new factors have been accumulated in Q_i with comparatively little effort. Analyzing the situation, we arrive at the figures indicated as (worst) and (best) respectively, in the table shown above. Considering the cost/efficiency function, measured by the number of operations required in order to accumulate a certain number of factors in Q_i, we find that this function can be calculated as # operations/$\sqrt{\text{# factors}}$, giving the values

$$4\sqrt{2} = 5.66 \text{ for Pollard's method}$$

and a number between

$$\frac{4}{3}\sqrt{\frac{27}{2}} = 4.90 \text{ and } \frac{3}{2}\sqrt{6} = 3.67$$

for Brent's modified algorithm. This explains why Brent's modification runs approximately 25% faster (note that 75% of 5.66 is 4.24, and the mean value between 4.90 and 3.67 is 4.29).

Searching for Factors of Certain Forms

It is sometimes known that the factors of a composite number are of a certain mathematical form. The number may, e.g. have the form $N = 6a^2 - b^2$ with $(a, b) = 1$, and hence we know that 6 is a quadratic residue of N, implying that all prime factors of N (if N can be factorized) take either the form $p = 24k \pm 1$ or $p = 24k \pm 5$. (See Table 32.) Several of the methods for factor search can quite easily be modified to take advantage of such a situation. In the method of trial division for instance, it is quite simple to generate trial divisors of a given linear form and no others. Just set the initial values p1:=-5; p2:=1; d1:=14; and d2:=2; and in the division loop write

$$\text{d1:=24-d1; d2:=24-d2; p1:=p1+d1; p2:=p2+d2;}$$

subsequently checking whether N is divisible, first by p1 and then by p2.

Exercise 5.4. An alternative to the above construction is to write *four* division statements in the program loop. Try this (compare with the program Trial division on p. 150).

Later we shall briefly hint at how several of the more important factorization methods can be modified to use shortcuts when the factor has the form $p = 2kn + 1$ and where n is a given integer, and $k = 1, 2, \ldots$ To indicate the importance of this case, we give

Legendre's Theorem for the Factors of $N = a^n \pm b^n$

It has long been known that, under certain conditions, the prime factors of $N = a^p \pm b^p$ (p being an odd prime), all have the form $2kp + 1$. This result has been generalized by Legendre, who proved

Theorem 5.7. Legendre's Theorem. All prime factors p of the number $N = a^n \pm b^n$, with $(a, b) = 1$, have the form $p = kn + 1$, apart from those which divide the algebraic factors of the form $a^m \pm b^m$, $m < n$, of N (see Appendix 6). Moreover, disregarding the prime $p = 2$ and the case $a^{2n} - b^{2n}$, the prime factors of N take the form $p = 2kn + 1$.

Proof. Suppose $a^n \pm b^n \equiv 0 \pmod{p}$, where p is prime and $(a, b) = 1$. First we deduce that $(a, p) = (b, p) = 1$, because if $a \equiv 0 \pmod{p}$, then the congruence would give $b \equiv 0 \pmod{p}$, and thus, contrary to the assumption, (a, b) would be divisible by p. Since p is a prime, a, b as well as a/b are in the field $\mathbf{Z}_p$ of primitive residue classes $\pmod{p}$. Thus, the congruence may be divided by b^n to give

$$\left(\frac{a}{b}\right)^n \equiv x^n \equiv \mp 1 \pmod{p},$$

with x a number in the field $\mathbf{Z}_p$. Suppose also that x belongs to the exponent $d \pmod{p}$. We know from Fermat's theorem that $d | p-1$, so we can put $d = (p-1)/k$. But $x^n \equiv \pm 1 \pmod{p}$ then implies that $n = l \cdot d = l(p-1)/k$, in the case $N = a^n - b^n$, and that $n = (2l-1)d/2 = (2l-1)(p-1)/(2k)$ in the case $N = a^n + b^n$. Here l is a positive integer. (In the latter case, d must obviously be an even number.) Hence, we immediately have

$$x^d \equiv \left(\frac{a}{b}\right)^d \equiv 1 \pmod{p}, \quad \text{i.e.} \quad a^d - b^d \equiv 0 \pmod{p}$$

and

$$x^{\frac{d}{2}} \equiv \left(\frac{a}{b}\right)^{\frac{d}{2}} \equiv -1 \pmod{p} \quad \text{i.e.} \quad a^{\frac{d}{2}} + b^{\frac{d}{2}} \equiv 0 \pmod{p}$$

respectively. If $l > 1$, i.e. if $d < n$, or $d/2 < n$, then the algebraic factor $a^d - b^d$ of $a^n - b^n$ or $a^{d/2} + b^{d/2}$ of $a^n + b^n$ contains p as prime factor, a case that has been excluded from the theorem. Thus, l must be $= 1$ and n must be $= (p-1)/k$ or $= (p-1)/(2k)$, in the respective cases, and thus we finally arrive at the result

$$\begin{cases} p = kn + 1 & \text{in the case} \quad N = a^n - b^n \\ p = 2kn + 1 & \text{in the case} \quad N = a^n + b^n. \end{cases}$$

If n is odd in the expression $kn+1$, then k must be even in order for $kn+1$ to be odd. Therefore, excepting the prime $p = 2$ and the case $a^{2n} - b^{2n} = (a^n - b^n)(a^n + b^n)$, the prime factors of $a^n \pm b^n$ have the form $2kn+1$.— The requirement of excluding those factors which are already factors of the algebraic factors of N does not reduce the practical applicability of the theorem. The algebraic factors can be considered in advance, and all the prime factors in these, which are called *algebraic* prime factors of N, divided out before what remains of N is searched for factors of the specific form $2kn + 1$. These last factors are called *primitive* prime factors of N.

Please note that the algebraic factors as such may not be the only divisors of N which are not of the form $2kn + 1$. *It could happen that a prime factor of an algebraic factor Q of N divides N with greater multiplicity than it divides Q.* We provide here a few simple examples of this phenomenon:

N	Q	N/Q
$2^3 + 1$	$2 + 1$	$3 \neq 6k + 1$
$10^3 - 1$	$10 - 1$	$3 \cdot 37$

This possibility renders it extremely important to check for multiple factors of any p found to divide N. Neglecting this could lead to mistakes concerning the form of the possible prime factors in the remaining cofactor.

When N is large, the amount of work required in searching for a factor is considerably reduced by applying Legendre's theorem. A famous illustration of this is Euler's factorization of the Fermat number $F_5 = 2^{2^5} + 1 = 2^{32} + 1 = 641 \cdot 6700417$. The Fermat numbers $F_n = 2^{2^n} + 1$ lack algebraic factors, a fact that was already recognized by Fermat himself, who invented them in the hope of constructing only primes. At the beginning of the series Fermat was fortunate, as F_0, F_1, F_2, F_3 and F_4 are all primes. For F_5 the theorem of Legendre states that all factors are of the form $64k + 1$. Dividing F_5 by these primes, which are 193, 257, 449, 577, 641, ..., the factorization of F_5 is very soon discovered. It is easy to verify that the cofactor 6700417 is also a prime! Merely extend the computation by dividing the cofactor by

$$641, 769, 1153, 1217, 1409, 1601 \text{ and } 2113$$

in turn, i.e. by all primes of the form $64k + 1$ below $\sqrt{6700417} = 2588.52$. Since none of these divisions results in a zero remainder, the cofactor must be a prime! (Actually it is possible to prove, by using the theory of quadratic residues, that every prime factor of F_5 has the form $128k + 1$, a fact which eliminates the divisions by 193, 449, 577, 1217, 1601 and 2113. Also the factor $257 = F_3$ can be eliminated, since no two Fermat numbers F_n and F_m have a common divisor, so the first trial is by 641.)

As another example on the use of Legendre's theorem, we shall factorize

$$N = 3^{105} - 2^{105}.$$

The algebraic factorization of $a^{105} - b^{105}$, according to formula (A6.29) on p. 322, is:

$$a^{105} - b^{105} = \frac{(a^{35} - b^{35})(a^{21} - b^{21})(a^{15} - b^{15})}{(a^7 - b^7)(a^5 - b^5)(a^3 - b^3)}(a - b) \times$$

$$\times \text{ an irreducible cofactor.}$$

Instead of making quite complicated algebraic deductions from this formula, test all the factors of

$$3 - 2, \, 3^3 - 2^3, \, 3^5 - 2^5, \, 3^7 - 2^7, \, 3^{15} - 2^{15}, \, 3^{21} - 2^{21}, \, 3^{35} - 2^{35}$$

for multiplicity in N. The remaining cofactor must possess factors all having the form $p = 210k + 1$. Here follows the computation:

$N = 125\ 23673753\ 78787534\ 01295235\ 32574262\ 84187176\ 25274211$,

$3^3 - 2^3 = 19, \, N/19 = N_1 =$
$\quad = 6\ 59140723\ 88357238\ 63226065\ 01714434\ 88641430\ 32909169$

$3^5 - 2^5 = 211, \, N_1/211 = N_2 =$
$\quad = 3123889\ 68665200\ 18309128\ 27022343\ 29330054\ 17217579$

$3^7 - 2^7 = 2059 = 29 \cdot 71, \, N_2/(29 \cdot 71) = N_3 =$
$\quad = 1517\ 18780313\ 35608697\ 97390491\ 66745667\ 82621281$

$3^{15} - 2^{15} = (3^3 - 2^3)(3^5 - 2^5) \cdot 3571, \, N_3/3571 = N_4 =$
$\quad = 42486356\ 85056177\ 17669389\ 66442661\ 90642011$

$3^{21} - 2^{21} = (3^3 - 2^3)(3^7 - 2^7) \cdot 267331$,
$\quad\quad p|267331 \Rightarrow p = 42k + 1, \, 267331 = 43 \cdot 6217$,
$\quad N_4/(43 \cdot 6217) = N_5 = 158\ 92790903\ 62201606\ 50799819\ 11062281$

$3^{35} - 2^{35} = (3^5 - 2^5)(3^7 - 2^7) \cdot 1151\ 60837611$,
$\quad p|1151\ 60837611 \Rightarrow p = 70k + 1, \, 1151\ 60837611 = 6091 \cdot 18906721$,
$\quad N_5/(6091 \cdot 18906721) = N_6 = 13800516\ 94075552\ 57322971$.

At this point we have eliminated all algebraic factors of N. Testing each of the primes found for multiplicity, we observe that there are no multiple factors in this case. Therefore, all the prime factors of the cofactor N_6 must be of the form $210k + 1$. Using this result, we obtain by trial division

$$N_6 = 11971 \cdot 722611 \cdot 159536\ 60849491,$$

thereby completing the prime factorization of N.

Remark. Another method of finding all algebraic factors of $N = a^n - b^n$ is to apply Euclid's algorithm in order to compute $(a^m - b^m, a^n - b^n)$ for all $m|n$. When a factor has been found and removed, the G.C.D. computation is *repeated with the same value of m until no more factors emerge*. This procedure will ensure that all multiple factors of N are discovered. Only after this has been accomplished for a certain value of m does the computation proceed with the next m. For the value of N considered above, this computation successively yields:

$$(3^3 - 2^3, N) = 19, \qquad N/19 = N_1, \qquad (19, N_1) = 1$$

$$(3^5 - 2^5, N_1) = 211, \qquad N_1/211 = N_2, \qquad (211, N_2) = 1$$

$$(3^7 - 2^7, N_2) = 2059, \qquad N_2/2059 = N_3, \qquad (2059, N_3) = 1$$

$$(3^{15} - 2^{15}, N_3) = 3571, \qquad N_3/3571 = N_4, \qquad (3571, N_4) = 1$$

$$(3^{21} - 2^{21}, N_4) = 267331, \qquad N_4/267331 = N_5, \quad (267331, N_5) = 1$$

$$(3^{35} - 2^{35}, N_5) = 1151\,60837611, \quad N_5/1151\,60837611 = N_6,$$

$$(1151\,60837611, N_6) = 1.$$

where the factors found, viz. 19, 211, 2059, 3571, 267331, 1151 60837611 and N_6 must be decomposed into prime factors or verified for primality in a manner similar to that demonstrated above.

Adaptation to Search for Factors of the Form $p = 2kn + 1$

Most factorization methods discussed in this book can easily be adapted to run faster on computers if the factor searched for is of the form $p = 2kn + 1$, $k = 1, 2, 3, \ldots$ However, Shanks' method, as well as Gauss' and the related continued fraction method do not take any advantage of the particular form of the divisors of N. This is of importance in the choice between the different factorization methods, as we shall see when discussing strategies for the factorization of large numbers.

Adaptation of Trial Division

There is a fairly obvious adaptation of the trial division to the case $p = 2kn + 1$. Just as in ordinary trial division, it is usually not worthwhile to first sieve out the *primes* $p = 2kn + 1$ and then divide only by these, but rather to divide N by *all* integers $2kn + 1$, except possibly those divisible by 3. Thus, if $3|n$ then divide N by all integers of the form $2kn + 1$. On the other hand if $3 \nmid n$ then, since $3 \nmid 6n + 1$, either $2n + 1$ or $4n + 1$ must be divisible by 3. Check which and denote the other of these two integers by p1. Denoting $6n$ by n6, the computer will generate the divisors which are of the form $2kn + 1$ and not divisible by 3 if we write the program code

$$\text{p:=1; \ d:=p1-1;}$$

prior to the division loop, and in the loop let

$$\text{p:=p+d; \ d:=n6-d;}$$

Taking an example, suppose we are searching for factors of the Mersenne numbers $M_p = 2^p - 1$ with p around 100,000. With the described shortcut we can then easily reach the search limit 2^{35} for p. The number of trial divisions will be $2^{35}/(3p) \approx 2^{35}/3 \cdot 10^5 \approx 115000$, taking a few seconds of computing time on a fast computer if the division is programmed as a reduction (mod p) of a power of 2, as in Fermat's compositeness test on p. 94.

Exercise 5.5. Factor search of $a^n \pm b^n$. Carry out the adaptation hinted at above of the computer program for trial division from Exercise 5.1 on p. 151. Decide on a reasonable value for the search limit G, depending on n. Check your program by applying it to some of the integers given in Tables 7–31. Use the program to search for factors of $11^n \pm 4^n$, for as high values of n as your program can handle.

Adaptation of Fermat's Factoring Method

One of the advantages of Fermat's method is that it can very easily be adapted to cases where different conditions are imposed on the possible prime factors of N. Like trial division, Fermat's method runs n times faster than usual if the form $p = 2kn + 1$ is introduced directly into the calculations. However, due to the simplicity of the method, it is actually possible to do much better than that. As a matter of fact, Fermat's method can be speeded up by a factor of $2n^2$ if all factors of N have the form $2kn + 1$. This is based on the following observation: Let $N = p \cdot q = (2kn + 1)(2ln + 1)$. Then $N + 1 = 4kln^2 + p + q$, and thus $(p + q)/2 \equiv (N + 1)/2 \pmod{2n^2}$. Now, since

$$N = pq = \left(\frac{p+q}{2}\right)^2 - \left(\frac{p-q}{2}\right)^2 = x^2 - y^2,$$

this implies that the feasible values of $x = (p + q)/2$ belong to precisely one residue class (mod $2n^2$), which reduces the amount of work required by a factor $2n^2$.

In the example quoted earlier in the text on p. 154, the 19-digit number $N = (10^{22} + 1)/(89 \cdot 101)$ was mentioned. As we know from Legendre's theorem, all prime factors of N are of the form $p = 44k + 1$. Introducing $a = 44k + 1$ and $b = 44l + 1$ in $N = x^2 - y^2$ we have, by the above reasoning, that

$$x \equiv \frac{N+1}{2} \ (\text{mod } 2 \cdot 22^2) \equiv 507 \ (\text{mod } 968).$$

Thus we find that it suffices to let m assume only values of the form $968t + 507$ in the factorization scheme. The computation proceeds as follows:

$$N = \frac{10^{22} + 1}{89 \cdot 101} = 1112\,47079\,76415\,61909,$$

$$\sqrt{N} = 10547\,37312.149\ldots, \quad [\sqrt{N} + 1]^2 - N = 17928\,98060,$$

$$[\sqrt{N} + 1] = 968 \cdot 1089604 + 641.$$

Hence, the first possible value which x can take, if x has to be of the form $968t + 507$, is $968 \cdot 1089605 + 507$, and the calculation runs

t	$x = 968t + 507$	$m = 1936(x + 484)$	$z = x^2 - N$
10 89605	10547 38147	204 19739 89616	176 10954 31700
06	10547 39115	204 19758 63664	380 30694 21316
			$= 19\,50146^2$

Here the third column, $(x + 968)^2 - x^2 = 1936(x + 484)$, contains the differences between consecutive values of z. The values of m are incremented by $1936 \cdot 968 = 1874048$.— This time the factorization is found in only 2 steps, or roughly 1/968 of the number required when no advantage was taken of the special form of the factors. This method of taking advantage of the form $2kn + 1$ of the factors of N is due to D. H. Lehmer, and is described in [13].

Adaptation of Euclid's Algorithm as an Aid to Factorization

The adaptation of Euclid's algorithm is also quite obvious. Instead of using the products P_s of all primes in the search intervals between g_s and G_s, simply accumulate the products of the primes of the given form(-s) between g_s and G_s and proceed as described previously.—Similarly as before, it is only in a rather specific situation that this method proves to be favourable.

Adaptation of the Pollard-Brent Method

The adaptation of this method is not as apparent as the earlier ones. If, however, we search for factors of the form $p = 2kn + 1$, then Legendre's Theorem 5.7 states that $a^n \pm b^n$ has all its divisors of the form $2kn + 1$, apart from those originating from algebraic factors. Therefore, if we use the recursion formula

$$x_{i+1} = x_i^n + a \,(\text{mod } N) \qquad (5.38)$$

instead of, as in the original method, $x_{i+1} = x_i^2 + a \,(\text{mod } N)$, then

$$x_i - x_j = x_{i-1}^n - x_{j-1}^n,$$

will, according to Legendre's theorem, accumulate mainly prime factors of the form $2kn+1$ rather than just any primes. (The other primes also sneak into the algebraic factors, but at a lower rate as they do in the original recursion formula.) This tends to reduce the number of cycles needed to discover a certain prime factor p of N by a factor of about $\sqrt{n-1}$ as compared to the original formula. This gain should be weighed against the extra labour demanded to compute x^n (mod N) instead of x^2 (mod N), but since the amount of work required to calculate x^n (mod N) with the power algorithm grows only logarithmically with n, the modification is often worthwhile, especially when n is large. By using the recursion formula $x_{i+1} \equiv x_i^{1024} + 1$ (mod N) with $x_0 = 3$, Brent and Pollard in 1980 managed to discover the factor

12389263 61552897 of the Fermat number F_8.

The computation took a couple of hours on a large computer [12].

Shanks' Factoring Method SQUFOF

Another recently discovered factoring method which we shall describe is Shanks' method "square forms factorization," SQUFOF for short. Gauss [2] was the first to apply systematically the theory of binary quadratic forms (expressions of the form $Ax^2 + Bxy + Cy^2$) to find factorizations of integers. This approach has been quite successful over the years and many variations on this theme can be found in the literature. Most of the methods are very complicated, especially when they are to be administered

191

on a computer, and not ideal for use with computers. However, around 1975 Daniel Shanks managed to construct a feasible algorithm. Unfortunately Shanks' theory, on which the algorithm is based, is too complicated to be given and discussed in detail here, but we shall describe how the algorithm works. For somewhat more detailed accounts, see [14]–[16]. The method uses the continued fraction expansion of $\sqrt{N}$, where N is the number to be factored. The reader who is unfamiliar with this subject should now consult Appendix 5 for more details before reading further.

Shanks' method makes use of the regular continued fraction expansion of $\sqrt{N}$ in the following way: The formula

$$A_{n-1}^2 \equiv (-1)^n Q_n \ (\text{mod } N),$$

proved as formula (A5.34) on p. 316, is applied to solve Legendre's congruence $x^2 \equiv y^2 \ (\text{mod } N)$. All that is necessary is to expand $\sqrt{N}$ until a *square number* $Q_n = R^2$ is found for an *even* value of n. Then Legendre's congruence has the solution $x \equiv A_{n-1}$, $y \equiv R$ and, if this is not one of the trivial solutions, the factors p and q of N can be obtained by Euclid's algorithm applied to N and $A_{n-1} \pm R$. The idea of looking for a square in the continued fraction expansion of $\sqrt{N}$ is actually an old one, but not much used before the advent of computers because of the many steps that are normally required to produce such a square.

Now, Shanks' algorithm differs in certain details from the short description just given. The most important distinction is that the numerators A_i of the convergents A_i/B_i to the continued fraction for $\sqrt{N}$ are *not* computed directly. This is because the continued fraction expansion itself (omitting the convergents (mod N) from the computation) can be effected, as demonstrated in Appendix 5, with all quantities involved $< 2\sqrt{N}$, whereas A_n (mod N) would require more than twice as many digits if the usual recursive formula $A_s = b_s A_{s-1} + A_{s-2}$ were used. The computation of A_{n-1} (mod N) is replaced by another, less time-consuming strategy for finding the factor of N as soon as some $(-1)^n Q_n = R^2$ has been obtained. This other computation has been included in the algorithm given below.

Shanks also managed to distinguish between those square $(-1)^n Q_n$'s which lead to the useful non-trivial solutions to Legendre's congruence and others which do not. The $(-1)^n Q_n$'s which may give rise to trivial solutions are stored in a list and can thereby be avoided before computing what would turn out to be only a trivial factorization of N.

Actually Shanks' algorithm is quite simple. We start by giving an example and then provide a more formalized description.

Example. Find the factors of $N = 1000009$.

Compute the regular continued fraction expansion of $\sqrt{N}$ until, after an *even number of steps*, a *square denominator* is found, the square root of which has never previously occurred as a denominator during the computation. This will guarantee a non-trivial solution to Legendre's congruence $A_{n-1}^2 \equiv (-1)^n Q_n \equiv R^2 \pmod{N}$. Starting by $\sqrt{N} = \sqrt{1000009} = 1000 + (\sqrt{N} - 1000)$, we successively find

$$\frac{1}{\sqrt{N} - 1000} = \frac{\sqrt{N} + 1000}{9} = 222 + \frac{\sqrt{N} - 998}{9} \tag{1}$$

$$\frac{9}{\sqrt{N} - 998} = \frac{9(\sqrt{N} + 998)}{4005} = \frac{\sqrt{N} + 998}{445} = 4 + \frac{\sqrt{N} - 782}{445} \tag{2}$$

$$\frac{445}{\sqrt{N} - 782} = \frac{\sqrt{N} + 782}{873} = 2 + \frac{\sqrt{N} - 964}{873} \tag{3}$$

$$\frac{873}{\sqrt{N} - 964} = \frac{\sqrt{N} + 964}{81} = 24 + \frac{\sqrt{N} - 980}{81}. \tag{4}$$

Here the square 81 occurs. It is ruled out, however, since the denominator $\sqrt{81} = 9$ has already appeared in step (1). Continuing the expansion, we find

$$\frac{81}{\sqrt{N} - 980} = \frac{\sqrt{N} + 980}{489} = 4 + \frac{\sqrt{N} - 976}{489} \tag{5}$$

$$\vdots$$

$$\frac{375}{\sqrt{N} - 997} = \frac{\sqrt{N} + 997}{16}. \tag{18}$$

Now the square 16 obtained can be used, and computing $A_{17} \pmod{N}$, we find

$$A_{17} \equiv 494881, \text{ and therefore } 494881^2 \equiv Q_{18} = 4^2 \pmod{1000009}.$$

Thus, $(494777, N) = 293$ and $(494885, N) = 3413$ are factors of N. However, in order to avoid the computation of $A_{17} \pmod{N}$, Shanks calculates instead what is called the square root of the corresponding quadratic form under the law of composition. This is done simply by altering a sign in the numerator and taking the square root of the denominator; yielding $(\sqrt{N} - 997)/4$ from $(\sqrt{N} + 997)/16$. Next, we expand this new number until the *coefficients in the numerators of two consecutive steps are equal*:

$$\frac{4}{\sqrt{N} - 997} = \frac{\sqrt{N} + 997}{1500} = 1 + \frac{\sqrt{N} - 503}{1500} \tag{1}$$

$$\frac{1500}{\sqrt{N} - 503} = \frac{\sqrt{N} + 503}{498} = 3 + \frac{\sqrt{N} - 991}{498} \tag{2}$$

$$\vdots$$

$$\frac{1410}{\sqrt{N} - 673} = \frac{\sqrt{N} + 673}{388} = 4 + \frac{\sqrt{N} - 879}{388} \tag{10}$$

$$\frac{388}{\sqrt{N} - 879} = \frac{\sqrt{N} + 879}{586} = 3 + \frac{\sqrt{N} - 879}{586}. \tag{11}$$

At this stage, in steps (10) and (11) the coefficients -879 in the numerators are the same, and therefore the last denominator 586 or $586/2 = 293$ is a factor of N. Thus $N = 293 \cdot 3413$.

Experience seems to indicate that the number of cycles required for the method is $O(\sqrt[4]{N})$, comparable with the number of cycles for the Pollard-Brent method in the case when N is composed of only two factors of about equal size (the difficult case within factorization). Each cycle runs faster, however, since it deals only with numbers of size approximately $\sqrt{N}$, while Pollard's methods works with numbers of magnitude about N^2.

Using the formulas for the continued fraction expansion of $\sqrt{N}$ from Appendix 5, we can describe this algorithm in the following way:

Let $\quad P_0 = 0, \quad Q_0 = 1, \quad Q_1 = N - P_1^2,$

$$q_i = \left[\frac{\sqrt{N} + P_i}{Q_i}\right], \quad P_{i+1} = q_i Q_i - P_i, \quad Q_{i+1} = Q_{i-1} + (P_i - P_{i+1})q_i.$$

$$(5.39)$$

If any Q_i is $< 2\sqrt{2\sqrt{N}}$, it is stored in a list. This list will contain all numbers which are useless for the factorization of N (as well as some others). Continue until some $Q_{2i} = R^2$. Then compare with the list:

1. If R (or $R/2$, if R is even) is included in the list, then continue the expansion.

2. Otherwise, a useful square has been found. In the expansion of $\sqrt{N}$, the expression now arrived at is

$$q_{2i} = \left[\frac{\sqrt{N} + P_{2i}}{Q_{2i}}\right] = \left[\frac{\sqrt{N} + P_{2i}}{R^2}\right]. \tag{5.40}$$

Next, continue by expanding the number $(\sqrt{N} - P_{2i})/R$. This is achieved by simply taking

$$P_0' = -P_{2i}, \quad Q_0' = R, \quad Q_1' = \frac{N - P_1'^2}{R}, \tag{5.41}$$

and using the same formulas for the recursion as above. Continue the expansion of this new number until some $P_{j+1} = P_j$. This will occur after about half the number of cycles required to find the square $Q_{2i} = R^2$. (In fact, by utilizing composition of quadratic forms, this second part of the computation could be performed in a logarithmic number of steps and could thus be programmed to take very little time compared with the first part of the computation, that of finding a square $(-1)^n Q_n$. However, we shall not elaborate on this here.) Then, finally Q_j (or $Q_j/2$ if Q_j is even) is a factor of N.

A Computer Program for Shanks' Method

A PASCAL program for Shanks' algorithm is given below:

```
PROGRAM SQUFOF
(Input,Output);
{Factorizes N < 10↑20 by Shanks' Square Forms Factorization
  method. Computer arithmetic allowing integers up to 2↑35
  is used. Warning: Make sure N is composite, otherwise the
  list could extend outside the specified index bounds!}
LABEL 1,2,3,4,5,6;
TYPE vector=ARRAY[0..30] OF INTEGER;
VAR List : vector;
    c,c2,sq,d,a,b,z,z1,z2,sq2sqN,Q,Q0,Q1,Q2,P1,P2,
    i,j,k,ks,u,r,w,s1,s2 : INTEGER;

PROCEDURE isqrtd(a,b : INTEGER; VAR sq,d : INTEGER);
```

195

```
{Computes n=10↑10*a+b, sq:=[sqrt(n)] and d:=n-sq↑2}
LABEL 1,2;
VAR c,c2,n,r,r1,r2,s,s1,s2,z,sw : INTEGER;
    rn,rsq,w,rs1,rs2 : REAL;
BEGIN
 c:=100000; c2:=10000000000; sw:=0;
 rn:=1E10*a+b; rsq:=sqrt(rn); r:=trunc(rsq);
 {r is a first approximation to sqrt(n)}
1: r1:=r DIV c; r2:=r MOD c; {r=10↑5*r1+r2}
 z:=2*r1*r2; s1:=a-r1*r1- z DIV c;
 s2:=b-r2*r2-(z MOD c)*c;
 {Here d=n-r↑2 has been computed as 10↑10s1+s2}
 IF sw=1 THEN GOTO 2 {d was <0 previously!};
 z:=2*r; rs1:=s1; rs2:=s2; w:=rs1*1E10+rs2;
 s:=trunc(w/z);
 {Here the correction s=(n-r↑2)/(2r) has been computed}
 IF s <> 0 THEN BEGIN r:=r+s; GOTO 1 END;
2: d:=s1*c2+s2; IF d<0 THEN BEGIN r:=r-1; sw:=1; GOTO 1 END;
 sq:=r;
END {isqrtd};

BEGIN c:=100000; c2:=10000000000;
1: write('Input N = 10↑10a+b as two integers a, b: ');
 read(a); IF a<0 THEN GOTO 6; read(b);
 isqrtd(a,b,sq,d); IF d=0 THEN
  BEGIN writeln('N is the square of',sq:11); GOTO 1 END;
 z:=2*sq; IF d>=sq THEN z:=z+1 {[2*sqrt(N)]};
 sq2sqN:=trunc(sqrt(z)) {sqrt(2*sqrt(N))};
 List[0]:=1; Q0:=1; P1:=sq; Q1:=d;
 {Here all initial values are set for the continued
 fraction expansion}
 FOR i:=1 TO 10000000 DO
  BEGIN {Continued fraction expansion starts}
   IF i MOD 50000=0 THEN writeln(i:8,' cycles passed');
   Q:=(sq+P1) DIV Q1; P2:=Q*Q1-P1; Q2:=Q0+Q*(P1-P2);
   u:=Q1; IF NOT odd(u) THEN u:=u DIV 2;
   Q0:=Q1; Q1:=Q2; P1:=P2;
   IF (u < sq2sqN) AND (u > 1) THEN
    BEGIN List[List[0]]:=u; List[0]:=List[0]+1 END;
    {Here a small denominator is placed in the list}
   IF odd(i) THEN
    BEGIN
```

196

```
      IF Q1 MOD 4 > 1 THEN GOTO 3 {No square, goto next i};
      r:=trunc(sqrt(Q1)); IF Q1=r*r THEN {A square!}
        BEGIN ks:=List[0]-1; FOR k:=1 TO ks DO
         IF r=List[k] THEN GOTO 3 {Square not useful, next i};
         IF r > 1 THEN GOTO 4 ELSE
           BEGIN
             write('The period has been searched (i=',i:7);
             writeln(') without finding any useful form');
             GOTO 1
           END
        END
     END;
3: END;
4: writeln('Number of steps to find a square was',i:8);
writeln('Number of elements in the list is',List[0]-1:3);

 {Here the computation of the square root of the
   square quadratic form found is started}
 z:=sq-(sq-P1) MOD r; z1:=z DIV c; z2:=z MOD c;
 {z=z1*10↑5+z2 for the computation of (n-z↑2)/r}
 w:=2*z1*z2; s1:=a-z1*z1-w DIV c; s2:=b-z2*z2-(w MOD c)*c;
5: IF s2<0 THEN
   BEGIN s2:=s2+c2; s1:=s1-1; GOTO 5 END;
   Q0:=r; P1:=z; z:=s1*c+s2 DIV c;
   Q1:=(z DIV r)*c+((z MOD r)*c+s2 MOD c)DIV r;
   FOR j:=1 TO 5000000 DO
     BEGIN {Here the expansion of the second form starts}
       Q:=(sq+P1) DIV Q1; P2:=Q*Q1-P1;
       Q2:=Q0+Q*(P1-P2); Q0:=Q1;
       u:=P1; P1:=P2; Q1:=Q2; IF u=P2 THEN
         BEGIN
           IF NOT odd(Q0) THEN Q0:=Q0 DIV 2;
           writeln('Factor=',Q0:11); GOTO 1
         END
       ELSE P1:=P2
     END;
6: END.
```

Please note that this program will operate only on a computer with a word length of at least 36 bits (or having an equivalent integer arithmetic). If your computer has a word length of s bits, then it will be convenient to use the largest power of 100 below 2^{s-2} as the constant c2, and the square root of this number as c in the program. The tricky part of the programming is at the very beginning and also at the start of the second part of the computation, where the square root of the quadratic form is taken. In these two places occasionally double precision arithmetic must be used. In the computer's hardware language there are normally certain operations which would be of use to us, such as division of a two-word integer by a one-word integer, giving a full quotient and remainder. Unfortunately, however, these devices are not easily accessible from high-level languages such as PASCAL, so we have to circumvent the problem of arithmetic overflow during this part of the computation.—Rather than using a pre-programmed package for this, it is shown above how it can be done in a reasonably simple manner. Nevertheless, a better solution for a calculation, involving more extensive double precision computations would undoubtedly have been to use a double precision integer arithmetic package such as the one described in Appendix 7.

Exercise 5.6. Shanks' method. Modify the PROGRAM SQUFOF above as to operate on your computer. This effort may be your easiest way of achieving a reasonably fast computer program for factorization of nearly double precision integers.— During the running-in phase of the program you might want to temporarily change the periodic printouts at each 50000th cycle to more frequent and more informative messages. Try your program on some of the numbers in the factor tables at the end of the book. Be careful to feed the program with *composite* numbers only, otherwise the ARRAY List may run out of bounds!

As has already been mentioned, the number of cycles in Shanks' method seems to be $< C\sqrt[4]{N}$. In some cases, however, the number of cycles is considerably smaller than this bound, so the computing time varies considerably for numbers N of approximately the same size. Thus, it is actually worth applying Shanks' algorithm simultaneously to e.g. the numbers N and $2N$ or to N and $3N$, because the probability of finding a square Q_{2i} early increases considerably when two expansions are available. The average benefit is not very great, but the trick helps to avoid the very long computer runs that may occur when the running time for N is large.— Besides, this strategy can prevent us from getting stranded on values of N for which the continued fraction expansion of $\sqrt{N}$ has so short a period that no useful square $(-1)^n Q_n$ occurs at all!

Comparison Between Pollard's and Shanks' Methods

Pollard's method requires approximately $C\sqrt{p}$ cycles to find a factor p of N, while Shanks' method needs about $C_1\sqrt[4]{N}$ cycles. Now, since the smallest factor p of N is $< \sqrt{N}$, the number of cycles in Pollard's method will in general be smaller than in Shanks' method *unless N is the product of two primes of about equal size.*—On the other hand, Pollard's method uses in each cycle 8 heavy operations (multiplications or divisions) performed on numbers of the size of about N^2, while Shanks' method demands 9 arithmetic operations $+ \frac{1}{3}$ square root extraction per cycle (or even fewer if the rejection of non-squares is effected by seeking quadratic non-residues of some small primes and the square-root extracted only when this fails), which in total equals about 15 simple arithmetic operations per cycle. However, these are performed *on numbers of size $< 2\sqrt{N}$.* (As shown above, in Shanks' algorithm only the computation of the numbers $P_1 = [\sqrt{N}]$, $Q_1 = N - P_1^2$ and $Q_1' = (N - P_{2i}^2)/R$ involve arithmetic operations on numbers $> 2\sqrt{N}$.) Since the labour of multiplying or dividing large numbers normally increases with the square of the length of the numbers involved, this means that one cycle of Shanks' algorithm is roughly about 10 times as fast as one cycle in Pollard's algorithm. This factor can sometimes be much larger. For instance, if Shanks' method due to the small size of N can be programmed to use directly the computer's hardware arithmetic, while Pollard's is restricted to using a pre-programmed multiple precision arithmetic package—quite a common situation—then the speed factor can easily exceed 100 or more per cycle.

Morrison and Brillhart's Continued Fraction Method

Morrison and Brillhart's method is one of the most efficient *general* factorization methods which has been put to extensive use on computers. One of the ideas behind the algorithm [17] is to find a non-trivial solution to the congruence $x^2 \equiv y^2 \pmod{N}$ and then compute a factor p of N by means of Euclid's algorithm applied to $(x+y, N)$. The technique of finding solutions to $x^2 \equiv y^2 \pmod{N}$ is inspired both by Legendre's factorization method, from which the idea of finding small quadratic residues from the continued fraction expansion of $\sqrt{N}$ is taken, and by an idea of Maurice Kraïtchik in which known residues are combined to form new ones, *in our case squares.* Historically the situation is much the same for this method as for Pollard's $(p-1)$-method—the underlying ideas have been known for quite a long time and have occasionally been applied to specific cases, in particular by D. H. Lehmer, R. E. Powers [18] and Kraïtchik [19]. The current version of the method is, however, due to Morrison and Brillhart, who have systematically explored the potentials of these ideas and have constructed a good algorithm. We shall now give a brief description of this algorithm.

First, part of the regular continued fraction expansion of $\sqrt{N}$ is computed, just as in Legendre's or Shanks' methods. The notations and formulas usually employed are listed in the second half of Appendix 5. Let us just mention here the two most important formulas, one of which is (A5.31) on p. 315:

$$A_{n-1}^2 - NB_{n-1}^2 = (-1)^n Q_n, \qquad (5.42)$$

immediately giving (A5.34):

$$A_{n-1}^2 \equiv (-1)^n Q_n \pmod{N}. \qquad (5.43)$$

The calculation of the expansion is not quite as fast as in Shanks' method, since Shanks uses *only* the quantities P_n and Q_n, while Morrison-Brillhart's method requires in addition $A_{n-1} \pmod{N}$, the numerator of the $(n-1)^{\text{th}}$ convergent A_{n-1}/B_{n-1} of the continued fraction for $\sqrt{N}$. However, while Shanks' algorithm involves *waiting until a perfect square shows up* among the $(-1)^n Q_n$, Morrison and Brillhart try to form *combinations which yield a square by multiplying together some of the quadratic residues generated*, and thereby may find a square with far fewer cycles than Shanks' method.

The Factor Base

The great improvement introduced in Morrison and Brillhart's method consists of the way of producing a combination of the quadratic residues found which is a square. Just as in Gauss' factorization method, an upper bound p_m is set on the small prime factors used and the Q_i's are searched for prime factors $\leq p_m$. The prime factorizations of those Q_i's which are completely factored in this way are retained, while the other Q_i's are rejected as soon as p_m is reached without Q_i having been completely factored. In order to render the method slightly more efficient and not discard any easily factored Q_i's, those factorizations are also kept for which the co-factor of Q_i after reaching the search limit p_m is larger than the search limit, but $< p_m^2$, in which case the co-factor must be a prime. The

m primes $p_1, p_2, p_3, \ldots, p_m$ form what is called the *factor base*. The number of *all* primes (below p_m^2) occurring in these factorizations is denoted by S.

The limitation of the Morrison-Brillhart method lies in the huge number of trials that must be carried out before a sufficient number of completely factored Q_i's can be collected. However, on a supercomputer possessing array processors one Q_i could be divided by 64 primes (or whatever the number of possible operations running in parallel might be) in the same operation, thus gaining a considerable speed factor over an ordinary computer.

For all the odd primes p_i in the above factorizations, the value of Legendre's symbol $(N/p_i) = 1$ (or, possibly, $(N/p_i) = 0$ which occurs when $p_i | N$, a case which, however, should be tested for and excluded at the start) because if $p_i | Q_n$, then $A_{n-1}^2 - NB_{n-1}^2 \equiv 0 \pmod{p_i}$, and therefore $N \equiv (A_{n-1}/B_{n-1})^2$ is a square $\pmod{p_i}$. Thus the primes in the factor base have about half the density of all primes. (Please note that p_i in this reasoning *does not* denote the ith prime!) Hence the value of S, the number of all primes found during these factorizations, is $\approx \pi(p_m)/2 +$ the number of primes between p_m and p_m^2 which happen to show up as factors.

The factorizations discovered are stored in the form of binary vectors Γ with $S + 1$ components $(\gamma_0, \gamma_1, \gamma_2, \ldots, \gamma_m, \ldots, \gamma_S)$, each element having the value 0 or 1. If the standard factorization of

$$(-1)^n Q_n = (-1)^{\alpha_0} \prod_{i=1}^{S} p_i^{\alpha_i},$$

then

$$\gamma_i \equiv \alpha_i \pmod{2}. \tag{5.44}$$

This means that it is really information about the *square-free part* $\prod p_i^{\gamma_i}$ of Q_n which is saved by storing the exponents γ_i *rather than the complete factorizations!*

The next step in the algorithm is to search for square combinations of the now completely factored Q_i's. A square combination

$$(-1)^{n_1} Q_{n_1} (-1)^{n_2} Q_{n_2} \cdots (-1)^{n_k} Q_{n_k} = (-1)^{\gamma} \prod_{i=1}^{S} p_i^{\sum_\nu \gamma_{i_\nu}}$$

is produced if *all the exponents* γ *and* $\sum \gamma_{i_\nu}$ *are even*. In order to find such a combination of the Q_i's, first generate $S + 1$ complete prime factorizations and then perform Gaussian elimination (mod 2) on the $(S+1) \times (S+1)$-matrix whose rows are the vectors Γ_ν. Even if this matrix *could*, in theory be non-singular, it is usual for a number of sets of linearly dependent rows to be discovered during the Gaussian elimination. Experience shows, that if S is large then a set of linearly dependent rows often is discovered already when the number of rows is only about 80–90% of S. If the matrix happens to be non-singular, then more Q_i's have to be found and factored.— Remember that a linear combination of binary vectors (mod 2) is just a sum (mod 2) of some of the vectors! Each of these sums leads to a factorization (trivial or non-trivial) of N. Experience has shown that about 5 or 6 such linear dependencies are generally enough to find a factorization, but nevertheless there have been examples reported of numbers defying factorization after as many as 25 square combinations have been found!

Before delving more deeply into the theoretical details of this method, let us consider

An Example of a Factorization

Let us take $N = 12007001$ (the same example as for Gauss' method). We choose the factor base 2, 5, 23, 31 (comprising the small quadratic residues of N, found by computing the value of Legendre's symbol (N/p) for all $p \leq p_m = 31$, with the upper bound used on the primes being $u_m = 97$). The continued fraction expansion of $\sqrt{N}$ and the subsequent factorization of the Q_n's found results in the following values of the vectors Γ:

n	$(-1)^n Q_n$	-1	2	5	23	31	97	59	71
1	$-2^3 \cdot 97$	1	1	0	0	0	1		
3	$-5 \cdot 97$	1	0	1	0	0	1		
6	$2^3 \cdot 5$	0	1	1	0	0	0		
10	$5^2 \cdot 59$	0	0	0	0	0	0	1	
12	$2^4 \cdot 71$	0	0	0	0	0	0	0	1
18	$2^4 \cdot 5^2$	0	0	0	0	0	0	0	0

Here we arrive at a square, $2^4 \cdot 5^2$. We need to test whether $A_{17}^2 \not\equiv 20^2 \pmod{N}$, which would lead to a non-trivial solution of Legendre's congruence! However, $A_{17} \equiv 20 \pmod{N}$, so that this square fails to factor N, just as it does in Shanks' method! Therefore we continue the expansion:

n	$(-1)^n Q_n$	-1	2	5	23	31	97	59	71	61
21	$-2^4 \cdot 61$	1	0	0	0	0	0	0	0	1
24	$2^3 \cdot 5 \cdot 59$	0	1	1	0	0	0	1	0	0
26	$2^3 \cdot 5 \cdot 71$	0	1	1	0	0	0	0	1	0
27	$-5 \cdot 31$	1	0	1	0	1	0	0	0	0
28	2^{11}	0	1	0	0	0	0	0	0	0

At this point we have 10 vectors Γ (discarding the useless row for $n = 18$) and 9 primes in the set, so that there is some chance of a square combination being found. Please note that it is not essential to have a *square* matrix at this stage!

The Gaussian elimination is performed by systematically adding two rows (mod 2) at a time, thereby creating zeros in the lower triangular part of the matrix. This is achieved by adding together (mod 2) rows having their left-most ONEs in the same column. In order to keep track of the additions carried out we extend the vectors to the right by means of rows of a unit matrix, and operate also on these (mod 2) when adding the first parts of the rows Γ.—For the sake of brevity in this pencil-and-paper example, we can, for the moment, discard the *columns* below the primes 23, 31 and 61, since at least *two* rows with $\gamma_i = 1$ must occur before the prime p_i can enter into a square combination! This also eliminates the rows for which $n = 21$ and $n = 27$. Thus we start with the matrix

n	-1	2	5	97	59	71								
1	1	1	0	1	0	0	1	0	0	0	0	0	0	0
3	1	0	1	1	0	0	0	1	0	0	0	0	0	0
6	0	1	1	0	0	0	0	0	1	0	0	0	0	0
10	0	0	0	0	1	0	0	0	0	1	0	0	0	0
12	0	0	0	0	0	1	0	0	0	0	1	0	0	0
24	0	1	1	0	1	0	0	0	0	0	0	1	0	0
26	0	1	1	0	0	1	0	0	0	0	0	0	1	0
28	0	1	0	0	0	0	0	0	0	0	0	0	0	1

The first step in the Gaussian elimination is to replace row 2 by row 1 + row 2 (mod 2):

New row 2: $0\ 1\ 1\ 0\ 0\ 0\ |\ 1\ 1\ 0\ 0\ 0\ 0\ 0\ 0$

Next, replace row 3 by new row 2 + row 3:

New row 3:   0 0 0 0 0 0 | 1 1 1 0 0 0 0 0.

Only zeros result! A square combination has been found! The corresponding row of the unit matrix tells us that it is the sum of the first, second and third rows which produced this result. We therefore form

$$A_0^2 A_2^2 A_5^2 \equiv (-Q_1)(-Q_3)Q_6 = 2^6 \cdot 5^2 \cdot 97^2 \ (\mathrm{mod}\ N),$$

having the solutions $A_0 A_2 A_5 \equiv \pm 2^3 \cdot 5 \cdot 97$. However, as $A_0 A_2 A_5 = 3465 \cdot 31186 \cdot 2228067 \equiv 3880 \ (\mathrm{mod}\ N)$ and $2^3 \cdot 5 \cdot 97$ is also $\equiv 3880$ $(\mathrm{mod}\ N)$ this square combination fails to factor N and we must resume the elimination. Since the linear combination in which row 3 is involved is useless, we discard it together with the corresponding 3rd column of the unit matrix and continue the elimination process. We have now arrived at

n	-1	2	5	97	59	71							
1	1	1	0	1	0	0	1	0	0	0	0	0	0
3	0	1	1	0	0	0	1	1	0	0	0	0	0
10	0	0	0	0	1	0	0	0	1	0	0	0	0
12	0	0	0	0	0	1	0	0	0	1	0	0	0
24	0	1	1	0	1	0	0	0	0	0	1	0	0
26	0	1	1	0	0	1	0	0	0	0	0	1	0
28	0	1	0	0	0	0	0	0	0	0	0	0	1

We successively obtain:

New row 5:   0 0 0 0 1 0 | 1 1 0 0 1 0 0

New row 6:   0 0 0 0 0 1 | 1 1 0 0 0 1 0

New row 7:   0 0 1 0 0 0 | 1 1 0 0 0 0 1

Now, new row 5 is identical to row 3, so that the sum of these (mod 2) produces zeros:

0 0 0 0 0 0 | 1 1 1 0 1 0 0.

The unit matrix part tells us which rows are involved in the square combination now detected. Hence, we compute

$$A_0 A_2 A_9 A_{23} \equiv \pm \sqrt{Q_1 Q_3 Q_{10} Q_{24}} \ (\mathrm{mod}\ N),$$

and find $A_0 A_2 A_9 A_{23} \equiv 3465 \cdot 31186 \cdot 668093 \cdot 7209052 \equiv 1144600 \pmod{N}$
and $\sqrt{Q_1 Q_3 Q_{10} Q_{24}} = 2^3 \cdot 5^2 \cdot 59 \cdot 97 = 1144600$. Once again, the
square combination found does not factor N. However, adding together the
remaining two identical rows, namely row 4 and the new row 6 yields

$$0\ 0\ 0\ 0\ 0\ 0\ |\ 1\ 1\ 0\ 1\ 0\ 1\ 0,$$

corresponding to

$$A_0^2 A_2^2 A_{11}^2 A_{25}^2 \equiv Q_1 Q_3 Q_{12} Q_{26} = 2^{10} \cdot 5^2 \cdot 71^2 \cdot 97^2 \pmod{N}.$$

This time we find $1101920 \equiv \pm 1101920 \pmod{N}$. Again no factorization
has been achieved! Therefore we continue the elimination! Now, using
pencil and paper once more we remove the two useless new combinations
(new rows 5 and 6) and are left with

n	-1	2	5	97	59	71				
1	1	1	0	1	0	0	1	0	0	0
3	0	1	1	0	0	0	1	1	0	0
10	0	0	0	0	1	0	0	0	1	0
28	0	1	0	0	0	0	0	0	0	1

Now, since the prime 59 occurs once only, row 3 cannot enter a zero sum.
Thus, discarding row 3 and replacing row 4 by row 2 + row 4 leads to

n	-1	2	5	97
1	1	1	0	1
3	0	1	1	0
28	0	0	1	0

which has no zero combinations (because the determinant of its leading
3×3-matrix has the value 1 (mod 2), and hence is non-singular).

In this unhappy situation we must return, expand more of $\sqrt{N}$ and
factor more Q_i's and start the elimination process all over again from the
beginning (provided some new small primes can be included below the factor
limit as a result of having factored further Q_i's). We find that the following
Q_i's possess only small prime factors:

n	$(-1)^n Q_n$	-1	2	5	23	31	97	59	71	61
30	$2^5 \cdot 5^3$	0	1	1	0	0	0	0	0	0
32	$31 \cdot 61$	0	0	0	0	1	0	0	0	1
33	$-2^3 \cdot 5 \cdot 61$	1	1	1	0	0	0	0	0	1
34	$31 \cdot 97$	0	0	0	0	1	1	0	0	0
36	$2^4 \cdot 59$	0	0	0	0	0	0	1	0	0
38	$5^2 \cdot 71$	0	0	0	0	0	0	0	1	0
39	$-2^3 \cdot 31$	1	1	0	0	1	0	0	0	0
41	$-31 \cdot 71$	1	0	0	0	1	0	0	1	0
45	$-5^2 \cdot 61$	1	0	0	0	0	0	0	0	1

By preliminary verification we observe that the easily obtained squares

$$A_{20}^2 A_{24}^2 \equiv (\pm 2^2 \cdot 5 \cdot 61)^2, \qquad A_9^2 A_{35}^2 \equiv (\pm 2^2 \cdot 5 \cdot 59)^2,$$

$$A_5^2 A_{29}^2 \equiv (\pm 2^4 \cdot 5^2)^2, \qquad A_{11}^2 A_{37}^2 \equiv (\pm 2^2 \cdot 5 \cdot 71)^2 \pmod{N}$$

all fail to factor N. Still working with pencil and paper, we discard the useless rows for $n = 30$, 36, 38 and 45. (In a computer program, these rows would in fact be included in the process of elimination, leading to the discovery of trivial factorizations as soon as the corresponding square combinations were found.) This time, the elimination proceeds from

n	$(-1)^n Q_n$	-1	2	5	23	31	97	59	71	61
1	$-2^3 \cdot 97$	1	1	0	0	0	1	0	0	0
3	$-5 \cdot 97$	1	0	1	0	0	1	0	0	0
10	$5^2 \cdot 59$	0	0	0	0	0	0	1	0	0
12	$2^4 \cdot 71$	0	0	0	0	0	0	0	1	0
24	$2^3 \cdot 5 \cdot 59$	0	1	1	0	0	0	1	0	0
26	$2^3 \cdot 5 \cdot 71$	0	1	1	0	0	0	0	1	0
28	2^{11}	0	1	0	0	0	0	0	0	0
32	$31 \cdot 61$	0	0	0	0	1	0	0	0	1
33	$-2^3 \cdot 5 \cdot 61$	1	1	1	0	0	0	0	0	1
34	$31 \cdot 97$	0	0	0	0	1	1	0	0	0
39	$-2^3 \cdot 31$	1	1	0	0	1	0	0	0	0
41	$-31 \cdot 71$	1	0	0	0	1	0	0	1	0

Performing the Gaussian elimination (mod 2), we find the following square combination:

$$A_0^2 A_{11}^2 A_{27}^2 A_{33}^2 A_{40}^2 \equiv 9815310^2 \equiv (\pm 2^9 \cdot 31 \cdot 71 \cdot 97)^2 \equiv (\pm 1247455)^2,$$

which leads to a non-trivial solution of Legendre's congruence and thus Euclid's algorithm yields $(12007001, 9815310 - 1247455) = 3001$ whereby N is finally factorized.

Further Details of the Method

Even if unrealistically small, the above example does illustrate fairly well the various situations that can arise in the application of Morrison and Brillhart's version of the continued fraction factorization method. We wish only to add certain details which were not expressly stated in the example. One of these concerns the computation of the square root of $Q_{i_1} Q_{i_2} \ldots Q_{i_s}$ (mod N). Since there is no known efficient algorithm for computing square roots modulo a *composite* number N without first factorizing N, the square root must be computed first, followed by the reduction (mod N). If the Q_i's were stored in their *factored* form the computation would be simple, since the number α_i of factors of the prime p_i could then be counted and the product $\prod_i p_i^{\alpha_i/2}$ (mod N) formed successively with all primes p_i participating in the square combination under consideration. However, since the binary vectors contain no information on the *multiple* prime factors of the Q_i's we would have to, at least partially, refactor the Q_i's a second time which would require much computer time, and so it would be convenient if some other method could be devised to effect this computation. (Of course, the straightforward way would be to first multiply together all the factors Q_i and then extract the square root, but since the product would be a very large number, this approch is highly unpractical!) An elegant algorithm to solve this problem is presented on p. 190 in Morrison and Brillhart's paper [17]. The idea is to successively accumulate the product of the Q_i's but, before involving a factor Q_s, to check by means of Euclid's algorithm whether Q_s has some factor X in common with the product so far accumulated. Suppose that we have arrived at the product Q and are about to include the factor Q_s. Then if $(Q, Q_s) = X$, instead of forming QQ_s calculate $(Q/X) \times (Q_s/X)$, thereby reducing the size of QQ_s. The factor X is accumulated in what will eventually be the square root (mod N) we are seeking. Let us demonstrate this technique on the above case, where we had to compute the square root of

$$Q_1 Q_{12} Q_{23} Q_{34} Q_{41} \text{ (mod } N) \equiv$$

$$\equiv 776 \cdot 1136 \cdot 2048 \cdot 3007 \cdot 2201 \text{ (mod } 12007001).$$

Denoting the accumulated product by Q and the accumulated square root

$(\bmod\ N)$ by $\sqrt{R}$, we have

$$Q = 1, \quad \sqrt{R} = 1.$$

$$(776, 1136) = 8, \quad Q = \frac{776}{8} \cdot \frac{1136}{8} = 13774, \quad \sqrt{R} = 8.$$

$$(13774, 2048) = 2, \quad Q = \frac{13774}{2} \cdot \frac{2048}{2} = 7052288,$$

$$\sqrt{R} = 8 \cdot 2 = 16.$$

$$(7052288, 3007) = 97, \quad Q = \frac{7052288}{97} \cdot \frac{3007}{97} = 2253824,$$

$$\sqrt{R} = 16 \cdot 97 = 1552.$$

$$(2253824, 2201) = 2201, \quad Q = \frac{2253824}{2201} \cdot \frac{2201}{2201} = 1024,$$

$$\sqrt{R} = 1552 \cdot 2201 = 3415952.$$

In this way only the square root of the final product $Q = 1024$ needs to be computed and this square root is accumulated in $\sqrt{R}$:

$$\sqrt{1024} = 32 \text{ and } \sqrt{R} \equiv 3415952 \cdot 32 \ (\bmod\ N) \equiv 1247455 \ (\bmod\ 12007001).$$

Another technical detail, which we wish to remark on here, involves the Gaussian elimination. Since subsequent columns of the matrix contain the less common prime divisors of the Q_n's and thus fewer ONEs, the entire process of elimination runs faster if, contrary to customary practice, the elimination of ONEs is started at the *right-most column and is run from right to left*.

A third detail we include here concerns how to proceed in case the period of the continued fraction expansion of $\sqrt{N}$ is too short to supply sufficient easily factorized quadratic residues. This problem can be countered just as in the other versions of the continued fraction method: Expand $\sqrt{kN}$ instead of $\sqrt{N}$, for some suitably chosen integer k.—In this situation, of course, the reductions are still performed $(\bmod\ N)$, and not $(\bmod\ kN)$.

For the following reason the value of k chosen will influence the primes which go into the factor base: (5.42) with kN instead of N gives

$$A_{n-1}^2 - kNB_{n-1}^2 = (-1)^n Q_n. \tag{5.45}$$

Assuming that $p|Q_n$, then $kN \equiv (A_{n-1}/B_{n-1})^2 \pmod{p}$, and kN is a quadratic residue of p. Thus the factor base will consist of those (small) primes p for which $(kN/p) = +1$. Also the prime factors of k and the prime 2 occur in the factor base and thus have to be included. Manipulating with k we can, to some extent, gain control over the primes in the factor base.

Exercise 5.7. Controlling the factor base. Utilizing the FUNCTION Jacobi on p. 286, write a computer program helping you, for any given number N, to find the most efficient (square-free) multiplier $k \leq 1000$, such that a maximal number of primes ≤ 100 will belong to the factor base.

The Early Abort Strategy

If a Q_n does not have any small prime factors it is not likely to factor at all before the search limit p_m of the factor base has been reached. Thus it may be advantageous to give up the trial division on Q_n after a while, and instead produce a new Q_n and work on that one. If, however, there are small factors, the remaining co-factor of Q_n might still be unlikely to factor within the factor base, if no more factors appear for a while. Taking this into consideration, a rather complicated strategy for giving up the Q_n's unlikely to factor has been developed by Pomerance (see [20]). This strategy indeed speeds up the Morrison-Brillhart algorithm considerably and is thus of advantage to include in the computer program. Before concluding the description of the continued fraction factorization method, we should like to give a few rules of thumb for the abortion of Q_n's and also mention some results achieved by the method.

In [21] Pomerance and Wagstaff describe experiments with the Early Abort Strategy. As a result of these experiments they give the following recommendations for the case when *two* abort points are chosen: divide Q_n by the first 15 primes in the factor base. If the unfactored portion of Q_n is $> \sqrt{N}/500$, give up this Q_n and find Q_{n+1}. If not, perform trial division by 80 more primes from the factor base. If the unfactored portion now is $> \sqrt{N}/2 \cdot 10^7$, then give up. If not, continue the divisions up to p_m. This recommendation applies for numbers N in the range from 10^{40} to 10^{54}. The factor base used by these tuning experiments contained 959 primes.

Results

We shall mention here only a few of the factorizations discovered with this method. One is the factorization of the Fermat number

$$F_7 = 2^{128} + 1 = 5\,96495891\,27497217 \cdot 570468\,92006851\,29054721,$$

found in 1970 by Morrison and Brillhart after having computed 1,330,000 Q_n's in the expansion in $\sqrt{257F_7}$. Of these, 2059 were completely factored into small primes. These factorizations were combined to provide the following non-trivial solution of Legendre's congruence:

$$2335036\,48380835\,85217723\,21436182\,27956476^2 \equiv$$
$$\equiv 2518647\,81457280\,41297312\,27193485\,20212223^2 \pmod{F_7},$$

from which the factorization was deduced. The Gaussian elimination used 1504K bytes of computer storage and the whole computation took less than 2 hours of computing time on a fast computer.

More recently, Thorkil Naur [22] has implemented the continued fraction method on what he describes as a "reasonably fast computer which can be used almost exclusively for factorization." The computing times are reported to range from "less than an hour for 35 digit numbers to about 24 hours for 45 digit numbers and about a week for 50 digit numbers." The most difficult number reported by Naur was a 56 digit number N'_{11} which, after 35×24 hours, was factorized as

$$5603023\,94853703\,82805887 \times 8\,89340324\,57788067\,00898245\,74922371.$$

(The numbers Naur had chosen to factor are defined recursively by A. A. Mullin [23] as $p_1 = 2$ and

$$N_i = p_1 \cdot p_2 \ldots p_{i-1} + 1, \quad \text{where} \quad p_j = \text{the largest prime factor of } N_j.$$

Before the next Mullin number p_i can be determined, the largest prime factor of N_i must be found. The numbers N_i grow quite rapidly:

$$N_2 = 3, \quad N_3 = 7, \quad N_4 = 43, \quad N_5 = 1807 = 13 \cdot 139$$

$$N_6 = 251085 = 5 \cdot 50207$$

$$N_7 = 126\,03664039 = 23 \cdot 1607 \cdot 340999$$

$$N_8 = 42978368\,33293963 = 23 \cdot 79 \cdot 23653\,47734339$$

$$N_9 = 10165\,87861619\,05754590\,68761119 =$$
$$= 17 \cdot 1277\,70091783 \cdot 4\,68022256\,41471129$$

$$N_{10} = 89 \cdot 839491 \cdot 5\,56266121 \cdot 8363\,12735653 \cdot 13688452\,06580129$$

$$N_{11} = 1307 \cdot N'_{11},$$

where N'_{11} is the 56 digit number referred to above. These numbers are, incidentally, precisely the numbers we employed on p. 44 in order to construct an infinite set of primes.)

Running Time Analysis

The Morrison-Brillhart version of the continued fraction algorithm has been subject to detailed heuristic theoretical running time analysis [20]. The conclusion is that running time, if the Early Abort Strategy is used, increases as

$$C \cdot N^{\sqrt{1.5 \ln \ln N / \ln N}}. \tag{5.46}$$

Experience with computer programs for the continued fraction method seems to confirm the increase with N in the running time predicted by the theoretical analysis.

The Quadratic Sieve

In the Morrison-Brillhart method most of the computing time is spent on factoring the quadratic residues. What is particularly disadvantageous is that most residues *do not factor completely* within the factor base (even if the technique described allowing for a special factor is used). As we have already pointed out, the 2059 factored residues, mentioned above in the factorization of F_7, were found after as many as 1,330,000 trials. Another disadvantage is that if N is large the trial divisions have to be carried out on multiple precision numbers.

If a procedure could be devised such that those trial divisions which would end in failure were never performed, and if those residues could be directly pointed out, which factor completely within the factor base, clearly a great gain in computational labour would result. Such a procedure has been found by Carl Pomerance [20]. It is called the *quadratic sieve* and can be described as follows.

With $[\sqrt{N}] = m$, small quadratic residues can be generated as

$$Q(x) = (x + m)^2 - N, \quad x = 0, \pm 1, \pm 2, \ldots \tag{5.47}$$

These numbers form an arithmetic series of the second order. Each of the residues may be identified by its argument x. Now it so happens that for each prime p in the factor base, the prime power $p^\alpha | Q(x)$ at points x, *which* are *evenly spaced* with difference p^α because

$$Q(x + kp^\alpha) = (x + kp^\alpha + m)^2 - N \equiv Q(x) \ (\text{mod } p^\alpha) \qquad (5.48)$$

for $k = 0, \pm1, \pm2, \ldots$ Thus, if only *one single value* of x can be located, for which $p^\alpha | Q(x)$, then other instances of this event can be found by a sieving procedure on x, similar to the sieve of Eratosthenes for locating multiples of p^α in an interval.

Smallest Solutions to $Q(x) \equiv 0 \ (\text{mod } p)$

The values of x for which $p^\alpha | Q(x)$ are falling into two series, corresponding to the two solutions of the quadratic congruence

$$(x + m)^2 - N \equiv 0 \ (\text{mod } p^\alpha). \qquad (5.49)$$

If p is an odd prime not dividing N and if *one* solution $x_{\alpha-1}$ to (5.49) for $(\text{mod } p^{\alpha-1})$, where $\alpha > 1$, is known, then a whole series of solutions can be found by putting $x_\alpha = x_{\alpha-1} + zp^{\alpha-1}$, yielding

$$(x_{\alpha-1} + zp^{\alpha-1} + m)^2 - N \equiv (x_{\alpha-1} + m)^2 - N + 2zp^{\alpha-1}(x_{\alpha-1} + m).$$

Dividing by $p^{\alpha-1}$ we get

$$\frac{(x_{\alpha-1} + m)^2 - N}{p^{\alpha-1}} + 2(x_{\alpha-1} + m)z \equiv 0 \ (\text{mod } p). \qquad (5.50)$$

This is a linear congruence for z, whose solution exists and solves (5.49) for $\alpha > 1$. (The solution exists because $(p, N) = 1$ implies $(x_{\alpha-1} + m, p) = 1$.) The problem of solving (5.49) is thus reduced to solving

$$(x + m)^2 - N \equiv 0 \ (\text{mod } p), \qquad (5.51)$$

which has solutions when $(N/p) = 1$ (or when $p|N$, a case which has, however, been ruled out in advance by trying N for all small factors). For $p = 4n + 3$ (5.51) has the two solutions

$$x \equiv -m \pm N^{(p+1)/4} \ (\text{mod } p). \qquad (5.52)$$

For a prime p of the form $4n + 1$ the two solutions of (5.51) are more complicated to find (see remark on p. 288), but can still be found in nearly polynomial time.

Remark. Since p is small (less than about 20000, if there are about 1000 primes in the factor base), (5.49) could also be solved by trial and error. Or, better still, since solutions are sought for *all* primes in the factor base, one could perform some trial divisions with these primes and note for which values of x there is a "hit" with the prime p.

The divisibility rules for $Q(x)$ by powers of 2 are a little bit more complicated than for odd prime powers, and we do not discuss this here.

In order to save still more computing time, the multiple precision divisions occurring in the quadratic sieve in the search for small factors can be replaced by single precision subtractions in the following manner. For each argument x in the sieving interval, start by loading the corresponding entry with an approximate value of $\log |Q(x)|$. When a location x is identified for which $p^\alpha | Q(x)$, subtract $\log p$ from what is in the corresponding entry. (Remember that if $p^\alpha | Q(x)$, it has already earlier during the sieving process been established that $p^{\alpha-1} | Q(x)$.) Choose some bound B and sieve for all $p_i^\alpha \le B$, where $p_i \le p_m$ is in the factor base. Those quadratic residues which after the sieving have their corresponding entry ≈ 0 can now be computed and factored by trial division—they are guaranteed to factor.

Special Factors

Those residues for which the entries lie between $\log p_m$ and $2 \log p_m$ factor completely within the factor base except for one possible additional special prime factor q. (It may happen that such a residue is instead divisible by a power $> B$ of a prime in the factor base.) Just as is the case in the Morrison-Brillhart method, the special factors greatly increase the efficiency of the process, when several completely factored residues containing the same special factor q can be obtained. This is achieved by sieving for small factors on a subsequence of residues, corresponding to values of $Q(x)$ all divisible by q. James Davis and Diane Holdridge report [24] that this technique may, for large numbers N, reduce the sieving time needed to produce a nearly square matrix to about one-sixth of the time needed without this device.

Results

In [24] the factorization of some of the most difficult numbers so far factored is reported. They vary in size from 51 to 63 digits. The factor bases used contained between 6485 and 6568 primes, and the number of initial residues sieved ranged from $1.5 \cdot 10^8$ to $5.5 \cdot 10^9$. The number of

213

residues with special factors q ranged from 10^9 to $1.7 \cdot 10^{10}$. The running time ranged from one hour for $N = 2^{193} - 1$ to 7.82 hours for $2^{198} + 1$ on a CRAY-1 computer.—Through this effort the old question about the structure of the original Mersenne numbers $M_n = 2^n - 1$ for n prime ≤ 257 has been completely answered. The most difficult of these factorizations turned out to be M_{253}, M_{193}, M_{211} and M_{251}.—Quite recently also the 71-digit number $(10^{71} - 1)/9$ was factored in 9.5 hours by aid of SANDIA's CRAY XMP computer. The result is reported on p. 399.

Running Time Analysis

A heuristic running time analysis of the quadratic sieve algorithm using Gaussian elimination indicates an asymptotic running time of

$$CN^{\sqrt{1.125 \ln\ln N / \ln N}}, \qquad (5.53)$$

asymptotically faster than (5.46) and with a cross-over point with (5.46), which is highly machine-dependent but may be as large as $\approx 10^{60}$. With a new idea by Wiedemann for the elimination, the (heuristic) running time comes down to

$$CN^{\sqrt{\ln\ln N / \ln N}}. \qquad (5.54)$$

Using the fact that a 70-digit number can be factorized in ten hours time with a supercomputer, we can estimate the running time for $N \approx 10^{100}$ in the following way. The exponent in (5.53) takes the value 0.1776 for $N = 10^{70}$ and the value 0.1537 for $N = 10^{100}$. Thus the estimated running time for $N = 10^{100}$ would be

$$10 \cdot 10^{100 \cdot 0.1537 - 70 \cdot 0.1776} \approx 10^{3.94} \text{ hours} = 1 \text{ year},$$

and thus out of reach at present. It is obvious that very much more powerful methods will be needed before numbers as large as the largest primes known today can be factorized.

Schroeppel's Method

By speeding up one of the time-consuming parts of the Morrison-Brillhart algorithm, which involves the factorization or rejection of the Q_i's, Richard Schroeppel claims to have reduced the algorithm's running time for very large N. Using Gaussian elimination, however, Schroeppel's original scheme had heuristic running time

$$CN^{1.5\sqrt{\ln\ln N/\ln N}},$$

which is worse than for the Morrison-Brillhart method. By using faster elimination methods now available, Schroeppel's algorithm could be brought down to (heuristically)

$$CN^{\sqrt{\ln\ln N/\ln N}}.$$

We shall not provide further details here, but refer the reader to [15] and [20].—There is also an algorithm with a *proven* running time

$$O\left(N^{\sqrt{2.5\ln\ln N/\ln N}}\right),$$

given by Dixon, see [25].

The Schnorr-Lenstra Method

The ideas published by Shanks in [14] have recently been taken up and developed further by Schnorr and Lenstra [26]. The resulting factorization algorithm, which is quite complicated, has been subject to computer implementation and testing. It seems to be faster than the Morrison-Brillhart method, having an asymptotic mean running time of only

$$o\left(N^{\sqrt{\ln\ln N/\ln N}}\right).$$

However, even if the *mean* running time is smaller than (5.53), the running times for numbers of approximately the same size vary enormously, and thus the running time for any particular number is quite unpredictable. This disadvantage can partly be remedied the same way as in Shanks' method SQUFOF, where one or more multipliers may be utilized. Applying the algorithm on several numbers kN simultaneously will greatly enhance the possibility of hitting upon a factorization rapidly and will thus help avoiding the very long running times that otherwise occasionally may occur.

Strategies in Factoring

The various factorization methods we have described are all useful in different situations. When factoring a large number, the method to be chosen must depend on knowledge about the factors of the number. To begin with you must make sure that the number is composite, so that you do not make a long computer run in vain resulting in nothing. Also, there is a difference in the applicability of on the one hand, those methods which yield the factors in approximate order of magnitude, such as trial division or Pollard's ρ-method, and on the other hand, those algorithms where running time is roughly independent of the size of the factors, for instance Shanks' method or the continued fraction method. It is frustrating to discover, after a very long run that N has the prime factorization $N = 97 \cdot p$, which could have been obtained almost immediately by using trial division.

Further, you could take a chance using Fermat's method in case N is the product of two almost equal factors, or Pollard's $(p-1)$-method (or the $(p+1)$-version of it) in the event of N having one factor p with $p-1$ or $p+1$ being a product of only small primes.—Some or all of these methods may be combined in such a way that, say, 10000 steps are run with each method to discover a possible "easy" factor. Only after all this fails, do you need to fall back on the "heavy artillery," i.e. the time consuming continued fraction method.

A well-balanced strategy, based on extensive computing experience, has been developed by Naur [22]. We take the liberty to summarize Naur's strategy as follows:

1. Make sure N is composite. Since very small divisors are quite common and are found very quickly by trial division, it is worthwhile attempting trial division up to, say, 100 or 1000 even before applying the strong pseudoprime test.

2. Perform trial division up to 10^5 or 10^6 (or even higher, if all factors of N are of the form $p = 2kn + 1$ with n large!). If Pollard's ρ-method is available then trial division need only be performed to a much lower search limit, e.g. 10^4, since the small divisors will fall out rapidly also with Pollard's method. One reason why trial division with the small primes is useful, despite the fact that Pollard's ρ-method finds small factors quickly, is that the small factors tend to appear multiplied together when found by Pollard's ρ-method, and thus have to be separated by trial division anyhow! A second reason for using trial division prior to applying Pollard's method is that although the running time for Pollard's ρ-method is only $C_1 \sqrt{p}$ while that for trial division is $C_2 p$, the constant C_1 in Pollard's method is much larger than C_2, so that it is advantageous to first find the very small divisors by trial division rather than by Pollard's

method, and to subsequently proceed with Pollard's method on a number which has been reduced in size by the removal of its small factors. This correspondingly reduces the size of the constant C_1.— Apply a compositeness test on what is left of N, each time factors have been found and removed.

3. At this point you need to "chance your arm," and with a little luck, shorten running time enormously—it could even be decisive for quick success or complete failure in the case when N is very large! The strategy to be employed is: Take the methods you have implemented on your computer covering various situations, which will mean one or more of the following: Pollard's ρ-method for finding "small" factors, Pollard's $(p-1)$-method, the $(p+1)$-method, Fermat's method for nearly equal factors and whatever else you may have at your disposal. (Also Shanks' method can be included here— its running time being very irregular and unpredictable it may yield a factorization quickly.) The programs ought to incorporate re-start points, allowing the runs to be suspended after a certain number of steps and possibly later continued from the same point. Try all these methods for a reasonable running time, depending upon your computer's capacity. Since you cannot possibly know in advance which of these methods will achieve a factorization (if a factorization will be found at all!), it is a good technique at this stage to run the program of each method in sequence for a predetermined number of steps, say 1000 or 10000, and breaking the runs off at re-start points in order to be able to proceed, if necessary. If N does not factorize during such a run you have to repeat the whole process from the re-start points of the previous run. The combination of programs should now be tuned, i.e. the proportion of computing time given to each program adjusted so as to optimize the chance of success for the total amount of computing time expended. This can be realised by applying theoretical running time analysis, as well as practical experience from running the various programs in the program combination. It is also important at this stage not to consume excessive amounts of computing time on methods which have a comparatively high average running time, such as Lehman's method ($O(N^{0.33})$) or Shanks' method ($O(N^{0.25})$), at least not if the continued fraction algorithm with or without the quadratic sieve ($O(N^{\approx 0.15})$) is also available.

4. If the number N has still not been factored, you will need to rely upon the "heavy artillery." Depending on the size of the number and on the capacity of your computer, this can be Pollard's ρ-method, Shanks' method or, preferably, the continued fraction method. Now you have only to sit down and wait; fairly good estimates of maximal running times are available for all these methods, so that you will know approximately how long the computer run could take. Of course you could have a pleasant surprise—for instance, if you are using Pollard's method and some factor is not too large, or if you apply Shanks' method and it just happens to terminate fast.

How Fast Can a Factorization Algorithm Be?

The author of this book believes that much faster factorization methods could be devised than those in existence today, in particular the continued fraction factorization method. Whilst there have been algorithms published displaying a more favourable order of magnitude for the growth with N of the labour required, they are at the moment much too complicated for practical implementation on computers. Some of them might be considered as proofs of existence rather than feasible algorithms.

However, there are probably factorization methods yet to be constructed which show only *polynomial time* growth with $\ln N$, at least that is the author's opinion. The argument in favour of this conjecture is the following:

A variation of the Prime Number Theorem (see p. 47) states that

(see p. 47)

$$\sum_{p_i \le x} \ln p_i \sim x. \tag{5.55}$$

The fact that (5.55) follows from the Prime Number Theorem is easy to prove, using Stieltjes integrals:

$$\sum_{p \le x} \ln p = \int_{2-0}^{x} \ln x \, d\pi(x) = \left[\pi(x) \ln x \right]_{2-0}^{x} - \int_{2-0}^{x} \frac{\pi(x) \, dx}{x} \sim$$

$$\sim x + O\!\left(\frac{x}{\ln x}\right) \sim x, \quad \text{as } x \to \infty.$$

218

Therefore, we expect the *length* (in digits) of the product of the primes between x and y to be

$$\approx \sum_{x \le p_i \le y} \log_{10} p_i \approx (y - x) \log_{10} e = 0.43(y - x) \text{ decimal digits.} \quad (5.56)$$

The reader should now once again consider the expressions on p. 152. Here $y - x = 100$ and thus the products given should contain about 43 decimal digits. Indeed, the length of these products varies between 35 and 46 digits. As a matter of fact, the re-formulation (5.55) of the Prime Number Theorem states that the primes on average thin out at precisely the rate required to allow the product of all primes, in an interval of fixed length, to remain itself of fixed length. If we choose the interval higher up in the number series, then the individual primes will be larger but more scarce, and their product will still have approximately the same length (at least unless we are going so high up in the number series that there are no or very few primes in the interval under consideration).—Hence the product of all primes between t and $2t$ is a number containing about $0.43t$ decimal digits. Next, what can be said about the product x_k of all primes between 2^k and 2^{k+1}? This product has approximately $0.43 \cdot 2^k$ digits. Thus, if we let k assume successively the values $k = 1, 2, 3, \ldots$ and form the numbers

$$x_1 = \prod_{2 \le p < 4} p, \quad x_2 = \prod_{4 < p < 8} p, \quad \ldots, \quad x_k = \prod_{2^k < p < 2^{k+1}} p, \quad (5.57)$$

then we can use Euclid's algorithm to find the factor p of N in $\log_2 N$ trials, since the prime p will be present in x_k if $2^k < p < 2^{k+1}$, i.e. for $k = [\log_2 p]$. What is wrong with this approach? Two things. Obviously one is the unmanageable size of the larger of the numbers x_k (already x_{10} has 435 digits!), and the other, the fact that all the primes are needed in order to construct the x_k's. In what way can these difficulties be overcome? Firstly, the large size of the x_k's can be dealt with if $x_k \pmod{N}$ is computed rather than x_k. Since the length of x_k approximately doubles in each step, it is somewhat akin to the squaring and reduction $\pmod{N}$ previously encountered in Pollard's ρ-method! But how can we avoid having to use all the primes? Actually we cannot, but possibly we could avoid employing each individual prime *explicitly*. If we allow a few multiple prime factors to creep into the numbers x_k, then these numbers will grow a little faster than necessary, but perhaps will be easier to compute. Thus we could, for instance, instead of x_k choose $y_k = 2^{k+1}!/2^k!$, a number that is obviously

219

a multiple of x_k (i.e. contains all primes between 2^k and 2^{k+1} as factors). Using Stirling's formula, we find that the number of decimal digits in y_k is approximately $0.3k \cdot 2^k$. However, we can do even better than this by utilizing well-known numbers, such as the binomial coefficients

$$\binom{2^{k+1}}{2^k} = \frac{(2^k+1)(2^k+2)\ldots(2^{k+1}-1) \cdot 2^{k+1}}{1 \cdot 2 \ldots (2^k-1) \cdot 2^k},$$

which are also multiples of x_k, as none of the primes between 2^k and 2^{k+1} in the numerator can cancel out, not being present as factors in the denominator. This number has only $0.6 \cdot 2^k$ digits, and is thus only slightly longer than the true x_k. Therefore, if we could discover a means of generating the binomial coefficients $\binom{2^{k+1}}{2^k}$ by using only the four elementary operations and possibly some other operation which can be performed (mod N), then we would finally be able to obtain the factor p of N in $\log_2 p$ steps! If we were able to generate $\binom{2^{k+1}}{2^k}$ (mod N) in $O(\log N)^k$ seconds, then, as we shall see in a moment, a factorization algorithm operating in polynomial time would actually be at hand.

We may now look afresh at Pollard's ρ-method. The reader ought, for the moment, to forget all about the corner-stones of this method, the birthday problem and Floyd's cycle-finding algorithm, and instead regard the algorithm as follows: The product Q_i in (5.33) will successively contain more and more of the small primes, just like $2^k!$ or the product of the binomial coefficients

$$\binom{2}{1}\binom{4}{2}\binom{8}{4}\binom{16}{8}\cdots\binom{2^{k+1}}{2^k}$$

Unfortunately, the very large primes in Q_i, which are most unlikely to be factors of N anyway, are present in "normal" quantities, as demonstrated earlier using the algebraic model of the ρ-method, and so Q_i contains, on average, all the primes up to only about Ci^2 for some constant C, which is poor considering the huge size of Q_i. The number of digits of each new factor of Q_i doubles twice in each step, and so does the number of digits of Q_i, which thus turns out to be about $C \cdot 4^i$ for some constant C. If *only* the small primes were present in Q_i, then its size would allow all primes up to $\ln Q_i \approx C_1 + 4^i \ln 2$ to appear and, if this could be achieved, only $O(\ln p)$ steps would be necessary to find the factor p by performing Euclid's algorithm, which is $O(\log N)$ in time, on Q_i and N.

Summarizing this line of thought: Pollard's ρ-method may be regarded as a technique for generating huge integers which, after i steps, contain all the small primes up to some limit Ci^2 as factors, and the amount of work needed to find a specific prime factor p of N is therefore $O(\sqrt{p})$ steps. An "ideal" prime generator could, after i steps, contain the primes up to about Cr^i and would thus require only $O(\ln p)$ steps, with each step performing in $O(\log N)^{k+1}$ seconds, in order to identify p as a factor of N. This corresponds to polynomial time performance.

Even if such an ideal prime generator cannot be constructed, the enormous gap between the orders of magnitude $\sqrt{p}$ and $(\log p)(\log N)^{k+1}$ means that there is certainly much room for improvement of existing algorithms. Ultimately, the ideal could be approximated so closely that in practice, a factorization algorithm performing in nearly polynomial time is achieved.

Bibliography

1. R. Sherman Lehman, "Factoring Large Integers," *Math. Comp.* **28** (1974) pp. 637–646.

2. C. F. Gauss, *Disquisitiones Arithmeticae*, Yale University Press, New Haven, 1966, Art. 329–332.

3. G. H. Hardy and E. M. Wright, *An Introduction to the Theory of Numbers*, Fifth edition, Oxford, 1979, pp. 354–359, 368–370.

4. Wladyslaw Narkiewicz, *Number Theory*, World Scientific, Singapore, 1983, pp. 251–259.

5. Mark Kac, *Statistical Independence in Probability, Analysis and Number Theory*, Carus Math. Monogr. no. 12, John Wiley and Sons, 1959, pp. 74–79.

6. Donald E. Knuth and Luis Trabb-Pardo, "Analysis of a Simple Factorization Algorithm," *Theoretical Computer Sc.* **3** (1976) pp. 321–348.

7. Karl Dickman, "On the Frequency of Numbers Containing Prime Factors of a Certain Relative Magnitude," *Ark. Mat. Astr. Fys.* **22A** #10 (1930), pp. 1–14.

8. J. M. Pollard, "Theorems on Factorization and Primality Testing," *Proc. Cambr. Philos. Soc.* **76** (1974) pp. 521–528.

9. H. C. Williams, "A $p + 1$ Method of Factoring," *Math. Comp.* **39** (1982) pp. 225–234.

10. J. M. Pollard, "A Monte Carlo Method for Factorization," *Nordisk Tidskrift för Informationsbehandling (BIT)* **15** (1975) pp. 331-334.

11. Richard P. Brent, "An Improved Monte Carlo Factorization Algorithm," *Nordisk Tidskrift för Informationsbehandling (BIT)* **20** (1980) pp. 176–184.

12. Richard P. Brent and J. M. Pollard, "Factorization of the Eighth Fermat Number," *Math. Comp.* **36** (1981) pp. 627–630.

13. John Brillhart and John L. Selfridge, "Some Factorizations of $2^n \pm 1$ and Related Results," *Math. Comp.* **21** (1967) pp. 87–96.

14. Daniel Shanks, "Class Number, A Theory of Factorization, and Genera," *Amer. Math. Soc. Proc. Symposia in Pure Math.* **20** (1971) pp. 415–440.

15. Louis Monier, *Algorithmes de Factorisations d'Entiers,* IRIA, Paris, 1980, pp. 3.13–3.24.

16. R. J. Schoof, "Quadratic Fields and Factorization," printed in H. W. Lenstra, Jr. and R. Tijdeman, *Computational Methods in Number Theory, Part II,* Mathematisch Centrum, Amsterdam 1982, pp. 235–286.

17. Michael A. Morrison and John Brillhart, "A Method of Factoring and the Factorization of F_7," *Math. Comp.* **29** (1975) pp. 183–205.

18. D. H. Lehmer and R. E. Powers, "On Factoring Large Numbers," *Bull. Am. Math. Soc.* **37** (1931) pp. 770–776.

19. Maurice Kraïtchik, *Théorie des Nombres. Tome II,* Gauthiers-Villars, Paris, 1926, pp. 195–208.

20. Carl Pomerance, "Analysis and Comparison of Some Integer Factoring Algorithms," printed in H. W. Lenstra, Jr. and R. Tijdeman, *Computational Methods in Number Theory, Part I,* Mathematisch Centrum Tract 154, Amsterdam 1982, pp. 89–139.

21. Carl Pomerance and Samuel S. Wagstaff, Jr., "Implementation of the Continued Fraction Integer Factoring Algorithm," *Congr. Num.* **37** (1983) pp. 99–118.

22. Thorkil Naur, *Integer Factorization,* DAIMI report, University of Aarhus, 1982.

23. A. A. Mullin, "Recursive Function Theory," *Bull. Am. Math. Soc.* **69** (1963) p. 737.

24. James A. Davis and Diane B. Holdridge, "Factorization Using the Quadratic Sieve Algorithm, "SANDIA report, SAND83-1346, SANDIA National Laboratories, Livermore, 1983.

25. John D. Dixon, "Asymptotically Fast Factorization of Integers," *Math. Comp.* **36** (1981) pp. 255–260.

26. C. P. Schnorr and H. W. Lenstra, Jr., "A Monte Carlo Factoring Algorithm with Linear Storage," *Math. Comp.* **43** (1984) pp. 289–311.

27. Dennis Parkinson and Marvin Wunderlich, "A Compact Algorithm for Gaussian Elimination over $GF(2)$ Implemented on Highly Parallel Computers," *Parallel Computing* **1** (1984) pp. 65–73.

CHAPTER 6

PRIME NUMBERS AND CRYPTOGRAPHY

Practical Secrecy

There is a remarkable disparity between the degree of difficulty of the task of multiplication and that of factorization. Multiplying integers together is a reasonable exercise for a young child if the integers are small, and it remains a very straightforward task even when the integers are very large. The reverse operation, however, that of resolving a given integer into factors, is cumbersome except for the very smallest integers and becomes nearto impossible for large numbers. This assymmetry is exploited in a new kind of cryptosystem, called RSA after its discoverers, Rivest, Shamir and Adleman. In the RSA system secrecy is provided by placing a would-be codebreaker in a situation where in principle he *commands all information necessary* for reading the protected message but is confronted with an arithmetic task which in practice is prohibitively time-consuming.

In this chapter we describe and discuss the basic algorithms underlying such cryptosystems.

Since these systems seem to achieve secrecy without keeping any key secret they are often referred to by the term *Public Key Cryptosystems* or *Open Key Cryptosystems*. We prefer the designation *Open Encryption-Key Systems*, since there also is a decryption key which, of course, must be kept secret.

Keys in Cryptography

Encryption may be seen as an operation on a segment *(coding unit)* of plaintext T yielding a corresponding segment of ciphertext C according to an encryption function f:

$$C = f(T). \tag{6.1}$$

Decryption is the inverse operation, performed by the function f^{-1}:

$$T = f^{-1}(C). \tag{6.2}$$

In most conventional cryptosystems, the mathematical relation between f and f^{-1} is trivial. For decryption, the encryption tables need only be read the other way round. If encryption is performed as word-by-word substitution of codes for plaintext words, then the writer will require, say, an English-to-Crypto code-book while the reader will find it convenient to have a Crypto-to-English dictionary, if the crypto vocabulary is at all large. Either of these books can be obtained from the other by a mere sorting process.

The encryption function f is traditionally defined by an algorithm and a parameter for that algorithm. The parameter *(key of the day)* is changed more often than the algorithm and different parameters may be chosen for different correspondents sharing the same algorithm (and communications network and crypto machinery). It is customary for cryptographers to assume that the algorithm is known to illegitimate readers, *the enemy*, but that the parameter is not, and the parameter is therefore called the *key*. Similarly, f^{-1} is described by a different algorithm with its own parameter. Since these parameters, if at all different, can easily be derived one from the other, the term key is often used loosely for the parameters for both encryption and decryption.

Now, the innovation introduced by RSA lies in the design of a system where f^{-1} may remain unknown to someone who possesses f. The existence of such a crypto opens up fascinating perspectives. Since f can be openly announced, crypto traffic is not hampered by the traditional difficulties of safely conveying the key from one place to another. The key for each addressee could in fact be published in a telephone directory or—to take a more practical example—in conjunction with the list of mailboxes in a computerized message system. The cryptosystem becomes less vulnerable, since every correspondent will have sole responsibility for the decryption key applicable to messages directed to him. Unauthorized readers will gain nothing if they come into the possession of what is otherwize lethal for most cryptosystems, a pair of plain and crypto texts. In fact, a sender of encrypted text will not be able to decode what he himself has written!

In this context we can no longer consider the distinction between the encryption and decryption keys as one involving merely a practical reformulation. The *practical* inconvenience for the unauthorized reader is precisely the heart of the matter. This is the reason why we shall prefer the term Open *Encryption*-Key Systems.

In a famous paper [1] Claude Shannon, the father of mathematical information theory, discusses the security of secrecy systems. He shows that the plaintext can be completely and unambiguously recovered if and only if the redundancy of the plaintext is at least as large as the sum of the noise and the information content of the key. Here redundancy and information content of the key have to be taken in the quantitative sense defined by Shannon in his paper. He hastens to add that this is a theoretical lower limit: where a cipher is breakable in principle according to this condition, it could still be practically secure because the number of operations required to determine the key might be excessively large. Cryptosystems, he points out, differ not only in the size of the information content of their keys but also in the amenability of the cipher text to effective tests.—The ciphers now under discussion represent an extreme case in the relation between key size and computational complexity for the codebreaker! There is no secret key information to detect whatsoever and "only" a formidable computational problem!—From this point of view, the RSA-system can be regarded as having a key with information content zero. Such cryptosystems have not been in use earlier.

Arithmetical Formulation

It is often convenient to describe the functions f and f^{-1} by using arithmetical formulas. To do so, we first need to represent the text, plaintext as well as cryptotext, as a series of digits which can be split into parts and each part interpreted as an integer. Thus, each segment of text to be encrypted or decrypted (i.e. each coding unit) has first to be expressed as an integer. A common way of deriving at such an integer representation of the text is to replace each letter and punctuation mark by a two-digit decimal number and splitting the resulting decimal digit string into integers of suitable size.—In order to simplify the terminology in the following, we shall use the words plaintext and cryptotext also for the integer representations of the coding units of plaintext and cryptotext, respectively.

Example. A large class of encryption algorithms may be described as addition of a number to the (integer representing the) plaintext of a coding unit:

$$C = T + k, \quad \text{yielding} \quad T = C - k,$$

which is the corresponding decryption algorithm (where k, of course, is the key).

RSA Cryptosystems

The encryption algorithm in the RSA type of ciphers can be described as exponentiation of the plaintext, followed by reduction (mod N):

$$C \equiv T^k \pmod{N}. \tag{6.3}$$

Decrypting, then, would entail extracting the kth root of C (mod N). Now, while it is possible with reasonable computational effort, to raise to a power (mod N), it is virtually impossible for very large C and N to extract the corresponding root (mod N) because of the absence of any (known) reasonably fast root extraction algorithm (mod N) for *composite* numbers N. The function f, having no easily computed inverse function f^{-1}, is therefore said to be a *trapdoor* function. See [3].

However, if k and N have been suitably chosen, it turns out that there exists an integer k' such that

$$T \equiv C^{k'} \pmod{N}. \tag{6.4}$$

This means that *decryption as well as encryption* is performed by the computationally acceptable operation of raising to a power (mod N). In order to be able to perform this, the *recovery exponent* k' needs to be known and, though determined by k and N, k' is in fact not easy to find.

Thus the basis of these procedures is that

1. $a^{kk'} \equiv a \pmod{N}$, for any a.

2. Given N and k, it is difficult to determine k' or any other integer s such that $a^{ks} \equiv a \pmod{N}$.

How to Find the Recovery Exponent

This section contains deductions making use of certain results from group theory and number theory. Therefore, readers unfamiliar with these topics can either skip this section or, before reading on, consult Appendix 1, in particular Lagrange's Theorem on pp. 248–249, and Appendix 2, especially "The Structure of the Group M_n" on pp. 272–277.

The connection between the encryption exponent k and the recovery exponent k' in an RSA cryptosystem may be found by considering the structure of the group M_N of primitive residue classes (mod N). From Lagrange's Theorem it follows that every element a of this group has an order e which is a factor of the order of the group, i.e. of the number of group elements, which is Euler's totient function $\varphi(N)$. Thus, for any plaintext T transformed to an integer a (mod N), we have

$$a^{\varphi(N)} \equiv 1 \;(\text{mod } N), \tag{6.5}$$

provided that $(a, N) = 1$. (The exception $(a, N) > 1$ will be discussed below.) Thanks to Carmichael's Theorem on p. 276, however, we can do a little better than this, namely

$$a^{\lambda(N)} \equiv 1 \;(\text{mod } N), \tag{6.6}$$

where $\lambda(N)$ is Carmichael's function. The reason why (6.6) is better than (6.5) is that $\lambda(N)$ is always a *proper* divisor of $\varphi(N)$ when N is the product of distinct odd primes, $\lambda(N)$ being smaller than $\varphi(N)$ in this instance. The choice of N as a product of precisely two primes turns out to be the principal case in RSA systems, a circumstance in which $\lambda(N)$ takes a particularly simple form, viz. $\lambda(pq) = (p-1)(q-1)/\text{G.C.D.}(p-1, q-1)$. Now, Theorem A2.12 immediately gives the relationship between k and k':

$$a^{kk'} \equiv a \;(\text{mod } N) \quad \text{if} \quad kk' \equiv 1 \;(\text{mod } \lambda(N)). \tag{6.7}$$

Note that provided N is a product of distinct primes, (6.7) holds for *all* a, thus dealing with the above-mentioned exception $(a, N) > 1$ in Euler's Theorem.

Further, the function $f(T)$ used in the RSA system is constructed as follows: Take a composite number N which is difficult to decompose into prime factors, for instance the product of two *large* primes $p \cdot q$. Compute first $\lambda(N)$, then choose an integer m such that you can find a factorization of

$$m\lambda(N) + 1 = k \cdot k'. \tag{6.8}$$

Now let the encryption trapdoor-function $f(T)$ be

$$f(T) = C \equiv T^k \;(\text{mod } N). \tag{6.9}$$

Knowing the factorization (6.8), by (6.7) we can now recover T from C by computing

$$C^{k'} \equiv T^{kk'} \equiv T^{m\lambda(N)+1} \equiv T \ (\text{mod } N). \tag{6.10}$$

The decryption is performed by aid of this formula. However, if the factorization (6.8) *is not known* then $f(T)$ can be considered to be a true trapdoor-function which is practically impossible to invert. Since the decomposition of large integers into prime factors is, at present, difficult to accomplish, it is likewise difficult to crack the code by first splitting N into $p \cdot q$, from this factorization computing $\lambda(N) = \text{L.C.M.}[p-1, q-1]$ and, finally, determining k' when N and k (the encryption keys) are known.—It is because of this relationship between k and k' that the security of the RSA cryptosystems stands and falls with the difficulty of practically factorizing large integers.

A Worked Example

An easy way of becoming acquainted with the details of the RSA method is to work out an example in full. Let us first select two primes p and q of reasonable size rather than very large primes, in order to get an easier-to-handle example. We have to start by computing N and $\lambda(N)$:

$$p = 4403346 \ 54777631, \quad q = 1452951 \ 43558111$$

$$N = 63978 \ 48687952 \ 71438588 \ 31415041$$

$$\lambda(N) = [p-1, q-1] = [4403346 \ 54777630, \ 1452951 \ 43558110].$$

Applying Euclid's algorithm to $p-1$ and $q-1$, we find the G.C.D. of $p-1$ and $q-1$ to be 90, and so

$$\lambda(N) = \frac{(p-1)(q-1)}{90} = 710 \ 87207643 \ 91839803 \ 22589770.$$

Next, we have the choice of two methods in order to find integers m, k and k' satisfying (6.8). *One alternative* is to try to factorize $m\lambda(N)+1$ for $m = 1, 2, 3, \ldots$ successively until we obtain a factorization that can be used. The factorizations of the first six numbers of this kind turn out to be:

$$\lambda(N) + 1 = \ 710\ 87207643\ 91839803\ 22589771 =$$
$$= 1193 \cdot 17107 \cdot 55511 \cdot 2990957 \cdot 2\ 09791523$$

$$2\lambda(N) + 1 = 1421\ 74415287\ 83679606\ 45179541 =$$
$$= 47 \cdot 131 \cdot 199 \cdot 1716499 \cdot 3322357 \cdot 2\ 03474209$$

$$3\lambda(N) + 1 = 2132\ 61622931\ 75519409\ 67769311 =$$
$$= 6257 \cdot 674683 \cdot 2\ 97801601 \cdot 16\ 96366781$$

$$4\lambda(N) + 1 = 2843\ 48830575\ 67359212\ 90359081 =$$
$$= 17 \cdot 53 \cdot 5605331 \cdot 56\ 30220352\ 11575351$$

$$5\lambda(N) + 1 = 3554\ 36038219\ 59199016\ 12948851 =$$
$$= 15647 \cdot 17450633 \cdot 1\ 30172483\ 87079301$$

$$6\lambda(N) + 1 = 4265\ 23245863\ 51038819\ 35538621 =$$
$$= 2293 \cdot 29581 \cdot 49864411 \cdot 12610\ 58128567$$

Please note that the example given is unrealistically small. If you assume that no-one can factorize your N, then it is probably about as difficult to find the complete prime factorizations of the numbers $m\lambda(N) + 1$ (which are of approximately the same size as N) as it would be to factorize N, unless you happen to be the *only* factorization champion around. For a much larger value of N than the one we have chosen, you would probably find only one or two small prime factors in attempting the factorization of the first numbers in the series $m\lambda(N) + 1$. Thus, unless particularly fond of factorizing, you might stop after having computed the first factor 1193 of the initial number $\lambda(N) + 1$ and then choose $k = 1193$ and $k' = (\lambda(N) + 1)/1193 = 59586930\ 12901793\ 63220947$.

The other alternative for obtaining valid combinations of m, k and k', which happens to be useful even if there is no factorization program at all available, is to compute m such that $m\lambda(N) + 1$ is divisible by some factor k *that has been chosen in advance*. This leads to a linear congruence (mod k) for m, and is therefore quite a simple problem. If, as an example, we pick $k = 101$, then $m\lambda(N) + 1 \equiv 0 \pmod{101}$, which leads to $11m \equiv -1 \pmod{101}$ and $m \equiv 55 \pmod{101}$, and thus we may choose $m = 55$, yielding $k' = (55\lambda(N) + 1)/101 = 387\ 10855647\ 67833556\ 21212251$.

Suppose in this example we decide to use the first possible combination found, $m = 1$, $k = 1193$, then the public keys for the cryptosystem in question will be

$$N = 63978\ 48687952\ 71438588\ 31415041 \text{ and } k = 1193.$$

Next, we consider the minor problem of converting the original text to integers (mod N). This is achieved by translating the characters of the text into digits by using some code like the following one:

A = 01	K = 11	U = 21	1 = 31
B = 02	L = 12	V = 22	2 = 32
C = 03	M = 13	W = 23	3 = 33
D = 04	N = 14	X = 24	4 = 34
E = 05	O = 15	Y = 25	5 = 35
F = 06	P = 16	Z = 26	6 = 36
G = 07	Q = 17	, = 27	7 = 37
H = 08	R = 18	. = 28	8 = 38
I = 09	S = 19	? = 29	9 = 39
J = 10	T = 20	0 = 30	! = 40

etc. until we reach

$$\text{blank space} = 00.$$

There is no need to construct a special conversion code since many codes for information interchange could be used that do already exist. If, working on a binary computer, you utilize the internal representation for the characters of the alphabet, the conversion of the plaintext to numerical representation is effected automatically when the computer reads the text.

After deciding how to encode the letters and other characters, group the resulting numbers in such a way that the final number is $< N$. In our example, with N a 29-digit number, we group together 14 2-digit numbers, corresponding to a string of text consisting of 14 letters, figures or punctuation marks. Thus the original text is, this case, split into groups of 14 characters and, if necessary, extended with blanks to complete the final group of 14 characters.

Suppose now that we wish to encrypt the following message:

THIS IS AN EXAMPLE OF THE RSA PUBLIC KEY CRYPTOSYSTEM.

First we apply the code above to transform the characters in the text into two-digit groups. These are at the same time grouped together into 28-digit groups and the last one is filled out with blanks at the end:

$$T_1 = 20\,08\,09\,19\,00\,09\,19\,00\,01\,14\,00\,05\,24\,01$$

$$T_2 = 13\,16\,12\,05\,00\,15\,06\,00\,20\,08\,05\,00\,18\,19$$

$$T_3 = 01\,00\,16\,21\,02\,12\,09\,03\,00\,11\,05\,25\,00\,03$$

$$T_4 = 18\,25\,16\,20\,15\,19\,25\,19\,20\,05\,13\,28\,00\,00.$$

Next, these numbers are transformed by means of the trapdoor-function $C \equiv T^{1193}$ (mod N), leading to 29 digits of ciphertext in each:

$$C_1 = 38582\ 33510426\ 58202111\ 82477004$$
$$C_2 = 34717\ 15763572\ 72707638\ 39377289$$
$$C_3 = 02867\ 37837457\ 86578878\ 47201742$$
$$C_4 = 08831\ 02912032\ 34436084\ 99714688.$$

Finally, these four 29-digit numbers, written in sequence, constitute the complete ciphertext:

3858233510426582021118247700434717157635727270763839377289
0286737837457865788784720174208831029120323443608499714688

Consider now the converse problem. In order to *decipher*, we first split the ciphertext into groups of 29 digits (since N contains 29 digits), resulting in, of course, the four numbers C_1, C_2, C_3 and C_4 above. The next step is to transform each of these numbers by applying

$$T \equiv C^{5958693012901793632209 47} \ (\text{mod } N),$$

yielding T_1, T_2, T_3 and T_4 as *29-digit* numbers, which means that each number begins with a redundant zero. Deleting this superfluous zero and using the code for transcribing the characters, we finally convert this sequence of two-digit groups back to the original plaintext.

Selecting Keys

In order to find valid combinations of N, k and k', the size of the coding unit must first be decided upon, and thereby the order of magnitude of N. Thus, if the coding unit is chosen to contain M decimal places, then N must be at least equal to 10^M in order that all distinct messages possible within the coding unit have different representations when reduced (mod N). The longer the coding unit, the greater the labour of encryption and ineffiency for short messages, but the higher the security. With the computational resources available today it is reasonable to set M to, say, 100–200.

Further, we would like N to be composed of two primes, p and q, which are *both* large, so that the enemy cannot use the relatively simple factorization procedures available based on finding small factors. To determine such numbers N, p and q we select p in the range of, say, $N^{0.4}$ to

$N^{0.45}$ and search for a value of q close to N/p. In the next subsection we describe how to find primes of a given approximate size.

After having chosen p and q, we can obtain a valid pair k and k' by following one of the procedures demonstrated in the worked example above. Please note that k should not be taken too small because it might occasionally happen that for a very short message T, its ciphertext $C \equiv T^k \pmod{N}$ is simply T^k and because T^k is smaller than N it is thus not affected by the reduction (mod N). (Very small decimal representations of units of plaintext can be avoided by placing these texts first in the coding unit; not last. Taking this precaution, a small value of k does no harm!)

Finding Primes

In order to design an RSA-system, we must be able to find a prime P of a certain approximate size. One way this could be done would be simply to test random numbers of the wanted size for primality and then pick a prime when one occurs. However, if P is rather large, say about 10^{60}, the scarcity of primes will force us to make in average $\ln P \approx 138$ trials before a prime is produced. This large number of trials can be reduced in several ways. One way to enhance the probability for P to be prime is to choose P of some special form which avoids many of the small primes as factors, such as $P = 2 \cdot 3 \cdot 5 \cdot 7 \cdot 11 \cdot 13 \cdot 17 \cdot 19k \pm 1 = 9699690 \pm 1$. These two particular forms would reduce the 138 average trials to 24. However, when using this technique, you have to be careful not to end up with a prime P having $P - 1$ or $P + 1$ composed of only small prime factors, because in such a case the factor P of N can easily be revealed by using the $(p - 1)$- or the $(p + 1)$-method, and the code cracked.—Furthermore, any P constructed in this way carries with it another disadvantage if used in an RSA-system: the code will expose what we shall call many fixed points of low order. This will be explained in detail in the next subsection.

A safer way of producing a "random" prime P is to apply the sieve of Eratosthenes as described on pp. 5–6. Again, looking for a prime near 10^{60}, just pick a "starting point" near 10^{60} for the interval to sieve and then sieve, say 5000 consecutive odd integers with all odd primes < 1000. This will leave approximately 81 numbers untouched by the sieve, among them approximately 72 primes in the interval chosen. Test these numbers for primality until a prime is found.—To check that the prime picked is suitable for use in an RSA-system, apply the rules given in the following two subsections.

232

Exercise 6.1. Sieving for a prime in $(10^{60}, 10^{60} + 10000)$. In sieving with one of the trial divisors p in the sieve above, *one operation only* needs to be performed in multiple precision arithmetic, namely the computation of the first odd multiple of p above the number chosen (10^{60}). To represent the sieving interval $(10^{60}, 10^{60} + 10000)$ more conveniently, first displace it to $(0, 10000)$ and then remove all even integers. After this transformation the sieving with the trial divisor p can proceed exactly as shown in the program Eratosthenes on p. 6. However, the statement start:=(m DIV q)*q+p in the program has to be changed to start:=((10↑60+1)DIV q)*q+p-10↑60 and carried out in multiple precision arithmetic. (If you have no multiple precision arithmetic available you may instead take advantage of the special form of the number 10^{60} and compute $10^{60}+1$ (mod q) by successively computing 10^7 (mod q), 10^8 (mod q), 10^{15} (mod q), 10^{30} (mod q) and, finally, $10^{60} + 1$ (mod q).)—After having sieved the interval with all small primes, test some of the remaining integers for pseudoprimality with several bases and pick one of the pseudoprimes found. (Note that a pseudoprime for the base a works as well as a genuine prime in (6.5)–(6.7) and thus works fine in an RSA-system *as far as encryption and decryption is concerned!* The main disadvantage of using pseudoprimes is that N will be composed of much smaller prime factors than necessary and thus the code more vulnerable to cracking by factoring N.) Check that $p - 1$ and $p + 1$ are not composed of *only* small prime factors by making a factor search up to 10^5. (If $P = 10^{60} + z$, looking for a factor $p < 10^5$ of $P \pm 1$ can, as above, be done without using multiple precision arithmetic, by first computing $r \equiv 10^{60}$ (mod p) in the way described earlier and then proceeding by computing $r + z$ (mod p) and checking if this is 1 or $p - 1$. Do not divide N by any factors found during the trial division, because N/p is unlikely to show any special form useful for shortcuts during the computations. Just store the factors and reduce N in the end by the factors found. Do not forget to test all factors for multiplicity in N.)

The Fixed Points of an RSA System

The following fact imposes a serious restriction on the choice of p and q: there exist plaintexts T which are revealed by the application of the *encryption* algorithm on the corresponding ciphertext C! This can be explained as follows: Although $\lambda(N) + 1$ is the smallest exponent s for which $a^s \equiv a$ (mod N) for *all* a, there do exist *particular values* of a satisfying $a^s \equiv a$ (mod N) for much smaller values of s than $\lambda(N) + 1$. If such values are abundant for some rather small values of s, then many units of ciphertext will be transformed into plaintext simply by reducing low powers of the ciphertext (mod N)! In some instances even the *ciphertext itself* is identical to the plaintext! Thus, we could have

$$f(T) = C \equiv T^k \equiv T \ (\text{mod } N). \tag{6.11}$$

233

Such a text T is called a *fixed point* of the transformation $f(T)$. In the case when the text is revealed by applying f n times, T is called a *fixed point of order n*. Mathematically, the problem of determining all the fixed points of order n is equivalent to finding all solutions to the congruence

$$T^{k^n} \equiv T \ (\mathrm{mod}\ N). \tag{6.12}$$

It can be shown that the number of solutions to this congruence is

$$g(n) = (1 + (k^n - 1, p - 1)) \times (1 + (k^n - 1, q - 1)). \tag{6.13}$$

Hence, since k is always odd, the total number of fixed points (of all orders) lies between at least $(1 + 2)(1 + 2) = 9$ and a number which depends upon the choice of p, q and k. It is obvious from (6.13) that if $p - 1$ and/or $q - 1$ contains many small prime factors, then the chance that $(k^n - 1, p - 1)$ and/or $(k^n - 1, q - 1)$ is large for small values of n increases. Therefore, such choices should be avoided in order to reduce the risk that a substantial portion of the message can be revealed simply by repeatedly using the known transformation f on the ciphertext C.

Example. Let us apply formula (6.13) to our earlier example

$$N = 4403346\ 54777631 \cdot 1452951\ 43558111$$

with the choice $k = 1193$. Putting the computer to work, we find that

$$g(1) = g(3) = g(5) = g(7) = g(9) = g(11) = \ldots = 9,$$
$$g(2) = g(10) = g(14) = g(22) = g(26) = \ldots = 49,$$
$$g(4) = g(8) = g(16) = g(20) = g(28) = \ldots = 961,$$

and the rather horrifying

$$g(6) = 31141,\ g(12) = 745381,\ g(36) = 2219761,\ g(84) = 5217121, \ldots$$

This result, which makes our choice of p, q and k somewhat unsecure, is due to the factorizations of $p - 1$ and $q - 1$:

$$p - 1 = 2 \cdot 3^3 \cdot 5 \cdot 31249 \cdot 52189481, \quad q - 1 = 2 \cdot 3^2 \cdot 5 \cdot 7^2 \cdot 13 \cdot 25\ 34364967,$$

containing too many small prime factors to be recommendable.—It is far better from this point of view to choose p and q such that

$$p = 2p' + 1, \quad q = 2q' + 1 \quad \text{with } p',\ q' \text{ primes.} \tag{6.14}$$

With this choice, the only fixed points of low order are *normally* (there are exceptions) the nine fixed points of order 1, three of which have the numerical values 0 and ± 1. We may now ask what about the remaining six fixed points of order 1? Should they be avoided as coding units? Ironically, they ought not to be because, if they are explicitly pointed out, then p and q could be deduced and hence the code broken! Thus, we are in the peculiar situation that we must not reveal which messages are not hidden by the encryption.

How Safe is an RSA Cryptosystem?

As we have seen above, the choice of safe values of p and q is not trivial. One pitfall is the choice of p and q such that N is comparatively easy to factorize. As a matter of fact, the problem of factoring N and that of cracking the RSA code based on N can be shown to be mathematically equivalent. Thus, the safest procedure to follow after p and q have been chosen is to try to factorize $N = p \cdot q$ with all known methods, just in case N is vulnerable to easy factorization by some algorithm. A knowledge of the theoretical background of the various factorization methods in existence can help us to avoid certain dangerous situations such as the case when $p - 1$ or $q - 1$ has only small prime factors, in which circumstance the factorization of N is found without difficulty by Pollard's $(p - 1)$-method. (By the way, this is the reason why also in such instances the ciphertext is easily revealed by repeated application of the trapdoor transformation f.)

Further, since it is possible to factorize 75-digit numbers, it is obvious that N must be chosen much larger in order to obtain a safe cryptosystem based on the RSA method. However, the larger the value of N, the longer the blocks used in the encryption and decryption algorithms will be and the more time-consuming and costly will the procedure become.

Superior Factorization

All the arguments above concerning the difficulty of factorizing N rest on the assumption that we can estimate the enemy's capacity to factorize large numbers. The general methods available today require an effort which is proportional to $N^{\approx 0.15}$. Increased computer power will therefore only marginally aid the enemy. However, when faster computers do become universally available we will, in turn, be able to afford to increase the length of the coding unit and hence N.

However, in theory there is no reason why it should not be possible one day to devise a factorization method with a time consumption which grows only as a polynomial function of ln N, in which case it might be meaningful to spend the computer power needed to compute the factorization of N and thereby reveal the secret message. Since this requires new intellectual insight into number theory, there is no way of estimating the probability of it happening—or even the probability that it has already happened in some taciturn enemy's headquarters somewhere. In this sense RSA cryptos are, and will remain, unsafe.

Bibliography

1. Claude Shannon, "A Mathematical Theory of Communication," *Bell System Technical Journal* **27** (1948) pp. 379–423 and 623–656.

2. Martin Gardner, "A New Kind of Cipher that Would Take Millions of Years to Break," *Scientific Am.* **237** (Aug. 1977) pp. 120–124.

3. W. Diffie and M. E. Hellman, "New Directions in Cryptography," *IEEE Trans. Inform. Theory* **22** (1976) pp. 644–654.

4. R. L. Rivest, A. Shamir and L. Adleman, "A Method For Obtaining Digital Signatures and Public-Key Cryptosystems," *Comm. ACM* **21** (1978) pp. 120–126.

5. Alan G. Konheim, *Cryptography: A Primer*, Wiley-Interscience, New York, 1981.

BASIC CONCEPTS IN HIGHER ALGEBRA

Introduction

The study of various algebraic structures is included in higher algebra. Some of these structures are important in number theory, and so we give a short account of these. The examples given will frequently be taken from higher arithmetic.

Modules

Let us start by giving the

Definition A1.1. A set of numbers M is called a module, if the following is true: If x and y belong to M, then $x + y$ and $x - y$ also belong to M.

This may be expressed differently: A module is closed under addition and subtraction. Actually, it suffices to require closure under *subtraction only,* because if $x - y$ always belongs to M, so does $x - x = 0$, and consequently $0 - y = -y$, and finally $x - (-y) = x + y$.—A module is thus a very simple structure.

Examples of modules.

1. The set of all integers n, because the sum and the difference of two integers are always integers.
2. The set of all (integer) multiples of an arbitrary number, α (which need not be an integer). Obviously $n\alpha \pm m\alpha = (n \pm m)\alpha$ also belongs to the set, which thus is a module.

If a module contains the number α, it contains all multiples $n\alpha$ of α. If the module consists only of integers, we have this wonderfully simple theorem, which leads to quite far-reaching consequences:

Theorem A1.1. All elements of a module M, containing only integers, are multiples of a certain integer d, which is the smallest positive integer of M. (Exception: the module consisting of only the integer 0.)

Proof. Suppose that d is the smallest positive integer of M. Now, suppose further that M contains some integer x, which *is not* a multiple of d. Since x is not a multiple of d, it must fall between two consecutive multiples of d: $nd < x < (n+1)d$. But in such a case the integer

$$y = x - d - d - \cdots - d = x - nd$$

is also a member of M, because of its construction. But then obviously $0 < x - nd = y < d$, which implies that there is a positive number $y < d$ in M, which contradicts the definition of d as the *smallest* positive integer of M. Hence it is impossible to have any integer $\neq nd$ in the module, which proves the theorem.

If a module contains the numbers α_i, $i = 1, 2, 3, \ldots$, it also contains all (integer) multiples of these numbers and all sums and differences of these multiples (this is what is termed all linear combinations with integer coefficients of the numbers α_i). If this completely exhausts the module, every element x of the module can be written in the form

$$x = \sum_i n_i \alpha_i, \tag{A1.1}$$

where the n_i's are arbitrary integers. If none of the numbers α_i can be written as a linear combination with integer coefficients of the other α_i's, then the α_i's are called *generators* of the module, and the module (A1.1) is also the smallest module containing these generators. Then also the representation (A1.1) for a general element of the module is the simplest possible.

Example. The module having the generators 1, $\sqrt{2}$ and π has the elements

$$x = m + n\sqrt{2} + p\pi,$$

where m, n and p are arbitrary integers.

Theorem A1.1 above can also be re-formulated as: Every integer module is generated by its smallest positive element.

Euclid's Algorithm

An important problem is: Given the integers a and b, what is the smallest module M containing a and b as elements? Since M is obviously an integer

238

module and since each integer module consists of the integral multiples of its generator d, the problem is to find d. Since both a and b must be multiples of d, $a = a_1 d$ and $b = b_1 d$, d certainly is a common divisor of a and b. Since we are looking for the *smallest* possible M, d obviously has to be as *large* as possible. Thus, d is the *greatest common divisor* (G.C.D.) of a and b, denoted by $d = (a, b)$.

Effective computation of d from a and b is carried out by Euclid's algorithm which, by repeated subtractions of integers known to belong to the module M, finally arrives at d. The technicalities of Euclid's algorithm may be described in this way:

If $a = b$, obviously $d = (a, b) = a$, and the problem is solved. If $a \neq b$, we can, if necessary, re-name the integers and call the larger of the two integers a. (Without loss of generality we consider only *positive* integers a and b, since $-n$ has the same divisors as n.) Now subtract b from a as many times as possible to find the least non-negative remainder:

$$a - b - b - \cdots - b = r.$$

These subtractions are performed as a division of a by b, giving the quotient q and the remainder r:

$$a = bq + r, \quad \text{with} \quad 0 \leq r < b. \tag{A1.2}$$

The integer r belongs to the module M by construction. If $r > 0$, a still smaller member r_1 of M can be found by a similar procedure from b and r, which both belong to M:

$$b = rq_1 + r_1, \quad \text{with} \quad 0 \leq r_1 < r. \tag{A1.3}$$

In this manner, a decreasing sequence of remainders can be found until some remainder r_n becomes $= 0$. We then have

$$a > b > r > r_1 > \ldots > r_{n-1} > r_n = 0. \tag{A1.4}$$

Some r_n will $= 0$ sooner or later, because every strictly decreasing sequence of positive integers can have only a *finite* number of elements. The integer r_{n-1} is the desired G.C.D. of a and b. This follows from two facts: Firstly every common divisor of a and b is found in each of the remainders r_i. Secondly all the integers a, b, r, r_1, r_2, $\ldots$, r_{n-2} are multiples of r_{n-1}.

Example. Find $(8991, 3293)$. The computation runs:

$$8991 = 2 \cdot 3293 + 2405$$
$$3293 = 1 \cdot 2405 + 888$$
$$2405 = 2 \cdot 888 + 629$$
$$888 = 1 \cdot 629 + 259$$
$$629 = 2 \cdot 259 + 111$$
$$259 = 2 \cdot 111 + 37$$
$$111 = 3 \cdot 37.$$

Thus $d = 37$, and we find $a = 37 \cdot 243$, $b = 37 \cdot 89$.

Since $-n$ has the same divisors as n, Euclid's algorithm operates equally well when using the smallest *absolute* remainder instead of using the smallest *positive* remainder, as we have done above. Normally this cuts down the number of divisions required. This variant, performed on the same integers as above, runs:

$$8991 = 3 \cdot 3293 - 888$$
$$3293 = 4 \cdot 888 - 259$$
$$888 = 3 \cdot 259 + 111$$
$$259 = 2 \cdot 111 + 37$$
$$111 = 3 \cdot 37.$$

The Labour Involved in Euclid's Algorithm

The worst possible case in Euclid's algorithm is when all successive quotients $= 1$, because in this case the remainders decrease as slowly as they possibly can, and the computation will take a maximal number of steps for integers of a given size. It can be shown that this case occurs when $a/b \approx \lambda = (1 + \sqrt{5})/2 = 1.618$, where the maximal number of steps is about $\log_\lambda a$ which is $4.8 \log_{10} a$. The *average* number of steps for randomly chosen integers a and b is much smaller and turns out to be $1.94 \log_{10} a$. This slow increase of the computational labour as a grows is very favourable indeed, because it implies that if we double the *length* (in digits) of a and b, the maximal number of steps needed to find (a, b) is only doubled. If we also take into account that the multiplication and division labour, using reasonably simple multiple-precision algorithms in a computer, grows *quadratically* with the length of the numbers involved, then the computing labour for Euclid's algorithm in total is at most

$$O(\log \max(a, b))^3. \tag{A1.5}$$

This growth of the labour involved in executing an algorithm is called *polynomial* growth (because it is of *polynomial* order of log N, N being the size of the number(s) involved). The best algorithms found in number theoretic computations are of polynomial order growth.

A Definition Taken From the Theory of Algorithms

The amount of computational labour required to solve a certain mathematical problem usually depends on the size of the problem given, i.e. on the number of variables or on the size of the numbers given, etc. For instance, the work needed to solve a system of n linear equations is proportional to n^3, and the work involved in computing the value of the product of two very large numbers, using the normal way to calculate products, is proportional to the product of the lengths of the numbers, which means that numbers twice as large will demand computations four times as laborious.

Within the theory of algorithms, it is important to study how much computational labour is required to carry out different existing algorithms. In this context, the size of a variable is usually defined using its *information content*, which is given by a number proportional to the length of the variable *measured in digits*. The amount of computational labour needed to carry out the algorithm is then deduced as a function of the information content of each of the variables occurring as input to the problem in question. That is the reason why the work involved in Euclid's algorithm is said to grow *cubically* (tacitly understood "with the length of the largest of the two numbers a and b given as input to the algorithm"). The most important algorithms in number theory are those which have only polynomial order of growth, since for these algorithms the work needed grows only in proportion to some power of log N. This is very favourable indeed, and such algorithms are applicable to very large numbers without leading to impossibly long execution times for computer runs. As examples, we may mention simple compositeness tests, such as Fermat's test on p. 91, and the solution of a quadratic congruence with prime modulus on p. 287.

For many algorithms there is a trade-off between the computing time and the storage requirements when implementing the algorithm. This aspect is most important in all practical work with many of the algorithms described in this book. As an example we give factor search by trial division in which very little storage is needed, if implemented as on p. 150. If much storage is available, however, the algorithm can be speeded up considerably by storing a large prime table in the computer, as described on p. 10. Another example is the computation of $\phi(x,a)$ on pp. 17–19, where the function values are taken from a table for all small values of a and have to be computed recursively for the larger values of a. The border line between the two cases depends on the storage available in the computer, and affects the speed of the computation in the way that a larger storage admits a faster computation of $\phi(x,a)$ for any fixed, large value of a.

A Computer Program for Euclid's Algorithm

A PASCAL function for the calculation of (a,b) by means of Euclid's algorithm is shown below

```
FUNCTION Euclid(a,b : INTEGER) : INTEGER;
{Computes (a,b) with Euclid's algorithm}
VAR m,n,r : INTEGER;
BEGIN m:=a; n:=b;
  WHILE n <> 0 DO BEGIN r:=m MOD n; m:=n; n:=r END;
  Euclid:=m
END {Euclid};
```

Exercise A1.1. Fast search for very small factors $\leq G \approx 30$. Write a program using the FUNCTION Euclid above to find all prime factors $\leq G$ of a single precision integer N. Choose the search limit G such that the product P of all primes $\leq G$ is as large as possible in single precision. (Thus, if the word length of your computer is 36 bits, $G = 29$, since $2 \cdot 3 \cdot 5 \cdots 29 = 6469693230 < 2^{35}$, while $2 \cdot 3 \cdot 5 \cdots 31 > 2^{35}$.) The result $d = (N,P)$ from Euclid's algorithm can now be used in various manners. If $(N,P) = 1$, there are no factors $\leq G$ and the factor search has to proceed with the larger primes. If $(N,P) > 1$, then N has one or more small prime factors. These can be found by trial division, but in many cases they are found much faster by a table look-up procedure. Write such a PROCEDURE factors(d), covering $d \leq 1000$, by utilizing a 2-dimensional ARRAY in which all integers ≤ 1000, which can be written as products of different prime factors $\leq G$ only, have their standard factorizations stored in the same form as in the PROCEDURE divide on p. 150. If $d > 1000$ (or whatever limit you have set for your PROCEDURE factors) the small factors can be found by trial division.

In order not to miss any multiple small prime factors, the whole algorithm so far described has to be repeated on N/d until Euclid's algorithm delivers the result $d = 1$.

Furthermore, in order to reduce the number of repetitions caused by the occurrence of multiple small prime factors, it might speed up the computer program to choose P not as the product of *primes* $\leq G$, but rather as the product of the *largest powers* $\leq G$ *of each prime* $\leq G$. (Again, if the computer's word length is 36 bits, then $P = 7 \cdot 9 \cdot 11 \cdot 13 \cdot 16 \cdot 17 \cdot 19 \cdot 23 \cdot 25 = 22544121600$.)—Make computer experiments with both versions of this approach. (Note that in the second variant the table utilized by the PROCEDURE factors has to be tailored according to the set of numbers ≤ 1000 which with this changed P can result from Euclid's algorithm on N and P.)

As test cases for numerical experiments, time the factorization of all integers between 10^6 and $10^6 + 10^3$ by the two variants. Also compare with the running times by direct factor search from the very beginning by trial division as exposed on p. 150 and also by instead utilizing a pre-stored table of primes.

The technique described may be used to speed up the factorization of quadratic residues in connection with the Morrison-Brillhart method of factorization given on p. 199 ff. In this case P is chosen as a product of powers of primes in the factor base.

Reducing the Labour

When applying Euclid's algorithm to very large integers demanding the use of multiple-precision arithmetic, much of the labour can be reduced. One way to achieve this is due to D. H. Lehmer, who observed that the quotient $q = a/b$ as well as the subsequent quotients q_i are seldom larger than one computer word. Thus, this quotient q can be computed almost always by dividing the *leading parts only* of the numbers a and b, while the remainder $a - bq$ will not require full multiple-precision multiplication but rather the much simpler operation of multiplication of a multi-precise integer by *one single word*. The number of division steps is unaffected by this simplification but each step will take only a time which is $O(\log \max(a, b))$, so that the time required for running the entire algorithm is thereby reduced to $O(\log \max(a, b))^2$ from $O(\log \max(a, b))^3$.

Binary Form of Euclid's Algorithm

Another way of avoiding division and multiplication of multi-precise integers is to apply the *binary form* of Euclid's algorithm. It uses subtractions only, coupled with the fact that it is particularly easy in a binary computer to divide integers by a power of 2. The integer has only to be shifted s positions to the right to effect division by 2^s. The scheme for finding $d = (a, b)$ according to this algorithm is:

1. Suppose that a ends in α binary zeros and b in β binary zeros. Let $d = 2^{\min(\alpha, \beta)}$.

2. Form $a' = a/2^\alpha$ and $b' = b/2^\beta$, i.e. shift a and b to the right until all binary zeros at the right ends are shifted away.

3. Now both a' and b' are *odd* numbers. Form $c = a' - b'$ which is *even*. Shift away the trailing zeros of c and repeat step 3, this time with the reduced c and $\min(a', b')$. When the result of the subtraction becomes 0, the current value of a' is the largest *odd* factor d' of (a, b).

4. Finally, put $(a, b) = dd'$.

It is quite easy to show that the binary form of Euclid's algorithm takes at most about $\log_2 \max(a, b) = 3.32 \log_{10} \max(a, b)$ steps. However, as in the case of the ordinary Euclidean algorithm, the *average* number of steps is smaller, approximately $2.35 \log_{10} \max(a, b)$ and thus, because of the simplicity of the operations performed in each step, the binary form competes very favourably with the standard version of the algorithm. For the deduction of this last result, see [1].

Example. Let us take the same example as above, $(8991, 3293)$. We shall write down all the numbers involved in decimal notation, so that the right shifting of numbers will be replaced here by division by some power of 2. We follow the algorithm constructed above and obtain:

1. Both $a = 8991$ and $b = 3293$ are odd, so $\alpha = \beta = 0$, and $d = 1$ initially.

2. $a' = 8991$, $b' = 3293$ give

3. $c = 8991 - 3293 = 5698 = 2 \cdot 2849$.

2a. $a'_1 = 3293$, $b'_1 = 2849$ give

3a. $c_1 = 3293 - 2849 = 444 = 4 \cdot 111$.

2b. $a'_2 = 2849$, $b'_2 = 111$ give

3b. $c_2 = 2849 - 111 = 2738 = 2 \cdot 1369$.

2c. $a'_3 = 1369$, $b'_3 = 111$ give

3c. $c_3 = 1369 - 111 = 1258 = 2 \cdot 629$.

2d. $a'_4 = 629$, $b'_4 = 111$ give

3d. $c_4 = 629 - 111 = 518 = 2 \cdot 259$.

2e. $a'_5 = 259$, $b'_5 = 111$ give

3e. $c_5 = 259 - 111 = 148 = 4 \cdot 37$.

2f. $a'_6 = 111$, $b'_6 = 37$ give

3f. $c_6 = 111 - 37 = 74 = 2 \cdot 37$.

2g. $a'_7 = 37$, $b'_7 = 37$ give

3g. $c_7 = 0$, and the algorithm terminates with $(a, b) = d' = 37$.

The binary version of Euclid's algorithm is particularly advantageous if implemented in assembly code combined with a multiple-precision arithmetic package.

The Diophantine Equation $ax + by = c$

Suppose that a, b and c are integers. When does the equation

$$ax + by = c \qquad (A1.6)$$

admit integer solutions x, y? Well, all integers of the form $ax + by$, according to theorem A1.1, constitute a module M with generator $d = (a, b)$, which can be found by Euclid's algorithm. Thus (A1.6) has integer solutions if and only if c belongs to M, i.e. if and only if c is a multiple of d. The notation for this is $d|c$ (d divides c). Hence, we have

Theorem A1.2. The equation $ax + by = c$ has integer solutions if and only if $(a, b)|c$.

Groups

We now turn our attention to another, most important, algebraic structure, namely the *groups*. A group is a set together with a composition rule, called group operation or group multiplication, by which any two group elements can be composed or multiplied, the result always being an element of the group. This is formalized in the following

Definition A1.2. A set G with a composition rule is called a group if

1. Every ordered pair A, B of elements of G has a uniquely determined "product" C, which also belongs to G (Notation: $AB = C$).

2. The associative law always holds for the group operation, i.e. if A, B and C are group elements, then $A(BC) = (AB)C$; this product may thus, without notational ambiguity, be denoted by ABC.

3. G must have a so-called neutral element I, satisfying $AI = IA = A$ for all A belonging to G.

4. Every element A of G must have an *inverse element* A^{-1} in G which satisfies $AA^{-1} = A^{-1}A = I$.

245

Examples. All rationals $\neq 0$ form a group, if ordinary multiplication of rationals is chosen as group operation. The number 1 is the neutral element, and the inverse element of a/b is b/a.—The set of *all* rationals forms another group, if the group operation is the usual *addition* of rationals. In this case 0 is the neutral element $(a + 0 = 0 + a = a)$, and the inverse of a is $-a$, since $a + (-a) = 0$.

What can be said about the elements i, a, b and c possessing the following multiplication table (in which each entry is formed by the composition of an element in the leftmost column and one in the top row)?

	i	a	b	c
i	i	a	b	c
a	a	c	i	b
b	b	i	c	a
c	c	b	a	i

Do they constitute a group? To investigate this, we need only to verify the requirements in the definition of a group given above. Let us test some cases. Is $(ab)c = a(bc)$? The multiplication table for the group gives $(ab)c = ic = c$, and $a(bc) = aa = c$, which agrees. Thus, in this particular case, associativity holds. By systematically working through all possible cases of products of 3 elements (64 cases), we can prove the associative law in general for this multiplication table. In addition, the element i acts as neutral element, since $i \cdot a = a$ etc. a and b are each other's inverse elements due to the fact that $ab = i$ and i and c are their own inverses (i always is, and $cc = i$).

The following group multiplication table shows the structure of the simplest non-commutative group, which is of order 6. In this table the neutral element is again denoted by i:

	i	a	b	c	d	e
i	i	a	b	c	d	e
a	a	b	i	e	c	d
b	b	i	a	d	e	c
c	c	d	e	i	a	b
d	d	e	c	b	i	a
e	e	c	d	a	b	i

Group theory is extremely important within many branches of algebra and number theory. In this appendix we shall develop it as far as necessary for our applications to number theory.

We commence by considering products of identical elements:

$$A \cdot A \cdots A \quad (n \text{ factors}).$$

We denote this product by A^n. Just as in elementary algebra, it is easy to verify the rules for power products

$$A^m \cdot A^n = A^{m+n} \tag{A1.7}$$

and

$$(A^m)^n = A^{mn}, \tag{A1.8}$$

if m and n are positive integers. Since the inverse of A, A^{-1}, exists by definition, these laws can be extended to all integers m and n; not necessarily positive. We have, as in elementary algebra, to identify A^0 with the neutral element I of the group and A^{-n} with $(A^{-1})^n$, which is the inverse element of A^n, as can readily be seen by forming

$$A^{-n}A^n = A^{-1}A^{-1}\cdots A^{-1}A \cdots AA = I.$$

(Due to associativity, we can group together the factors of this product pairwise, starting at any point, e.g. in the middle. There $A^{-1} \cdot A = I$ cancel and this continues, until we finally arrive at the first and last factors: $A^{-1}IA = A^{-1}A = I$.)—The elements A^2, A^3, ... are said to be generated by A. Before we can proceed, some further definitions are necessary.

Definition A1.3. If the number of elements of a group G is finite and $= n$, G is said to be a *finite* group, and n is termed the *order* of the group.

Definition A1.4. If AB always $= BA$, i.e. if commutativity always holds for the group operation, then G is called an *abelian* group.

Example. The group studied above, having elements i, a, b, c, is finite with order $= 4$. Also, it is easily seen from its multiplication table, that it is an abelian group. (This follows because the table is symmetric with respect to its "main diagonal" $i - c - c - i$).

Now consider the successive powers A^n, $n = 1, 2, 3, \ldots$, for an element A of a *finite* group G. Obviously all the powers A^n cannot be distinct because only a finite number of possibilities exist for elements of G. Hence, sooner or later some power A^r will coincide with an earlier power A^s, which implies that $A^{r-s} = I$. Some particular power A^e of A will thus be equal to the neutral element I of G.

Lagrange's Theorem. Cosets

Which values are possible for the exponent e, encountered in the previous paragraph? The answer to this question is provided by a beautiful theorem of Lagrange:

Theorem A1.3. Lagrange's Theorem. If A is an element of a finite group G of order n, then $A^n = I$.

Lagrange's theorem implies that $e|n$. This is because if the smallest positive exponent s for which $A^s = I$ is called e, then *all* exponents for which $A^s = I$ constitute a module M generated by e. e is also termed the *order* of the element A in the group G. The elements of the module M are all the multiples of e. However, since $A^n = I$ then n belongs to this module, and thus n must be a multiple of e.

Now let us prove Lagrange's theorem. If e is the order of the element A, then all the elements $A, A^2, A^3, \ldots, A^e = I$ are necessarily distinct. (If they were not, then some power A^r would equal A^s, i.e. $A^{r-s} = I$ for $r - s < e$, contradicting the definition of e as the *smallest* positive exponent for which $A^e = I$.) In fact, the elements $A, A^2, \ldots, I$ constitute an abelian group G_1 of their own, generated by A, and of order e. This group is called a *subgroup* of G, since it is entirely contained in G. Now, if e chances to be $= n$, the subgroup G_1 exhausts all the elements of G and $A^e = A^n = I$, so that Lagrange's theorem follows immediately. If this is not the case, we must take any element B of G outside G_1, and consider the set of elements

$$AB, A^2B, A^3B, \ldots, A^eB = IB = B. \tag{A1.9}$$

None of these elements of G belongs to G_1, and they are all distinct. This is because if any of (A1.9) were in G_1, we would have $A^rB = A^s$, i.e. $B = A^{-r}A^s = A^{s-r}$, contrary to the choice of B which, being outside G_1, is certainly not a power of A. (Note that even if $s - r < 0$, A^{s-r} is an element of G_1, namely the element A^{e+s-r}.) Further, all of (A1.9) are distinct since $A^rB = A^sB$ would once again imply $A^{r-s} = I$ for $r - s < e$, a contradiction. The set (A1.9) is said to be a *coset* of the subgroup G_1.

Now, if G_1 and the coset (A1.9) together exhaust G, then $n = 2e$, and again the proof is completed. If not, we can, however, choose a new element C of G, which is neither in G_1 nor in the coset (A1.9), and form a second coset to G_1, consisting of the elements

$$AC, A^2C, A^3C, \ldots, A^eC = C. \tag{A1.10}$$

Following a similar reasoning, we find that this coset (A1.10) has no elements in common with the subgroup G_1, and that all its elements are distinct. The important property is that *the elements of this new coset (A1.10) are all dinstinct from the elements of the first coset (A1.9).* A proof of this fact is as follows: If $A^r B = A^s C$, then $C = A^{r-s} B$ for some value of $r - s$ which is $< e$. This is contrary to the assumption that C is chosen outside the coset (A1.9). (Again, if $r - s < 0$, then $e + r - s$ is > 0 and $< e$.) This proves that any cosets of G_1 are either composed of entirely different elements, or are identical. From this it follows that all elements of G may be organized in a number of cosets, mutually distinct, and each containing e elements:

$$
\begin{aligned}
&A, \quad A^2, \quad \ldots, \ A^e \ = I \\
&AB, A^2 B, \ \ldots, \ A^e B = B \\
&AC, A^2 C, \ \ldots, \ A^e C = C \\
&\qquad \vdots \\
&AH, A^2 H, \ \ldots, \ A^e H = H.
\end{aligned}
\tag{A1.11}
$$

But then $n = e \times$ (the number of cosets h), $A^n = A^{eh} = (A^e)^h = I^h = I$, and the proof is complete. We shall encounter Lagrange's theorem in several disguises in the theory of numbers, namely as Fermat's theorem and as Euler's theorem, two special cases of Lagrange's theorem discovered prior to the general theorem.—We conclude this subsection by giving an example of a group divided into cosets. Let us take the non-commutative group of order 6, whose multiplication table we have given above. The element c has order 2, because $cc = i$, and thus the aggregate (i, c) constitutes a subgroup. Pick an element $\neq i$ or c, e.g. a. The coset will have the two elements $(ia, ca) = (a, d)$. Now there only remain the elements b and e, which so far have not been included in any coset. Pick one of these, say b, and form the next coset: $(ib, cb) = (b, e)$ which finally exhausts the group. So the result is that if we start with the subgroup of order 2, (i, c), the group is split into three disjoint cosets of two group-elements each:

$$
(i, c), \quad (a, d) \quad \text{and} \quad (b, e).
$$

Abstract Groups. Isomorphic Groups

As long as we are merely interested in the intrinsic structure of a group, and do not wish to take into account the particular mathematical type of its elements, the notation we choose for the group elements is irrelevant. Such a group is called an *abstract group*. Of course, such a group is as "real" as any of the groups which we recognize in mathematics, such as the group mentioned previously of all non-zero rational numbers with the group operation being ordinary multiplication. Moreover, the group mentioned earlier of order 4 is well-known in mathematics, and it is called the multiplicative group of primitive residue classes (mod 5). See Appendix 2, p. 270. If the primitive residue classes (mod 5) are denoted by 1, 2, 3 and 4, the multiplication table of the group takes the form

	1	2	3	4
1	1	2	3	4
2	2	4	1	3
3	3	1	4	2
4	4	3	2	1

By comparing this table with the first one given on p. 246, we find them identical, *apart from the notation of the group-elements*, $i = 1$, $a = 2$, $b = 3$ and $c = 4$. When two groups have the same structure, such as these two, they are called *isomorphic groups*. If some property, depending only on the group-structure, is known for the elements of a given group, the corresponding property also holds for the elements of any group isomorphic to the given group.

It now appears that the group we are studying is also isomorphic to another well-known group, namely the group of *all* residue classes (mod 4), if the group operation is chosen as addition (mod 4). In this case, the group operation table is actually an "addition table" with the following appearance:

	0	1	2	3
0	0	1	2	3
1	1	2	3	0
2	2	3	0	1
3	3	0	1	2

By once again, comparing the table with the multiplication table of the abstract group, we recognize the two groups as being isomorphic when choosing $i = 0$, $a = 1$, $b = 3$ and $c = 2$. It is, however, also possible to select $i = 0$, $a = 3$, $b = 1$ and $c = 2$. Thus, two groups may be isomorphic in several different ways.

In an abelian group it is often convenient to call the group operation addition instead of multiplication. In so doing the isomorphism found above can be expressed thus: the multiplicative group of primitive residue classes (mod 5) is isomorphic to the additive group of all residue classes (mod 4). In an additively written abelian group, the power A^n is replaced by the multiple nA. The inverse of A is denoted by $-A$ and the neutral element is denoted by 0. We recognize this structure as a module, apart from the fact that the group-elements are not always real or complex numbers, but may be other mathematical objects. Thus, an additively written abelian group may be regarded as a generalization of a module.

The Direct Product of Two Given Groups

Definition A1.5. Let two groups be given, G of order g and with neutral element I_G, and H of order h and with neutral element I_H. Consider all pairs (G_i, H_j) of elements, one from G and the other from H. These gh pairs form a new group denoted by $G \times H$ and called the *direct product* of G and H, if composition in the new group is defined by

$$(G_i, H_j) \cdot (G_k, H_l) = (G_i G_k, H_j H_l).$$

It is easy to prove that this rule of composition conforms with the requirements for a group. Firstly, (I_G, I_H) behaves as the neutral element of $G \times H$, and the inverse element of (G_i, H_j) is obviously (G_i^{-1}, H_j^{-1}). Secondly, since associativity holds in each of G and H, it also holds in $G \times H$, since

$$\{(G_i, H_j) \cdot (G_k, H_l)\} \cdot (G_m, H_n) =$$
$$= (G_i G_k, H_j H_l) \cdot (G_m, H_n) = ((G_i G_k) G_m, (H_j H_l) H_n) =$$
$$= (G_i(G_k G_m), H_j(H_l H_n)) = (G_i, H_j) \cdot (G_k G_m, H_l H_n) =$$
$$= (G_i, H_j) \cdot \{(G_k, H_l) \cdot (G_m, H_n)\}.$$

In addition, if G and H are both *abelian* groups then so is $G \times H$, because then $(G_i, H_j) \cdot (G_k, H_l) = (G_i G_k, H_j H_l) = (G_k G_i, H_l H_j) = (G_k, H_l) \cdot (G_i, H_j)$. We shall now state, without proof, the following theorem concerning the structure of all finite abelian groups:

251

Theorem A1.4. Every finite abelian group can be written as a direct product of abelian groups of prime power orders.

Such representations will be discussed on pp. 272–276 for the important groups M_n whose elements are the primitive residue classes (mod n).

Cyclic Groups

According to Lagrange's theorem, the order e_A of the element A of a group divides the order n of that group. If a group has one or more elements of order n, then the powers $A, A^2, A^3, \ldots, A^n = I$ of any of these elements will exhaust the whole group. All elements of the group are then *generated* by the element A. Such a group is called a *cyclic* group of order n. We express this formally:

Definition A1.6. A group which is generated by a single group element is called a cyclic group.

Rings

The simple algebraic structures treated so far, modules and groups, may be characterized by the fact that only one pair of inverse operations is allowed in each of these structures. In a module this pair is always called addition-subtraction. In a group, the notion of group operation is a little bit more general and we have so far given examples of the pair multiplication-division as well as of the pair addition-subtraction of group elements. In elementary arithmetic, however, we are accustomed to having *two* pairs of inverse operations available simultaneously, both the pair addition-subtraction and the pair multiplication-division. However, the main reason why it is useful to have *both* these pairs at our disposal is not that we are used to that kind of algebraic structure, but rather that multiplication and addition of ordinary numbers are not independent of each other. Since $a + a + \cdots + a$ (to n terms) $= n \times a$, multiplication by a positive integer may be regarded as repeated addition. Thus, there exist more complicated structures than modules and groups. We shall now describe two such structures, starting with

Definition A1.7. A set which always allows addition, subtraction and multiplication of its elements, without ever producing a result outside the set, is called a ring. Multiplication has to be distributive with respect to

addition, i.e. $a(b + c) = ab + ac$ must hold for all a, b and c in the set.—
If the multiplication is commutative, a very common situation, then the
structure is said to be a commutative ring.

This means that a ring is a kind of module, the elements of which
may also be multiplied, *with the products always remaining in the module.*
Many modules actually are rings, but by no means all. An example of a
module that is also a ring is the set of all integers under ordinary addition.
This well-known module is also a ring if ordinary multiplication is introduced as multiplicative composition rule. An example of a module that is
not a ring is that consisting of the numbers $n\sqrt{2}$, with n an integer. The
product of two of its elements is $a\sqrt{2} \cdot b\sqrt{2} = 2ab$, which is not a member
of the module unless $a = 0$ or b is $= 0$.

In a ring all additions, subtractions and multiplications of elements
of the ring may be performed, without exception. Thus, in any ring, it is
possible to form polynomials of one or more variables:

$$P(x) = \sum_{i=0}^{n} a_i x^i, \quad P(x, y) = \sum_{i,j} a_{ij} x^i y^j, \quad \text{and so on.} \quad \text{(A1.12)}$$

Here the a_i's and the a_{ij}'s are fixed elements of the ring, and x and y
denote arbitrary elements of the ring.

Zero Divisors

There exist rings *in which* $ab = 0$, with neither $a = 0$ nor $b = 0$. a and b
are in such a case called *zero divisors.* In rings without zero divisors, the
theorems of algebra are very similar to those of elementary algebra. If the
ring has zero divisors, modifications will occur, of which one of the most
important is that an *equation of degree n may have more than n roots.*
This can never happen in a ring without zero divisors due to the following
Theorem A1.5. In a ring without zero divisors the equation of degree n

$$P(x) = a_n x^n + a_{n-1} x^{n-1} + \cdots + a_1 x + a_0 = 0, \quad a_n \neq 0, \quad \text{(A1.13)}$$

has as its roots at most n (distinct) elements of the ring.

Proof. We shall not here prove this theorem for the most general ring,
but only for the most common case, which is applicable to the solution
of higher degree congruences. Thus, we will assume a little more than is

stated in the theorem, namely that the ring has a unit element, denoted by 1. In this case we certainly can also construct a "polynomial division algorithm" for effectively calculating the quotient $Q(x)$ and the remainder $R(x)$, if we wish to divide $P(x)$ by an arbitrary polynomial $A(x)$ of degree $\leq n$ with its leading coefficient $=1$:

$$P(x) = Q(x)A(x) + R(x). \tag{A1.14}$$

The degree of $R(x)$ can be made less than the degree of $A(x)$. That a polynomial division algorithm exists in our case follows from the fact that a polynomial of degree less than that of $P(x)$ can be found by subtracting the leading term of $P(x)$, $a_n x^n$, in the form $a_n x^k A(x)$ for a suitable value of k. By performing a new subtraction in order to remove the new leading term of what now remains of $P(x)$, we can successively reduce $P(x)$ to a polynomial $R(x)$ of a degree less than the degree of $A(x)$.

After having identified the existence of a polynomial division algorithm in our ring, we are now in a position to prove the theorem by repeating the standard proof of the corresponding theorem of elementary algebra. Firstly, if x_1 is an arbitrary element of the ring, then $P(x)$ may be written in the form

$$P(x) = Q(x)(x - x_1) + P(x_1), \tag{A1.15}$$

where $Q(x)$ will be of degree $n - 1$. This follows if we first apply the polynomial division algorithm to $P(x)$ and $x - x_1$, yielding $P(x) = Q(x)(x - x_1) + R(x)$, where $R(x)$ is a polynomial of degree zero, i.e. a constant. Putting $x = x_1$, we find the value of the constant to be $P(x_1)$, which proves (A1.15). Secondly, if x_1 is a root of the equation $P(x) = 0$ in the ring, then certainly

$$P(x) = Q(x)(x - x_1), \quad \text{since} \quad P(x_1) = 0. \tag{A1.16}$$

However, and now comes the crucial point, if $x_2 \neq x_1$ is another root of $P(x) = 0$ in the ring, and *if there are no zero divisors*, then $Q(x_2)(x_2 - x_1) = 0$ implies $Q(x_2) = 0$. Thus, $Q(x) = 0$ has the root x_2, and if we repeat the reasoning a number of times we will arrive at

$$P(x) = (x - x_1) \dots (x - x_k) Q_k(x) \tag{A1.17}$$

where, finally, $Q_k(x) = 0$ has no further roots in the ring. From this representation of $P(x)$ it follows that $P(x) = 0$ cannot have roots other than precisely $x_1, x_2, \dots, x_k$ (again because there are no zero divisors). Now, the number of linear factors in (A1.17) can never exceed n, since the degrees of the polynomials in both sides of (A1.17) must be identical. This completes the proof.

Exercise A1.2. Wilson's Theorem. Fermat's Theorem (A2.14) states that if p is a prime, then the congruence $x^{p-1} - 1 \equiv 0 \pmod{p}$ has the solutions $x \equiv 1, 2, 3, \ldots, p-1 \pmod{p}$. Using this fact, write $P(x) = x^{p-1} - 1$ as a product of linear factors (mod p) as indicated in (A1.17). Comparing the constant terms on both sides, which formula do you find?

Fields

Is it possible to construct a domain in which *all* divisions by arbitrary elements can be performed? What about our "everyday" domain, the real numbers? At first you might be inclined to answer: "Yes, the division a/b of two real numbers is always possible!" But at second thought there is one exception to this, *you must not divide by 0.* This exception is quite general. There are many domains in which the result of all four rules of arithmetic can always be uniquely defined, *with the exception of the division by 0.* Such a domain is called a *number field* or, simply, a *field.* Thus, we arrive at the following

Definition A1.8. A domain in which the four simple arithmetic operations can always be carried out is called a field. The only forbidden operation is division by 0.

How is a field structured? First of all, it is a module, since all additions and subtractions are possible. Thus, the field must always contain the zero element, because the element $a - a = 0$ belongs to every module. This is the forbidden divisor. All elements $\neq 0$ now constitute a group (frequently abelian), in which the group operation and its inverse operation are the multiplication and the division, respectively, of elements of the field. This requires that the neutral element of the group, $1 = a/a$, also belongs to the field.

Examples of fields.

1. The set of all rational numbers, **Q**.
2. The set of all real numbers, **R**.
3. The set of all complex numbers $a + b\sqrt{-1}$, where a and b are real numbers. This set is denoted by **C**.
4. The group of order 4 studied above, being the group of primitive residue classes (mod 5), can be extended to a field if it is enlarged by the residue class 0 (mod 5). The addition and multiplication tables for this field are the following

$a+b$					
	0	1	2	3	4
0	0	1	2	3	4
1	1	2	3	4	0
2	2	3	4	0	1
3	3	4	0	1	2
4	4	0	1	2	3

ab					
	0	1	2	3	4
0	0	0	0	0	0
1	0	1	2	3	4
2	0	2	4	1	3
3	0	3	1	4	2
4	0	4	3	2	1

As the reader may verify, the addition table shows that the structure is a module (or if you prefer: an additively written cyclic group) of order 5. All numbers $\neq 0$ of the field constitute, as we have shown earlier, a multiplicatively written cyclic group of order 4.

5. The set of all numbers of the form $a + b\sqrt{D}$, with a and b rationals and D a positive or a negative non-square integer. The four arithmetic operations are defined by

$$(a + b\sqrt{D}) \pm (c + d\sqrt{D}) = a \pm c + (b \pm d)\sqrt{D} \qquad \text{(A1.18)}$$

$$(a + b\sqrt{D})(c + d\sqrt{D}) = ac + bdD + (ad + bc)\sqrt{D} \qquad \text{(A1.19)}$$

$$\frac{a + b\sqrt{D}}{c + d\sqrt{D}} = \frac{ac - bdD}{c^2 - d^2 D} + \frac{(bc - ad)\sqrt{D}}{c^2 - d^2 D} \qquad \text{(A1.20)}$$

As D is not a square, $c^2 - d^2 D$ will always be $\neq 0$, except when $c = d = 0$. Thus all divisions by $c + d\sqrt{D}$ can be carried out in (A1.20), except the division by $0 + 0\sqrt{D} = 0$, the zero element of the field. Since the result of any arithmetic operation can always be reduced to $p + q\sqrt{D}$, with p and q rational numbers, the domain is closed under arithmetic operations, and it is thus a field. This kind of field is called a *quadratic field*, and is denoted by $K(\sqrt{D})$. In Appendix 4 we present several arithmetic properties of the numbers of $K(\sqrt{D})$. These are utilized in Chapter 4.

6. Cyclotomic fields (see Appendix 6). Let m be a positive integer and ς a primitive m^{th} root of unity. The smallest field $R_m(\varsigma)$ containing ς is called the mth *cyclotomic field*. The cyclotomic fields have only recently become of interest in the study of primes, due to their use in fast primality proofs for numbers not possessing a special form [2].—However, the field $R_p(\varsigma)$ where p is an odd prime, has been indispensible in number theory for a long time because of its

connection with Fermat's conjecture that $x^p + y^p = z^p$ has no solutions in integers all $\neq 0$ for $p > 2$. In fact $R_p(\varsigma)$ is useful because the polynomial $x^p + y^p$ has *linear* factors in $R_p(\varsigma)$:

$$x^p + y^p = \prod_{k=1}^{p} (x + \varsigma^k y). \tag{A1.21}$$

Note that the cyclotomic fields $R_3(\varsigma)$, $R_4(\varsigma)$ and $R_6(\varsigma)$ are identical to the quadratic fields $K(\sqrt{-3})$, $K(\sqrt{-1})$ and $K(\sqrt{-3})$ respectively. For all other values of m, $R_m(\varsigma)$ is a more complicated structure than a quadratic field. The general element of $R_m(\varsigma)$ can be written in the form

$$a_0 + a_1\varsigma + a_2\varsigma^2 + \cdots + a_{\varphi(m)-1}\varsigma^{\varphi(m)-1}, \tag{A1.22}$$

where the a_i's are rational numbers. This is because ς satisfies an equation with rational coefficients of degree $\varphi(m)$, and thus $\varsigma^{\varphi(m)}$ and all higher powers of ς can be replaced by polynomials in ς of degree lower than $\varphi(m)$.—The rules for multiplication and, in particular, division are in general far more cumbersome than for quadratic fields. For instance the multiplication rule in $R_5(\varsigma)$, in which $\varsigma^4 = -\varsigma^3 - \varsigma^2 - \varsigma - 1$, is given by

$$(a_0 + a_1\varsigma + a_2\varsigma^2 + a_3\varsigma^3)(b_0 + b_1\varsigma + b_2\varsigma^2 + b_3\varsigma^3) =$$
$$= a_0b_0 - a_1b_3 - a_2b_2 + a_2b_3 - a_3b_1 + a_3b_2 +$$
$$+ (a_0b_1 + a_1b_0 - a_1b_3 - a_2b_2 - a_3b_1 + a_3b_3)\varsigma +$$
$$+ (a_0b_2 + a_1b_1 - a_1b_3 + a_2b_0 - a_2b_2 - a_3b_1)\varsigma^2 +$$
$$+ (a_0b_3 + a_1b_2 - a_1b_3 + a_2b_1 - a_2b_2 + a_3b_0 - a_3b_1)\varsigma^3.$$

The division rule in $R_5(\varsigma)$ is even more complicated.

Mappings. Isomorphisms and Homomorphisms

We have already encountered the concept isomorphism, on discovering two groups of identical internal structure. We shall now study this concept in more detail. We begin with two sets, M and M_1. Let every element of M correspond to a certain element of M_1. (It is allowed to use the same element of M_1 several times, but every element of M_1 must be used at least once.) We now say that M has been mapped onto M_1, denoted by $M \to M_1$.

257

Example. The set of all *positive* real numbers is mapped onto the set of *all* real numbers by the function $y = \log x$.

If a certain mapping $M \to M_1$ also preserves some algebraic structure, then the mapping is called homomorphic (with respect to the structure). We formalize this in

Definition A1.9. Suppose we map the set M with the elements $a, b, c, \ldots$ onto the set M_1 with the elements $a_1, b_1, c_1, \ldots$ in such a way that $a \to a_1, b \to b_1, c \to c_1, \ldots$ If in this mapping $a + b \to a_1 + b_1$, and $ab \to a_1 b_1$ for arbitrary a and b, the mapping is called a homomorphism.

Example. The ring of *all* rational integers can be mapped homomorphically onto the field of all residue classes (mod 5). (See 4 in the examples above.) The mapping is carried out by mapping the integer $5n + a$ for $0 \le a \le 4$ onto the residue class a (mod 5), which is an element of the field. Homomorphisms like this one will be described in greater detail in Appendix 2. (See pp. 263–264.)

If, in addition to the homomorphism from M to M_1, the elements of M_1 can be mapped back to M uniquely, so that there is a one-to-one correspondence between the elements of the two sets M and M_1: $a \leftrightarrow a_1, b \leftrightarrow b_1, c \leftrightarrow c_1, \ldots$, then the mapping is called an *isomorphism*, on condition that *all additive and multiplicative relations between elements are preserved* under the mapping.

Group Characters

Definition A1.10. Let G be a finite abelian group. A *group character* χ is a complex-valued, multiplicative function $\neq 0$, defined on the elements of G, i.e.

$$\chi(AB) = \chi(A)\chi(B) \quad \text{for all} \quad A, B \quad \text{in} \quad G. \tag{A1.23}$$

If I is the neutral element of the group, then $\chi(I) = 1$, because from $I^2 = I$ it follows that $\{\chi(I)\}^2 = \chi(I)$, having as only solution $\neq 0$ the possibility $\chi(I) = 1$. If A is a group element for which $A^2 = I$, then $\chi(A) = \pm 1$, since $\chi(A)^2 = \chi(A^2) = \chi(I) = 1$. Moreover, if the order of the group is n, then all values of any group character are n^{th} roots of unity, since it from Lagrange's Theorem $A^n = I$ follows that $\{\chi(A)\}^n = \chi(I) = 1$.—Choosing $\chi(A) = 1$ for all A of G we get what is called the *principal character* of G.

Example 1. Let G be a cyclic group of order n with generator A. Let $\varsigma = e^{2\pi i k/n}$ be any one of the nth roots of unity. Then obviously the function defined by $\chi(A^s) = e^{2\pi i k s/n}$ is a group character on G, since $\chi(A^s A^t) = \chi(A^{s+t}) = e^{2\pi i k(s+t)/n} = e^{2\pi i k s/n} \cdot e^{2\pi i k t/n} = \chi(A^s)\chi(A^t)$.

Example 2. The quadratic character. Consider M_p, the group of primitive residue classes modulo an odd prime p, with generator a. The number of group elements, n, is $=p-1$. Choosing $k = (p-1)/2$ in Example 1 above, we find $\chi(a^s) = e^{\pi i s} = (-1)^s$. The character takes the value $+1$ if s is even, and the value -1 if s is odd, i.e. the character takes the same value as Legendre's symbol (a^s/p) for any group element a^s.—This character is called the *quadratic character* of M_p. It can be utilized to write down formulas analogous to Euler's product (2.7) for certain Dirichlet series, such as the L-function shown below for $p = 5$:

$$L_{-5}(s) = 1 - \frac{1}{2^s} - \frac{1}{3^s} + \frac{1}{4^s} + \frac{1}{6^s} - \frac{1}{7^s} - \frac{1}{8^s} + \frac{1}{9^s} + \frac{1}{11^s} - \frac{1}{12^s} - \cdots =$$

$$= \sum_{n=1}^{\infty} \left(\frac{5}{n}\right) \frac{1}{n^s} = \prod_{p \neq 5} \left(1 - \left(\frac{5}{p}\right)\frac{1}{p^s}\right)^{-1}.$$

(The coefficients in the series show the periodic pattern $1, -1, -1, 1, 0, 1, -1, -1, 1, 0, \ldots$ because the Jacobi symbol $(5/n)$ is periodic (mod 5).)

The study of group characters is useful since information on the structure of a group can frequently be drawn from properties of one or more of its characters. To this end we shall prove a few theorems in this area. The technique used in these proofs will also be of help in deducing one of the modern primality tests, which depend heavily on the structure of the group M_N, where the primality of N is to be proved. See p. 139–143 for an example of this.

Theorem A1.6. Let G be a finite, abelian group of order n. Then

$$\sum_A \chi(A) = \begin{cases} n, & \text{if } \chi \text{ is the principal character} \\ 0, & \text{otherwise.} \end{cases} \tag{A1.24}$$

Proof. Let $S = \sum_A \chi(A)$. Then, for any element B of G we have

$$S\chi(B) = \sum_A \chi(A)\chi(B) = \sum_A \chi(AB) = \sum_A \chi(A) = S,$$

since AB runs through the whole group G as A does. Such an element B can always be found in case χ is not the principal character. Now, $S(\chi(B) - 1) = 0$ implies that $S = 0$, if $\chi(B) \neq 1$. This proves the second part of (A1.24). If, on the other hand, χ is the principal character, then every term in the sum is $= 1$, and the first part of (A1.24) follows immediately.

The Conjugate or Inverse Character

Taking the complex conjugate of (A1.23) we have $\overline{\chi}(AB) = \overline{\chi}(A)\overline{\chi}(B)$, which shows that if χ is a group character, so is its conjugate, $\overline{\chi}$. Now, since the value of $\chi(A)$ is always a root of unity ς_n, we have

$$\overline{\chi}(A) = \overline{\varsigma}_n = \frac{1}{\varsigma_n} = \chi^{-1}(A) = \chi(A^{-1}). \tag{A1.25}$$

Because of the relation $\overline{\chi}(A) = \chi^{-1}(A)$ this character is also called the *inverse* character of $\chi(A)$.

Before we proceed with the next theorem we have to give

Definition A1.11. Two complex-valued functions ϕ and ψ defined on a group G are said to be *orthogonal* on G if $\sum_A \phi(A)\overline{\chi}(A) = 0$, where the summation is extended over all elements A of G.

Theorem A1.7. Two distinct characters $\chi_1(A)$ and $\chi_2(A)$ defined on the same group are always orthogonal, i.e.

$$\sum_A \chi_1(A)\overline{\chi}_2(A) = 0, \qquad \text{if} \qquad \chi_1 \neq \chi_2. \tag{A1.26}$$

Proof. The product $\chi(A)$ of two characters $\chi_1(A)$ and $\overline{\chi}_2(A)$ is again a character, since

$$\chi(AB) = \chi_1(AB)\overline{\chi}_2(AB) = \chi_1(A)\chi_1(B)\overline{\chi}_2(A)\overline{\chi}_2(B) = \chi(A)\overline{\chi}(B).$$

Thus, from Theorem A1.6 above, we find that

$$\sum_A \chi(A) = \sum_A \chi_1(A)\overline{\chi}_2(A) = 0,$$

unless $\chi(A)$ is the principal character, which would imply $\chi_1(A)\overline{\chi}_2(A) = 1$ for all A of G. But this would lead to $\chi_2(A) = 1/\overline{\chi}_1(A) = \chi_1(A)$, contrary to the assumption that χ_1 and χ_2 be distinct characters.

Homomorphisms and Group Characters

Let p be a prime and ς a pth root of unity $\neq 1$. The group S generated by ς under complex multiplication is a cyclic group of order p. Suppose another group G has S as a subgroup and that there exists a homomorphism which maps G onto S in such a way that the elements A and B of G are mapped onto ς^j and ς^k, respectively, say. *This mapping induces a group character* $\chi(A)$ *defined on* G, since the homomorphism forces AB to be mapped onto $\varsigma^j \cdot \varsigma^k$, which means that $\chi(AB) = \varsigma^j \varsigma^k = \chi(A)\chi(B)$.

Example. Let $G = M_{31}$, the group of *primitive* residue classes (mod 31) and S be the group generated by $\varsigma = e^{2\pi i/5}$. M_{31} is cyclic of order 30 and has $g = 3$ as a generator. Its elements can thus be written as 3^j (mod 31), $j = 1, 2, 3, \ldots, 30$. Since $5|30$ and G is cyclic, G has a cyclic subgroup of order 5, i.e. a subgroup isomorphic to S. The mapping from the element 3^j of G onto ς^j of S is thus a homomorphism from G to S and the function $\chi(3^j) = \varsigma^j$ defines a group character on M_{31}.—Mapping 3^j onto ς^{js}, $1 \leq s \leq 4$, $(s, 5) = 1$, defines $\varphi(5) = 4$ different group characters for M_{31}. Considering, more generally, a cyclic subgroup of any order $d|30$ of M_{31}, we can in a similar way construct $\varphi(d)$ group characters for M_{31}. Doing this for all divisors d of 30 we will in total find $\sum_{d|30} \varphi(d) = 30$ different group characters on M_{31}.—In general, a finite, abelian group of order n possesses exactly n different group characters.

Bibliography

1. Richard P. Brent, "Analysis of the Binary Euclidean Algorithm," Printed in J. F. Traub, *New Directions and Recent Results in Algorithms and Complexity*, Academic Press, N. Y., 1976, pp. 321–355.

2. Carl Pomerance, "Recent Developments in Primality Testing," *The Mathematical Intelligencer*, 3, 1981, pp. 97–105.

BASIC CONCEPTS IN HIGHER ARITHMETIC

Divisors. Common Divisors

Let us commence with

Definition A2.1. d is called a divisor of n, if the division n/d leaves no remainder.

Example. The integer 15 has the divisors $-15, -5, -3, -1, 1, 3, 5$ and 15.

It is usually sufficient to consider only the *positive* divisors. If d is a divisor of m as well as of n, d is called a *common divisor* of m and n. The greatest common divisor (G.C.D.) d of m and n has been accorded a notation of its own: $d = (m, n)$. (Cf. p. 239.)

Example. 9 possesses the divisors 1, 3 and 9, and 15 has the divisors 1, 3, 5 and 15. Thus, $(9, 15) = 3$.

Euclid has proved a fundamental property of divisors of integers, namely

Theorem A2.1. If $c|ab$ and $(c, a) = 1$, then $c|b$.

Proof. Consider the diophantine equation $ax + cy = 1$. Since $(c, a) = 1$, according to theorem A1.2, this equation has integer solutions x, y. Therefore, there exists (at least) one pair of integers x, y such that

$$b = b(ax + cy) = (ab)x + c \cdot by = \left(\frac{ab}{c}x + by\right)c.$$

Since it is assumed that $c|ab$ the number in the larger parenthesis is an integer, and thus c must also divide b, proving the theorem.

The Fundamental Theorem of Arithmetic

We have now sufficient tools at our disposal for proving the fundamental theorem of arithmetic:

Theorem A2.2. If p_i and q_j are positive primes, and if

$$a = \prod_{i=1}^{n} p_i^{\alpha_i} = \prod_{j=1}^{m} q_j^{\beta_j} , \qquad (A2.1)$$

where

$$1 < p_1 < p_2 < \cdots < p_{n-1} < p_n$$

and

$$1 < q_1 < q_2 < \cdots < q_{m-1} < q_m,$$

then $n = m$, $p_i = q_i$ and $\alpha_i = \beta_i$, i.e. every positive integer has a composition into positive prime factors, which is unique apart from the order of the factors.

Proof. If p and q are distinct primes, then $p|q$ is never true since if this were the case then q would be composite, $q = p \cdot (q/p)$. Now, $p_1|a$, i.e. $p_1| \prod q_j^{\beta_j}$. Theorem A2.1 then tells us that p_1 divides at least one of the factors q_j. But since both p_1 and all the q_j's are assumed prime, this is only possible if $p_1 = q_j$. If we now divide the whole equation (A2.1) by $p_1 = q_j$ and repeat the procedure, we obtain the same result for another pair of prime factors p_k and q_l. Continuing in this way, we will finally arrive at the equation $1 = 1$. Thus, assuming that the prime factors in the two representations (A2.1) are given in increasing order of magnitude and counting the multiplicity of each factor, the theorem follows.

Remark. The representation of a positive integer n as a product of positive prime powers, where the primes are given in increasing order of magnitude, is called *the standard factorization* of n.

Congruences

The notion of congruence was introduced by Gauss, who gave the following

Definition A2.2. If $a - b$ is divisible by n, we write $a \equiv b \,(\mathrm{mod}\ n)$ (read: a is congruent to b modulo n).

Since the remainder r on division of a by n can always be chosen in such a way that $0 \leq r \leq n - 1$, any integer is congruent to one of the integers 0, 1, 2, $\ldots, n-1 \,(\mathrm{mod}\ n)$. Thus, what a congruence relation $(\mathrm{mod}\ n)$ actually defines is a mapping of *all* integers onto only precisely n integers, and the mapping is such that any two integers, which have a difference divisible by n, are always mapped upon the same integer. We shall show that this mapping is homomorphic. The integers 0, 1, 2, $\ldots, n - 1$ represent what is called the n different residue classes $(\mathrm{mod}\ n)$.

Example. If $n = 5$, the integers

$\ldots - 15, - 10, -5, 0, 5, 10, 15, \ldots$ are mapped onto 0

$\ldots - 14, - 9, -4, 1, 6, 11, 16, \ldots$ are mapped onto 1

$\ldots - 13, - 8, -3, 2, 7, 12, 17, \ldots$ are mapped onto 2

$\ldots - 12, - 7, -2, 3, 8, 13, 18, \ldots$ are mapped onto 3

$\ldots - 11, - 6, -1, 4, 9, 14, 19, \ldots$ are mapped onto 4.

The relation $a \equiv b \pmod{n}$ is equivalent to

$$a = b + nk, \quad \text{with integer } k. \tag{A2.2}$$

By using this rule, it is easy to deduce the following arithmetic rules for congruences:

Theorem A2.3. If $a \equiv b \pmod{n}$ and $c \equiv d \pmod{n}$, then also $a + c \equiv b + d \pmod{n}$, $a - c \equiv b - d \pmod{n}$ and $ac \equiv bd \pmod{n}$, i.e. the mapping of all integers onto the residue classes $\pmod{n}$ is homomorphic.

Thus, it is allowed to add, subtract and multiply congruences $\pmod{n}$. This means that the residue classes $\pmod{n}$ constitute a ring, called the ring of (all) residue classes $\pmod{n}$.

Example. $100 \equiv 4 \pmod{12}$, since $100 = 12 \cdot 8 + 4$, and $51 \equiv 3 \pmod{12}$, because $51 = 12 \cdot 4 + 3$. Performing addition, subtraction and multiplication of these two congruences we find that $151 \equiv 7 \pmod{12}$, $49 \equiv 1 \pmod{12}$ and $5100 \equiv 12 \equiv 0 \pmod{12}$, as is also easily verified by direct calculation.

Proofs of these rules are based on simple algebra: $a = b + kn$ and $c = d + ln$ imply that $a \pm c = b \pm d + (k \pm l)n$, i.e. $a \pm c \equiv b \pm d \pmod{n}$, and that $ac = (b + kn)(d + ln) = bd + (kd + bl + kln)n$, i.e. $ac \equiv bd \pmod{n}$.

The rule for *cancelling* a congruence by an integer is a little more complicated:

Theorem A2.4. If $ac \equiv bc \pmod{n}$, then

$$a \equiv b \left(\bmod \frac{n}{(c, n)} \right).$$

Hence, if the G.C.D. $d = (c, n)$ of c and n is > 1, the resulting congruence *is not valid* $\pmod{n}$, *but only* $\pmod{n/d}$.

Proof. Since $n|(a-b)c$, certainly $\frac{n}{d}\left|(a-b)\frac{c}{d}\right.$. Since $\left(\frac{n}{d},\frac{c}{d}\right)=1$, theorem A2.1 gives

$$\frac{n}{d}\left|a-b\right. \quad\text{i.e.}\quad a\equiv b\left(\bmod\ \frac{n}{d}\right).$$

Both sides of a congruence can thus be divided by the same integer c, only if the modulus n is in the same operation divided by (c,n). This is the *cancellation rule for congruences.*

Example. $3500\equiv 70\ (\bmod\ 245)$. When cancelling the factor 70, we must not forget to divide the modulus 245 by $(70,245)=35$; the result of the cancellation is therefore $50\equiv 1\ (\bmod\ 7)$.

The rules given for operating on congruences are, as a matter of fact, extremely efficient when an algebraic expression needs to be evaluated $(\bmod\ n)$. We give some examples. What is the result of $3^{100}\ (\bmod\ 101)$? In order to compute a power efficiently, we use the method of successive squarings, followed by occasional multiplications (cf. p. 93). In this particular case, we find, successively

$$
\begin{array}{llll}
3^5 = & & 243 = 202+41\equiv & 41\ (\bmod\ 101)\\[4pt]
3^{10} = & (3^5)^2\equiv & 41^2 = \quad 1681\equiv & 65\ (\bmod\ 101)\\[4pt]
3^{20}\equiv & & 65^2 = \quad 4225\equiv-17 & (\bmod\ 101)\\[4pt]
3^{40}\equiv & & (-17)^2 = \quad 289\equiv-14 & (\bmod\ 101)\\[4pt]
3^{50} = 3^{40}\cdot 3^{10}\equiv & -14\cdot 65 = & -910\equiv -1 & (\bmod\ 101)\\[4pt]
3^{100} = & (3^{50})^2\equiv & (-1)^2 = \quad 1\equiv & 1\ (\bmod\ 101)
\end{array}
$$

Another example is: What is the value of $4^n\ (\bmod\ 12)$? Since $4^2=16\equiv 4\ (\bmod\ 12)$, $4^3=4\cdot 4^2\equiv 4\cdot 4=4^2\equiv 4\ (\bmod\ 12)$ etc. obviously, $4^n\equiv 4\ (\bmod\ 12)$ for $n>0$.

Linear Congruences

If a and b are known residue classes, and x is an unknown residue class such that

$$ax\equiv b\ (\bmod\ n), \tag{A2.3}$$

then x is said to satisfy the *linear congruence* (A2.3). This congruence is equivalent to the diophantine equation

$$ax = b + ny \tag{A2.4}$$

which, according to theorem A1.2, is solvable if and only if

$$(a, n) | b. \tag{A2.5}$$

If $(a, n) = d$, (A2.3) can, in this case, be cancelled by d to give

$$\frac{a}{d} x \equiv \frac{b}{d} \left(\bmod \frac{n}{d} \right), \tag{A2.6}$$

a congruence with $(a/d, n/d) = 1$. Thus, if a linear congruence (A2.3) has any solutions, it can be transformed to the congruence (A2.6) or, with changed notation, to

$$ax \equiv b \,(\bmod\, n), \quad \text{with } (a, n) = 1. \tag{A2.7}$$

The situation is described by

Theorem A2.5. The linear congruence (A2.7) is always satisfied by precisely one residue class $x \,(\bmod\, n)$. This solution is sometimes denoted by $x \equiv a^{-1}b \,(\bmod\, n)$ or, alternatively, by $x \equiv b/a \,(\bmod\, n)$.

Proof. Let x extend through all the residue classes $(\bmod\, n)$. Then all the values of $y = ax \,(\bmod\, n)$ are distinct, since $ax_1 \equiv ax_2 \,(\bmod\, n)$ would imply $x_1 \equiv x_2 \,(\bmod\, n)$ due to the assumption that $(a, n) = 1$ and the cancellation rule for congruences. Since there are precisely n residue classes, the n different values of $ax \,(\bmod\, n)$ must each fall in one residue class. Thus for precisely one value of $x \,(\bmod\, n)$, a specific value b for which $ax \equiv b \,(\bmod\, n)$ holds, as assumed.

Linear Congruences and Euclid's Algorithm

A linear congruence can be solved by applying Euclid's algorithm. We demonstrate this with the example

$$3293x \equiv 222 \,(\bmod\, 8991)$$

which is equivalent to

$$3293x = 222 + 8991y.$$

Reduce this diophantine equation (mod 3293):

$$222 + 8991y \equiv 0 \ (\text{mod } 3293).$$

Now one line of Euclid's algorithm comes into effect (compare with the example on p. 240):

$$222 + (2 \cdot 3293 + 2405)y \equiv 222 + 2405y \equiv 0 \ (\text{mod } 3293).$$

Continuing in this way, we obtain successively:

$$222 + 2405y = 3293z, \qquad 222 \equiv 888z \ (\text{mod } 2405),$$
$$222 = 888z + 2405u, \qquad 222 \equiv 629u \ (\text{mod } 888),$$
$$222 = 629u + 888v, \qquad 222 \equiv 259v \ (\text{mod } 629),$$
$$222 = 259v + 629w, \qquad 222 \equiv 111w \ (\text{mod } 259),$$
$$222 = 111w + 259t, \qquad 222 \equiv 37t \ (\text{mod } 111),$$

$$222 = 37t + 111s.$$

At this stage $37|111$, and Euclid's algorithm terminates. The result is that the integers 3293 and 8991 have the G.C.D. 37. Now, check whether $37|222$. It does, and therefore the given congruence is solvable. Proceed as follows: First divide the last equation by the G.C.D. to obtain

$$6 = t + 3s, \quad \text{i.e.} \quad t = 6 - 3s,$$

where s is an arbitrary integer, a so-called *parameter*, used as an aid to express all solutions x, y. To achieve this goal, repeat the above calculation backwards, expressing everything in terms of s:

$$w = \frac{222 - 259t}{111} = \frac{222 - 259(6 - 3s)}{111} = -12 + 7s$$

$$v = \frac{222 - 629w}{259} = \frac{222 - 629(-12 + 7s)}{259} = 30 - 17s$$

$$u = \frac{222 - 888v}{629} = \frac{222 - 888(30 - 17s)}{629} = -42 + 24s$$

$$z = \frac{222 - 2405u}{888} = \frac{222 - 2405(-42 + 24s)}{888} = 114 - 65s$$

$$-y = \frac{222 - 3293z}{2405} = \frac{222 - 3293(114 - 65s)}{2405} = -156 + 89s$$

and, finally

$$x = \frac{222 - 8991y}{3293} = \frac{222 - 8991(-156 + 89s)}{3293} = 426 - 243s$$

i.e.

$$x \equiv 426 \,(\text{mod } 243) \equiv 183 \,(\text{mod } 243).$$

A more elegant way to find the value of x, making use of continued fractions, is presented in Appendix 5, pp. 308–311, where the reader can also find a computer program for solving a linear congruence.

Systems of Linear Congruences

Consider a system of simultaneous linear congruences,

$$a_i x \equiv b_i \,(\text{mod } n_i) \quad \text{with} \quad (a_i, n_i) = 1, \tag{A2.8}$$

each of them assumed solvable. Furthermore, suppose that no two n_i's have common divisors > 1, so that $(n_i, n_j) = 1$ if $i \neq j$. Then we have the following

Theorem A2.6. The Chinese Remainder Theorem. The system of simultaneous congruences (A2.8) is solvable, and the congruences have precisely one common solution $x(\text{mod } \prod n_i)$, if $(n_i, n_j) = 1$ for all pairs i, j.

Proof. Start by considering just two simultaneous congruences

$$a_1 x \equiv b_1 \,(\text{mod } n_1) \tag{A2.9}$$

and

$$a_2 x \equiv b_2 \,(\text{mod } n_2). \tag{A2.10}$$

We shall prove that these two congruences have precisely one common solution $x \,(\text{mod } n_1 n_2)$. Firstly, (A2.9) implies that $a_1 x$ can be written as $a_1 x = b_1 + n_1 t$ for some integer t, or equivalently,

$$n_2 a_1 x = n_2 b_1 + n_1 n_2 t, \tag{A2.11}$$

and secondly, (A2.10) implies that $a_2 x = b_2 + n_2 u$ or

$$n_1 a_2 x = n_1 b_2 + n_1 n_2 u. \tag{A2.12}$$

Subtracting (A2.11) and (A2.12), we have

$$(n_2 a_1 - n_1 a_2)x \equiv n_2 b_1 - n_1 b_2 \ (\text{mod } n_1 n_2). \tag{A2.13}$$

However, since the solvability test is satisfied, this new congruence is, according to Theorem A2.5, always solvable (mod $n_1 n_2$). It is easily observed that this is the case:

$$(n_2 a_1 - n_1 a_2, \ n_1) = (n_2 a_1, n_1) = 1$$

since both $(n_2, n_1) = 1$ and $(a_1, n_1) = 1$, and analogously for

$$(n_2 a_1 - n_1 a_2, \ n_2) = (n_1 a_2, n_2) = 1.$$

The fact that the solution of (A2.13) really satisfies both (A2.9) and (A2.10), and not only the combination (A2.13) of these two, can be verified by reducing (A2.13) first to (mod n_1) and then to (mod n_2). These reductions lead back to precisely (A2.9) and (A2.10).

The solution of the general case with any number of simultaneous congruences, with pairwise relatively prime moduli, can now be built up step-wise by successively applying the result for two congruences and then appending the remaining congruences, one at a time. (Two integers m and n are said to be *relatively prime*, if $(m, n) = 1$.) This proves the general theorem.

The Residue Classes (mod p) Constitute a Field

Theorem A2.4 yields a particularly simple result if the module n happens to be a prime number p. Always, in such a case, $(c, p) = 1$ or p. However, if $(c, p) = p$, on cancelling the congruence by p, we obtain $a \equiv b \ (\text{mod } 1)$, which is a trivial congruence, since the difference $a - b$ of any two integers is always divisible by 1. Thus, if we disregard this trivial case and forbid cancellation by p, we may state that a congruence (mod p) can always be cancelled without affecting the module, since the module $p/(c, p)$ remains $= p$. Hence, divisions by all integers except p, acting as the zero element, are allowed in the ring of residue classes (mod p). This ring thus constitutes a field, since according to Theorem A2.5, all these divisions give unique results. This conclusion can be re-phrased as

Theorem A2.7. If p is a prime, then the ring of all residue classes (mod p) is a field. The $p-1$ residue classes $\not\equiv 0$ (mod p) constitue a (multiplicative) abelian group of order $p-1$. This group is denoted by M_p.

Example. The structure we studied in Appendix 1, p. 250, is the field corresponding to $p = 5$. The division a/b has to be carried out as a multiplication ab^{-1}, which is always possible to perform, since the multiplicative group contains the inverse b^{-1} of every element $b \neq 0$.

From this general theory, the very important theorem of Fermat now follows easily as a special case of Lagrange's Theorem A1.3 on p. 248:

Theorem A2.8. Fermat's Theorem. If p is a prime, and $a \not\equiv 0$ (mod p), then

$$a^{p-1} \equiv 1 \text{ (mod } p\text{).} \tag{A2.14}$$

Multiplication by a yields

$$a^p \equiv a \text{ (mod } p\text{),} \tag{A2.15}$$

a variation of the theorem which holds for *all* a, i.e. also when $a \equiv 0$ (mod p).

The Primitive Residue Classes (mod p)

If n is a composite number, the ring of all residue classes (mod n) cannot also be considered to be a field. This is because cancellation of a congruence (mod n) by *any of the divisors* d of n also requires the corresponding cancellation of n, and thus carries us from the ring (mod n) to another ring, viz. (mod n/d). In this case, d is said to be a zero divisor of the ring, since $d|n$ and $n \equiv 0$ (mod n).— However, if we avoid the zero divisors of n, i.e. if we consider only the so-called *primitive* residue classes a (mod n) with $(a, n) = 1$ then, according to Theorem A2.5, all divisions by *these* elements can be uniquely performed. The primitive residue classes (mod n) actually constitute a multiplicative group M_n of order $\varphi(n)$. This important number-theoretic function $\varphi(n)$ is called Euler's totient function. In the next section we shall describe the structure of M_n in some detail, but for the moment we shall concentrate on the following important question: How many primitive residue classes are there (mod n), i.e. of which order $\varphi(n)$ is the group M_n or which is the same thing, how can we enumerate all a's satisfying

$$(a, n) = 1 \quad \text{and} \quad 1 \leq a \leq n? \tag{A2.16}$$

Consider first the simplest case, when $n = p^\alpha$, a prime power. Then $(a, n) > 1$ if and only if $p | a$, which is true if a is one of the $p^{\alpha-1}$ multiples of p between 1 and p^α, i.e. for the integers

$$a = p, \ 2p, \ 3p, \ldots, p^{\alpha-1} \cdot p.$$

The remaining $p^\alpha - p^{\alpha-1}$ integers below p^α represent the different *primitive* residue classes (mod p^α), and thus we find the value of Euler's function in this case to be

$$\varphi(p^\alpha) = p^{\alpha-1}(p-1), \quad \text{if } p \text{ is a prime.} \tag{A2.17}$$

In particular $\alpha = 1$ gives

$$\varphi(p) = p - 1. \tag{A2.18}$$

Taking this simple result as a starting point, we now have to deduce $\varphi(n)$ when $n = \prod p_i^{\alpha_i}$. This can be done by using the relation

$$\varphi(mn) = \varphi(m)\varphi(n), \quad \text{if } (m, n) = 1. \tag{A2.19}$$

The relation (A2.19) asserts that Euler's function is multiplicative; a fact that can be proved as follows: If $(m, n) = 1$, each residue class a (mod mn) with $(a, mn) = 1$ simultaneously satisfies both $(a, m) = 1$ and $(a, n) = 1$, and conversely. But from Theorem A2.6, we know that every combination of primitive residue classes

$$\left. \begin{array}{l} x \equiv b_1 \ (\text{mod } m) \\ x \equiv b_2 \ (\text{mod } n) \end{array} \right\} \tag{A2.20}$$

corresponds to precisely one primitive residue class (mod mn). Thus, if $(m, n) = 1$, $\varphi(m)$ primitive residue classes (mod m) can be combined with $\varphi(n)$ primitive residue classes (mod n) to give $\varphi(m)\varphi(n)$ primitive residue classes (mod mn). This is the reason why Euler's function is multiplicative, and (A2.19) is immediate.

Using the basic results (A2.17) and (A2.19) we now obtain

$$\varphi(n) = \prod_i p_i^{\alpha_i-1}(p_i - 1) = n \prod_i \left(1 - \frac{1}{p_i}\right), \quad \text{if } n = \prod_i p_i^{\alpha_i}. \tag{A2.21}$$

As usual, the primes p_i must all be distinct. Lagrange's theorem A1.3 now leads to the following generalization of Fermat's theorem A2.8 for the case of composite modules n:

271

Theorem A2.9. Euler's Theorem. If $(a.n) = 1$ and $n = \prod p_i^{\alpha_i}$, then

$$a^{\varphi(n)} \equiv 1 \ (\text{mod } n), \qquad (\text{A2.22})$$

where $\varphi(n) = \prod p_i^{\alpha_i - 1}(p_i - 1)$ is the number of primitive residue classes (mod n).

Using Euler's Theorem, we can find a way of explicitly writing down the quotient of two residue classes, a/b (mod n). The formula is

$$\frac{a}{b} = ab^{-1} \equiv ab^{\varphi(n)}b^{-1} = ab^{\varphi(n)-1} \ (\text{mod } n). \qquad (\text{A2.23})$$

The Structure of the Group M_n

The multiplicative group of all primitive residue classes (mod n), M_n, contains precisely $\varphi(n)$ elements. Gauss proved that M_n is a cyclic group if and only if $n = 4, n = p^\alpha$ or $n = 2p^\alpha$, and p is an odd prime. A proof can be found in [1]. The result by Gauss is a special case of a more general theorem on the structure of M_n, which we shall not prove in this book, see [2]. We shall however describe the structure of M_n in the general case.

According to Theorem A1.4 on p. 252, every finite abelian group can be written as a direct product of cyclic groups of prime power orders. Moreover, the direct product of two cyclic groups of orders m and n, if $(m, n) = 1$, is itself a cyclic group. Therefore, in general, there is no need to reduce an abelian group into group factors of *prime power orders* in order to be able to express the group as a product of cyclic groups—decomposition can be effected with certain suitably chosen larger cyclic group factors. We shall now describe the possible ways in which this can be accomplished for the group M_n.

If $n = \prod p_i^{\alpha_i}$, with p_i all distinct primes, then the order of M_n is, as we have just seen, $\varphi(n) = \prod p_i^{\alpha_i - 1}(p_i - 1)$. Next, assume that

$$p_i - 1 = \prod_j q_{ij}^{\beta_{ij}}, \qquad (\text{A2.24})$$

where q_{ij}, $j = 1, 2, \ldots$, are distinct primes. Then $\varphi(n)$ takes the form

$$\varphi(n) = \prod_{i,j} p_i^{\alpha_i - 1} q_{ij}^{\beta_{ij}}. \qquad (\text{A2.25})$$

The representation (A2.25) will, with a minor modification, give rise to those prime powers which actually are the searched for orders of the cyclic groups in the decomposition of M_n as a direct product of groups in accordance with Theorem A1.4. The modification is that if $2^\alpha | n$ for $\alpha \geq 3$, then a factor 2 has to be separated from $\varphi(2^\alpha)$, and thus $\varphi(2^\alpha) = 2^{\alpha-1}$ has to be written as $2 \cdot 2^{\alpha-2}$ in such a case.

Examples.

$$
\begin{array}{lll}
n = 4 = 2^2 & \text{has } \varphi(n) = 2 & \text{(A2.26)} \\
n = 8 = 2^3 & \text{has } \varphi(n) = 2 \cdot 2 & \text{(A2.27)} \\
n = 15 = 3 \cdot 5 & \text{has } \varphi(n) = 2 \cdot 4 & \text{(A2.28)} \\
n = 16 = 2^4 & \text{has } \varphi(n) = 2 \cdot 4 & \text{(A2.29)} \\
n = 17 & \text{has } \varphi(n) = 16 & \text{(A2.30)} \\
n = 24 = 2^3 \cdot 3 & \text{has } \varphi(n) = 2 \cdot 2 \cdot 2 & \text{(A2.31)} \\
n = 63 = 3^2 \cdot 7 & \text{has } \varphi(n) = 2 \cdot 3 \cdot 2 \cdot 3 = 6 \cdot 6 & \text{(A2.32)} \\
n = 65 = 5 \cdot 13 & \text{has } \varphi(n) = 4 \cdot 4 \cdot 3 = 12 \cdot 4 & \text{(A2.33)} \\
n = 104 = 2^3 \cdot 13 & \text{has } \varphi(n) = 2 \cdot 2 \cdot 4 \cdot 3 = 12 \cdot 2 \cdot 2 & \text{(A2.34)} \\
n = 105 = 3 \cdot 5 \cdot 7 & \text{has } \varphi(n) = 2 \cdot 4 \cdot 2 \cdot 3 = 12 \cdot 2 \cdot 2 & \text{(A2.35)}
\end{array}
$$

In each instance, the factorization of $\varphi(n)$ given first indicates the order of the cyclic groups referred to in Theorem A1.4. Using the just mentioned possibility of multiplying together those cyclic groups which have orders without common divisors, we can multiply occurring powers of different primes, and thus sometimes find several different representations of M_n as a product of cyclic groups. Among all these possibilities, it is interesting to select the one which will give the *least* number of group factors. This is accomplished by first multiplying all the *highest* prime powers of each individual prime in the given representation of $\varphi(n)$. The procedure is then repeated on the prime powers remaining after the first round, and so on. Whenever this manner of writing $\varphi(n)$ differs from the one arrived at in the first place, it has been indicated in the examples above by giving a second representation of $\varphi(n)$. (This occurs in (A2.32)–(A2.35).) Thus e.g. both M_{104} and M_{105} can be written as a direct product of three cyclic groups, of orders 12, 2 and 2, respectively, and thus M_{104} and M_{105} are isomorphic.

In order to clarify this discussion, let us take a close look at an example, $n = 15$. The group M_{15} has $\varphi(15) = 8$ elements a. It has two generators,

g_1 and g_2 of orders 2 and 4, respectively, so that the group elements a can be written as

$$a = g_1^{e_1} g_2^{e_2}, \tag{A2.36}$$

with $0 \le e_1 \le 1$ and $0 \le e_2 \le 3$. A possible choice of g_1 and g_2 is $g_1 = 11$ and $g_2 = 2$. It is easy to verify the following congruences (mod 15):

$$1 \equiv 11^0 2^0, \quad 2 \equiv 11^0 2^1, \quad 4 \equiv 11^0 2^2, \quad 7 \equiv 11^1 2^1,$$
$$8 \equiv 11^0 2^3, \quad 11 \equiv 11^1 2^0, \quad 13 \equiv 11^1 2^3, \quad 14 \equiv 11^1 2^2,$$

giving the representation of all 8 elements of M_{15} in the form (A2.36).

The table on the next page contains representations, analogous to (A2.26)–(A2.35), for all groups M_n with $n \le 50$, as products of cyclic groups with generators g_i of the cyclic group factors.

We observe that if $\varphi(m)$ and $\varphi(n)$ have identical factorizations under the rules established, then the groups M_m and M_n are isomorphic. The table shows that, among others, the groups M_{15}, M_{16}, M_{20} and M_{30} all are isomorphic (of type $2 \cdot 4$). So are M_{35}, M_{39} and M_{45} (of type $2 \cdot 12$). On the other hand, M_{16} (of type $2 \cdot 4$) and M_{24} (of type $2 \cdot 2 \cdot 2$), both of order 8, show different structure.

Homomorphisms of M_q when q is a Prime

Suppose n is a prime q such that $q - 1 = \prod_i p_i$ is a *square-free number*. Since M_q is cyclic of order $\prod p_i$, M_q contains cyclic subgroups C_p of order p for each of the primes $p = p_i$. If g is a generator of M_q then $g^{(q-1)/p}$ is a generator of C_p. The mapping of the element g^j in M_q on $g^{(q-1)j/p}$ in C_p is a homomorphism. Let ς_p be a pth root of unity. Then this homomorphism induces a group character $\chi_p(g^j) = \varsigma_p^j$. All those elements A of M_q, for which $\chi(A) = 1$ form a subgroup of M_q, called the *kernel* of the homomorphism. The kernel is composed of the elements $g^p, g^{2p}, g^{3p}, \ldots, g^{q-1} = I$.—If this construction is carried through for *all* prime factors p_i of $q - 1$, one kernel is found for each of the mappings. For these kernels we have the following theorem which we shall need in the proof of Lenstra's primality testing algorithm on p. 142.

Theorem A2.10. If $q - 1$ is square-free, then all these kernels have precisely one element in common, the neutral element of the group M_q.

Proof. An element of one particular of the kernels can be written in the form g^{jp_i}, where p_i is the prime used in the corresponding homomorphism.

Thus, in order to belong to *all* of the kernels simultaneously, an element must be of the form $g^j \prod p_i$. But the only element of this form in M_q is $g\prod p_i = g^{q-1} = I$.

n	$\varphi(n)$	$\lambda(n)$	g_i	n	$\varphi(n)$	$\lambda(n)$	g_i
3	2	2	2	27	18	18	2
4	2	2	3	28	$2 \cdot 6$	6	13, 3
5	4	4	2	29	28	28	2
6	2	2	5	30	$2 \cdot 4$	4	11, 7
7	6	6	3	31	30	30	3
8	$2 \cdot 2$	2	7, 3	32	$2 \cdot 8$	8	31, 3
9	6	6	2	33	$2 \cdot 10$	10	10, 2
10	4	4	3	34	16	16	3
11	10	10	2	35	$2 \cdot 12$	12	6, 2
12	$2 \cdot 2$	2	5, 7	36	$2 \cdot 6$	6	19, 5
13	12	12	2	37	36	36	2
14	6	6	3	38	18	18	3
15	$2 \cdot 4$	4	14, 2	39	$2 \cdot 12$	12	38, 2
16	$2 \cdot 4$	4	15, 3	40	$2 \cdot 2 \cdot 4$	4	39, 11, 3
17	16	16	3	41	40	40	6
18	6	6	5	42	$2 \cdot 6$	6	13, 5
19	18	18	2	43	42	42	3
20	$2 \cdot 4$	4	19, 3	44	$2 \cdot 10$	10	43, 3
21	$2 \cdot 6$	6	20, 2	45	$2 \cdot 12$	12	44, 2
22	10	10	7	46	22	22	5
23	22	22	5	47	46	46	5
24	$2 \cdot 2 \cdot 2$	2	5, 7, 13	48	$2 \cdot 2 \cdot 4$	4	47, 7, 5
25	20	20	2	49	42	42	3
26	12	12	7	50	20	20	3

Carmichael's Function

The technique described for finding a group decomposition of M_n, that of first multiplying the largest powers of each prime in the modified factorization of $\varphi(n)$, and then repeating this procedure with what remains of $\varphi(n)$, leads to a result in which the first factor found is an integer multiple of all factors subsequently discovered. Those cyclic groups G_i, the direct products of which give M_n, thus have orders d_i which all divide the largest of these orders, called Carmichael's function $\lambda(n)$. Referring back to (A2.25), we see that $\lambda(n)$ is the least common multiple of all the factors $\varphi(p_i^{\alpha_i})$ of $\varphi(n)$, with a modification similar to the one made above, i.e. if $8|n$ then the power $2^{\alpha-2}$ is regarded as a factor instead of $\varphi(2^\alpha) = 2^{\alpha-1}$. Formally, this rule for computing $\lambda(n)$ can be stated as

$$\begin{cases} \lambda(p^\alpha) = \varphi(p^\alpha), & \text{when } p = 2 \text{ and } \alpha \le 2, \text{ and when } p \ge 3 \\[2mm] \lambda(2^\alpha) = \dfrac{1}{2}\varphi(2^\alpha), & \text{when } \alpha \ge 3 \\[2mm] \lambda(n) \ = \text{L.C.M.}[\lambda(p_i^{\alpha_i})]_i, & \text{if } n = \displaystyle\prod_i p_i^{\alpha_i}. \end{cases} \qquad (\text{A2.37})$$

If

$$M_n = G_1 \times G_2 \times \cdots \times G_k, \qquad (\text{A2.38})$$

and d_i is the order of the group G_i, then every element a of M_n can be written in the form

$$a = g_1^{e_1} g_2^{e_2} \ldots g_k^{e_k}, \quad \text{with} \quad 1 \le e_i \le d_i. \qquad (\text{A2.39})$$

Furthermore, for every i,

$$g_i^{d_i} \equiv 1 \ (\text{mod } n), \quad \text{with} \quad d_i|\lambda(n). \qquad (\text{A2.40})$$

Carmichael's Theorem

An immediate consequence of (A2.39) and (A2.40) is

Theorem A2.11. Carmichael's Theorem. If $(a, n) = 1$, then

$$a^{\lambda(n)} \equiv 1 \ (\text{mod } n). \qquad (\text{A2.41})$$

This is a very useful, but often forgotten generalization of Euler's theorem A2.9.

276

Example. For $n = 65520 = 2^4 \cdot 3^2 \cdot 5 \cdot 7 \cdot 13$, Euler's function $\varphi(n)$ assumes the value $8 \cdot 6 \cdot 4 \cdot 6 \cdot 12 = 13824$, while $\lambda(n) = [4, 6, 4, 6, 12] = 12$. For all a with $(a, n) = 1$ we thus have

$$a^{12} \equiv 1 \pmod{65520}.$$

In the small table above, we have given $\lambda(n)$ as well as $\varphi(n)$.

In the chapter on cryptography we require a slight generalization of (A2.41) to include values of a with $(a, n) > 1$. This is provided by

Theorem A2.12. If n is a product of distinct primes then, for all a,

$$a^{\lambda(n)+1} \equiv a \pmod{n}. \tag{A2.42}$$

Proof. Let $n = \prod p_i$. By Fermat's Theorem,

$$a^{p_i-1} \equiv 1 \pmod{p_i}, \quad \text{yielding} \quad a^{m(p_i-1)} \equiv 1 \pmod{p_i}.$$

Multiplying by a, we infer that

$$a^{s_i} \equiv a \pmod{p_i} \quad \text{if} \quad s_i \equiv 1 \pmod{p_i - 1} \tag{A2.43}$$

for all a satisfying $(a, p_i) = 1$. It is an interesting fact that this congruence holds also when $(a, p_i) = p_i$, because then both sides of (A2.43) are $\equiv 0 \pmod{p_i}$. Choosing a value of $s \equiv s_i \equiv 1 \pmod{p_i - 1}$ for every value of i, we arrive at the following congruences:

$$a^s \equiv a \pmod{p_i}, \quad i = 1, 2, \ldots, \quad \text{for all } a. \tag{A2.44}$$

Now, by the Chinese remainder theorem, these congruences may be combined into one congruence (mod n) which obviously is $a^s \equiv a \pmod{n}$ provided that the exponent $s \equiv 1 \pmod{p_i - 1}$ for every prime factor p_i of n. However, since $\lambda(n)$ is the L.C.M.$[p_i - 1]$, the choice $s = \lambda(n) + 1$ satisfies the conditions imposed on s, thus proving (A2.42).

Remark. Since $\lambda(n)$ is the *least* common multiple of all the $p_i - 1$, $\lambda(n) + 1$ is also the *smallest* value of $s > 1$ satisfying $s \equiv 1 \pmod{p_i - 1}$ for all i. In general $a^s \equiv a \pmod{n}$ holds for *all* values of a if and only if $s \equiv 1 \pmod{\lambda(n)}$. For *particular* values of a, a^s could, of course, be $\equiv a$ for much smaller values of s than $\lambda(n) + 1$.—Moreover, if n contains multiple prime factors, then (A2.42) need no longer be true, as can be seen from the example $n = 12$ with $\lambda(n) = 2$ and $2^3 = 8 \not\equiv 2 \pmod{12}$.

Bibliography

1. O. Ore, *Number Theory and Its History*, McGraw-Hill, New York, 1948.

2. D. Shanks, *Solved and Unsolved Problems in Number Theory*, Second edition, Chelsea, New York, 1978.

QUADRATIC RESIDUES

Legendre's Symbol

Let us commence with

Definition A3.1. If $(a, n) = 1$ and if the congruence

$$x^2 \equiv a \ (\text{mod } n) \tag{A3.1}$$

has solutions x, then a is called a *quadratic residue* of n. If (A3.1) has no solutions, a is a *quadratic non-residue* (mod n).

For the case when n is an odd prime p, Legendre introduced a special symbol (a/p) which is given the value $+1$ if a is a quadratic residue and the value -1 if a is a quadratic non-residue of n.

Examples. Calculating $(\pm 1)^2$, $(\pm 2)^2$ and $(\pm 3)^2$ (mod 7) we find that $a \equiv 1, 2$ and 4 are the quadratic residues (mod 7). The remaining residue classes, i.e. $a \equiv 3, 5$ and 6 are the quadratic non-residues (mod 7).—Analogously, it can be shown that $a \equiv 1$ is the only quadratic residue (mod 12).—Please note that $a \equiv 4$ and $a \equiv 9$ *do not count* as quadratic residues (mod 12), since $(a, 12) > 1$ for these values of a.

Arithmetic Rules for Residues and Non-Residues

In Definition A3.1 a condition is that $(a, n) = 1$, and thus we consider only those residue classes which belong to the group M_n of primitive residue classes (mod n). As a matter of fact, definition A3.1 divides M_n into two parts. One part consists of all quadratic residues (mod n), and is actually a subgroup of M_n. The other part consists of all quadratic non-residues (mod n), and is composed of one or several cosets to this subgroup. (The other residue classes (mod n), with $(a, n) > 1$, which are not in M_n at all, are sometimes called semi-residues of n. The quantity (a/n) is, in these cases, given the value 0.) Now quadratic residuacity is governed by the following two simple rules:

Theorem A3.1. The product of two quadratic residues is always a quadratic residue, and the product of a residue and a non-residue is a non-residue.

Proof. $x^2 \equiv a \pmod{n}$ and $y^2 \equiv b \pmod{n}$ imply $(xy)^2 \equiv ab \pmod{n}$, which proves the first half of the theorem. However, if the product of some quadratic residue x^2 with some quadratic non-residue z were to result in another residue y^2, then this would imply $x^2 z \equiv y^2 \pmod{n}$, which would immediately give $z \equiv (y/x)^2 \pmod{n}$, i.e. contrary to the assumption, z would actually be a quadratic residue. (Note that y/x is an integer $\pmod{n}$, since we are allowed to form yx^{-1} without leaving the group M_n of *primitive* residue classes.) This completes the proof.

In general, no simple rule can be applied in order to determine the quadratic character of the product of two quadratic non-residues. Only in the case when M_n is a *cyclic* group (and $n > 2$), and in particular when n is an odd prime, are there equal numbers of residues and non-residues, and in that case the product of two non-residues is always a residue.

Proof. If M_n is a cyclic group, then all the group elements can be written as

$$g^s, \quad \text{for} \quad 1 \leq s \leq \varphi(n), \tag{A3.2}$$

where g is a generator of M_n. Obviously the $\frac{1}{2}\varphi(n)$ group-elements

$$g^2, g^4, g^6, \ldots, g^{\varphi(n)} \tag{A3.3}$$

with *even* exponents are all quadratic residues. Furthermore g must be a quadratic non-residue, because if g were $= x^2$, then Lagrange's theorem A1.3 would give $g^{\frac{1}{2}\varphi(n)} = x^{\varphi(n)} = 1$. But then, contrary to the assumption, g would not be a generator of the cyclic group M_n. Now, since g is a non-residue, all the residues (A3.3) multiplied by g are also non-residues:

$$g, g^3, g^5, \ldots, g^{\varphi(n)-1}. \tag{A3.4}$$

But this exhausts all elements of M_n, and so we find the same number of residues and non-residues, both $= \frac{1}{2}\varphi(n)$.—Besides, the product of two non-residues is, in this simple case, a residue, since

$$g^{\text{odd number}} \cdot g^{\text{odd number}} = g^{\text{even number}}.$$

The result can be summed up in

Theorem A3.2. If M_n is a cyclic group, i.e. if $n = 4, n = p^\alpha$ or $n = 2p^\alpha$ for some odd prime p, then there are $\frac{1}{2}\varphi(n)$ quadratic residues and $\frac{1}{2}\varphi(n)$ quadratic non-residues (mod n). The product of two non-residues is always a residue in such a case, which means that Legendre's symbol always obeys the rule

$$\left(\frac{a}{p}\right)\left(\frac{b}{p}\right) = \left(\frac{ab}{p}\right), \tag{A3.5}$$

if p is an odd prime and if $(a, n) = (b, n) = 1$, since M_p is a cyclic group.

Example. We have discussed above the example $p = 7$, with $(a/p) = +1$ for $a \equiv 1, 2$ and 4 (mod 7), and $(a/p) = -1$ for $a \equiv 3, 5$ and 6 (mod 7). The multiplication table of M_7 can be arranged in the following manner with the quadratic residues boxed:

	1 2 4	3 5 6
1	1 2 4	3 5 6
2	2 4 1	6 3 5
4	4 1 2	5 6 3
3	3 6 5	2 1 4
5	5 3 6	1 4 2
6	6 5 3	4 2 1

Euler's Criterion for the Computation of (a/p)

Euler discovered a method of finding out if a is a quadratic residue or not of p without having to calculate all the quadratic residues of p as we did above for $p = 7$:

Theorem A3.3. Euler's Criterion. If $(a, p) = 1$ and if p is an odd prime, then

$$\left(\frac{a}{p}\right) \equiv a^{(p-1)/2} \text{ (mod } p). \tag{A3.6}$$

Proof. Let g be a generator of the cyclic group M_p. For which values of s is $g^s \equiv -1$ (mod p)? Obviously g^{2s} must then be $\equiv 1$ (mod p). This can only be the case if $2s$ is a multiple of $p - 1$, since $p - 1$ is the smallest positive exponent e, making $g^e \equiv 1$ (mod p). Therefore s must assume one of the values $0, (p - 1)/2, p - 1, \ldots$ Now $g, g^2, \ldots, g^{p-1} \equiv 1$ constitute all the different elements of M_p (in some order), and so $g^{(p-1)/2}$ must be that power of g which is $\equiv -1$ (mod p).—Now, if a can be represented as

280

$a \equiv g^t$, say, then

$$a^{(p-1)/2} \equiv g^{t(p-1)/2} \equiv (-1)^t \pmod{p}, \qquad (A3.7)$$

i.e. $a^{(p-1)/2} \equiv +1$ if t is even, which means that a is one of the quadratic residues, listed in (A3.3). Alternatively $a^{(p-1)/2} \equiv -1$ if t is odd, in which case a is one of the quadratic non-residues of (A3.4). This proves Euler's Criterion.

The Law of Quadratic Reciprocity

One of the highlights of number theory is the law of quadratic reciprocity, known to Euler and Legendre and proved by Gauss. It relates the two Legendre symbols (q/p) and (p/q) by

Theorem A3.4. The quadratic reciprocity law. If p and q are odd primes, then

$$\left(\frac{p}{q}\right) = \left(\frac{q}{p}\right)(-1)^{\frac{1}{2}(p-1)\frac{1}{2}(q-1)}. \qquad (A3.8)$$

We shall not here prove the theorem, but rather refer the reader to textbooks in number theory, such as [1] or [2].

If Theorem A3.4 is supplemented with a few of the cases which it does not cover, then it can be used for fast computation of the value of Legendre's symbol (a/p), when p is a prime. The required supplements are provided in

Theorem A3.5. If p is an odd prime, then

$$\left(\frac{-1}{p}\right) = (-1)^{\frac{1}{2}(p-1)}, \quad \text{which is} \begin{cases} +1, & \text{if } p = 4k+1 \\ -1, & \text{if } p = 4k-1 \end{cases} \qquad (A3.9)$$

and

$$\left(\frac{2}{p}\right) = (-1)^{(p^2-1)/8}, \quad \text{which is} \begin{cases} +1, & \text{if } p = 8k \pm 1 \\ -1, & \text{if } p = 8k \pm 3. \end{cases} \qquad (A3.10)$$

The proofs may also be found in [1] or [2].

Example 1. Computation of (123/4567), where the integer 4567 is known to be a prime.

$$\left(\frac{123}{4567}\right) = \left(\frac{3 \cdot 41}{4567}\right) = \left(\frac{3}{4567}\right) \cdot \left(\frac{41}{4567}\right) =$$

$$= \left(\frac{4567}{3}\right)(-1)^{2283} \times \left(\frac{4567}{41}\right)(-1)^{20 \cdot 2283} = \left(\frac{1}{3}\right)(-1)\left(\frac{16}{41}\right) = -1.$$

Using Euler's criterion we have been able to answer the fairly simple question: "Which a's are quadratic residues (mod p)?" Using the quadratic reciprocity law we can answer the much more difficult question: "For which primes p is a a quadratic residue?"

Example 2. When is $(10/p) = +1$? The deduction runs as follows:

$$\left(\frac{10}{p}\right) = \left(\frac{2}{p}\right)\left(\frac{5}{p}\right) = \left(\frac{2}{p}\right)\left(\frac{p}{5}\right).$$

But

$$\left(\frac{2}{p}\right) = +1, \quad \text{if } p = 8k \pm 1 \text{ and}$$

$$\left(\frac{2}{p}\right) = -1, \quad \text{if } p = 8k \pm 3 \text{ according to (A3.10).}$$

Also,

$$\left(\frac{p}{5}\right) = +1, \quad \text{if } p = 5m \pm 1 \text{ and}$$

$$\left(\frac{p}{5}\right) = -1, \quad \text{if } p = 5m \pm 2 \text{ according to (A3.6).}$$

Thus, $(2/p)(p/5)$ is evidently a periodic number-theoretic function with its period-length (at most) $= 5 \cdot 8 = 40$. Thus it suffices to study all integers p in the interval $[1, 40]$ with $(p, 2) = 1$ and $(p, 5) = 1$.

These different values of p lead to the following values of the Legendre-symbols $(2/p)$ and $(p/5)$. Below the line the product of these two Legendre-symbols has been calculated.

$p \pmod{40}$	1	3	7	9	11	13	17	19	21	23	27	29	31	33	37	39
$\left(\dfrac{2}{p}\right)$	1	−1	1	1	−1	−1	1	−1	−1	1	−1	−1	1	1	−1	1
$\left(\dfrac{5}{p}\right)$	1	−1	−1	1	1	−1	−1	1	1	−1	−1	1	1	−1	−1	1
$\left(\dfrac{2}{p}\right)\left(\dfrac{p}{5}\right)$	1	1	−1	1	−1	1	−1	−1	−1	−1	1	−1	1	−1	1	1

Therefore, $(10/p) = +1$, if p is given by one of the forms

$$p = 40k + 1, 3, 9, 13, 27, 31, 37, 39 \qquad (A3.11)$$

and $(10/p) = -1$, if p satisfies one of

$$p = 40k + 7, 11, 17, 19, 21, 23, 29, 33. \qquad (A3.12)$$

Remark. The reciprocal of the prime p has a periodic decimal expansion (except for the two primes $p = 2$ and 5). The period obviously contains d digits if the order of the residue class 10 of the group M_p is d, because then $10^d - 1 \equiv 0 \pmod{p}$, while $10^s - 1 \not\equiv 0 \pmod{p}$ for any positive s smaller than d. The maximal period-length $d = p - 1$ is obtained if 10 is one of the group elements that can be used as a generator of M_p. Now, if $(10/p) = +1$, Euler's criterion rules out a maximal period-length, since then already $10^{(p-1)/2} \equiv 1 \pmod{p}$, and the period-length d must, in such a case, be a divisor of $(p-1)/2$. Hence the primes with maximally long periods for their reciprocals are all to be found among the numbers (A3.12). The period-lengths for all primes below 101 except 2 and 5 are given in the following table:

The period-length d of the decimal fraction $1/p$											
p	3	7	11	13	17	19	23	29	31	37	41	43
d	1	6	2	6	16	18	22	28	15	3	5	21
p	47	53	59	61	67	71	73	79	83	89	97	101
d	46	13	58	60	33	35	8	13	41	44	96	4

It is seen from the table that all p's with maximal period, 7, 17, 19, 23, 29, 47, 59, 61 and 97, belong to the set (A3.12), while all p's belonging to (A3.11) have shorter periods.

The calculation we completed above to find all primes with $(10/p) = 1$ can, of course, be easily done for any positive or negative value of a instead of 10. The result will be analogous: $(a/p) = 1$ (or $= -1$) for all p's falling within certain arithmetic series with the difference $= 4a$.

For certain computations it is convenient to have access to these arithmetic series. At the end of this book, tables are provided in which all such series can be found for all square-free a with $|a| \leq 101$ (Table 32).

Jacobi's Symbol

Originally the Legendre symbol (a/p) was only defined for the case when p is a prime. As we have seen, it obeys the rules (A3.5) and (A3.8)–(A3.10). Jacobi found the following definition to be of value:

Definition A3.2. Jacobi's symbol. Let n be an odd integer with $(a, n) = 1$. Define

$$\left(\frac{a}{n}\right) \quad \text{as} \quad \prod_i \left(\frac{a}{p_i}\right)^{\alpha_i}, \quad \text{if} \quad n = \prod_i p_i^{\alpha_i}. \tag{A3.13}$$

This extended symbol satisfies the same formal rules as does the original Legendre symbol.

Thus, we have

Theorem A3.6. If m and n are odd positive integers, and if $(a, n) = 1$ and $(b, n) = 1$, then

$$\left(\frac{a}{n}\right)\left(\frac{b}{n}\right) = \left(\frac{ab}{n}\right) \tag{A3.14}$$

$$\left(\frac{-1}{n}\right) = (-1)^{\frac{1}{2}(n-1)} \tag{A3.15}$$

$$\left(\frac{2}{n}\right) = (-1)^{(n^2-1)/8} \tag{A3.16}$$

and

$$\left(\frac{m}{n}\right) = \left(\frac{n}{m}\right)(-1)^{\frac{1}{2}(m-1)\frac{1}{2}(n-1)}. \tag{A3.17}$$

A proof is given in [3].

At the beginning of this appendix, we mentioned that it could happen that the product of two quadratic non-residues is another *non-residue*. This may occur if M_n is not a cyclic group. However, since (A3.14) always holds, the Jacobi symbol (a/n) *occasionally assumes, in such a case, the value* $+1$ *for a quadratic non-residue* a (mod n). As an example we give

$$\left(\frac{2}{15}\right) = \left(\frac{2}{3}\right)\left(\frac{2}{5}\right) = (-1)(-1) = +1,$$

although 2 is a quadratic non-residue (mod 15).

On the other hand, it is easy to prove that (a/n) always takes the value $+1$, if a is a quadratic residue (mod n), because since $x^2 \equiv a$ (mod n) is solvable, this congruence must also be solvable (mod p_i) for each factor p_i of n. As a consequence, each of the factors (a/p_i) in (A3.13) takes the value $+1$, implying that their product (a/n) also takes the same value $+1$ in this case.

The conclusion to be drawn from all this is that Jacobi's symbol (a/n) cannot be of direct use to decide whether or not a is a quadratic residue (mod n). Nonetheless, since Jacobi's symbol obeys the same formal laws as Legendre's symbol, its value can be computed in the same fast way without first establishing whether n is prime or not. If it subsequently emerges that n actually is a prime, then the value of the symbol provides information about the quadratic character of a (mod n). The fact that $(a/n) = +1$ (or $= -1$) may also be interpreted thus: n falls within one of those arithmetic series which contain all primes p with $(a/p) = +1$ (or $= -1$, respectively).

Example. In (A3.11) the following assertion is made:

$$\left(\frac{10}{p}\right) = +1, \quad \text{if } p = 40k + 9 \quad \text{is a prime.}$$

From this fact we also deduce that

$$\left(\frac{10}{n}\right) = +1, \quad \text{if } n \text{ is any number of the form } n = 40k + 9,$$

even if n happens to be composite. We have e.g.

$$\left(\frac{10}{49}\right) = \left(\frac{10}{7}\right)\left(\frac{10}{7}\right) = \left(\frac{3}{7}\right)\left(\frac{3}{7}\right) = (-1)(-1) = +1.$$

Obviously, 10 is a quadratic non-residue (mod 49), since $10 \equiv 3$ (mod 7), and 3 is a non-residue (mod 7). In Chapter 4 we sometimes use Jacobi's symbol in this way.

A PASCAL Function for Computing (a/n)

In the following function Jacobi (d,n) (d/n) is computed recursively according to formulas (A3.14)–(A3.17). If n is even or if $(d,n) > 1$, then an error message is delivered. In such a case the variable Jacobi is given the value 0.

```
FUNCTION Jacobi(d,n : INTEGER) : INTEGER;
{Computes Jacobi's symbol (d/n) for odd n}
LABEL 1,2,3,4;
VAR d1,n1,i2,m,n8,u,z,u4 : INTEGER;
BEGIN
   d1:=d; n1:=abs(n); m:=0; n8:=n1 MOD 8;
   IF odd(n8+1) THEN
     BEGIN writeln(tty,'n even in (d/n) is not allowed');
       GOTO 3 END;
   IF d1<0 THEN
     BEGIN d1:=-d1; IF (n8=3) OR (n8=7) THEN m:=m+1 END;
1: IF d1=0 THEN
     BEGIN writeln(tty,'(d,n)>1 in (d/n) is not allowed');
       GOTO 3 END;
   i2:=0;
2: u:=d1 DIV 2; IF d1=u*2 THEN
     BEGIN i2:=i2+1; d1:=u; GOTO 2 END;
   IF odd(i2) THEN m:=m+(n8*n8-1) DIV 8;
   u4:=d1 MOD 4; m:=m+(n8-1)*(u4-1) DIV 4; z:=n1 MOD d1;
   n1:=d1; d1:=z; n8:=n1 MOD 8; IF n1>1 THEN GOTO 1;
   m:=m MOD 2; IF m=0 THEN Jacobi:=1 ELSE Jacobi:=-1; GOTO 4;
3: Jacobi:=0;
4: END {Jacobi};
```

Exercise A3.1. The number of quadratic residues. Deduce a formula for the number of quadratic residues of a *composite* number $n = 2^\alpha \prod p_i^{\alpha_i}$. Hints

1. Remember that (a, n) has to be $= 1$.

2. Any residue (mod n) has to be a residue of 2^α and of *all* $p_i^{\alpha_i}$.

3. The residues of $p_i^{\alpha_i}$ are easy to find if you make use of the representation of the group $M_{p_i^{\alpha_i}}$, which is cyclic if p_i is odd and if $p_i^{\alpha_i} = 2$ or 4, and is of type $2 \times 2^{\alpha-2}$ if 2^α for $\alpha \geq 2$ is considered.

4. The residues of the various moduli 2^α and $p_i^{\alpha_i}$ may be combined by multiplication to give all residues (mod n).

Check your formula on $n = 15$ and $n = 24$. How can all residues (mod 15) be found from the representations (A2.36) of the elements of M_{15}?

The Quadratic Congruence $x^2 \equiv c \pmod{p}$

We conclude this appendix by demonstrating how to compute efficiently the solutions to the congruence

$$x^2 \equiv c \pmod{p}, \quad p \text{ prime and } \left(\frac{c}{p}\right) = +1. \tag{A3.18}$$

Since Euler's criterion gives $c^{(p-1)/2} \equiv 1 \pmod{p}$, we have

$$c^{(p+1)/2} \equiv c \pmod{p}. \tag{A3.19}$$

Thus, if $(p+1)/2$ is *even*, i.e. when p is of the form $4k - 1$, then we immediately find the solution

$$x \equiv \pm c^{(p+1)/4} \pmod{p}. \tag{A3.20}$$

The Case $p = 4k + 1$

The situation when $p = 4k + 1$ is more difficult but, contrary to what is frequently believed, there does exist an effective algorithm working in polynomial time only, see [4]. It is certainly not surprising that this algorithm is based on the theory of Lucas sequences, because in Lucas primality tests $p+1$ frequently plays the same rôle as does $p-1$ in ordinary Fermat tests. Starting with two Lucas sequences U_n and V_n, defined as usual by (4.31), with the characteristic equation $\lambda^2 - P\lambda + c = 0$ and P chosen so that $P^2 - 4c$ is a quadratic non-residue of p, we have, by (4.53)

$$U_{p+1} \equiv 0 \pmod{p}. \tag{A3.21}$$

Thus, according to (4.38),

$$U_{p+1} = U_{(p+1)/2} V_{(p+1)/2} \equiv 0 \pmod{p}. \tag{A3.22}$$

Further, using the same notation as on p. 113:

$$V_n^2 - (a-b)^2 U_n^2 = (a^n + b^n)^2 - (a^n - b^n)^2 = 4a^n b^n = 4Q^n.$$

Since $(a-b)^2 = P^2 - 4Q = P^2 - 4c$ in our case, we obtain

$$V_{(p+1)/2}^2 - (P^2 - 4c)U_{(p+1)/2}^2 = 4c^{(p+1)/2} \equiv 4c \pmod{p}. \tag{A3.23}$$

At this point we ask the question: Which of the two factors $U_{(p+1)/2}$ and $V_{(p+1)/2}$ in (A3.22) is divisible by p? *Certainly not $V_{(p+1)/2}$, because then* (A3.23) would imply that $-(P^2 - 4c)$ and, since p is of the form $4k + 1$, also $P^2 - 4c$, are quadratic residues of p, contrary to our choice of P. Thus, $U_{(p+1)/2}$ must be divisible by p, and hence

$$V_{(p+1)/2}^2 \equiv 4c \pmod{p}, \tag{A3.24}$$

providing a solution to the congruence $x^2 \equiv c \pmod{p}$ in this case. Since the terms of a Lucas sequence can be computed in a logarithmic number of steps, the algorithm is of polynomial time. Besides, the amount of work required to compute $V_n \pmod{p}$ is about three times that needed to find $a^n \pmod{p}$.

Remark. *Only if the Lucas sequence has already been found* can the computation of $V_{(p+1)/2}$ be accomplished in polynomial time. Such a situation occurs e.g. when p is fixed and many square roots (mod p) have to be computed. If, however, a square root has to be computed modulo a *different* p each time, then also the time to find a suitable Lucas sequence has to be taken into account. Since there is no known *deterministic* algorithm for this search (there are only algorithms based on trial and error) the running time may in this case be little worse than polynomial time.

Exercise A3.2. Square roots (mod p). Write a computer program for the computation of $\sqrt{c} \pmod{p}$, p odd, if the square root exists. (If not, print an error message.) Use the PROGRAM Fermat on p. 92 for the case $p = 4k - 1$ and the PROGRAM Fibonacci in exercise 4.2 on p. 115 for the case $4k + 1$.

Bibliography

1. G. H. Hardy and E. M. Wright, *An Introduction to the Theory of Numbers*, Fifth edition, Oxford, 1979, pp. 73–78.

2. Loo Keng Hua, *Introduction to Number Theory*, Springer, New York, 1982, pp. 37–42.

3. Trygve Nagell, *Introduction to Number Theory*, Almqvist & Wiksell, Uppsala, 1951, pp. 145–149.

4. D. H. Lehmer, *Computer Technology Applied to the Theory of Numbers*. Printed in W. J. LeVeque, *Studies in Number Theory*, Prentice-Hall, N. J., 1969, pp. 117–151.

THE ARITHMETIC OF QUADRATIC FIELDS

Integers of $K(\sqrt{D})$

We shall define the concept of *integers in a quadratic field*. In order to distinguish these generalized integers from the ordinary integers $0, \pm 1, \pm 2, \ldots$, we shall call the ordinary integers *rational integers*.— Consider all numbers z of the form

$$z = r + s\sqrt{D}, \qquad (A4.1)$$

with r and s rational numbers and D a square-free integer. (A square-free integer is an integer with no square factor, i.e. an integer with its standard factorization containing no multiple prime factors.) As we have demonstrated in Appendix 1, formulas (A1.18)–(A1.20) on p. 256, these numbers constitute a (quadratic) field, since all four arithmetic operations can be performed on any two numbers of the field, except division by 0, and give as result a number which is also of the form (A4.1). In order to discuss quadratic fields in greater detail, we require the following

Definition A4.1. The number z is said to be an *integer* of the quadratic field $K(\sqrt{D})$ if it satisfies a quadratic equation of the form

$$z^2 + pz + q = 0, \qquad (A4.2)$$

with coefficients p and q that are rational integers.

Since the number $z = r + s\sqrt{D}$ is a solution of the quadratic equation of the form

$$(z - r)^2 = Ds^2, \quad \text{i.e.} \quad z^2 - 2rz + r^2 - Ds^2 = 0, \qquad (A4.3)$$

the following conditions are necessary and sufficient in order that $r + s\sqrt{D}$ be an integer of the field $K(\sqrt{D})$:

$$-2r \quad \text{and} \quad r^2 - Ds^2 \quad \text{must both be integers.}$$

Putting $r = a/2$ for integral a, $2r$ is always a rational integer and $r^2 - Ds^2 = a^2/4 - Ds^2$ is a rational integer if either

289

1. a is even (i.e. if r is a rational integer) and s is a rational integer.

or

2. a is odd (in which case $a^2/4 \equiv 1/4 \pmod 1$), and $Ds^2 \equiv 1/4$ (mod 1). The latter case occurs if and only if $s \equiv 1/2 \pmod 1$ and at the same time $D \equiv 1 \pmod 4$.

These two separate cases can be reformulated as

Theorem A4.1. The integers of $K(\sqrt{D})$ are

$$x = r + s\sqrt{D} = r + s\rho, \quad \text{if } D \equiv 2 \text{ or } \equiv 3 \text{ (mod 4)} \tag{A4.4}$$

$$x = r + s \cdot \frac{-1 + \sqrt{D}}{2} = r + s\rho, \quad \text{if } D \equiv 1 \text{ (mod 4)}, \tag{A4.5}$$

r and s being arbitrary rational integers.

It is easy to verify that the given definition of integers of $K(\sqrt{D})$ complies with the normal rules for integers, namely that the sum, difference and product of any two integers is also an integer:

$$(r_1 + s_1\rho) \pm (r_2 + s_2\rho) = (r_1 \pm r_2) + (s_1 \pm s_2)\rho \tag{A4.6}$$

and

$$(r_1 + s_1\rho)(r_2 + s_2\rho) = r_1r_2 + s_1s_2\rho^2 + (r_1s_2 + s_1r_2)\rho =$$
$$= \begin{cases} r_1r_2 + Ds_1s_2 + (r_1s_2 + s_1r_2)\rho, & \text{if } \rho = \sqrt{D} \\ r_1r_2 + \dfrac{D-1}{4}s_1s_2 + (r_1s_2 + s_1r_2 - s_1s_2)\rho, & \text{if } \rho = \dfrac{-1+\sqrt{D}}{2}. \end{cases}$$
$$\tag{A4.7}$$

Definition A4.2. The numbers $x = r + s\sqrt{D}$ and $\bar{x} = r - s\sqrt{D}$ are called *conjugate numbers* in $K(\sqrt{D})$. The rational number

$$x\bar{x} = N(x) = N(\bar{x}) = \begin{cases} r^2 - Ds^2, & \text{if } D \equiv 2 \text{ or } \equiv 3 \text{ (mod 4)} \\ r^2 - rs - \dfrac{D-1}{4}s^2, & \text{if } D \equiv 1 \text{ (mod 4)} \end{cases} \tag{A4.8}$$

is termed the *norm* of x in $K(\sqrt{D})$.

Note that

$$N(x) = 0 \text{ if and only if } x = 0, \tag{A4.9}$$

and that the norm of an integer of $K(\sqrt{D})$ is a rational integer.

Examples. The set of ordinary complex numbers $x + iy$ (corresponding to $D = -1$), contains the so-called Gaussian integers $a + ib$, a and b being rational integers, as a subset. The norm is $N(a + ib) = a^2 + b^2$. The integers of $K(\sqrt{-1})$ are the so-called lattice points of the coordinate system. See figure A4.1 on the next page.

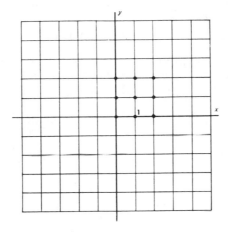

Figure A4.1. The Gaussian integers $a + ib$, of $K(\sqrt{-1})$

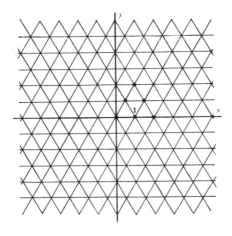

Figure A4.2. The integers $a + b\rho$, $\rho = (-1 + i\sqrt{3})/2$, of $K(\sqrt{-3})$

291

Choosing $D = -3$ instead, the numbers of the field $K(\sqrt{-3})$ are $x + y\sqrt{-3}$ with their integers being of the form $a + b\rho$, where $\rho = (-1 + \sqrt{-3})/2$. If depicted by points in the complex plane, the integers of $K(\sqrt{-3})$ are the lattice points of the hexagonal lattice drawn in figure A4.2.—For other negative values of D, we arrive at situations analogous to the two described above; the integers of $K(\sqrt{D})$ being the lattice points of a particular lattice in the complex plane.

The most important property of the norm of a number is that the norm is a completely multiplicative function. Thus, we have

Theorem A4.2. $N(\alpha\beta) = N(\alpha)N(\beta).$ $\hspace{2cm}$ (A4.10)

Proof. If $D \equiv 2$ or $3 \pmod 4$, (A4.7) and (A4.9) imply that

$$N(r_1 + s_1\sqrt{D}) \cdot N(r_2 + s_2\sqrt{D}) = (r_1^2 - Ds_1^2)(r_2^2 - Ds_2^2) =$$
$$= (r_1r_2 + Ds_1s_2)^2 - D(r_1s_2 + s_1r_2)^2 =$$
$$= N\{(r_1 + s_1\sqrt{D})(r_2 + s_2\sqrt{D})\}.$$

If $D \equiv 1 \pmod 4$, we have instead, putting $(D-1)/4 = T$:

$$N(r_1 + s_1\rho) \cdot N(r_2 + s_2\rho) = (r_1^2 - r_1s_1 - Ts_1^2) \cdot (r_2^2 - r_2s_2 - Ts_2^2) =$$
$$= (r_1r_2 + Ts_1s_2)^2 - (r_1r_2 + Ts_1s_2)(r_1s_2 + s_1r_2 - s_1s_2) -$$
$$-T(r_1s_2 + s_1r_2 - s_1s_2)^2 = N\{(r_1 + s_1\rho)(r_2 + s_2\rho)\}.$$

Units of $K(\sqrt{D})$

We begin by making

Definition A4.3. If

$$N(x) = \pm 1, \hspace{2cm} \text{(A4.11)}$$

then x is called a *unit* of $K(\sqrt{D})$.

Examples. Among the Gaussian integers there are four units, ± 1 and $\pm i$, corresponding to all solutions in rational integers of the equation $r^2 + s^2 = 1$. (Note that $r^2 + s^2 = -1$ has no solutions!) In $K(\sqrt{-3})$ the units are the integer solutions of

$$N(r + s\rho) = r^2 - rs + s^2 = \pm 1,$$

which are

$$\begin{cases} r = 0 \\ s = \pm 1 \end{cases} \qquad \begin{cases} r = \pm 1 \\ s = 0 \end{cases} \qquad \begin{cases} r = 1 \\ s = 1 \end{cases} \qquad \begin{cases} r = -1 \\ s = -1, \end{cases}$$

leading to the six units

$$x = \pm 1, \qquad x = \pm \frac{-1 \pm \sqrt{-3}}{2}. \qquad \text{(A4.12)}$$

Real quadratic fields $(D > 0)$ have infinitely many units. An example is the field $K(\sqrt{2})$, in which

$$x = (1 + \sqrt{2})^n = a_n + b_n\sqrt{2} \qquad \text{(A4.13)}$$

is a unit for every integral value of n, since

$$x\bar{x} = (a_n + b_n\sqrt{2})(a_n - b_n\sqrt{2}) = (1 + \sqrt{2})^n(1 - \sqrt{2})^n = (-1)^n = \pm 1.$$

Associated Numbers in $K(\sqrt{D})$

The concept of associated numbers, well-known from the rational integers, see p. 1 (n and $-n$ are associated in the ring of rational integers), can also be generalized, according to the following

Definition A4.4. Two integers α and β of $K(\sqrt{D})$ are said to be *associated*, if their quotient is a unit.

Examples. $2 + 3i$ and $i(2 + 3i) = -3 + 2i$ are associated integers in $K(\sqrt{-1})$.—In $K(\sqrt{2})$ the integers $5 + \sqrt{2}$ and $11 - 7\sqrt{2}$ are associated, because

$$\frac{11 - 7\sqrt{2}}{5 + \sqrt{2}} = \frac{(11 - 7\sqrt{2})(5 - \sqrt{2})}{23} = \frac{69 - 46\sqrt{2}}{23} = 3 - 2\sqrt{2}$$

and

$$N(3 - 2\sqrt{2}) = 3^2 - 2 \cdot 2^2 = 1.$$

Associated numbers always have the same norm (or the same norm with a sign change). This is due to the fact that if $\alpha = \beta x$, then also $N(\alpha) = N(\beta)N(x)$ with $N(x) = \pm 1$, which proves the assertion. However, the converse is not always true: *There are integers in $K(\sqrt{D})$ with the same norm, which are not associated.* As an example, consider $2 + i$ and $2 - i$ in $K(\sqrt{-1})$, both of norm 5, but nevertheless $(2 + i)/(2 - i) = (2 + i)^2/5 = \frac{3}{5} + \frac{4}{5}i$ is not a unit in $K(\sqrt{-1})$.

Divisibility in $K(\sqrt{D})$

If $\alpha = \beta\gamma$, with α, β and γ integers of $K(\sqrt{D})$, then α is said to be divisible by β (and by γ). Notation: $\beta|\alpha$ (β divides α). Now, if $\beta|\alpha$, then $N(\beta)|N(\alpha)$ must necessarily hold in the field of *rational* numbers, because $N(\alpha) = N(\beta)N(\gamma)$ and the norms are *rational* integers.—The arithmetic of congruences (mod n) in the field $K(\sqrt{D})$ is very similar to the arithmetic of congruences (mod n) in the field of rational numbers, since every *rational* integer n is also an integer of $K(\sqrt{D})$ with the norm $N(n) = n^2$.—However, special care must be taken when applying Theorem A2.4 on p. 264, the cancellation rule for congruences. Whilst it *is* true that the cancellation rule

$$dx \equiv dy \ (\text{mod } dn) \Rightarrow x \equiv y \ (\text{mod } n)$$

is valid also in the field $K(\sqrt{D})$, it is easy to overlook a common factor of the modulus and the two sides of the congruence. This stems from the fact that such a common factor must be an *integer* of $K(\sqrt{D})$, but the factor, if any, is not quite as easily recognizable as is a common factor in the field of rationals. If, for instance,

$$(1 + \sqrt{3})x \equiv (1 + \sqrt{3})^2 \ (\text{mod } 10),$$

then it *does not follow* by cancellation of the easily recognizable factor $1 + \sqrt{3}$ that

$$x \equiv 1 + \sqrt{3} \ (\text{mod } 10),$$

but only that

$$x \equiv 1 + \sqrt{3} \left(\text{mod } \frac{10}{1 + \sqrt{3}} \right) = 1 + \sqrt{3} \ (\text{mod } 5(\sqrt{3} - 1)).$$

If for some reason we are interested only in rational modules, than it even follows less, only that

$$x \equiv 1 + \sqrt{3} \ (\text{mod } 5).$$

Exercise A4.1. Arithmetic operations (mod 7) **in the quadratic field** $K(\sqrt{6})$. Write down all residue classes (mod 7) in $K(\sqrt{6})$. (There are 49 of them.) Find out how to compute an expression like $(a + b\sqrt{6})/(c + d\sqrt{6})$ (mod 7). How much is $(3 + 5\sqrt{6})/(8 - \sqrt{6})$ (mod 7)? (Answer: $6 + 4\sqrt{6}$.)

Fermat's Theorem in $K(\sqrt{D})$

A most interesting fact is that Fermat's theorem A2.8 on p. 270 has a nice generalization to the field $K(\sqrt{D})$. Before we are able to expand on this, we need to establish when an integer $\alpha = r + s\rho$ of $K(\sqrt{D})$ is $\equiv 0 \pmod{n}$. This means that we would like n to divide α. Hence there must be an integer $\beta = t + u\rho$ of $K(\sqrt{D})$ such that $\alpha = n\beta$, i.e. such that $r + s\rho = n(t + u\rho)$, which implies that $r = nt$ and $s = nu$. Obviously, then, the condition that $n \mid r + s\rho$ is equivalent to $r \equiv s \equiv 0 \pmod{n}$. Applying this result, we are now in a position to generalize Fermat's theorem

$$a^p \equiv a \pmod{p} \tag{A4.14}$$

to integers of $K(\sqrt{D})$. Consider first the case with $D \equiv 2$ or $\equiv 3 \pmod 4$. Suppose p is an odd rational prime which does not divide D. Then, the binomial theorem yields

$$a^p = (r + s\sqrt{D})^p = r^p + \binom{p}{1} r^{p-1} s\sqrt{D} + \binom{p}{2} r^{p-2} s^2 D + \cdots +$$

$$+ \binom{p}{p-1} r s^{p-1} D^{\frac{1}{2}(p-1)} + s^p D^{\frac{1}{2}(p-1)}\sqrt{D} \equiv$$

(since all the binomial coefficients $\binom{p}{k}$ are $\equiv 0 \pmod{p}$)

$$\equiv r^p + s^p D^{\frac{1}{2}(p-1)}\sqrt{D} \pmod{p} \equiv$$

(according to Fermat's theorem in the field of rationals)

$$\equiv r + s \cdot D^{\frac{1}{2}(p-1)}\sqrt{D} \equiv$$

(using Euler's criterion for quadratic residuacity on p. 280)

$$\equiv r + s\left(\frac{D}{p}\right)\sqrt{D} \equiv \begin{cases} a \pmod{p}, & \text{if } \left(\dfrac{D}{p}\right) = +1 \\[2mm] \bar{a} \pmod{p}, & \text{if } \left(\dfrac{D}{p}\right) = -1. \end{cases}$$

In the case when $D \equiv 1 \pmod 4$, an analogous computation gives

$$a^p = \left(r + s \cdot \frac{-1 + \sqrt{D}}{2}\right)^p = 2^{-p}\left((2r - s) + s\sqrt{D}\right)^p \equiv$$

295

$$\equiv 2^{-p}\left((2r-s)^p + s^p D^{\frac{1}{2}(p-1)}\sqrt{D}\right) \pmod{p} \equiv$$

$$\equiv \frac{1}{2}\left((2r-s) + s\sqrt{D}\left(\frac{D}{p}\right)\right) =$$

$$= r + s \cdot \frac{-1 + \left(\dfrac{D}{p}\right)\sqrt{D}}{2} \equiv \begin{cases} a \pmod{p}, & \text{if } \left(\dfrac{D}{p}\right) = +1 \\[2ex] \bar{a} \pmod{p}, & \text{if } \left(\dfrac{D}{p}\right) = -1. \end{cases}$$

All these cases can be collected as

Theorem A4.3. Fermat's theorem in $K(\sqrt{D})$. If a is an integer of $K(\sqrt{D})$, and if p is an odd rational prime, then

$$a^p \equiv a \pmod{p}, \quad \text{if } \left(\frac{D}{p}\right) = +1 \tag{A4.15}$$

or

$$a^p \equiv \bar{a} \pmod{p}, \quad \text{if } \left(\frac{D}{p}\right) = -1. \tag{A4.16}$$

This theorem can be applied in various ways: of particular importance for us is (A4.16) which, when multiplied by a factor a, yields

$$a^{p+1} \equiv a\bar{a} = N(a) \pmod{p}, \quad \text{if } \left(\frac{D}{p}\right) = -1. \tag{A4.17}$$

In addition, (A4.15) implies that if $n > 2$, then

$$a^{p^n} = (a^p)^{p^{n-1}} = (a + kp)^{p^{n-1}} \equiv a^{p^{n-1}} \pmod{p^n},$$

since all the terms in the binomial expansion of $(a + kp)^{p^{n-1}}$, apart from the term $a^{p^{n-1}}$, are divisible by a factor p^n. Now, if $(a, p) = 1$ in $K(\sqrt{D})$ then, on cancelling the congruence by $a^{p^{n-1}}$, we arrive at

$$a^{p^{n-1}(p-1)} \equiv 1 \pmod{p^n}, \quad \text{if } \left(\frac{D}{p}\right) = 1. \tag{A4.18}$$

In an analogous manner, (A4.17) leads to

$$a^{p^{n-1}(p+1)} = (a^{p+1})^{p^{n-1}} = (N(a) + kp)^{p^{n-1}} \equiv$$

$$\equiv (N(a))^{p^{n-1}} \pmod{p^n}, \quad \text{if } \left(\frac{D}{p}\right) = -1. \tag{A4.19}$$

The congruences (A4.18) and (A4.19), which are a generalization of Euler's Theorem A2.9 on p. 272 to $K(\sqrt{D})$ in the case when the module n is of the form p^n, were used in Chapter 4.

Primes in $K(\sqrt{D})$

Every integer α of $K(\sqrt{D})$ can be written as a product $\alpha = \beta\gamma$ of integers of $K(\sqrt{D})$. In order to see this, we need only to choose β as a suitable unit of $K(\sqrt{D})$ and γ as one of the integers associated with α. Normally $K(\sqrt{D})$ has *several different units*, e.g. ± 1. A useful definition of primes of $K(\sqrt{D})$ is, as a consequence, not as obvious as for the ring of rationals. However, the following definition is suitable:

Definition A4.5. If the integer α of $K(\sqrt{D})$ cannot be decomposed as $\alpha = \beta\gamma$, with β and γ integers of $K(\sqrt{D})$, unless one of β or γ is a *unit* of $K(\sqrt{D})$, then α is said to be a prime of $K(\sqrt{D})$. Otherwise α is said to be *composite* in $K(\sqrt{D})$.

Examples. It follows that the primes in $K(\sqrt{-1})$ are the following Gaussian integers:

1. The numbers $1+i$, $2\pm i$, $3\pm 2i$, $4\pm i$, $\ldots$, i.e. all the factors $a\pm bi$ of $p = a^2 + b^2$, where $p = 2$ or p is a rational prime of the form $4k + 1$.
2. The rational integers $3, 7, 11, 19, \ldots$, i.e. all rational primes q of the form $q = 4k - 1$.
3. The primes associated to these primes must be added.

Exercise A4.2. Machin's formula for π. Write $239 + i$ as a product of Gaussian primes. Hint: $N(239 + i) = 57122 = 2 \cdot 13^4$. Start by looking for primes with norms 2 and 13 and try them as factors!

Use the prime factorization found to prove

$$\left(\frac{5+i}{5-i}\right)^4 \left(\frac{239+i}{239-i}\right)^{-1} = \frac{1+i}{1-i} = i.$$

Prove that this formula can be re-written as

$$4 \arctan\frac{1}{5} - \arctan\frac{1}{239} = \arctan 1 = \frac{\pi}{4}$$

(Machin's formula for $\pi/4$). Hint:

$$\arctan\frac{1}{z} = \text{arccot}\, z = \frac{1}{2i} \ln\frac{z+i}{z-i},$$

and thus $\sum_k a_k \, \text{arccot}\, u_k = \pi/4$ is equivalent to

$$\prod_k \left(\frac{u_k + i}{u_k - i}\right)^{a_k} = e^{\frac{2i\pi}{4}} = i.$$

Find more formulas of Machin's kind by factorizing all Gaussian integers of the form $a \pm i$ for $a \leq 57$ and looking for combinations of factors satisfying the product formula above. (You should write some computer FUNCTIONs for the arithmetic operations of Gaussian integers and let the computer print out a list of factorizations for you.) Did you find Euler's $\operatorname{arccot} 2 + \operatorname{arccot} 3 = \pi/4$? Gauss' $12 \operatorname{arccot} 18 + 8 \operatorname{arccot} 57 - 5 \operatorname{arccot} 239 = \pi/4$? Störmer's $6 \operatorname{arccot} 8 + 2 \operatorname{arccot} 57 + \operatorname{arccot} 239 = \pi/4$?

Factorization of Integers in $K(\sqrt{D})$

Using the prime numbers defined above, every integer α of $K(\sqrt{D})$ can be decomposed into prime factors belonging to $K(\sqrt{D})$:

$$\alpha = \prod_i \pi_i^{a_i} \tag{A4.20}$$

At first sight the situation appears very similar to the corresponding result concerning factorization in the field of rationals. This is, however, not at all the case! It is only in a few exceptional cases that the decomposition (A4.20) is unique! As a matter of fact, we have the following

Theorem A4.4. If $D = -163, -67, -43, -19, -11, -7, -3, -2, -1,$ $2, 3, 5, 6, 7, 11, 13, 17, 19, 21, 29, 33, 37, 41, 57$ and 73, then the theorem of unique factorization into prime factors is valid in $K(\sqrt{D})$. The negative values of D mentioned are all *negative* values for which the theorem holds.

A proof can be found in [1].

According to Theorem A4.4 there exist quadratic fields in which the *fundamental theorem of arithmetic does not hold*. A simple example is when $D = -6$. In the field $K(\sqrt{-6})$, it can be proved in the following manner that the integers $2, 3$ and $\sqrt{-6}$ all are primes. If a number α is composite in $K(\sqrt{-6})$, then

$$\alpha = \beta\gamma = (a + b\sqrt{-6})(c + d\sqrt{-6}),$$

and

$$N(\alpha) = (a^2 + 6b^2)(c^2 + 6d^2).$$

However, the norms of the integers $2, 3$ and $\sqrt{-6}$ are $4, 9$, and -6, respectively, and none of these numbers admits a decomposition in the form $\pm(a^2 + 6b^2)(c^2 + 6d^2)$, unless one of the factors is a unit thus proving the primality of $2, 3$, and $\sqrt{-6}$ in $K(\sqrt{-6})$.

Now, what can be said about the factorization of the integer 6? Obviously

$$6 = 2 \cdot 3 = -\sqrt{-6}\sqrt{-6} \qquad (A4.21)$$

yielding *two essentially distinct prime factorizations* of the integer 6. How can this be possible? The phenomenon may be explained as follows: The number field $K(\sqrt{-6})$ can be extended to a larger field K_1, consisting of all numbers of the form

$$a + b\sqrt{-2} + c\sqrt{3} + d\sqrt{-6}.$$

In the field K_1 both $\sqrt{-2}$ and $\sqrt{3}$ are primes, and therefore the integer 6 possesses *in this field* a unique factorization (apart from the usual possibility to use associated primes):

$$6 = -\sqrt{-2}\sqrt{-2}\sqrt{3}\sqrt{3}. \qquad (A4.22)$$

Combining the factors of this factorization in different groupings, it appears that several groupings give rise to primes in the smaller field $K(\sqrt{-6})$:

$$(-\sqrt{-2}\sqrt{-2}) \cdot (\sqrt{3}\sqrt{3}) \quad \text{and} \quad (-\sqrt{-2}\sqrt{3}) \cdot (\sqrt{-2}\sqrt{3}).$$

This is how the two essentially different prime factorizations of 6 arise.— By introducing what is now called *ideals*, Kummer succeeded in restoring the fundamental theorem of arithmetic to cyclotomic rings $Z_p(\varsigma)$, which correspond to the cyclotomic fields $R_p(\varsigma)$ mentioned on p. 256. The theory of ideals has later been perfected by Dedekind. For a detailed account of these theories, see [2].—The fundamental theorem of arithmetic in its naïve form (i.e. without using ideals) once again breaks down in a general cyclotomic ring. The theorem does hold in $Z_p(\varsigma)$ for the first seven odd primes $p = 3, 5, 7, 11, 13, 17$ and 19, but is no longer valid for $p = 23$.

Bibliography

1. G. H. Hardy and E. M. Wright, *An Introduction to the Theory of Numbers,* Fifth edition, Oxford, 1979, pp. 211–215.

2. Z. I. Borevich and I. R. Shafarevich, *Number Theory*, Academic Press, New York, 1966, pp. 155–250.

APPENDIX 5

CONTINUED FRACTIONS

Introduction

In this appendix we shall provide the reader with a glimpse at the theory of continued fractions. This is a fascinating and very useful theory, both theoretically and computationally. It is, however, a little neglected in mathematics courses, and there are only a few elementary books [1], [2] on the subject.

What Is a Continued Fraction?

We start by making

Definition A5.1. A continued fraction is an expression of the form

$$b_0 + \cfrac{a_1}{b_1 + \cfrac{a_2}{b_2 + \cfrac{a_3}{b_3 + \cdots + \cfrac{a_n}{b_n}}}} \tag{A5.1}$$

The numbers a_i are termed *partial numerators* and the numbers b_i (apart from b_0) are called *partial denominators*. If all partial numerators a_i are equal to 1, and if b_0 is an integer and all partial denominators b_i are *positive* integers, then the continued fraction is said to be *regular*.

Before we proceed we shall introduce a more convenient notation for the cumbersome (A5.1):

$$b_0 + \frac{a_1 \rfloor}{\lceil b_1} + \frac{a_2 \rfloor}{\lceil b_2} + \cdots + \frac{a_n \rfloor}{\lceil b_n}$$

A (finite) continued fraction (with rational a_i's and b_i's) represents a rational number. The significance of *regular* continued fractions is that if such an expansion is truncated at any point, then this shorter expansion

300

(called *convergent*) will represent a number that approximates the value of the complete expansion. The greater the portion of the complete expansion used, the more accurate will the approximation be. Thanks to this property, regular continued fractions are of the utmost importance in the theory of approximation of real numbers by rational numbers.

Regular Continued Fractions. Expansions

Regular continued fractions are the simplest ones to deal with. They possess certain very simple and useful properties, which we now present.

To begin with, we show how to find the regular continued fraction expansion of a real number x. We would like to write x in the form

$$x = b_0 + \cfrac{1}{b_1 + \cfrac{1}{b_2 + \cdots}} = b_0 + \frac{1|}{|b_1} + \frac{1|}{|b_2} + \cdots, \qquad (A5.2)$$

where b_0 is an integer and the partial denominators $b_1, b_2, \ldots$ all *positive* integers. This is achieved by successively computing the numbers given by the following algorithm:

$$b_0 = [x], \quad x_1 = \frac{1}{x - b_0}, \quad b_1 = [x_1], \quad x_2 = \frac{1}{x_1 - b_1},$$

$$b_2 = [x_2], \quad x_3 = \frac{1}{x_2 - b_2}, \cdots, \quad b_n = [x_n], \quad x_{n+1} = \frac{1}{x_n - b_n} \cdots \quad (A5.3)$$

The computation will *terminate* with one of the numbers x_i being exactly an integer, if and only if x is a *rational* number. If x is an *irrational* number, then the expansion is *infinite*.

Example 1. The regular continued fraction expansion of $\sqrt{2}$ is computed as follows:

$$b_0 = [\sqrt{2}] = 1, \qquad x_1 = \frac{1}{\sqrt{2} - 1} = \sqrt{2} + 1$$

$$b_1 = [\sqrt{2} + 1] = 2, \quad x_2 = \frac{1}{\sqrt{2} + 1 - 2} = \sqrt{2} + 1.$$

From this point on the calculation is repeated periodically. Thus, b_2 and all subsequent b_i's equal $b_1 = 2$, and the (infinite) expansion becomes

$$\sqrt{2} = 1 + \frac{1|}{|2} + \frac{1|}{|2} + \frac{1|}{|2} + \cdots, \qquad (A5.4)$$

Example 2. The base of the natural logarithms, $e = 2.71828182\ldots$, yields

$$
\begin{aligned}
b_0 &= 2, & x_1 &= 1/0.71828182\ldots = 1.39221119\ldots \\
b_1 &= 1, & x_2 &= 1/0.39221119\ldots = 2.54964677\ldots \\
b_2 &= 2, & x_3 &= 1/0.54964677\ldots = 1.81935024\ldots \\
b_3 &= 1, & x_4 &= 1/0.81935024\ldots = 1.22047928\ldots \\
b_4 &= 1, & x_5 &= 1/0.22047928\ldots = 4.53557347\ldots \\
b_5 &= 4, & x_6 &= 1/0.535573\ldots \quad = 1.867157\ldots \\
b_6 &= 1, & x_7 &= 1/0.867157\ldots \quad = 1.153193\ldots \\
b_7 &= 1, & x_8 &= 1/0.153193\ldots \quad = 6.527707\ldots \\
b_8 &= 6, & x_9 &= 1/0.5277\ldots \quad\;\; = 1.8949\ldots \\
b_9 &= 1, & x_{10} &= 1/0.8949\ldots \quad\;\; = 1.1173\ldots \\
b_{10} &= 1, & x_{11} &= 1/0.1173\ldots \quad\;\; = 8.5226\ldots \\
b_{11} &= 8, & &\ldots
\end{aligned}
$$

This computation produces the famous expansion

$$
e = 2 + \frac{1|}{|1} + \frac{1|}{|2} + \frac{1|}{|1} + \frac{1|}{|1} + \frac{1|}{|4} + \frac{1|}{|1} + \frac{1|}{|1} + \frac{1|}{|6} + \cdots \qquad (A5.5)
$$

Further, it is useful to observe that the expanded number x can always be recovered by first truncating the expansion at a partial denominator b_k and then replacing b_k by x_k:

$$
x = b_0 + \frac{1|}{|b_1} + \frac{1|}{|b_2} + \cdots + \frac{1}{|b_{k-1}} + \frac{1|}{|x_k}. \qquad (A5.6)
$$

This fact is an immediate consequence of algorithm (A5.3) above for the computation of the b_i's, since x_k has precisely the expansion

$$
b_k + \frac{1}{|b_{k+1}} + \frac{1}{|b_{k+2}} + \cdots
$$

This is actually identical to the part of the expansion following the partial denominator b_{k-1} in the expansion of the original number x.—Thus, from the expansion (A5.4) above of $\sqrt{2}$ we find

$$
\sqrt{2} = 1 + \frac{1|}{|2} + \frac{1|}{|2} + \cdots + \frac{1|}{|2} + \frac{1}{|\sqrt{2}+1|}, \qquad (A5.7)
$$

the dots representing an arbitrary number of partial denominators $= 2$.

Evaluating a Continued Fraction

How can the value of a regular continued fraction be calculated? The first time you see a continued fraction written down, such as

$$\frac{243}{89} = 2 + \frac{1|}{|1} + \frac{1|}{|2} + \frac{1|}{|1} + \frac{1|}{|2} + \frac{1|}{|2} + \frac{1|}{|3}$$

you may think that the only way to evaluate this expression is to start at the bottom and compute backwards through the fractions, as follows: $2 + \frac{1}{3} = \frac{7}{3}$, $2 + \frac{3}{7} = \frac{17}{7}$, $1 + \frac{7}{17} = \frac{24}{17}$, $2 + \frac{17}{24} = \frac{65}{24}$, $1 + \frac{24}{65} = \frac{89}{65}$ and, finally, $2 + \frac{65}{89} = \frac{243}{89}$.

However, there is another algorithm which actually runs *forwards* through the partial denominators. It is based on the following

Theorem A5.1. Let b_0 be an integer and let all partial denominators of the regular continued fraction

$$\frac{A_n}{B_n} = b_0 + \frac{1|}{|b_1} + \frac{1|}{|b_2} + \cdots + \frac{1|}{|b_n} \tag{A5.8}$$

be integers > 0. Let the so-called *partial quotients* or *convergents* A_s/B_s be given in their lowest terms. If we define $A_{-1} = 1$, $B_{-1} = 0$, $A_0 = b_0$ and $B_0 = 1$, then A_n/B_n can be computed recursively by the formulas

$$\begin{cases} A_s = b_s A_{s-1} + A_{s-2} \\ B_s = b_s B_{s-1} + B_{s-2}. \end{cases} \tag{A5.9}$$

Proof. $A_1/B_1 = b_0 + 1/b_1 = (b_1 b_0 + 1)/b_1$, which gives $A_1 = b_1 b_0 + 1 = b_1 A_0 + A_{-1}$ and $B_1 = b_1 = b_1 B_0 + B_{-1}$, so that the recursion formulas hold for $s=1$. If they hold for s, how can they be proved for $s+1$? Well, the partial quotient A_{s+1}/B_{s+1} can be constructed from A_s/B_s simply by writing $b_s + 1/b_{s+1}$ for b_s. Thus, if (A5.9) holds for s, then also

$$\frac{A_{s+1}}{B_{s+1}} = \frac{(b_s + \frac{1}{b_{s+1}})A_{s-1} + A_{s-2}}{(b_s + \frac{1}{b_{s+1}})B_{s-1} + B_{s-2}} = \frac{b_s A_{s-1} + A_{s-2} + \frac{A_{s-1}}{b_{s+1}}}{b_s B_{s-1} + B_{s-2} + \frac{B_{s-1}}{b_{s+1}}} =$$

$$= \frac{A_s + \frac{A_{s-1}}{b_{s+1}}}{B_s + \frac{B_{s-1}}{b_{s+1}}} = \frac{b_{s+1}A_s + A_{s-1}}{b_{s+1}B_s + B_{s-1}},$$

303

which agrees with (A5.9) if we choose

$$\begin{cases} A_{s+1} = b_{s+1}A_s + A_{s-1} \\ B_{s+1} = b_{s+1}B_s + B_{s-1}. \end{cases}$$

But how can we ensure that the fraction A_{s+1}/B_{s+1} will be given in its *lowest terms*? This is guaranteed by another important theorem, viz.

Theorem A5.2. The A_s and B_s as defined in (A5.9) satisfy

$$A_{s-1}B_s - A_sB_{s-1} = (-1)^s. \tag{A5.10}$$

Proof. $A_{-1}B_0 - A_0B_{-1} = 1 \cdot 1 - b_0 \cdot 0 = 1$, so that the relation holds for $s = 0$. Suppose it holds for a particular value of s. Then for $s+1$ we obtain

$$A_sB_{s+1} - A_{s+1}B_s = A_s(b_{s+1}B_s + B_{s-1}) - (b_{s+1}A_s + A_{s-1})B_s =$$
$$= -(A_{s-1}B_s - A_sB_{s-1}) = -(-1)^s = (-1)^{s+1}.$$

From this theorem it follows that all partial quotients A_s/B_s are in their lowest terms, because if A_s and B_s had a common factor $d > 1$, then this factor would certainly appear as a divisor of $A_{s-1}B_s - A_sB_{s-1}$. This result concludes the proof of Theorem A5.1.

The recursion formulas (A5.9) immediately give rise to the following tabular scheme for the efficient calculation of the value of a regular continued fraction:

	b_1	b_2	...	b_{s-1}	b_s	b_{s+1}	...
1	b_0	$b_1 \cdot b_0 + 1$	...	A_{s-2}	A_{s-1}	$b_s \cdot A_{s-1} + A_{s-2}$	...
0	1	$b_1 \cdot 1 + 0$		B_{s-2}	B_{s-1}	$b_s \cdot B_{s-1} + B_{s-2}$	

The method is as follows: First, write down the two left-most partial quotients $A_{-1}/B_{-1} = 1/0$ and $A_0/B_0 = b_0/1$. Next introduce the successive partial denominators $b_1, b_2, b_3, \ldots$, at the head of the table. Finally, calculate A_s/B_s recursively by means of (A5.9), i.e. multiply the next b_i by the term below it and then add the term immediately to the left on the same line. Calculate first the numerator and then the denominator in each of the convergents by following this rule twice.

304

Example. The fraction 243/89 earlier developed can be re-calculated from its regular continued fraction expansion in the following way:

	1	2	1	2	2	3	
$\frac{1}{0}$	$\frac{2}{1}$	$\frac{3}{1}$	$\frac{8}{3}$	$\frac{11}{4}$	$\frac{30}{11}$	$\frac{71}{26}$	$\frac{243}{89}$

Formulas (A5.9) may surprise you at first sight, particularly the fact that both the numerator and the denominator satisfy the *same* recursion formula, but they arise as a natural consequence of the convergent A_n/B_n being a fractional linear function of b_n. Write

$$\frac{A(t)}{B(t)} = b_0 + \frac{1}{\lfloor b_1} + \frac{1}{\lfloor b_2} + \cdots + \frac{1}{\lfloor b_{n-1}} + \frac{1}{\lfloor t} \qquad (A5.11)$$

with the variable t replacing b_n in (A5.8). Then

$$R(t) = \frac{A(t)}{B(t)} = \frac{a + bt}{c + dt} \qquad (A5.12)$$

for some suitably chosen constants a, b, c and d. Now, $t \to \infty$ in (A5.11) gives $R(\infty) = A_{n-1}/B_{n-1}$, and $t \to 0$ in (A5.11) yields

$$\frac{1}{b_{n-1} + \dfrac{1}{t}} = \frac{1}{b_{n-1} + \infty} = 0,$$

so that $R(0) = A_{n-2}/B_{n-2}$. Thus we observe that

$$\frac{A(t)}{B(t)} = \frac{A_{n-2} + kA_{n-1}t}{B_{n-2} + kB_{n-1}t},$$

must hold for some constant k in order for $R(0)$ and $R(\infty)$ to assume correct values. Inserting $t = b_n$, $R(b_n) = A_n/B_n$, and thus

$$\frac{A_n}{B_n} = \frac{A_{n-2} + kA_{n-1}b_n}{B_{n-2} + kB_{n-1}b_n}.$$

This finally gives $k = 1$ due to (A5.9), and we obtain the following value for the continued fraction (A5.11):

$$\frac{A(t)}{B(t)} = \frac{A_{n-2} + A_{n-1}t}{B_{n-2} + B_{n-1}t}. \qquad (A5.13)$$

Continued Fractions as Approximations

As we have already pointed out at the beginning of this appendix, if we truncate a regular continued fraction expansion of an irrational number x after the nth partial denominator b_n, then the truncated fraction, the convergent A_n/B_n, approximates to the number x. We shall now study this approximation in some detail. Firstly, if x is an irrational number > 0, then the first approximation b_0 will be too small, the second approximation $b_0 + 1/b_1$ too large, the third too small and so on. This is obvious if you consider the construction of the continued fraction. Developing these ideas, let x be an irrational number whose third approximation is given by

$$\xi = b_0 + \cfrac{1}{b_1 + \cfrac{1}{b_2}},$$

This means that

$$x = b_0 + \cfrac{1}{b_1 + \cfrac{1}{b_2 + \theta}}, \quad \text{with } 0 < \theta < 1,$$

since b_2 has been computed as the integral part of some number $b_2 + \theta$ between b_2 and $b_2 + 1$. Hence, ξ is an approximation to x with b_2 too small, and thus $1/b_2$ and also $b_1 + 1/b_2$ will be too large. Finally, ξ will be smaller than x. Whether the approximation ξ emerges smaller or greater than x will depend upon whether an even or an odd number of divisions are involved in the above reasoning, proving our assertion.

Introducing, as in Theorem A5.1, the convergent A_n/B_n, we can express x in the form of a convergent alternating series:

$$
\begin{aligned}
x &= \frac{A_0}{B_0} + \left(\frac{A_1}{B_1} - \frac{A_0}{B_0}\right) + \left(\frac{A_2}{B_2} - \frac{A_1}{B_1}\right) + \cdots + \left(\frac{A_s}{B_s} - \frac{A_{s-1}}{B_{s-1}}\right) + \cdots \\
&= \frac{A_0}{B_0} + \frac{1}{B_0 B_1} - \frac{1}{B_1 B_2} + \frac{1}{B_2 B_3} - \cdots + \frac{(-1)^{s-1}}{B_{s-1} B_s} + \cdots, \quad (A5.14)
\end{aligned}
$$

since by Theorem A5.2, $A_{s-1} B_s - A_s B_{s-1} = (-1)^s$. The fact that the series in (A5.14) converges is easy to prove. The B_i's must grow at least as fast as the Fibonacci numbers U_n (see p. 114) since, according to Theorem A5.1, $B_s = b_s B_{s-1} + B_{s-2} \geq B_{s-1} + B_{s-2}$ because $b_s \geq 1$. However, the Fibonacci number U_n is $\approx ((1 + \sqrt{5})/2)^n/\sqrt{5}$, and thus the series must converge at least as rapidly as a geometric series with its quotient $= -((1 + \sqrt{5})/2)^{-2} = -0.382$.

We shall now consider the following important question: How close will the convergent A_n/B_n approach to x? This is answered by

Theorem A5.3. Let the regular continued fraction expansion of the irrational number x be

$$x = b_0 + \frac{1}{|b_1} + \frac{1}{|b_2} + \cdots + \frac{1}{|b_n} + \cdots$$

Then the nth convergent,

$$\frac{A_n}{B_n} = b_0 + \frac{1}{|b_1} + \frac{1}{|b_2} + \cdots + \frac{1}{|b_n},$$

expressed in its lowest terms, approximates x and we have

$$0 < (-1)^n \left(x - \frac{A_n}{B_n} \right) < \frac{1}{B_n^2 \sqrt{5}}. \tag{A5.15}$$

Proof. Since we will need only the weaker result

$$0 < (-1)^n \left(x - \frac{A_n}{B_n} \right) < \frac{1}{B_n^2}, \tag{A5.16}$$

we do not prove (A5.15) but refer the reader to [3] for its proof. (A5.16) is proved in the following way:

$$(-1)^n \left(x - \frac{A_n}{B_n} \right) = \frac{1}{B_n B_{n+1}} - \frac{1}{B_{n+1} B_{n+2}} + \cdots < \frac{1}{B_n B_{n+1}} < \frac{1}{B_n^2}.$$

Euclid's Algorithm and Continued Fractions

The sequence of computations performed when developing a/b as a regular continued fraction is identical to that when (a, b) is calculated by Euclid's algorithm. The reader should compare the example on Euclid's algorithm for (8991,3293), given on p. 240 with the example given above in which we developed $243/89 = 8991/3293$ as a continued fraction. The sequence of quotients occurring in Euclid's algorithm is exactly the same as the sequence of partial denominators in the continued fraction expansion, in our example 2, 1, 2, 1, 2, 2, 3. This observation is most useful for the efficient solution of linear diophantine equations (or linear congruences, which is almost the same thing).

Linear Diophantine Equations and Continued Fractions

In Appendix 2, pp. 266–268, we presented an algorithm for solving the linear congruence $ax \equiv c \pmod{b}$, which is equivalent to the diophantine equation $ax - by = c$. This algorithm starts by first applying Euclid's algorithm to calculate (a, b), followed by a calculation running backwards through the chain of equations in order to find the desired values of x and y. Now, by employing continued fractions, we can obtain all this in one step!

Let us begin by considering the simplest case, the diophantine equation $ax - by = 1$, with $(a, b) = 1$ and $b > a$. First calculate the continued fraction expansion of b/a. Essentially this means performing Euclid's algorithm, since the sequence of divisions on performing Euclid's algorithm on the integers a and b is precisely the same as in carrying out the regular continued fraction expansion of b/a. Subsequently, compute the successive partial quotients by means of the scheme we demonstrated on p. 304. When the final partial quotient A_s/B_s before b/a is reached, the solution to $ax - by = 1$ is immediate, since $aA_s - bB_s = \pm 1$ (Theorem A5.2).

If we wish to solve $ax - by = c$ instead of $ax - by = 1$, still with $(a, b) = 1$, we need only to multiply the values of x and y found as the solution of $ax - by = 1$ by c.—If $(a, b) > 1$, then this will be evident at the end of the continued fraction expansion, just as it becomes in Euclid's algorithm. In this case we must verify that $(a, b) | c$ and, if so, when cancelling the whole diophantine equation by (a, b), which does not affect the continued fraction expansion, we end up having to solve a diophantine equation with $c/(a, b)$ rather than the original c.—Let us now once again consider the example given in Appendix 2

$$3293x \equiv 222 \pmod{8991}.$$

We have already on p. 240 performed a calculation which is equivalent to the continued fraction expansion of $8991/3293 = 243/89$ and on p. 305 we have calculated the penultimate partial quotient of this expansion. It was found to be $71/26$, and thus $89 \cdot 71 - 243 \cdot 26 = 1$. We can always proceed this far without any difficulty. But one may ask: "Why 243 and 89 rather than 8991 and 3293?" The reason is that the final two integers possess a common divisor $= 3293/89 = 37$. Now, we must check whether this common divisor $37 | 222$. It does, and therefore the congruence we are considering is solvable. Hence, we can cancel the factor 37, yielding

$89x \equiv 6 \pmod{243}$, corresponding to $89x - 243y = 6$. However, we have just found that $89x - 243y = 1$ has the solution $x = 71$, $y = 26$. Thus our congruence has the solution $x \equiv 6 \cdot 71 \equiv 426 \equiv 183 \pmod{243}$.

A Computer Program

On p. 242 in Appendix 1 we have seen a computer program for Euclid's algorithm. This program can easily be modified to provide the solution of a linear congruence $ax \equiv c \pmod{b}$. To this end we must incorporate the recursion (A5.9) into the program. Since the recursion runs forwards, just as Euclid's algorithm does, *it is not necessary to save all the partial denominators!* As soon as a new partial denominator is calculated, it is used at once to compute the values of the next A_s and B_s and can then be discarded. A flow-chart of the program is shown below:

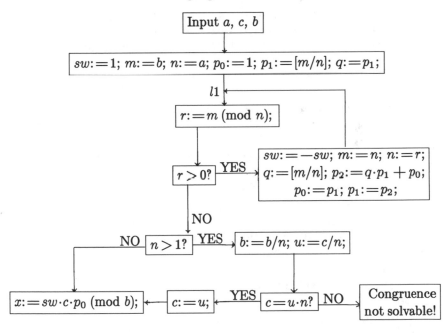

The following is a PASCAL program Dioeqv for the solution of a linear diophantine equation with two unknowns. It is based on a PASCAL function lincon which solves the linear congruence $ax \equiv c \pmod{b}$, and is an implementation of the flow-chart given above.—The variable sw contains a switch having the value $(-1)^{s+1}$.

```
PROGRAM Dioeqv;
{Solves the diophantine equation ax-by=c}
(Input,Output);
LABEL 1;
VAR a,b,c,x : INTEGER;

FUNCTION lincon(a,b,c : INTEGER) : INTEGER;
{Solves the linear congruence ax = c (mod b)}
LABEL 1,2;
VAR  sw,m,n,p0,p1,p2,q,r,u : INTEGER;
BEGIN
  sw:=1; m:=abs(b); n:=a; IF n<0 THEN BEGIN c:=-c; n:=-n END
  {Here signs are adjusted if a and b are not positive};
  p0:=1; p1:=m DIV n; q:=p1 {Starting values!};
1: r:=m MOD n;
  IF r>0 THEN
    BEGIN sw:=-sw; m:=n; n:=r; q:=m DIV n;
      p2:=q*p1+p0; p0:=p1; p1:=p2; GOTO 1
      {This constitutes the continued fraction algorithm}
    END
  ELSE
    BEGIN IF n>1 THEN
      BEGIN b:=b DIV n; u:=c DIV n;
        IF c <> u*n
          THEN
            BEGIN writeln(tty,'Congruence not solvable!');
              sw:=0; GOTO 2
            END
          ELSE
            BEGIN c:=u;
              writeln(tty,'Modulus reduced to',b:8)
            END
      END;
    2:lincon:=sw*c*p0 MOD b
    END
END {lincon};

BEGIN
1:write(tty,'Input a,b and c in the congruence ax=c (mod b)');
  write(tty,'and ax-by=c: '); read(tty,a,b,c);
  IF a <> 0 THEN
    BEGIN x:=lincon(a,b,c);
```

310

```
        writeln(tty,'x=',x:8,' y=',(a*x-c)DIV b:8);
        GOTO 1
    END
END.
```

Exercise A5.1. Division (mod n). Use the PROGRAM Dioeqv above to compute

$$\frac{1}{2}+\frac{1}{3}+\frac{1}{5}+\frac{1}{7}+\frac{1}{11}+\frac{1}{13}\ (\text{mod } 29).$$

Hint: Compute each of the fractions $x_p \equiv 1/p$ (mod 29) separately by solving $px_p - 1 = 29y$. (Answer: 15.)

Continued Fraction Expansions of Square Roots

We now turn our attention to a subject of great interest to number theory, namely the regular continued fraction expansion of the square root of a non-square integer N. The expansion can formally be defined as follows:

$$x_0 = \sqrt{N}, \quad b_i = [x_i], \quad x_{i+1} = 1/(x_i - b_i), \tag{A5.17}$$

where $[x]$, as usual, denotes the integer part of x. The definition implies that

$$\sqrt{N} = b_0 + \frac{1}{\lfloor b_1} + \frac{1}{\lfloor b_2} + \frac{1}{\lfloor b_3} + \cdots + \frac{1}{\lfloor b_n} + \cdots \tag{A5.18}$$

with all b_i's positive integers. The expansion is infinite, since $\sqrt{N}$ is irrational when N is not a perfect square. We shall prove that the expansion is always periodic.

The partial denominators b_i are best computed as in the following example: $N = 69$ gives $x_0 = \sqrt{69}$, $b_0 = [\sqrt{69}] = 8$ and

$$x_1 = \frac{1}{\sqrt{69}-8} = \frac{\sqrt{69}+8}{5} = 3 + \frac{\sqrt{69}-7}{5}$$

$$x_2 = \frac{5}{\sqrt{69}-7} = \frac{5(\sqrt{69}+7)}{20} = \frac{\sqrt{69}+7}{4} = 3 + \frac{\sqrt{69}-5}{4}$$

$$x_3 = \frac{4}{\sqrt{69}-5} = \frac{4(\sqrt{69}+5)}{44} = \frac{\sqrt{69}+5}{11} = 1 + \frac{\sqrt{69}-6}{11}$$

$$x_4 = \frac{11}{\sqrt{69}-6} = \frac{11(\sqrt{69}+6)}{33} = \frac{\sqrt{69}+6}{3} = 4 + \frac{\sqrt{69}-6}{3}$$

$$x_5 = \frac{3}{\sqrt{69}-6} = \frac{3(\sqrt{69}+6)}{33} = \frac{\sqrt{69}+6}{11} = 1 + \frac{\sqrt{69}-5}{11}$$

$$x_6 = \frac{11}{\sqrt{69}-5} = \frac{11(\sqrt{69}+5)}{44} = \frac{\sqrt{69}+5}{4} = 3 + \frac{\sqrt{69}-7}{4}$$

$$x_7 = \frac{4}{\sqrt{69}-7} = \frac{4(\sqrt{69}+7)}{20} = \frac{\sqrt{69}+7}{5} = 3 + \frac{\sqrt{69}-8}{5}$$

$$x_8 = \frac{5}{\sqrt{69}-8} = \frac{5(\sqrt{69}+8)}{5} = \sqrt{69}+8 = 16 + (\sqrt{69}-8)$$

$$x_9 = \frac{1}{\sqrt{69}-8} = 3 + \frac{\sqrt{69}-7}{5}.$$

Here, the partial denominators b_i in the expansion are the initial integers of the right-most expressions on each line. Since $x_9 = x_1$, the previous calculation will be repeated from this point onwards, so that the entire expression is periodic from x_9 and on. We thus have the expansion

$$\sqrt{69} = 8 + \overline{\frac{1|}{|3} + \frac{1|}{|3} + \frac{1|}{|1} + \frac{1|}{|4} + \frac{1|}{|1} + \frac{1|}{|3} + \frac{1|}{|3} + \frac{1|}{|16}} + \frac{1|}{|3} + \frac{1|}{|3} + \cdots,$$

which is periodic, with the first period marked by a bar.—Therefore, at least in this particular example, the regular continued fraction expansion of $\sqrt{N}$ is periodic. Note also that, excepting the last term 16, the period is symmetric, the point of symmetry being the partial denominator 1 in this case.

Proof of Periodicity

We shall now prove that the continued fraction expansion of $\sqrt{N}$ is always periodic. Since, in Chapter 5, we needed one more relation between the variables introduced in the following deduction, we shall provide this at the same time. We write a general line in the calculation above as (assuming the quantities P_i and Q_i to be > 0)

$$x_i = \frac{Q_{i-1}}{\sqrt{N}-P_i} = \frac{\sqrt{N}+P_i}{Q_i} = b_i + \frac{\sqrt{N}-P_{i+1}}{Q_i}. \qquad \text{(A5.19)}$$

with

$$b_i = [x_i] = \left[\frac{\sqrt{N} + P_i}{Q_i}\right], \quad P_{i+1} = b_i Q_i - P_i. \qquad (A5.20)$$

We will now show that $P_i < \sqrt{N}$ and $Q_i < 2\sqrt{N}$ limits the number of possibilities for the pairs (P_i, Q_i) to at most $2N$, and thus induces periodicity in the sequence $\{b_i\}$. The crucial point is that the expression

$$\frac{Q_{i-1}}{\sqrt{N} - P_i} = \frac{Q_{i-1}(\sqrt{N} + P_i)}{N - P_i^2} \qquad (A5.21)$$

can always be reduced to $(\sqrt{N} + P_i)/Q_i$, implying that

$$Q_i = \frac{N - P_i^2}{Q_{i-1}} \qquad (A5.22)$$

must always be an integer. We prove this by mathematical induction. In the first line of the computation above we have

$$x_1 = \frac{1}{\sqrt{N} - b_0} = \frac{\sqrt{N} + b_0}{N - b_0^2} = \frac{\sqrt{N} + P_1}{Q_1}, \qquad (A5.23)$$

where $Q_1 = N - P_1^2$. Here, the requirement that the right-hand-side has the desired form is fulfilled. Assuming that

$$x_i = \frac{\sqrt{N} + P_i}{Q_i}, \quad \text{with} \quad Q_i | N - P_i^2, \qquad (A5.24)$$

we obtain

$$x_i = b_i + \frac{\sqrt{N} - (b_i Q_i - P_i)}{Q_i} = b_i + \frac{\sqrt{N} - P_{i+1}}{Q_i}, \qquad (A5.25)$$

$$x_{i+1} = \frac{Q_i}{\sqrt{N} - P_{i+1}} = \frac{Q_i(\sqrt{N} + P_{i+1})}{N - P_{i+1}^2}. \qquad (A5.26)$$

Now, because $(N - P_i^2)/Q_i$ is an integer, the number $(N - P_{i+1}^2)/Q_i = (N - (b_i Q_i - P_i)^2)/Q_i = (N - P_i^2)/Q_i - b_i^2 Q_i + 2b_i P_i$ is also an integer, proving that (A5.22) is always an integer.

313

The choice of b_i implies that $P_i < \sqrt{N}$ and, since $P_{i+1} = b_i Q_i - P_i$, we find

$$Q_i = \frac{P_i + P_{i+1}}{b_i} < 2\sqrt{N}. \tag{A5.27}$$

Thus, the total number of different fractions $(\sqrt{N}+P_i)/Q_i$ in the expansion is at most $[\sqrt{N}] \cdot 2[\sqrt{N}] < 2N$. This means that after at most $2N$ steps we will find a fraction $(\sqrt{N} + P_k)/Q_k$ which has occurred earlier in the expansion, and therefore the expansion must be periodic with a period-length of at most $2N - 1$.

The relation mentioned above, which we needed in Chapter 5, is

$$Q_{i+1} = Q_{i-1} + (P_i - P_{i+1})b_i, \tag{A5.28}$$

which may be deduced as follows: From (A5.22) we get

$$Q_i Q_{i-1} = N - P_i^2 \quad \text{and} \quad Q_{i+1} Q_i = N - P_{i+1}^2. \tag{A5.29}$$

Subtracting these two, we find

$$Q_i(Q_{i+1} - Q_{i-1}) = P_i^2 - P_{i+1}^2 = (P_i - P_{i+1})(P_i + P_{i+1}).$$

Writing (A5.27) as $P_i + P_{i+1} = Q_i b_i$, we can cancel Q_i and arrive at (A5.28).

Exercise A5.2. Continued fraction expansion of $\sqrt{N}$. Starting the recursion (A5.19)–(A5.20) with $P_0 = 0$, $Q_0 = 1$, write a FUNCTION cfracsqrt(N), giving the expansion of $\sqrt{N}$. Stop when the first period has been found ($b_n = 2b_0$ at this point). Test your programming on some simple cases, such as $\sqrt{a^2+1}$ and $\sqrt{a^2+2}$ and on $N = 69$.

The Maximal Period-Length

In fact, the maximal period-length can never attain the value $2N - 1$. This is because the number of possible combinations of P_i and Q_i is limited also by the condition that Q_i must be a divisor of $N - P_i^2$ and not simply *any* number $< 2N$.

The number of divisors of a number n is called the *divisor function* $d(n)$, and is a very irregular function of n. However, its *average order of magnitude* is known and is, for numbers of size n, approximately equal to $\ln n$. Thus, *on the average*, we may expect the number of possible values

314

of Q_i i.e. the number of divisors of $N - P_i^2$ not exceeding $2\sqrt{N}$, to be about $\frac{1}{2}\ln n$. (Most of the values of $N - P_i^2$ are just below N, and for these values only about *half* of the divisors of $N - P_i^2$ are also smaller than $2\sqrt{N}$, which is the reason why a factor $\frac{1}{2}$ must be included.) This leads to the expectation that the period for the expansion of $\sqrt{N}$ is shorter than $\frac{1}{2}\sqrt{N}\ln N$. It has, in fact, been proved that, for $N > 7$, the period is always shorter than $0.72\sqrt{N}\ln N$, and more elaborate arguments suggest that the period-length actually is smaller than approximately

$$\begin{cases} \sqrt{D}\ln\ln D, & \text{for } D \equiv 1 \ (\text{mod } 8) \\ \sqrt{D}\ln\ln 4D, & \text{otherwise.} \end{cases} \tag{A5.30}$$

See [4] for further details.

Short Periods

The discussion above concerns the *longest possible* periods. Frequently, the periods are considerably shorter. For instance, if $N = \sqrt{a^2 + 1}$, we have

$$\sqrt{N} = \sqrt{a^2 + 1} = a + \frac{1}{\lfloor 2a} + \frac{1}{\lfloor 2a} + \cdots,$$

so that the period has only one element. Also $\sqrt{a^2 + 2}$ has a short period:

$$\sqrt{a^2 + 2} = a + \frac{1}{\lfloor a} + \frac{1}{\lfloor 2a} + \frac{1}{\lfloor a} + \frac{1}{\lfloor 2a} + \cdots,$$

with period-length $= 2$.

Continued Fractions and Quadratic Residues

We shall now demonstrate how the regular continued fraction expansion of $\sqrt{N}$ can be used to find small quadratic residues of N. The method is dependent on the following formula

$$A_{n-1}^2 - NB_{n-1}^2 = (-1)^n Q_n, \tag{A5.31}$$

valid for all $n \geq 0$. Here A_n/B_n is the nth convergent of the expansion of $\sqrt{N}$, and x_n and Q_n are defined as in (A5.19).

315

Proof. Substituting $t = x_n$ in (A5.13), and expanding $x = \sqrt{N}$ according to (A5.12), we obtain

$$\sqrt{N} = \frac{A_{n-2} + A_{n-1}x_n}{B_{n-2} + B_{n-1}x_n}. \qquad (A5.32)$$

Next, inserting $x_n = (\sqrt{N} + P_n)/Q_n$, we find

$$\sqrt{N} = \frac{Q_n A_{n-2} + P_n A_{n-1} + A_{n-1}\sqrt{N}}{Q_n B_{n-2} + P_n B_{n-1} + B_{n-1}\sqrt{N}},$$

or

$$(Q_n B_{n-2} + P_n B_{n-1})\sqrt{N} + N B_{n-1} = Q_n A_{n-2} + P_n A_{n-1} + A_{n-1}\sqrt{N}.$$

Assuming that N is not a square, $\sqrt{N}$ is irrational and we can identify the rational and the irrational terms in the equation:

$$\begin{cases} Q_n A_{n-2} + P_n A_{n-1} = N B_{n-1} \\ Q_n B_{n-2} + P_n B_{n-1} = A_{n-1}. \end{cases} \qquad (A5.33)$$

Eliminating P_n yields

$$(A_{n-2}B_{n-1} - A_{n-1}B_{n-2})Q_n = N B_{n-1}^2 - A_{n-1}^2,$$

or, using (A5.10),

$$(-1)^{n-1}Q_n = N B_{n-1}^2 - A_{n-1}^2,$$

which gives us (A5.31).

We now return to the problem of finding small residues of N. Reducing (A5.31) (mod N), we find that

$$A_{n-1}^2 \equiv (-1)^n Q_n \pmod{N}, \qquad (A5.34)$$

and hence $(-1)^n Q_n$ certainly is a quadratic residue (mod N). How large can it be? In the proof of the periodicity of the regular continued fraction expansion of $\sqrt{N}$, we found in (A5.27) that $Q_n < 2\sqrt{N}$. Thus, the continued fraction expansion of $\sqrt{N}$ can provide us with a fast way of obtaining lots of small quadratic residues of N. The method is rapid, because both the P_i's and the Q_i's are less than $2\sqrt{N}$, with the result that the whole computation runs much faster for large values of N, which require multiple precision arithmetic in a computer, than would the method of squaring numbers close to $\sqrt{kN}$, and then reducing these (mod N). How about *lots* of quadratic residues? Since the expansion is periodic, all the Q_i's will repeat periodically, and in the case when the period of the expansion happens to be short this will severely limit the number of quadratic residues accessible by this approach. (As a matter of fact, the period is always symmetric, so that at best only half the number of steps in the period yield a new Q_i!) Well, this sad situation can be remedied by expanding $\sqrt{kN}$ with k suitably chosen such that the period of this new continued fraction is long. The success of this trick is due to the fact that each quadratic residue of kN must also be a quadratic residue of N.—The device is employed in Shanks' method as well as in Morrison and Brillhart's method, both of which are described in Chapter 5.

Bibliography

1. Oskar Perron, *Die Lehre von den Kettenbrüchen*, Band I, Third edition, Teubner, Stuttgart, 1954.

2. A. Ya. Khintchine, *Continued Fractions*, P. Noordhoff, Groningen, 1963.

3. G. H. Hardy and E. M. Wright, *An Introduction to the Theory of Numbers*, Fifth edition, Oxford, 1979, pp. 164–165.

4. H. C. Williams, "A Numerical Investigation Into the Length of the Period of the Continued Fraction Expansion of $\sqrt{D}$," *Math. Comp.* **36** (1981), pp. 593–601.

ALGEBRAIC FACTORS

Introduction

Sometimes an integer has a particular form, such as $N = 10^{21} + 7^{21}$. This can be considered a special case of the polynomial $P(x, y) = x^{21} + y^{21}$ for $x = 10$, $y = 7$. Knowing that $P(x, y)$ has the polynomial factors $x + y$, $x^3 + y^3$ and $x^7 + y^7$, we immediately obtain the corresponding factors of N: $10 + 7$, $10^3 + 7^3$ and $10^7 + 7^7$. Thus in this case the complete factorization of N is very much facilitated by its special form. In this appendix we shall very briefly present some results from the theory of polynomials which are of considerable help in the factorization of certain integers.—A few fundamental results from group theory and from higher arithmetic will be required here; these can be found in Appendix 1 and 2.

Factorization of Polynomials

Certain polynomials can be factorized, such as

$$x^2 - 3x + 2 = (x - 1)(x - 2)$$

or

$$x^6 - y^6 = (x - y)(x + y)(x^2 + xy + y^2)(x^2 - xy + y^2).$$

Other polynomials are *irreducible*, for instance

$$x^2 + y^2 + 1,$$

which means that they cannot be factorized. Polynomials $P(x)$ of *only one variable* x are governed by a rather straightforward theory. Whether or not $P(x)$ can be factorized depends upon which numbers we allow as the coefficients of a factorization. According to Theorem A1.5 on p. 253 we have: If $P(x)$ has a zero a, then $P(x) = (x - a)Q(x)$, with $Q(x)$ another polynomial. If the zero a is a *rational number*, then obviously the factorization $(x - a)Q(x)$ will have rational coefficients (supposing that $P(x)$ has rational coefficients to begin with). If a is an irrational or a complex number and $P(x)$ is assumed to have rational coefficients, then the coefficients of $Q(x)$ are expressible in terms of a.

Examples.

$$x^2 - 2 = (x - \sqrt{2})(x + \sqrt{2})$$

and

$$x^2 + 1 = (x - i)(x + i).$$

Of course, it can occur that a polynomial with rational coefficients has several irrational zeros and that when the corresponding linear factors are multiplied, the result is a polynomial factor with rational coefficients, e.g.

$$x^4 - x^2 - 2 = (x - i)(x + i)(x - \sqrt{2})(x + \sqrt{2}) = (x^2 + 1)(x^2 - 2).$$

Homogeneous polynomials $P_n(x, y)$ of two variables are, as a matter of fact, just as simple to deal with as polynomials of one variable. (In a homogeneous polynomial all the terms are of the same degree.) This is because the substitution $x/y = z$, followed by a multiplication by y^{-n} will yield a one-variable polynomial, as in the example

$$x^6 - y^6 = y^6 \left\{ \left(\frac{x}{y} \right)^6 - 1 \right\} = y^6(z^6 - 1),$$

or, in general

$$y^{-n} P_n(x, y) = P_n \left(\frac{x}{y}, 1 \right) = P_n^*(z). \tag{A6.1}$$

In this manner the treatment of $P_n(x, y)$ is reduced to the study of the one-variable polynomial $P_n^*(z)$.—In this appendix we shall present *rational* factorizations of certain homogeneous polynomials of two variables with rational coefficients. The most important of these are

The Cyclotomic Polynomials

Consider

$$x^n - y^n = y^n \left\{ \left(\frac{x}{y} \right)^n - 1 \right\} = y^n(z^n - 1). \tag{A6.2}$$

The nth degree polynomial $F_n(z) = z^n - 1$ has zeros which are called n^{th} roots of unity,

$$z_k = \cos\frac{2\pi k}{n} + i\sin\frac{2\pi k}{n} = e^{\frac{2\pi ki}{n}}, \quad \text{for} \quad k = 1, 2, \ldots, n. \quad \text{(A6.3)}$$

Every z_k can be written as a power of a so-called *primitive* n^{th} root of unity. (These, which we shall refer to as z_l's, are precisely those z_k's which can be used as generators of the cyclic group of order n formed by the nth roots of unity under multiplication.) This is due to the fact that

$$z_k = z_1^k, \quad k = 1, 2, \ldots, n, \quad \text{with} \quad z_1 = e^{\frac{2\pi i}{n}}. \quad \text{(A6.4)}$$

It is easy to prove that z_l is a primitive n^{th} root of unity if and only if $(l, n) = 1$. Thus, there are precisely $\varphi(n)$ primitive n^{th} roots of unity.

Some of the nth roots of unity are also roots of unity of a degree lower than n. For instance z_1^d, with $d|n$, is a root of unity of degree n/d, because $(z_1^d)^{n/d} = z_1^n = 1$. In this manner, all the n roots of unity of degree n can be distributed into classes, each class consisting of the $\varphi(d)$ *primitive* roots of unity of degree d, with d a divisor of n. Counting the number of roots in each class and observing that there are n roots of unity in all, we arrive at the following well-known formula for Euler's φ-function:

$$\sum_{d|n} \varphi(d) = n. \quad \text{(A6.5)}$$

It is possible to prove that the *primitive* n^{th} roots of unity are the zeros of a polynomial $\Phi_n(z)$, irreducible in the field of rationals and having integer coefficients. The polynomial $\Phi_n(z)$ is called the nth cyclotomic polynomial. Its degree is $\varphi(n)$ and it is a so-called *primitive factor* of $F_n(z)$. It follows that the polynomial $F_n(z) = z^n - 1$ can be written as the product of all the cyclotomic polynomials $\Phi_d(z)$ with $d|n$, or

$$F_n(z) = z^n - 1 = \prod_{d|n} \Phi_d(z). \quad \text{(A6.6)}$$

Please note the difficulty in this notation: $\Phi_n(z)$ *does not* have, as might be expected from the notation, degree n but degree $\varphi(n)$.—Formula (A6.6) provides the (rational) algebraic factors of $z^n - 1$ if the polynomials $\Phi_d(z)$ are known, and thus also of $x^n - y^n$. As an example we take

$$x^{12} - y^{12} = y^{12}\,\Phi_1\!\left(\frac{y}{x}\right)\Phi_2\!\left(\frac{y}{x}\right)\Phi_3\!\left(\frac{y}{x}\right)\Phi_4\!\left(\frac{y}{x}\right)\Phi_6\!\left(\frac{y}{x}\right)\Phi_{12}\!\left(\frac{y}{x}\right) =$$

$$= (x-y)(x+y)(x^2+xy+y^2)(x^2+y^2)(x^2-xy+y^2)(x^4-x^2y^2+y^4).$$

If p is an odd prime, we can easily obtain Φ_d for $d = p$, $2p$ and $4p$:

$$\Phi_p = \frac{z^p-1}{z-1} = z^{p-1} + z^{p-2} + \cdots + z + 1 \tag{A6.7}$$

$$\Phi_{2p} = \frac{z^{2p}-1}{z^p-1}\frac{z-1}{z^2-1} = z^{p-1} - z^{p-2} + \cdots - z + 1 \tag{A6.8}$$

$$\Phi_{4p} = \frac{z^{4p}-1}{z^{2p}-1}\frac{z^2-1}{z^4-1} = z^{2p-2} - z^{2p-4} + \cdots - z^2 + 1. \tag{A6.9}$$

The first polynomials $\Phi_n(z)$ (for $n \le 20$) which do not satisfy the formulas (A6.7)–(A6.9) are

$$\Phi_1 = z - 1 \tag{A6.10}$$

$$\Phi_2 = \frac{z^2-1}{z-1} = z + 1 \tag{A6.11}$$

$$\Phi_4 = \frac{z^4-1}{z^2-1} = z^2 + 1 \tag{A6.12}$$

$$\Phi_8 = \frac{z^8-1}{z^4-1} = z^4 + 1 \tag{A6.13}$$

$$\Phi_9 = \frac{z^9-1}{z^3-1} = z^6 + z^3 + 1 \tag{A6.14}$$

$$\Phi_{15} = \frac{z^{15}-1}{z^5-1}\frac{z-1}{z^3-1} = z^8 - z^7 + z^5 - z^4 + z^3 - z + 1 \tag{A6.15}$$

$$\Phi_{16} = \frac{z^{16}-1}{z^8-1} = z^8 + 1 \tag{A6.16}$$

$$\Phi_{18} = \frac{z^{18}-1}{z^9-1}\frac{z^3-1}{z^6-1} = z^6 - z^3 + 1. \tag{A6.17}$$

When p is an odd prime we observe from the above formulas that always

$$z^p - 1 = \Phi_1\Phi_p \tag{A6.18}$$

$$z^{2p} - 1 = \Phi_1\Phi_2\Phi_p\Phi_{2p} \tag{A6.19}$$

$$z^{4p} - 1 = \Phi_1\Phi_2\Phi_4\Phi_p\Phi_{2p}\Phi_{4p}. \tag{A6.20}$$

For the other values of $n \leq 20$, (A6.6) leads to the following formulas:

$$z - 1 = \Phi_1 \tag{A6.21}$$

$$z^2 - 1 = \Phi_1 \Phi_2 \tag{A6.22}$$

$$z^4 - 1 = \Phi_1 \Phi_2 \Phi_4 \tag{A6.23}$$

$$z^8 - 1 = \Phi_1 \Phi_2 \Phi_4 \Phi_8 \tag{A6.24}$$

$$z^9 - 1 = \Phi_1 \Phi_3 \Phi_9 \tag{A6.25}$$

$$z^{15} - 1 = \Phi_1 \Phi_3 \Phi_5 \Phi_{15} \tag{A6.26}$$

$$z^{16} - 1 = \Phi_1 \Phi_2 \Phi_4 \Phi_8 \Phi_{16} \tag{A6.27}$$

$$z^{18} - 1 = \Phi_1 \Phi_2 \Phi_3 \Phi_6 \Phi_9 \Phi_{18}. \tag{A6.28}$$

Further, if the standard factorization of n is $\prod p_i^{\alpha_i}$, we have the following general expression for $\Phi_n(z)$:

$$\Phi_n(z) = \frac{(z^n - 1)\prod(z^{n/(p_i p_j)} - 1) \cdots}{\prod(z^{n/p_i} - 1)\prod(z^{n/(p_i p_j p_k)} - 1) \cdots} = \prod_{d|n}(z^d - 1)^{\mu(n/d)}, \tag{A6.29}$$

where μ is the Möbius function, defined on p. 53. So, as an example, for $n = 105 = 3 \cdot 5 \cdot 7$, we have:

$$\Phi_{105}(z) = \frac{(z^{105} - 1)(z^7 - 1)(z^5 - 1)(z^3 - 1)}{(z^{35} - 1)(z^{21} - 1)(z^{15} - 1)(z - 1)}.$$

Exercise A6.1. Factors of $x^n - y^n$. Express $x^{21} - y^{21}$ as a product of primitive cyclotomic polynomials. Use this composition and Legendre's Theorem 5.7 on p. 184 to factorize some integers of this form. Check your results with the corresponding items in Table 7–Table 31.

The Polynomial $x^n + y^n$

Since

$$x^n + y^n = \frac{x^{2n} - y^{2n}}{x^n - y^n}, \tag{A6.30}$$

the properties of $x^n + y^n$ can be deduced from those of the two polynomials F_{2n} and F_n. The relation to be utilized is

$$z^n + 1 = \frac{F_{2n}(z)}{F_n(z)} = \frac{\prod_{d|2n} \Phi_d(z)}{\prod_{d|n} \Phi_d(z)} = \prod_{d'} \Phi_{d'}(z), \tag{A6.31}$$

where d' extends through the values of those divisors of $2n$ which are not at the same time also divisors of n.

The Polynomial $x^n + ay^n$

Oddly enough, there exist binomials other than $x^n \pm y^n$ which admit rational factorizations. The simplest case is

$$x^4 + 4y^4 = (x^2 - 2xy + 2y^2)(x^2 + 2xy + 2y^2). \qquad (A6.32)$$

Choosing suitable numerical values of x and y in (A6.32), the resulting formula can be used to factor certain numbers $a^n + b^n$ whose factorization is not immediately covered by the theory of cyclotomic polynomials hitherto discussed. Thus, for instance, the choice $x = 1$ and $y = 2^{k-1}$ leads to Aurifeuille's formula

$$2^{4k-2} + 1 = (2^{2k-1} - 2^k + 1)(2^{2k-1} + 2^k + 1). \qquad (A6.33)$$

Aurifeuillian Factorisations

There are many formulas similar to (A6.33) which have been investigated by Lucas and others. We shall call these formulas Aurifeuillian-type factorizations. Lucas discovered the following

Theorem A6.1. Lucas' Theorem. The primitive factor $Q_n(x, y)$ of $x^n + y^n$ can, for a square-free n, be written in the form

$$Q_n(x, y) = U^2(x, y) + nxyV^2(x, y), \qquad (A6.34)$$

where U and V are homogeneous polynomials, and with the following rules for the sign:

If $n = 4l + 1$, then $Q_n = U^2 + nxyV^2$.
If $n = 4l + 3$, then $Q_n = U^2 - nxyV^2$.
If $n = 4l + 2$, then either sign may be chosen.

Examples.

$$n = 2 \text{ gives } x^2 + y^2 = (x - y)^2 + 2xy = (x + y)^2 - 2xy. \qquad (A6.35)$$

$$n = 3 \text{ gives } \frac{x^3 + y^3}{x + y} = (x + y)^2 - 3xy. \qquad (A6.36)$$

323

$$n = 5 \text{ gives } \frac{x^5 + y^5}{x + y} = (x^2 - 3xy + y^2)^2 + 5xy(x - y)^2. \qquad \text{(A6.37)}$$

$$n = 6 \text{ gives } \frac{x^6 + y^6}{x^2 + y^2} = (x^2 - 3xy + y^2)^2 + 6xy(x - y)^2 =$$

$$= (x^2 + 3xy + y^2)^2 - 6xy(x + y)^2. \qquad \text{(A6.38)}$$

$$n = 7 \text{ gives } \frac{x^7 + y^7}{x + y} = (x + y)^6 - 7xy(x^2 + xy + y^2)^2. \qquad \text{(A6.39)}$$

$$n = 10 \text{ gives } \frac{x^{10} + y^{10}}{x^2 + y^2} = (x^4 \mp 5x^3 y + 7x^2 y^2 \mp 5xy^3 + y^4)^2 \pm$$

$$\pm 10xy(x^3 \mp 2x^2 y + 2xy^2 \mp y^3)^2. \qquad \text{(A6.40)}$$

$$n = 15 \text{ gives } \frac{x^{15} + y^{15}}{x^5 + y^5} \frac{x + y}{x^3 + y^3} =$$

$$= (x^4 + 8x^3 y + 13x^2 y^2 + 8xy^3 + y^4)^2 - 15xy(x + y)^6. \qquad \text{(A6.41)}$$

In (A6.40) either the upper or the lower signs have to be taken.—Note that, if n is odd, the primitive factor of $x^n - y^n$, which is $y^{\varphi(n)}\Phi_n(y/x)$, can be represented by (A6.34) simply by changing the sign of y. Thus, for odd n the primitive factor of $x^n - y^n$ is $= y^{\varphi(n)}\Phi_n(-y/x)$.—If n is even, the primitive factor of $x^n - y^n$ is identical to the primitive factor of $x^{n/2} + y^{n/2}$ and thus the primitive factor of $x^{n/2} + y^{n/2}$ is $y^{\varphi(n)}\Phi_n(y/x)$. In Table 34, at the end of the book, formulas of the same kind as (A6.35)–(A6.41) are given for n below 119.

Exercise A6.2. Deduction of Lucas' formulas. Table 34 at the end of the book has been computed in the following way (for notation, see p. 452):

$$\Phi_n(z^2) = U_n^2(z^2) - (-1)^{\frac{n-1}{2}} nz^2 V_n^2(z^2) =$$

$$= \left(U_n(z^2) - (-1)^{\frac{n-1}{4}}\sqrt{n}\, zV_n(z^2)\right)\left(U_n(z^2) + (-1)^{\frac{n-1}{4}}\sqrt{n}\, zV_n(z^2)\right).$$

The polynomial $\Phi_n(z^2)$ has the $2\varphi(n)$ zeros $z_k = e^{\pi ik/n}$, $1 \le k \le 2n$, $(k, n) = 1$. Find a rule to decide which of these numbers are zeros of the first factor and which are zeros of the second factor of $\Phi_n(z^2)$. Use complex multiplication of the appropriate zeros, written in the form $z_k = \cos(\pi k/n) + i\sin(\pi k/n)$, to form the polynomial $\prod(z - z_k)$, identical to the second factor. The even-degree terms of this polynomial now form $U_n(z^2)$ and the odd-degree terms form $\sqrt{n}\, zV_n(z^2)$, if $n \equiv 1 \pmod 4$, and $i\sqrt{n}\, zV_n(z^2)$, if $n \equiv 3 \pmod 4$. Check the results of your computations with some of the entries in Table 34.

Factorization Formulas

Now, we may ask how Lucas' formula can help us in the factorization of numbers. Well it can in the following way. If x and y are properly chosen, Lucas' Teorem A6.1 will yield algebraic factors of certain binomials $x^n \pm ay^n$. This is because if we have a *minus-sign* and choose x and y such that nxy becomes a square, i.e. such that $xy = nc^2$, then we find

$$Q_n(x, y) = U^2 - n^2c^2V^2 = (U - ncV)(U + ncV). \qquad \text{(A6.42)}$$

In Lucas' theorem, a minus-sign occurs in the cases

$$x^{4l+1} - y^{4l+1}, \quad x^{4l+2} + y^{4l+2}, \quad x^{4l+3} + y^{4l+3}, \qquad \text{(A6.43)}$$

where the exponents must be square-free integers. Putting

$$x = n_1 a^2, \ y = n_2 b^2, \quad \text{where} \quad n_1 n_2 = n, \qquad \text{(A6.44)}$$

the formulas lead to the following Aurifeuillian-type algebraic factorizations:

$$4a^4 + b^4 = (2a^2 - 2ab + b^2)(2a^2 + 2ab + b^2). \qquad \text{(A6.45)}$$

$$27a^6 + b^6 = (3a^2 + b^2)(3a^2 - 3ab + b^2)(3a^2 + 3ab + b^2). \qquad \text{(A6.46)}$$

$$5^5 a^{10} - b^{10} = (5a^2 - b^2)(25a^4 - 25a^3b + 15a^2b^2 - 5ab^3 + b^4) \times$$
$$\times (25a^4 + 25a^3b + 15a^2b^2 + 5ab^3 + b^4). \qquad \text{(A6.47)}$$

$$6^6 a^{12} + b^{12} = (36a^4 + b^4)(36a^4 - 36a^3b + 18a^2b^2 - 6ab^3 + b^4) \times$$
$$\times (36a^4 + 36a^3b + 18a^2b^2 + 6ab^3 + b^4). \qquad \text{(A6.48)}$$

$$3^6 a^{12} + 2^6 b^{12} = (9a^4 + 4b^4)(9a^4 - 18a^3b + 18a^2b^2 - 12ab^3 + 4b^4) \times$$
$$\times (9a^4 + 18a^3b + 18a^2b^2 + 12ab^3 + 4b^4). \qquad \text{(A6.49)}$$

$$7^7 a^{14} + b^{14} = (7a^2 + b^2) \times$$
$$\times (343a^6 - 343a^5b + 147a^4b^2 - 49a^3b^3 + 21a^2b^4 - 7ab^5 + b^6) \times$$
$$\times (343a^6 + 343a^5b + 147a^4b^2 + 49a^3b^3 + 21a^2b^4 + 7ab^5 + b^6). \text{(A6.50)}$$

325

$$10^{10}a^{20} + b^{20} = (100a^4 + b^4) \times$$
$$\times (10000a^8 - 10000a^7b + 5000a^6b^2 - 2000a^5b^3 +$$
$$+ 700a^4b^4 - 200a^3b^5 + 50a^2b^6 - 10ab^7 + b^8) \times$$
$$\times (10000a^8 + 10000a^7b + 5000a^6b^2 + 2000a^5b^3 +$$
$$+ 700a^4b^4 + 200a^3b^5 + 50a^2b^6 + 10ab^7 + b^8). \quad \text{(A6.51)}$$

$$5^{10}a^{20} + 2^{10}b^{20} = (25a^4 + 4b^4) \times$$
$$\times (625a^8 - 1250a^7b + 1250a^6b^2 - 1000a^5b^3 +$$
$$+ 700a^4b^4 - 400a^3b^5 + 200a^2b^6 - 80ab^7 + 16b^8) \times$$
$$\times (625a^8 + 1250a^7b + 1250a^6b^2 + 1000a^5b^3 +$$
$$+ 700a^4b^4 + 400a^3b^5 + 200a^2b^6 + 80ab^7 + 16b^8). \quad \text{(A6.52)}$$

$$15^{15}a^{30} + b^{30} = (15a^2 + b^2)(225a^4 - 15a^2b^2 + b^4) \times$$
$$\times (50625a^8 - 3375a^6b^2 + 225a^4b^4 - 15a^2b^6 + b^8) \times$$
$$\times (50625a^8 - 50625a^7b + 27000a^6b^2 - 10125a^5b^3 +$$
$$+ 2925a^4b^4 - 675a^3b^5 + 120a^2b^6 - 15ab^7 + b^8) \times$$
$$\times (50625a^8 + 50625a^7b + 27000a^6b^2 + 10125a^5b^3 +$$
$$+ 2925a^4b^4 + 675a^3b^5 + 120a^2b^6 + 15ab^7 + b^8). \quad \text{(A6.53)}$$

$$5^{15}a^{30} + 3^{15}b^{30} = (5a^2 + 3b^2)(25a^4 - 15a^2b^2 + 9b^4) \times$$
$$\times (625a^8 - 375a^6b^2 + 225a^4b^4 - 135a^2b^6 + 81b^8) \times$$
$$\times (625a^8 - 1875a^7b + 3000a^6b^2 - 3375a^5b^3 +$$
$$+ 2925a^4b^4 - 2025a^3b^5 + 1080a^2b^6 - 405ab^7 + 81b^8) \times$$
$$\times (625a^8 + 1875a^7b + 3000a^6b^2 + 3375a^5b^3 +$$
$$+ 2925a^4b^4 + 2025a^3b^5 + 1080a^2b^6 + 405ab^7 + 81b^8). \quad \text{(A6.54)}$$

For the specific case $a = n^{k-1}$ and $b = 1$, these formulas lead amongst others, to (A6.33) and its analogues

$$3^{6k-3} + 1 = (3^{2k-1} + 1)(3^{2k-1} - 3^k + 1)(3^{2k-1} + 3^k + 1). \quad \text{(A6.55)}$$

$$5^{10k-5} - 1 = (5^{2k-1} - 1)L_{10k-5}M_{10k-5}, \quad \text{where}$$
$$L_{10k-5}, \ M_{10k-5} = 5^{4k-2} + 3 \cdot 5^{2k-1} + 1 \mp 5^k(5^{2k-1} + 1). \quad \text{(A6.56)}$$

$$6^{12k-6} + 1 = (6^{4k-2} + 1)L_{12k-6}M_{12k-6}, \quad \text{where}$$
$$L_{12k-6}, \ M_{12k-6} = 6^{4k-2} + 3 \cdot 6^{2k-1} + 1 \mp 6^k(6^{2k-1} + 1). \quad \text{(A6.57)}$$

$$7^{14k-7} + 1 = (7^{2k-1} + 1)L_{14k-7}M_{14k-7}, \quad \text{where}$$
$$L_{14k-7}, \ M_{14k-7} = (7^{2k-1} + 1)^3 \mp 7^k(7^{4k-2} + 7^{2k-1} + 1). \quad \text{(A6.58)}$$

$$10^{20k-10} + 1 = (10^{4k-2} + 1)L_{20k-10}M_{20k-10}, \quad \text{where}$$
$$L_{20k-10}, \ M_{20k-10} = 10^{8k-4} + 5 \cdot 10^{6k-3} + 7 \cdot 10^{4k-2} +$$
$$+5 \cdot 10^{2k-1} + 1 \mp 10^k(10^{6k-3} + 2 \cdot 10^{4k-2} + 2 \cdot 10^{2k-1} + 1). \quad \text{(A6.59)}$$

In all these formulas, the minus-sign is associated with the L-factor and the plus-sign with the M-factor.

Finally, putting $a = 2 \cdot 12^{k-1}$ and $b = 1$ in (A6.46), we have

$$12^{6k-3} + 1 = (12^{2k-1} + 1)(12^{2k-1} - 2^{2k-1}3^k + 1)(12^{2k-1} + 2^{2k-1}3^k + 1).$$
$$\text{(A6.60)}$$

As has already been pointed out, these formulas are all generalizations of Aurifeuille's formula (A6.33), well-known from the factorization of the Mersenne numbers:

$$M_{8k-4} = 2^{8k-4} - 1 = (2^{4k-2} - 1)(2^{4k-2} + 1) =$$
$$= (2^{2k-1} - 1)(2^{2k-1} + 1)(2^{2k-1} - 2^k + 1)(2^{2k-1} + 2^k + 1).$$

To show how powerful Lucas' formulas are, when applicable, we give just one

Example. Factor $10^{30} + 1$. Putting $k = 2$ in (A6.59), we find

$$10^{30} + 1 = 1000001 \cdot 904806804901 \cdot 1105207205101. \quad \text{(A6.61)}$$

Furthermore, we can use one of the "rules of cubes":

$$10^{30} + 1 = (10^{10} + 1)(10^{20} - 10^{10} + 1). \quad \text{(A6.62)}$$

The substitution $k = 1$ in (A6.59) gives

$$10^{10} + 1 = 101 \cdot 3541 \cdot 27961,$$

with all the factors primes. Moreover, the first factor of (A6.61) is

$$10^6 + 1 = (10^2 + 1)(10^4 - 10^2 + 1) = 101 \cdot 9901,$$

where both factors are primes. Thus, 3541 and 27961 must divide the large factors of (A6.61). It is easily verified that

$$904806804901 = 3541 \cdot 255522961$$

and

$$1105207205101 = 27961 \cdot 39526741.$$

The cofactors obtained, 39526741 and 255522961, have no factors in common with the primes in $10^6 + 1$ or $10^{10} + 1$. Legendre's Theorem 5.7 on p. 184 tells us that all prime factors of these two numbers are then of the form $60k + 1$. Performing trial divisions by these factors, we easily establish the primality of the smaller of these two numbers, and find the prime factorization

$$255522961 = 61 \cdot 4188901.$$

Thus the complete prime factorization of $10^{30} + 1$ is

$$61 \cdot 101 \cdot 3541 \cdot 9901 \cdot 27961 \cdot 4188901 \cdot 39526741.$$

Exercise A6.3. Simplifying some surds. Use (A6.45)–(A6.48) to find simpler expressions for the (complex) values of the surds $(-4)^{\frac{1}{4}}$, $(-27)^{\frac{1}{6}}$, $(5^5)^{\frac{1}{10}}$ and $(-6^6)^{\frac{1}{12}}$. (All of these can be expressed with the aid of square-roots only!).

The Algebraic Structure of Aurifeuillian Numbers

Since all numbers for which Aurifeuillian-type factorizations can be performed are of the form $x^n \pm y^n$, either (A6.6) or (A6.31) holds in the case when n is composite. Using this fact, we can find *algebraic* factors of the Aurifeuillian factors of $x^n \pm y^n$. As an example we give

$$2^{210} + 1 = (2^{105} - 2^{53} + 1)(2^{105} + 2^{53} + 1). \qquad \text{(A6.63)}$$

Here $2^{210} + 1$ possesses the algebraic factors $2^{70} + 1$, $2^{42} + 1$, $2^{30} + 1$,

$2^{14}+1$, $2^{10}+1$, 2^6+1 and 2^2+1, having the Aurifeuillian factorizations:

$$2^{70}+1 = (2^{35}-2^{18}+1)(2^{35}+2^{18}+1)$$
$$2^{42}+1 = (2^{21}-2^{11}+1)(2^{21}+2^{11}+1)$$
$$2^{30}+1 = (2^{15}-2^8+1)(2^{15}+2^8+1)$$
$$2^{14}+1 = (2^7-2^4+1)(2^7+2^4+1)$$
$$2^{10}+1 = (2^5-2^3+1)(2^5+2^3+1)$$
$$2^6+1 = (2^3-2^2+1)(2^3+2^2+1)$$
$$2^2+1 = (2^1-2^1+1)(2^1+2^1+1).$$

Thus each of these factors must divide one of the factors in the right-hand-side of (A6.63). Without entering into further details we state only the results:

$$2^{105}-2^{53}+1 =$$
$$= \frac{(2^{35}-2^{18}+1)(2^{21}+2^{11}+1)(2^{15}-2^8+1)(2^1-2^1+1)}{(2^7-2^4+1)(2^5+2^3+1)(2^3+2^2+1)} Q_{210}(2)$$

$$2^{105}+2^{53}+1 =$$
$$= \frac{(2^{35}+2^{18}+1)(2^{21}-2^{11}+1)(2^{15}+2^8+1)(2^1+2^1+1)}{(2^7+2^4+1)(2^5-2^3+1)(2^3-2^2+1)} Q'_{210}(2).$$

These show great structural resemblance to (A6.29) for $n = 105$. $Q_{210}(z)$ and $Q'_{210}(z)$ are (primitive) polynomial factors of $2z^{104}-2z^{52}+1$ and $2z^{104}+2z^{52}+1$, respectively.—The only difficulty in computing such algebraic Aurifeuillian factorizations involves choosing, in each position within the factorizations, the correct one of the two Aurifeuillian factors L or M possible. The problem is not too difficult theoretically and a solution can be found in [1], pp. lv–lvii, but if in doubt about which one of the two alternatives to choose, you can always verify numerically whether or not any particular factor divides the left-hand-side.—Larger tables of factors of $a^n \pm b^n$, such as [1], are usually organized so as to keep track of Aurifeuillian factors, and of course this information is made heavy use of in the factorization process.

Exercise A6.4. Aurifeuillian factorizations. Factorize $3^{87}+1$, $5^{55}-1$, $6^{54}+1$, $7^{49}+1$ and $12^{39}+1$.

A Formula by Gauss for $x^n - y^n$

There is another very interesting way of writing the cyclotomic polynomial, which is related to the theory of quadratic residues. This theory was established by Gauss in his Disquisitiones Arithmeticæ [3]. The theorem proved by Gauss is that for prime exponents p

$$4\frac{x^p - y^p}{x - y} = R^2(x, y) - (-1)^{\frac{p-1}{2}} pS^2(x, y), \qquad (A6.64)$$

or

$$4\Phi_p(z) = R^{*\,2}(z) - (-1)^{\frac{p-1}{2}} pS^{*\,2}(z), \qquad (A6.64^*)$$

if we prefer the more compact one-variable notation. Here, R and S are homogeneous polynomials in the variables x and y with integer coefficients. As an example we consider

$$4\frac{x^7 - y^7}{x - y} = 4(x^6 + x^5y + x^4y^2 + x^3y^3 + x^2y^4 + xy^5 + y^6) =$$
$$= (2x^3 + x^2y - xy^2 - 2y^3)^2 + 7(x^2y + xy^2)^2.$$

Kraïtchik [4] generalized Gauss' formula to odd composite square-free values of the exponent n. Also in this case the cyclotomic polynomial $\Phi_n(z)$ comes into play. The simplest case is $n = 15$:

$$4\frac{x^{15} - y^{15}}{x^5 - y^5}\frac{x - y}{x^3 - y^3} = 4(x^8 - x^7y + x^5y^3 - x^4y^4 + x^3y^5 - xy^7 + y^8) =$$

$$= (2x^4 - x^3y - 4x^2y^2 - xy^3 + 2y^4)^2 + 15(x^3y - xy^3)^2.$$

In Table 33 at the end of this book, the reader will find the coefficients of the polynomials R^* and S^* for all square-free odd n up to $n = 149$.— We mention these formulas because they may be of use in connection with Euler's method, discussed on p. 158.

Exercise A6.5. Deduction of Gauss' formulas. Table 33 at the end of the book has been computed in the following way (for notation, see p. 445):

$$4\Phi_n(z) = A_n^2(z) - (-1)^{\frac{n-1}{2}} nz^2 B_n^2(z) =$$

$$= \left(A_n(z) - (-1)^{\frac{n-1}{4}}\sqrt{n}\, zB_n(z)\right)\left(A_n(z) + (-1)^{\frac{n-1}{4}}\sqrt{n}\, zB_n(z)\right).$$

The polynomial $\Phi_n(z)$ has the $\varphi(n)$ zeros $z_k = e^{2\pi ik/n}$, $1 \leq k \leq n$, $(k, n) = 1$. Find a rule to decide which of these numbers are zeros of the first factor and which are zeros of the second factor of $\Phi_n(z)$. Use complex multiplication of the appropriate zeros, written in the form $z_k = \cos(2\pi k/n) + i\sin(2\pi k/n)$, to form the polynomial $2\prod(z - z_k)$, identical to the second factor. The real parts of the terms of this polynomial now form $A_n(z)$ and the imaginary parts form $i\sqrt{n}\, zB_n(z)$, if $n \equiv 3 \pmod 4$. How can A_n and B_n be found, if $n \equiv 1 \pmod 4$? Check the results of your computations with some of the entries in Table 33.

Bibliography

1. John Brillhart, D. H. Lehmer, J. L. Selfridge, Bryant Tuckerman and S. S. Wagstaff, Jr., *Factorisations of $b^n \pm 1$, $b = 2, 3, 5, 6, 7, 10, 11, 12$ up to High Powers,* American Mathematical Society, Providence R. I., 1983.

2. A. Schinzel, "On Primitive Factors of $a^n - b^n$," *Proc. Cambr. Philos. Soc.* **57** (1962).

3. C. F. Gauss, *Untersuchungen über höhere Arithmetik,* Chelsea Publishing Company, New York, 1965, p. 425–428 (articles 356–357).

4. Maurice Kraïtchik, *Recherches sur la Théorie des Nombres,* Tome II *Factorization,* Paris, 1929, p. 1–7.

MULTIPLE-PRECISION ARITHMETIC

Introduction

As the reader must certainly be aware of by now, it is virtually impossible to do computer work on large primes and the factorization of large numbers without having a *multiple-precision arithmetic package* available. Unfortunately, however, the performance of exact computations on large integers has a limited appeal, and computer manufacturers do not find it profitable to include such facilities in the software that goes with their hardware. This means that the reader may have to construct such a package himself for the computer he is using. In this appendix we shall discuss ways in which this can be done.

Various Objectives for a Multiple-Precision Package

Before attempting to construct a multiple-precision package for his computer, the reader should make clear what are the main objectives of the package. A choice can be made from the following list of important aspects:

1. The package itself should be easy to program and to check out.

2. The package should be easy to use.

3. Computations with the package should run at optimal speed.

4. Easy input and output should be favoured.

5. The package should require minimal computer storage.

6. The package should be portable between different types of computer.

There may be other factors in addition to the ones listed above that the reader might wish to consider.

Unfortunately, *all* the virtues of an ideal multiple-precision package cannot possibly be attained simultaneously. If, for instance, high speed is desirable, then the package must be programmed in machine language (or assembly language, which is machine language with mnemonic instruction codes and some additional options) in order to achieve the required speed. But in that case all the peculiarities of the computer's built-in arithmetic have to be taken into account, complicating the programming considerably.—It is obvious that we cannot, in the space available here, discuss all aspects of multiple-precision arithmetic, but have to limit ourselves to providing just a few clues as to how it can be implemented.

How to Store Multi-Precise Integers

Since the essence of multiple-precision arithmetic is to work with numbers that *cannot* be stored in one computer word, a multi-precise integer has to be stored as a one-dimensional (or linear) array. Further, because the integers occurring in a particular computation mostly are of various sizes it is practical to build a package that allows for integers of various lengths. It certainly saves storage not having to occupy memory space for the maximal allowable length of every number occurring in the computation and, generally, also saves computing time, since a full-length computation on a number preceded by one or more zero words is inefficient. Thus unless some very special computation, requiring all intermediate results to be of approximately the same size, is needed, a package allowing variable size integers is strongly recommended. However, with such an approach, some information about how many components (computer words) k each number occupies, must also be stored. Thus, if the number a is stored in a[0:k], then it is suitable to store k or sign$(a) \times k$ in a[0]. (Note that if using a computer with negative integers stored as $(2^t - 1)$-complements of the corresponding positive integers—so-called *complementary arithmetic* using the radix $B = 2^t$—then the minus sign of a negative integer may be stored in the sign position of *every* component, with no need for storing it also in a[0]. However, even in this case, it can be practical to keep the sign in a[0] as well!)

The representation, in which *positive* integers are stored in the components of a and in which the sign is attached to a[0] only, will be termed *sign arithmetic*. The number of components in sign arithmetic can be obtained simply by evaluating abs(a[0]), where the number 0 is held just as a[0]:=0.—If the number N is stored, its components are the digits of the number N in an enumeration system to some radix B:

$$N = \sum_{i=1}^{k} a_i B^{i-1}. \qquad (A7.1)$$

The radix B may be any integer up to the limit for single-precision integers (or for those single-precision reals which can be stored *exactly*, if the computer's built-in floating-point arithmetic is to be used.) In assembly code, B is usually precisely this limit 2^t, where $t + 1$ is the number of bits in one computer word (or where t is the number of bits in the mantissa of a floating-point number). In multiple-precision packages written in a high-level language, B may be chosen as some power of 10 or in some other way, depending on the application.—The order in which the components are stored does not really matter much, but the author has found some advantage in using the arrangement indicated in (A7.1), storing the *least significant B-digit first*, i.e. in a[1]. The advantage of using this order is that it becomes easy to compute certain frequently occurring expressions such as N (mod 8), if $N \neq 0$. This is simply a[1]MOD 8 or sign(a[0])*a[1]MOD 8 (depending upon how the sign of N is stored), as long as the radix B is divisible by 8.—Having introduced variable-length integers, it is of course advantageous to use dynamic arrays whenever possible.

Addition and Subtraction of Multi-Precise Integers

The addition and subtraction of multi-precise integers is almost self-evident. In *complementary arithmetic* (and assembly code), to add or subtract the numbers stored in a[0:k] and b[0:l], simply add or subtract the corresponding components a[i] and b[i] for i:-1,...,min(k,l), and take care of the overflow between successive components. The addition of all the carry digits could in fact induce new overflows, but the addition of these second order carry digits can never cause a third overflow.—Because of possible overflow in the leading component you must also allow for one component extra in the result $c = a \pm b$; thus c should be declared as c[0:max(k,l)+1].

In order to compute $c = a \pm b$ in *sign arithmetic* you must first verify which of the three possible values $|a| + |b|$, $|a| - |b|$ or $|b| - |a|$ it is that equals $|c|$. After this the computation can proceed as above, performed on the appropriate of these three expressions. Finally, the correct sign has to be given c.

334

Reduction in Length of Multi-Precise Integers

The result of an addition or a subtraction may need less space than the largest number involved, for example $(10^{16} + 1) + (-10^{16}) = 1$, thus allowing for the result of the operation to be stored in fewer components than $\max(k, l) + 1$. The reduction to fewer components is easily performed by repeatedly checking whether the final component equals zero and, if this happens, reducing c[0] by 1 as shown by the following program fragment:

```
m:=abs(c[0]);
WHILE (m>0) AND (c[m]=0) DO m:=m-1;
c[0]:=sign(c[0])*m;
```

Multiplication of Multi-Precise Integers

The multiplication of multi-precise integers is slightly more complicated than addition or subtraction. The logically simplest way of multiplying two radix-B numbers

$$M = \sum_{i=1}^{m} a_i B^{i-1} \quad \text{and} \quad N = \sum_{j=1}^{n} b_j B^{j-1} \tag{A7.2}$$

is to consider M and N as the values $M(B)$ and $N(B)$, respectively, of the following two polynomials:

$$M(x) = \sum_{i=1}^{m} a_i x^{i-1} \quad \text{and} \quad N(x) = \sum_{j=1}^{n} b_j x^{j-1}.$$

Then apply the usual multiplication formula for polynomials

$$M(x)N(x) = P(x) = \sum_{k=1}^{m+n-1} c_k x^{k-1}, \tag{A7.3}$$

where

$$c_k = \sum_{i} a_i b_{k+1-i}, \quad k = 1, 2, \ldots, m+n-1, \tag{A7.4}$$

subject to the conditions that

$$1 \leq i \leq m \quad \text{and} \quad 1 \leq k+1-i \leq n. \tag{A7.5}$$

335

Subsequently put $x = B$. This yields the components of the product $M \cdot N$, except in the case when $c_k \geq B$, when c_k has to be written in the form $c_k = Q \cdot B + c_k'$, and Q must be transferred as an overflow to the next (higher) radix-B digit c_{k+1}.

The accumulation of the sum of products for c_k in (A7.4) is performed by the following program loop:

```
c[k]:=0;
FOR i:=max(1,k+1-n) TO min(k,m) DO c[k]:=c[k]+a[i]*b[k+1-i];
```

There are, however, two problems involved in this computation. First, each product is a double-precision integer, so that *either B* must be chosen below *the square root* of the upper limit for integers, *or* a section of assembly code, taking care of the double-length product generated by the computer's hardware, must be incorporated.—The second problem is the overflow which may occur due to the addition of the products in (A7.4). This necessitates the accumulation of the c_k's as *triple* length integers, all of which are finally used in establishing the result $M \cdot N$. Next, in order not to render our description of the situation too abstract, we demonstrate all this in an

Example. Suppose you have a microcomputer allowing integers in the range $-32768 \leq N \leq 32767$, and that you choose $B = 10000$ and sign arithmetic. Then you need to store large integers in units of 4 decimal digits, e.g. $M = 123456789$ will be stored as

```
a[0]:=3; a[1]:=6789; a[2]:=2345; a[3]:=1;
```

and $N = 87654321$ in the form

```
b[0]:=2; b[1]:=4321; b[2]:=8765;
```

In order to form $M \cdot N$ using *single precision arithmetic only*, we first split the components of M and N into 2-digit numbers, corresponding to $B = 100$. M will then (temporarily for the multiplication) be stored as

```
aa[0]:= 5; aa[1]:=89; aa[2]:=67; aa[3]:=45; aa[4]:=23;
aa[5]:= 1;
```

and N as

```
bb[0]:= 4; bb[1]:=21; bb[2]:=43; bb[3]:=65; bb[4]:=87;
```

The calculation of the products of the radix-B digits, one from M and one from N, is then carried out in much the same way as in the usual algorithm for the multiplication of two integers using pencil and paper. Only the order by which these products are formed differs:

```
                1  23 45 67 89
                   87 65 43 21
            ───────────────────────
                         18 69    21·89
                      14 07       21·67
                      38 27       43·89
                    9 45          21·45
                   28 81          43·67
                   57 85          65·89
                 4 83             21·23
                19 35             43·45
                43 55             65·67
                77 43             87·89
                   21             21·1
                 9 89             43·23
                29 25             65·45
                58 29             87·67
                   43             43·1
             14 95               65·23
             39 15               87·45
                65               65·1
          20 01                  87·23
          87                     87·1
            ───────────────────────
        1  08 21 52 10 12 63 52 69
```

In the last line the products have been added in their appropriate positions, and the carry digits taken care of, exactly as in the ordinary pencil-and-paper method.—Finally, the result is assembled in the form

c[0]:=5; c[1]:=5269; c[2]:=1263; c[3]:=5210;
c[4]:=821; c[5]:=1;

Division of Multi-Precise Integers

Division is by far the most complicated of the four elementary operations. We shall only sketch how the division of two multi-precise integers M and N can be effected.

Just as when dividing two integers by ordinary pencil-and-paper arithmetic, a *provisory quotient* $q \approx M/N$ is set up yielding the beginning of the full quotient Q. A reduced numerator $M' = M - N \cdot q$ is then formed

337

and a new provisory quotient $q' \approx M'/N$, giving more digits of Q, is calculated, etc., until the number M has been exhausted. The only problem involved is to accomplish all this in such a manner that every conceivable case is covered by the algorithm used. Here are some hints on how to achieve this:

1. Divide only *positive* integers, even if you are using complementary arithmetic elsewhere.

2. The provisory quotient could be formed as a[m]DIV b[n]. However, if a[m] and/or b[n] has only a few digits, then the approximation of the first digits of Q will generally be poor. Using the computer's built-in floating-point arithmetic, use instead the approximation

$$\frac{a_m B + a_{m-1}}{b_n B + b_{n-1}}.$$

(Note that appropriate conversions from INTEGER TYPE variables to REAL TYPE variables may first have to be explicitly made. Moreover, it must be verified that both M and N have at least two components before the above expression is formed.)

3. In order to secure that the algorithm always operates with *positive* integers, be sure to always round off a provisory quotient *downward*.

4. $M' = M - N \cdot q$ can be constructed by means of the multiplication and addition/subtraction routines of the package or, if higher efficiency is desired, by writing a portion of code taking care of the reduction of M.

5. Next, check whether $M' < N$, in which case the division is complete. If not, return to step 2 above and calculate a new provisory quotient. Because of the downward roundings, it may occasionally happen that q becomes 0 instead of 1. In that case, reduce M' to $M' - N \cdot B^u$, and add B^u to Q. (The factor B^u corresponds to the position where the provisory quotient should be added in the final quotient Q.)

Input and Output of Multi-Precise Integers

Since in- and output invariably take only a small portion of computing time in applications of multi-precision arithmetic, it does not really matter if they are slightly inefficient. In the case where sign arithmetic has been chosen with a radix $B = 10^s$, the routines for in- and output are very simple indeed, involving only the reading or printing of the components of a vector (in reverse order), using the I/O-instructions of the computer's ordinary software. If, however, a little neater printout is preferred, check whether any components (after the most significant one has been printed) are $< 10^{s-1}$, in which case you may print the required zeros before the usual printing instruction is applied to print the digits of the component.

A Complete Package For Multiple-Precision Arithmetic

In order to give the reader at least some opportunity of trying computations with large numbers on his computer, we provide here a complete package in sign arithmetic, written in PASCAL. Since PASCAL lacks dynamic arrays, every array must be assigned a fixed length, called vsize in the package, corresponding to the maximal length of the integers occurring in the computation at hand. The package will allow for integers of at most vsize components to be processed. However, the value of the constant vsize can easily be changed by the reader, if necessary, in which case the constant lvsize, which has to be twice vsize, should also be adjusted.

The radix B must be a power of 100, and less than half the value of the largest integer allowed by the computer's fixed-point arithmetic. The radix B as well as all the constants dependent upon B are computed by the program from the key value sqrtB, which is $\sqrt{B}$ and has to be supplied in the list of constants.

The following PASCAL program includes the Multiple-Precision Package in the context of a complete computer program which tests all the procedures of the package on a few very simple cases:

```
PROGRAM Multiple-Precision Package
{Tests the procedures in the Package}
(Input,Output);
LABEL 1,2;
CONST sqrtB=100000; vsize=20; lvsize=40;
TYPE vector    = ARRAY[0..vsize] OF INTEGER;
     string    = PACKED ARRAY[1..8] OF CHAR;
VAR a,b,c,d          : vector;
    Base,B1,B101     : INTEGER;

FUNCTION sign(x : INTEGER) : INTEGER;
```

```
BEGIN IF x=0 THEN sign := 0 ELSE sign := abs(x) div x
END {sign};

PROCEDURE putin(txt : string; VAR a : vector);
{Prints a text string of 8 characters and reads one
  multiple-precision integer which is stored in a}
VAR a0,m,i : INTEGER;

BEGIN write(txt); read(a0); m:=abs(a0); a[0]:=a0;
  FOR i:=m DOWNTO 1 DO read(a[i]);
END {putin};

PROCEDURE putout(txt : string; VAR a : vector);
{Prints a text string of 8 characters followed by one
  multiple-precision integer a}
CONST l=0.434294481 {const=1/ln(10)};
VAR a0,i,j,s,sa,t,u,v : INTEGER;

BEGIN write(txt); a0:=a[0]; sa:=sign(a0); a0:=abs(a0);
  IF sa<0 THEN write('-') ELSE write(' ');
  v:=trunc(l*ln(Base)+0.5) {v is the number of digits
    allowed in each component of a};
  IF a0=0 THEN write(0:v) ELSE
    BEGIN write(a[a0]:v);
      FOR i:=a0-1 DOWNTO 1 DO BEGIN t:=a[i];
        IF t=0 THEN s:=1 ELSE
          IF t=B101 THEN s:=v-1 ELSE
            IF t>=Base div 10 THEN s:=v ELSE
              s:=trunc(1+l*ln(t));
        u:=v-s; write(' ');
        FOR j:=1 TO u DO write('0');
        write(t:s) END END;
  writeln('') END {putout};

PROCEDURE red(VAR c : vector);
{Reduces a multiple-precision integer to standard form
  in call of procedure addsub}
VAR c0,s,i,j : INTEGER;

BEGIN c0:=c[0]; s:=sign(c0); c0:=abs(c0); c[c0+1]:=0;
  WHILE (c0>0) AND (c[c0]=0) DO c0:=c0-1;
    {Here the leading zeros of c are removed}
```

340

```
    IF c[c0]<0 THEN FOR i:=0 TO c0 DO c[i]:=-c[i];
  FOR j:=1 TO 2 DO BEGIN
    FOR i:=1 TO c0 DO BEGIN IF c[i]>=Base
      THEN BEGIN c[i+1]:=c[i+1]+1; c[i]:=c[i]-Base END
        ELSE IF c[i]<0 THEN BEGIN
          c[i+1]:=c[i+1]-1; c[i]:=c[i]+Base END END END;
  c0:=c0+1; WHILE (c0>0) AND (c[c0]=0) DO c0:=c0-1;
  c[0]:=sign(c[0])*c0
END {reduction};

PROCEDURE let(VAR a : vector; b : vector);
{Puts a:=b for multiple-precision integers a and b}
VAR s,i : INTEGER;

BEGIN s:=abs(b[0]);
  FOR i:=0 TO s DO a[i]:=b[i]
END {let};

PROCEDURE addsub(a,b : vector; sgn : INTEGER; VAR c: vector);
{With a=sign(a[0])*sum a[i]*B↑(i-1),
  b=sign(b[0])*sum b[j]*B↑(j-1), and sgn=+1 or -1,
  c:=a+sgn*b. c may be one of a or b}
VAR sa,sb,q,r,max,min,i,s : INTEGER;

BEGIN sa:=a[0]; sb:=b[0]*sgn; q:=abs(sa); max:=q; r:=abs(sb);
  min:=r;
  IF min>max THEN BEGIN max:=r; min:=q END;
  s:=sign(sa)*sign(sb);
  FOR i:=1 TO min DO c[i]:=a[i]+s*b[i];
  FOR i:=min+1 TO max DO
    IF max=q THEN c[i]:=a[i] ELSE c[i]:=s*b[i];
  c[0]:=sa; c[max+1]:=0;
  IF sa=0 THEN let(c,b); {c:=b in case a=0}
  IF s<>0 THEN c[0]:=max*sign(sa);
    {Here the components of c may be >B or <0}
  red(c) END {addsub};


PROCEDURE mul(a,b : vector; VAR c : vector);
{Computes c:=a*b for multiple-precision integers.
  c may be one of a or b}
TYPE vector    = ARRAY[0..vsize] OF INTEGER;
```

```
        longvector = ARRAY[0..lvsize] OF INTEGER;
 VAR aa,bb : vector;
     cc    : longvector;
     m,n,p,sa,sb,sc,i,k,s,m2,n2,p2,m21,n21,mn : INTEGER;

BEGIN m:=a[0]; sa:=sign(m); m:=abs(m); n:=b[0]; sb:=sign(n);
 n:=abs(n); p:=m+n; sc:=sa*sb; m2:=m*2; n2:=n*2; p2:=p*2;
 m21:=m2-1; n21:=n2-1; mn:=m2+n21;

 FOR i:=1 TO m DO BEGIN s:=a[i]; aa[2*i]:=s div sqrtB;
  aa[2*i-1]:=s mod sqrtB END;

 FOR i:=1 TO n DO BEGIN s:=b[i]; bb[2*i]:=s div sqrtB;
  bb[2*i-1]:=s mod sqrtB END;

 FOR i:=0 TO lvsize DO cc[i]:=0;
 IF m<=n THEN

   BEGIN

    FOR i:=1 TO m21 DO BEGIN s:=0;
     FOR k:=1 TO i DO BEGIN s:=s+aa[k]*bb[i+1-k];
      IF s>=Base THEN BEGIN s:=s-Base; cc[i+2]:=cc[i+2]+1 END
     END; cc[i]:=cc[i]+s END;

    FOR i:=m2 TO n2 DO BEGIN s:=0;
     FOR k:=1 TO m2 DO BEGIN s:=s+aa[k]*bb[i+1-k];
      IF s>=Base THEN BEGIN s:=s-Base; cc[i+2]:=cc[i+2]+1 END
     END; cc[i]:=cc[i]+s END;

    FOR i:=n2+1 TO mn DO BEGIN s:=0;
     FOR k:=i-n21 TO m2 DO BEGIN s:=s+aa[k]*bb[i+1-k];
      IF s>=Base THEN BEGIN s:=s-Base; cc[i+2]:=cc[i+2]+1 END
     END; cc[i]:=cc[i]+s END;

   END

 ELSE BEGIN

   FOR i:=1 TO n21 DO BEGIN s:=0;
    FOR k:=1 TO i DO BEGIN s:=s+bb[k]*aa[i+1-k];
     IF s>=Base THEN BEGIN s:=s-Base; cc[i+2]:=cc[i+2]+1 END
```

```
      END; cc[i]:=cc[i]+s END;

    FOR i:=n2 TO m2 DO BEGIN s:=0;
     FOR k:=1 TO n2 DO BEGIN s:=s+bb[k]*aa[i+1-k];
        IF s>=Base THEN BEGIN s:=s-Base; cc[i+2]:=cc[i+2]+1 END
      END; cc[i]:=cc[i]+s END;

    FOR i:=m2+1 TO mn DO BEGIN s:=0;
     FOR k:=i-m21 TO n2 DO BEGIN s:=s+bb[k]*aa[i+1-k];
        IF s>=Base THEN BEGIN s:=s-Base; cc[i+2]:=cc[i+2]+1 END
      END; cc[i]:=cc[i]+s END;

  END;

 FOR i:=1 TO p2 DO BEGIN cc[i+1]:=cc[i+1]+cc[i] div sqrtB;
  cc[i]:=cc[i] mod sqrtB END;
 FOR i:=1 TO p DO c[i]:=cc[2*i-1]+sqrtB*cc[2*i];
 IF c[p]=0 THEN p:=p-1; c[0]:=p*sc
END {mul};

PROCEDURE quot(a,b : vector; VAR q,r : vector);
{Computes q and r satisfying a=b*q+r, 0<=r<b for multiple
   precision integers a and b. r may be either of a or b}
LABEL 1,2,3;
TYPE vector = ARRAY[0..vsize] OF INTEGER;
VAR one,qk,aa,bb,cc,qq              : vector;
    m,n,s,t,i,j,k,l,p,t1,j1,sa,sb,u : INTEGER;
    ar,br,qr                        : REAL;

BEGIN FOR i:=0 TO vsize DO qq[i]:=0; m:=a[0]; n:=b[0];
   sa:=sign(m); sb:=sign(n); s:=abs(m); t:=abs(n); l:=s-t+1;
   IF l<=0 THEN l:=1;
   IF s<t THEN u:=t ELSE u:=s;
   IF n=0 THEN BEGIN
     writeln('Division by zero attempted in quot');
     l:=0; GOTO 3 END;
   one[0]:=1; one[1]:=1; qk[0]:=1; t1:=t+1; b[0]:=t;
   let(aa,a); aa[0]:=s;
   IF t=1 THEN br:=b[1] ELSE br:=b[t]+b[t-1]/Base;
     {Here br := the leading part of b}
1: IF s<t THEN GOTO 3;
   IF s=t THEN BEGIN
```

343

```
   addsub(aa,b,-1,cc); IF cc[0]<0 THEN GOTO 3 END;
  j:=aa[0]-t;
  IF s<=1 THEN ar:=aa[s] ELSE ar:=aa[s]+aa[s-1]/Base;
    {Here ar := the leading part of a}
  qr:=ar/br; {qr := the approximate quotient a/b}
  IF (qr<1) AND (j=0) THEN GOTO 3 ELSE
    IF qr<1 THEN BEGIN qr:=qr*Base; j:=j-1;
      IF qr<1 THEN qr:=1 END;
  k:=trunc(qr); qk[1]:=k; qq[j+1]:=qq[j+1]+k;
2: mul(qk,b,bb); FOR i:=t1 DOWNTO 1 DO bb[i+j]:=bb[i];
  FOR i:=1 TO j DO bb[i]:=0;
  bb[0]:=bb[0]+j; addsub(aa,bb,-1,cc); s:=cc[0]; p:=s;
  IF p<0 THEN
    BEGIN qk[1]:=qk[1]-1; qq[j+1]:=qq[j+1]-1;
      IF qk[1]=0 THEN
        BEGIN qk[1]:=B1; qq[j]:=qq[j]+qk[1]; j:=j-1 END;
      GOTO 2 END;
  FOR i:=0 TO p DO aa[i]:=cc[i]; GOTO 1;

3: b[0]:=n; IF qq[l]=0 THEN l:=l-1;
  qq[0]:=l*sa*sb; let(q,qq);
  IF l=0 THEN aa[0]:=m ELSE aa[0]:=aa[0]*sa*sb;
  IF (sa*sb<0) AND (aa[0]<>0) THEN BEGIN
    addsub(q,one,-1,q); addsub(aa,b,sb,aa) END;
let(r,aa) END {quot};

PROCEDURE Euclid(a,b : vector; VAR d : vector);
{Computes d=G.C.D.(a,b) for multiple-precision integers
  with Euclid's algorithm}
TYPE vector = ARRAY[0..vsize] OF INTEGER;
VAR aa,bb,q,r          : vector;
    a0,b0,i,r0,aa0,bb0 : INTEGER;

BEGIN a0:=abs(a[0]); b0:=abs(b[0]);
  let(aa,a); aa[0]:=a0; let(bb,b); bb[0]:=b0;
  WHILE bb[0]>0 DO
    BEGIN quot(aa,bb,q,r); let(aa,bb); let(bb,r) END;
  let(d,aa)
END {Euclid};

PROCEDURE apowerb(a,b,n : vector; VAR r : vector);
{Computes r = a↑b (mod n) for multiple-precision
```

```
   integers, b>0}
LABEL 1;
TYPE vector = ARRAY[0..vsize] OF INTEGER;
VAR two,par,aa,bb,p,s,gb : vector;

BEGIN p[0]:=1; p[1]:=1; two[0]:=1; two[1]:=2; let(bb,b);
   IF bb[0]<=0 THEN BEGIN
     writeln('b<=0 attempted in apowerb'); GOTO 1 END;
   quot(a,n,gb,aa);  WHILE bb[0]>0 DO
     BEGIN quot(bb,two,gb,par); let(bb,gb);
       IF par[0]=1 THEN BEGIN mul(aa,p,s); quot(s,n,gb,p) END;
       mul(aa,aa,s); quot(s,n,gb,aa) END;
   let(r,p);
1:  END {apowerb};

BEGIN {The test program for the package starts here}

   Base:=sqr(sqrtB); B1:=Base-1; B101:=Base div 10-1;
   writeln('Base=10↑',trunc(0.434294481*ln(Base)+0.5):2);
1: putin('Give a: ',a); putin('Give b: ',b);
   IF (a[0]=0) AND (b[0]=0) THEN GOTO 2;
   addsub(a,b,1,c); putout('a+b =    ',c);
   addsub(a,b,-1,c); putout('a-b =   ',c);
   mul(a,b,c); putout('a*b =   ',c);
   quot(a,b,c,d); putout('a/b =    ',c); putout('remaind=',d);
   Euclid(a,b,c); putout('(a,b)=',c);
   putin('Give c: ',c); apowerb(a,b,c,d); putout('a↑b (c)=',d);
   GOTO 1;

2: END.
```

A Computer Program for Pollard's ρ-Method

As an example of the use of our multiple-precision package, we shall now modify the program supplied on p. 178 for Pollard's ρ-method so that N can be a multi-precise integer.

```
PROGRAM Multi Pollard
{Searches for factors with Pollard's rho-method}
(Input,Output);
```

```
LABEL 1;
CONST sqrtB=100000; vsize=10; lvsize=20;
TYPE vector    = ARRAY[0..vsize] OF INTEGER;
     string    = PACKED ARRAY[1..8] OF CHAR;
VAR a,x1,x,y,Q,N,z,w,p    : vector;
    i,Base,B1,B101        : INTEGER;

FUNCTION sign(x : INTEGER) : INTEGER;
BEGIN IF x=0 THEN sign := 0 ELSE sign := abs(x) div x
END {sign};
   .
   .
   .
1:  END {apowerb};

BEGIN  Base:=sqr(sqrtB); B1:=Base-1; B101:=Base div 10-1;
  {Pollard's method begins here}
  putin('Give a: ',a); putin(Give x1:',x1); let(x,x1);
  let(y,x1); Q[0]:=1; Q[1]:=1; putin('Give N: ',N);
  FOR i:=1 TO 10000 DO BEGIN
    mul(x,x,z); addsub(z,a,-1,z); quot(z,N,w,x);
      {x:=x↑2-a mod N}
    mul(y,y,z); addsub(z,a,-1,z); quot(z,N,w,y);
      {y:=y↑2-a mod N}
    mul(y,y,z); addsub(z,a,-1,z); quot(z,N,w,y);
      {y:=y↑2-a mod N}
    addsub(y,x,-1,z); mul(Q,z,Q); quot(Q,N,w,Q);
      {Q:=Q*(y-x)mod N}
    IF i mod 20 = 0 THEN
      BEGIN Euclid(Q,N,p);
        IF ((p[0]=1) AND (p[1]>1)) OR (p[0]>1) THEN
          BEGIN quot(N,p,z,w); IF w[0]=0 THEN
            BEGIN putout('factor= ',p);
              writeln(' discovered for i=',i:5);
                {Here a factor of N is found and divided out}
              let(N,z); IF (N[0]=1) AND (N[1]=1) THEN GOTO 1
            END
          END
      END
  END;
  writeln('No factor discovered in 10000 cycles!');
1: END.
```

Exercise A7.1. Multiple-precision computations. Adapt the multiple-precision package above to your computer. Use it in a factorization program including computation of $N = a^n \pm b^n$, a, b and n parameters, division of N by a given factor, one of the probable prime tests mentioned in Definition 4.4 on p. 99, Pollard's ρ-method, and search for small factors by trial division. Write the program in such a form that it is easy to maintain a dialogue, choosing one option or another from a menu. Use your program to factorize some numbers of reasonable size, depending on the number crunching power of your computer. Test values can be found in Table 7–Table 31.

FAST MULTIPLICATION OF LARGE INTEGERS

The Ordinary Multiplication Algorithm

In Appendix 7 we have described the most frequently used algorithm for multiplication of integers M and N which are so large that multiple precision arithmetic is required. In this algorithm the numbers M and N are first written in a radix system with a suitable radix B and are then multiplied according to the following formula for multiplication of numbers to radix B. This is essentially the formula for multiplication of two polynomials of one variable B:

$$M \cdot N = \sum_{i=1}^{m} a_i B^{i-1} \sum_{j=1}^{n} b_j B^{j-1} = \sum_{k=1}^{m+n-1} c_k B^{k-1}, \qquad \text{(A8.1)}$$

where the "radix-B digit" is

$$c_k = \sum_{i} a_i b_{k+1-i}, \quad k = 1, 2, \ldots, m+n-1. \qquad \text{(A8.2)}$$

A minor modification has to be made in order to allow for "overflow" between adjacent digits as soon as some digit exceeds $B - 1$ in size. The summation index i has to be varied in such a way that these relationships always hold:

$$1 \le i \le m \quad \text{and} \quad 1 \le k+1-i \le n. \qquad \text{(A8.3)}$$

These conditions ensure that all the a_i's and b_j's used in (A8.2) do exist. The radix B is usually chosen in such a way that any integers $\le B - 1$, which are the digits to radix B, can be stored in one computer word. Hence the total number of "elementary operations" to perform the multiplication of M and N, i.e. the number of computer multiplications and additions, is proportional to $m \cdot n \approx \log_B M \cdot \log_B N$. If both numbers are of approximately the same size N then the number of elementary operations and thus also the execution time in a computer for this algorithm, will be

$$O(\log N)^2. \qquad \text{(A8.4)}$$

At first sight it might seem difficult to improve on the speed of this algorithm. However, astonishingly enough, there exist algorithms which work considerably faster for very large values of N, namely running in only

$$O(\log N)^{1+\epsilon} \tag{A8.5}$$

elementary steps, where ϵ is any positive number which can be chosen as close to zero as we might wish. We shall not go very deep into details here, but only hint at some ideas involved in this improvement of the multiplication algorithm. For a more detailed account we refer the reader to [1].

Double Length Multiplication

The principles discussed in the preceding subsection would give the formula

$$M \cdot N = (a_2 B + a_1)(b_2 B + b_1) = a_2 b_2 B^2 + (a_2 b_1 + a_1 b_2)B + a_1 b_1 \tag{A8.6}$$

for the ordinary multiplication of two double length integers. (With double, triple, etc. length integers we mean integers which because of their size must be stored in two, three, etc. computer words.) In this formula we see that all *four* combinations $a_i b_j$ of components a_i from the first integer M and b_j from the second integer N occur: $a_2 b_2$, $a_2 b_1$, $a_1 b_2$, and, finally, $a_1 b_1$. But there exists another formula for the right-hand-side of (A8.6):

$$a_2 b_2(B^2 + B) + (a_2 - a_1)(b_1 - b_2)B + a_1 b_1(B + 1). \tag{A8.7}$$

In this formula only three multiplications of numbers to radix B are performed, the rest is just simple additions in different positions of the products formed! In order to understand (A8.7) better let us compare the ordinary multiplication formula (A8.6) with (A8.7) in an

Example. Suppose $B = 100$ and that we have to multiply the two 4-digit numbers $M = 4711$ and $N = 6397$. By (A8.6) we first have to compute

$$a_2 b_2 = 47 \cdot 63 = 2961$$
$$a_2 b_1 = 47 \cdot 97 = 4559$$
$$a_1 b_2 = 11 \cdot 63 = 693$$
$$a_1 b_1 = 11 \cdot 97 = 1067$$

349

and then perform the additions

```
29 61
   45 59
    6 93
      10 67
─────────────
30 13 62 67
```

In a general case, we need *four* multiplications and *four* additions of two-digit numbers plus the adding of overflow, should any overflow occur in the additions.

Performing the same multiplication by use of (A8.7), we have to form only *three* products of two-digit numbers:

$$a_2 b_2 = 47 \cdot 63 = 2961$$
$$(a_2 - a_1)(b_1 - b_2) = (47 - 11)(97 - 63) = 36 \cdot 34 = 1224$$
$$a_1 b_1 = 11 \cdot 97 = 1067.$$

Next we have to form

$$2961 \cdot 10100 + 1224 \cdot 100 + 1067 \cdot 101,$$

which is done by the following additions:

```
29 61
   29 61
   12 24
   10 67
      10 67
─────────────
30 13 62 67
```

In a general case a computation like this requires *three* multiplications and *eight* additions of two-digit numbers (plus the adding of possible overflow between adjacent digits).

The reader who is familiar with the execution times on modern computers may now ask: Since the multiplication time and the addition time in the computer are approximately equal, how can it speed up the algorithm to trade one multiplication for *four* additions? The answer is the following: It is only when addition and multiplication of *one-word numbers* are compared, that the execution times in the computer's arithmetic unit are about equal. This has not always been the case. Because the multiplication algorithm used in a computer's hardware is more complicated than the addition algorithm, in the early computers, there was a very marked difference in these two execution times, one multiplication then taking between 5 and 10 times as long as one addition. Because it was felt that multiplication was too slow as compared to addition, extra hardware was added to speed it up. And this is why multiplications run as fast as additions on modern computers.—But if multiple precision integers are considered, then the ordinary multiplication algorithm is much slower than the obvious addition algorithm for multi-precise integers, so in such a case the trade of one multiplication for four additions pays off well! (The reader may easily convince himself of this fact by performing some multiplication of eight-digit numbers such as

$$12345678 \cdot 98765432$$

by the two methods discussed above, and actually performing the multiplications involved by pencil and paper and not by a hand-held calculator!)

Recursive Use of Double Length Multiplication Formula

The significance of formula (A8.7) lies not in the fact that it might run slightly faster than (A8.6) does on a computer, but that it can be used recursively in the case when two very large integers are to be multiplied. In precisely the same way as (A8.7) takes care of one double-length multiplication by 3 single-length multiplications and some additions, the same formula with B changed to B^2,

$$(U_2 B^2 + U_1)(V_2 B^2 + V_1) =$$

$$= U_2 V_2 (B^4 + B^2) + (U_2 - U_1)(V_1 - V_2) B^2 + U_1 V_1 (B^2 + 1), \quad \text{(A8.8)}$$

takes care of one quadruple-length multiplication by performing instead 3 double-length multiplications and some double-length additions. (Please note that the numbers $U_2 B^2 + U_1$ and $V_2 B^2 + V_1$ denote *quadruple* length integers, if U_2, U_1, V_2 and V_1 are *double-length* integers!) Using formula (A8.8) recursively, a multiplication of two very large integers is first replaced by 3 multiplications of integers of about half the number of digits, which in turn are replaced by 9 multiplications of integers of only $\frac{1}{4}$ of the length of the original integers and so on, plus a lot of additions.

351

How fast would this algorithm be? If we don't count the additions at all (this makes the argument simpler, and is actually safe when we only want to know the *asymptotic* behaviour of the algorithm, i.e. its speed for *very* large integers N), we find that

Integers of the size 2^k words require 3^k computer multiplications.

But since

$$3^k = (2^{\log_2 3})^k = (2^k)^{\log_2 3}$$

and 2^k is the length in computer words of the numbers to be multiplied, which is proportional to $\log N$, we find that the execution time for the recursive use of (A8.7) is

$$O(\log N)^{\log_2 3} = O(\log N)^{1.585}, \tag{A8.9}$$

which is considerably better than (A8.4).

Now, since we have neglected the effect of the larger amount of additions occurring in the recursive use of (A8.7), we might expect the gain to be somewhat smaller than the formula shows. But for very large numbers N formula (A8.9) holds and the additions effect only the constant hidden in the O-notation of (A8.9). As a result, the break-even point between (A8.9) and (A8.4) is not reached until for very large numbers N. First when numbers of several hundred or even thousands of digits are involved, the more complicated organization for carrying out multiplications pays off. Remember that the recursive use of a procedure requires much administrative overhead which also costs computing time!—But since the very large primes known have *tens of thousands* of digits, this technique decidedly is favourable in performing Lucas tests on such huge numbers.

A Recursive Procedure for Squaring Large Integers

We shall not discuss the use of (A8.7) and (A8.8) in their full generality, but in order to give at least some example of how to program the algorithm just discussed, we take the slightly simpler case of *squaring* a large integer. Since this is what is needed in the Lucas test of the Mersenne numbers, at the same time it gives a glimpse of how these very large primes were found.

In order to square a very large integer, we first have to find a way to express the square in (at most) three smaller squares. This is done simply by putting $M = N = U_2 B + U_1$ in (A8.7):

$$(U_2 B + U_1)^2 = U_2^2(B^2 + B) - (U_2 - U_1)^2 B + U_1^2(B + 1). \tag{A8.10}$$

A PASCAL procedure which implements this formula by recursive calls takes the form

```
PROGRAM Squaretest
{Checks out the procedure square}
(Input,Output);
CONST sqrtR=100; vsize=33; lvsize=66;
TYPE vector     = ARRAY[0..vsize] OF INTEGER;
     string     = PACKED ARRAY[1..8] OF CHAR;
VAR a,b                   : vector;
    Radix,B1,B101         : INTEGER;

FUNCTION sign(x : INTEGER) : INTEGER;
BEGIN IF x=0 THEN sign := 0 ELSE sign := abs(x) DIV x
END {sign};
```

Here the PROCEDUREs putin, putout, red, let, addsub and mul from the Multiple Precision Package must be inserted

```
PROCEDURE expand(s : INTEGER; VAR u : vector);
{Multiplies a multiple precision integer u by Radix↑s, s>0}
VAR u0,i,k    : INTEGER;

BEGIN u0:=abs(u[0]);
  FOR i:=1 TO u0 DO
    BEGIN k:=u0+1-i; u[k+s]:=u[k]; u[k]:=0 END;
  FOR i:=u0+1 TO s DO u[i]:=0;
  u[0]:=sign(u[0])*(u0+s)
END {expand};

PROCEDURE square(p : vector; VAR q : vector);
{Computes q=p↑2 recursively by using (Bu+v)↑2 =
  = u↑2*B↑2 + [u↑2 + (u-v)↑2 +v↑2]*B + v↑2}
LABEL 1;
TYPE vector     = ARRAY[0..vsize] OF INTEGER;
VAR u,v,z,u2,v2,uv,s    : vector;
    p0,ap0,ap02,i,k     : INTEGER;

BEGIN p0:=p[0]; ap0:=abs(p0); IF ap0<=1 THEN
  BEGIN q[0]:=0; IF ap0=1 THEN
    BEGIN i:=sqr(p[1]); q[1]:=i MOD Radix; q[2]:=i DIV Radix;
      IF q[2]=0 THEN q[0]:=1 ELSE q[0]:=2 END;
```

```
{Here a piece of assembler code should replace the last
  two lines to compute the square of a single-word integer}
  GOTO 1
END;
ap02:=(ap0+1)DIV 2; p[0]:=ap02; let(v,p); p[0]:=p0;
  {Here v has been formed}
u[0]:=ap0-ap02; FOR i:=1 TO u[0] DO u[i]:=p[i+ap02];
  {Here u has been formed}
square(u,u2); addsub(u,v,-1,z); square(z,uv); square(v,v2);
addsub(u2,uv,-1,z); addsub(z,v2,1,s);
  {Here u↑2 - (u-v)↑2 + v↑2 has been formed}
expand(ap02,s); expand(2*ap02,u2);
addsub(u2,v2,1,z); addsub(z,s,1,q);
1: END {square};

BEGIN {Here BEGINs the test program for the procedure square}
  Radix:=sqr(sqrtR); B1:=Radix-1; B101:=Radix DIV 10-1;
  writeln('Radix=10↑',trunc(0.434294481*ln(Radix)+0.5):2);
putin('Give a= ',a); WHILE a[0]<>0 DO BEGIN
  square(a,b); putout('sqr(a)= ',b); putin('Give a= ',a) END
END.
```

Here a new procedure **expand** has been written in order to facilitate the multiplication of the multiple precision integers occurring in the computation by a power of the radix used. This operation is simply performed by adding a number of zero words at the end of the integer in question, which, because of our "the other way round" representation, has to be done by inserting these zero words at the beginning of the corresponding vector.

When the procedure Square is called to compute the square of a very large integer N, the integer is first split into two parts, U and V. Then Square is called again *three times* for the squaring of the smaller integers occurring in (A8.10). If any of *these* integers is still larger than one computer word, the procedure Square is again called three times to perform the squaring of this particular integer and so on, until the whole computation is reorganized to contain the squaring of single precision integers only and some(!!) additions. In order to keep track of what has to be done during this process, key information for each of the procedure calls is automatically being stored in the computer.

The substitution of the squaring of a large integer by three squarings of integers of about half the length may be symbolized as follows.

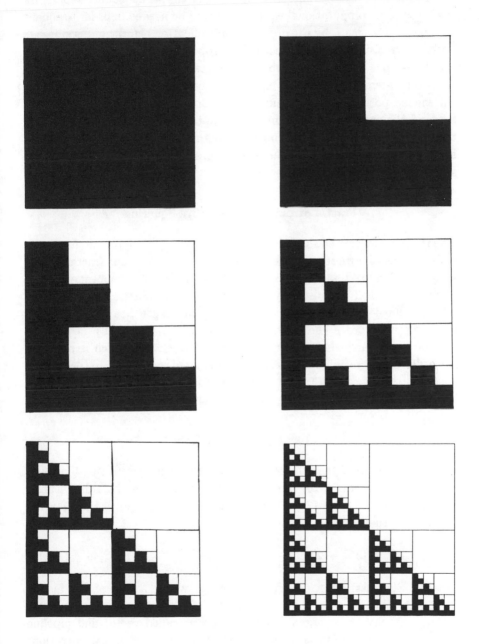

Figure A8.1. Fractal structure of recursive squaring. $D = 1.585$

In the first of these pictures, the area of the whole square represents the amount of computational work that would be needed in order to square an integer of length L by the ordinary multiplication algorithm, used in our Multiple Precision Package, described in Appendix 7. In the second picture, the integer has been split in two equal halves, and now only *three* of these shorter integers of length $L/2$ have to be squared. This is represented by the shaded portion of the second picture. Continuing in the same way, we successively arrive at an amount of computational labour, corresponding to the shaded portions of the following pictures. The dark portions of the squares are thinned out, the further we break down the computation in smaller and smaller parts.

Fractal Structure of Recursive Squaring

The thinning out of the shaded parts of the squares in the picture is what is called a *fractal structure*. More precisely, it is a self-similar fractal, called a square Sierpiński gasket, with inner cut-off. For this concept, see [2]. (The inner cutoff is caused by the fact that the splitting into halves is not continued indefinitely, but ends when the length has come down to one computer word.) The *fractal dimension* that characterizes the limiting case of the squares shown is $d = \log 3 / \log 2 = 1.585$, which is precisely the order of magnitude of the algorithm depicted, according to (A8.9). We thus find the concept of fractal dimension to be at the heart of the matter in the study of graphs showing the execution of computational schemes.

Large Mersenne Primes

In [3] it is reported that the technique described has been used by David Slowinski in his search for large Mersenne primes with the aid of Lucas' Test, Theorem 4.9 on p. 126. Choosing the radix $B = 2^n$ gives the following formula for the square of $U = U_2 B + U_1$:

$$U^2 = (2^{2n} + 2^n)U_2^2 - 2^n(U_1 - U_2)^2 + (2^n + 1)U_0^2, \qquad \text{(A8.11)}$$

used by Slowinski on a CRAY-1 computer, which is one of the fastest computers built to date. The high speed of the CRAY computer depends to large extent on its organization. It is a so-called vector and pipeline machine, containing an arithmetic unit in which 64 identical operations can be performed on 64 items of data simultaneously. By this means a speed factor of 64 is gained, *assuming that the computation at hand can be organized to fit this design.* Thus, in such a computer, the smallest "unit" to work with, in a multiple-precision arithmetic package, is an integer of 64 (or possibly 62) computer words. This concept replaces the single-word integer used in the assembly code within the PROCEDURE square in the program above, and the arithmetic package will employ sections of assembly code written to handle such entities. Of course, the entire computation is effected in binary arithmetic, which is extremely advantageous because of the simple binary structure (ONEs only) of the Mersenne numbers.

Bibliography

1. Donald E. Knuth, *The Art of Computer Programming*, Vol. 2, Second edition, Addison-Wesley, Reading, Massachusetts (1981) pp. 278–301.

2. Benoit B. Mandelbrot, *The Fractal Geometry of Nature*, San Francisco 1983.

3. John Ewing, "$2^{86243} - 1$ is Prime", *The Mathematical Intelligencer* **5** (1983) p. 60.

THE STIELTJES INTEGRAL

Introduction

In this appendix we give a short account of the theory of Riemann-Stieltjes integration. This concept is a generalization of the familiar theory of Riemann integration as extended to integrands with jump discontinuities. Since many of the fundamental number-theoretic functions, for instance $\pi(x)$, exhibit this type of behaviour, Riemann-Stieltjes integration is the appropriate tool to apply when such functions need to be integrated.— Moreover, it is an elegant little theory, also useful in other contexts, mainly because it admits sums to be written in a simple way as integrals.

Functions With Jump Discontinuities

A jump is the simplest sort of discontinuity a function can have. A jump discontinuity can be described as follows: At some point x_i the *limit when approaching x_i from the right,*

$$\lim_{x \to x_i + 0} \alpha(x) = \alpha(x_i + 0), \tag{A9.1}$$

and the *limit from the left,*

$$\lim_{x \to x_i - 0} \alpha(x) = \alpha(x_i - 0), \tag{A9.2}$$

both exist, but have different values. The magnitude of the jump is the difference between these two limiting values, $\alpha(x_i + 0) - \alpha(x_i - 0)$. A very simple example of a function with such discontinuities is offered by $y = [x]$ (the integer part of x), which has jumps of size 1 for each integer value of x. The structure of a function having only finite jump discontinuities is described by

Theorem A9.1. If the function $\alpha(x)$ has only finite jump discontinuities, then $\alpha(x)$ can be written as a sum of two terms

$$\alpha(x) = g(x) + w(x). \tag{A9.3}$$

Here $g(x)$ is a continuous function and $w(x)$ a step-function having the same points of discontinuity and jumps of the same magnitudes as $\alpha(x)$.

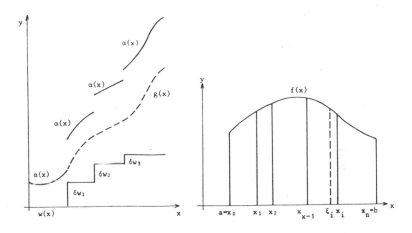

Figure A9.1. $\alpha(x) = g(x) + w(x)$ Figure A9.2. Riemann integration

Proof. The theorem is obvious if we consider fig. A9.1. All that is needed to construct $g(x)$ is to remove each jump of $\alpha(x)$ by a vertical movement of the parts of its graph between two adjacent jumps by a distance chosen such that the entire jump is spanned. Finally $w(x)$ is given by $\alpha(x) - g(x)$.

This fundamental theorem provides us with a tool which greatly simplifies the calculus of functions with jump discontinuities. Writing $\alpha(x)$ as $g(x) + w(x)$, in order to generalize the differential and integral calculus for continuous functions to this slightly more general class, we need only to study how the calculus operates on step-functions.

The Riemann Integral

Before stating the definition of the Riemann-Stieltjes integral, we recollect the definition of the somewhat simpler Riemann integral for a continuous function $f(x)$, integrated over a finite interval $[a, b]$:

Divide the given interval into sub-intervals by introducing the points $a = x_0, x_1, x_2, \ldots, x_{n-1}, x_n = b$, see figure A9.2. Choose an arbitrary point ξ_i in the sub-interval

$$x_{i-1} \leq \xi_i \leq x_i, \tag{A9.4}$$

and consider the limit

$$\lim \sum_{i=1}^{n} f(\xi_i)\Delta x_i, \quad \text{with} \quad \Delta x_i = x_i - x_{i-1}. \tag{A9.5}$$

Here the lengths Δx_i of all the sub-intervals must tend to zero. This expression then converges to Riemann's integral

$$\int_a^b f(x)\,dx = \lim_{\Delta x_i \to 0} \sum_{i=1}^{n} f(\xi_i)\Delta x_i. \tag{A9.6}$$

Please note that the limiting value is completely independent of the particular way in which the subdivision points are chosen, provided that the lengths of all the sub-intervals $[x_{i-1}, x_i]$ tend to zero.—We also remind the reader of the fairly obvious generalizations of (A9.6) to cases where one or both of the limits of integration tend to infinity:

$$\int_a^\infty f(x)\,dx = \lim_{X \to \infty} \int_a^X f(x)\,dx, \tag{A9.7}$$

$$\int_{-\infty}^b f(x)\,dx = \lim_{X \to -\infty} \int_X^b f(x)\,dx, \tag{A9.8}$$

and of the instance when $f(x)$ has but one single point of discontinuity, $x = c$ in the interval $[a, b]$:

$$\int_a^b f(x)\,dx = \lim_{\epsilon \to +0} \int_a^{c-\epsilon} f(x)\,dx + \lim_{\epsilon \to +0} \int_{c+\epsilon}^b f(x)\,dx. \tag{A9.9}$$

In case $f(x)$ has a finite number of discontinuities in $[a, b]$, the interval is split into sub-intervals, each sub-interval containing but one discontinuity of $f(x)$, and (A9.9) applied to each sub-interval. If each of the resulting limits exists, then $f(x)$ is said to be Riemann integrable over the interval $[a, b]$.

Definition of the Stieltjes Integral

The Stieltjes integral is a generalization of (A9.9) to integrands of a form other than that we have so far considered. Instead of (A9.5), consider the limit

$$\lim_{\Delta x_i \to 0} \sum_{i=1}^{n} f(\xi_i)(\alpha(x_i) - \alpha(x_{i-1})), \qquad (A9.10)$$

with x_i and ξ_i as in (A9.5). If $\alpha(x)$ is a *differentiable* function then, by the mean-value theorem,

$$\alpha(x_i) - \alpha(x_{i-1}) = \alpha'(\xi_i^*)\Delta x_i \qquad (A9.11)$$

for some suitably chosen ξ_i^* within the sub-interval $[x_{i-1}, x_i]$. Now, choosing ξ_i in (A9.5) to be ξ_i^*, it follows that the limit (A9.10) converges towards

$$\int_a^b f(x)\alpha'(x)\,dx = \int_a^b f(x)\,d\alpha(x). \qquad (A9.12)$$

Note here that even if $\alpha(x)$ is *not* a *differentiable function*, the limiting value (A9.10) will exist, provided that $\alpha(x)$ is continuous and the lengths of all the sub-intervals tend to zero. However, in such a case it is not possible in general to form $\alpha'(x)$, so that the notation $\int f(x)\,d\alpha(x)$ must be retained.

Next, if $\alpha(x)$ is not continuous but has jumps of magnitude δw_k at the points of discontinuity x_k, we may utilize Theorem A9.1 and write $\alpha(x)$ as $g(x) + w(x)$; that is to accumulate all the jumps in the step-function $w(x)$ while $g(x)$ is continuous. With this notation (A9.10) becomes

$$\lim_{\Delta x_i \to 0} \sum_{i=1}^{n} f(\xi_i)(g(x_i) - g(x_{i-1})) + \lim_{\Delta x_i \to 0} \sum_{i=1}^{n} f(\xi_i)(w(x_i) - w(x_{i-1})).$$
$$(A9.13)$$

The first limit is obviously well-defined and $= \int_a^b f(x)\,dg(x)$, as $g(x)$ is assumed to be continuous. Calculating the second limiting value, contributions are obtained only from those sub-intervals $[x_{i-1} - x_i]$ in which $w(x)$ jumps. Since the jumps of $w(x)$ are identical to those of $\alpha(x)$—which we have assumed to be δw_k at $x = x_k$—we find that the second limit equals

$$\sum_k f(\xi_i)\,\delta w_k = \sum_k f(x_k)\,\delta w_k, \qquad (A9.14)$$

if we suppose $f(x)$ to be continuous. Thus in this case the value of the Stieltjes integral is found to be

$$\int\limits_a^b f(x)\,d\alpha(x) = \int\limits_a^b f(x)\,dg(x) + \sum_k f(x_k)\,\delta w_k, \qquad \text{(A9.15)}$$

where $g(x)$ is the "continuous part" of $f(x)$, and the "step-function part" $\alpha(x)$ of $f(x)$ has jumps of sizes δw_k at $x = x_k$ in the interval $[a, b]$.

Generalizations of (A9.15) to an infinite range of integration, and to functions $f(x)$ with jump discontinuities, are achieved in a manner completely analogous to the corresponding generalizations of the ordinary Riemann integral discussed earlier. The only situation to be avoided is when $f(x)$ has a jump at one of the points x_k, a point of jump discontinuity for $\alpha(x)$, because in such a case the sum (A9.14) does not necessarily tend to a limit.

If the function $\alpha(x)$ has a jump at either $x = a$ or $x = b$, then the range of integration must be slightly adjusted depending on whether a contribution to the integral from the jump is desired or not. If no contribution is required, we write

$$\int\limits_{a+0}^{b-0} f(x)\,d\alpha(x), \quad \text{whereas} \quad \int\limits_{a-0}^{b+0} f(x)\,d\alpha(x) \qquad \text{(A9.16)}$$

is used, if both jumps are to be included in the evaluation of the integral.

Examples.

$$\sum_{x=m}^n f(x) = \int\limits_{m-0}^{n+0} f(x)\,d[x], \qquad \text{(A9.17)}$$

since $d[x]$ equals 1 for $x = m, m + 1, \ldots, n$, and is 0 elsewhere.

$$\sum_{p \text{ prime}} f(p) = \int\limits_1^\infty f(x)\,d\pi(x), \qquad \text{(A9.18)}$$

since $d\pi(x)$ is 1 whenever x is a prime p, and is 0 otherwise.

362

Rules of Integration for Stieltjes Integrals

The simplest rules of integration can be proved in much the same way as in the case of ordinary Riemann integrals. Directly from the definition we have

$$\int_a^b f(x)\, d\alpha(x) = \int_a^c f(x)\, d\alpha(x) + \int_c^b f(x)\, d\alpha(x) \qquad (A9.19)$$

for any number c between a and b. However, care must be taken if c is a point of discontinuity of $\alpha(x)$. The contribution to the integral from such a jump is $f(c)(\alpha(c+0) - \alpha(c-0))$, and it is important to ensure that this term is neither lost nor counted twice when rewriting the integral. This can be achieved by avoiding the point c itself as a limit of integration, i.e. to calculate either

$$\int_a^{c-0} f(x)\, d\alpha(x) + \int_{c-0}^b f(x)\, d\alpha(x) \qquad (A9.20)$$

or

$$\int_a^{c+0} f(x)\, d\alpha(x) + \int_{c+0}^b f(x)\, d\alpha(x) \qquad (A9.21)$$

Integration by change of variable follows the same formal rule as for Riemann integrals, i.e.

$$\int_a^b f(x)\, d\alpha(x) = \int_A^B f(\phi(t))\, d\alpha(\phi(t)), \qquad (A9.22)$$

where the new limits of integration are given by the well-known rules

$$\phi(A) = a \quad \text{and} \quad \phi(B) = b, \qquad (A9.23)$$

respectively.

Integration by Parts of Stieltjes Integrals

Integration by parts demands a closer scrutiny. Assume that $\alpha(x)$ does not jump at either $x = a$ or $x = b$. Write, as usual, $\alpha(x) = g(x) + w(x)$ with

363

$g(x)$ a continuous and $w(x)$ a piecewise constant function. Then

$$\int_a^b \alpha(x)\,df(x) = \int_a^b g(x)\,df(x) + \int_a^b w(x)\,df(x) =$$

$$= \int_a^b g(x)\,df(x) + \sum_k w_k \int_{x_k}^{x_{k+1}} df(x) =$$

$$= \int_a^b g(x)\,df(x) + \sum_k w_k(f(x_{k+1}) - f(x_k)), \qquad (A9.24)$$

where, as above,

$$w_k = w(x_k + 0) = \sum_{i=1}^k \delta w_i. \qquad (A9.25)$$

Here w_k is the accumulated sum of jumps of $\alpha(x)$ taken from left to right, i.e. the level which the step-function $w(x)$ has attained at $x = x_k + 0$. Adding (A9.15) and (A9.24) we have

$$\int_a^b f(x)\,d\alpha(x) + \int_a^b \alpha(x)\,df(x) = \int_a^b f(x)\,dg(x) + \int_a^b g(x)\,df(x) +$$

$$+ \sum_k \{f(x_{k+1})(w_{k+1} - w_k) + w_k(f(x_{k+1}) - f(x_k))\} =$$

(since telescopic cancellation reduces the sum completely; apart from its first and last terms)

$$= \int_a^b d(f(x)g(x)) + f(b)w(b) - f(a)w(a) = \Big[f(x)\alpha(x)\Big]_a^b. \qquad (A9.26)$$

We therefore conclude that the familiar formula for integration by parts of an ordinary Riemann integral

$$\int_a^b f(x)\,d\alpha(x) = \Big[f(x)\alpha(x)\Big]_a^b - \int_a^b \alpha(x)\,df(x) \qquad (A9.27)$$

is valid without any change also for Riemann-Stieltjes integrals.

In the case when $\alpha(x)$ jumps at $x = a$ and/or $x = b$, formula (A9.27) still holds if care is taken to count the contributions from the jumps equally on both sides of (A9.26).

Remark. If $g(x)$ is chosen identically 0, then (A9.26) yields

$$\sum_{k=0}^{s-1} f(x_{k+1})(w_{k+1} - w_k) = f(x_s)w_s - f(x_0)w_0 -$$

$$- \sum_{k=0}^{s-1} w_k(f(x_{k+1}) - f(x_k)). \qquad (A9.28)$$

This is Abel's celebrated formula for the partial summation of series.

The Mean Value Theorem

We shall also need the Mean Value Theorem for Stieltjes integrals. We first remind the reader of the Mean Value Theorem for ordinary Riemann integrals:

Theorem A9.2. The Mean Value Theorem for Riemann integrals. If $f(x)$ is a continuous function and $h(x) \geq 0$, then

$$\int_a^b f(x)h(x)\, dx = f(\xi) \int_a^b h(x)\, dx, \qquad (A9.29)$$

where ξ is a number between a and b.

The Mean Value Theorem is useful when it comes to estimation of an integral which is otherwise difficult to evaluate. For Stieltjes integrals this goal is achieved by

Theorem A9.3. The Mean Value Theorem for Riemann-Stieltjes Integrals. If $f(x)$ is a continuous and $\alpha(x)$ a non-decreasing function, then

$$\int_a^b f(x)\, d\alpha(x) = f(\xi) \int_a^b d\alpha(x) = f(\xi)(\alpha(b) - \alpha(a)), \qquad (A9.30)$$

where ξ again is a number between a and b.

Proof. Suppose, as is usual in the proof of the Mean Value Theorem, that $m \le f(x) \le M$ in the interval $[a, b]$, where m and M are the lower and upper bounds, respectively, of $f(x)$. Since $\alpha(x)$ is supposed to be non-decreasing, all the differences $\alpha(x_i) - \alpha(x_{i-1})$ in (A9.10) are non-negative, and thus

$$m(\alpha(x_i) - \alpha(x_{x-1})) \le f(\xi_i)(\alpha(x_i) - \alpha(x_{i-1})) \le M(\alpha(x_i) - \alpha(x_{i-1})).$$

Summing up all these inequalities for $i = 1, 2, \ldots, n$ and passing to the limit, we obtain

$$m \int_a^b d\alpha(x) \le \int_a^b f(x)\, d\alpha(x) \le M \int_a^b d\alpha(x).$$

Now, since $f(x)$ is continuous we can find a number ξ in $[a, b]$ such that the middle term

$$\int_a^b f(x)\, d\alpha(x) = f(\xi) \int_a^b d\alpha(x),$$

which proves the teorem.

Applications

It is much easier to transform integrals by means of the rules of integration than it is to carry through the corresponding transformations on sums. Consequently a large field of application of Stieltjes integrals is the transformation of various sums after having first converted them into integrals by introducing some suitable step-function, as done in the case of formulas (A9.17) and (A9.18) above.

Example 1. The famous Euler-MacLaurin summation formula can easily be deduced by this technique. We indicate here the first step of the deduction. Starting from (A9.17), we find

$$\sum_{k=m}^{n} f(k) = \int_{m-0}^{n+0} f(x)\, d[x] = \int_{m-0}^{n+0} f(x)\, dx - \int_{m-0}^{n+0} f(x)\, d(x - [x]) =$$

$$= \int_{m}^{n} f(x)\, dx - \int_{m-0}^{n+0} f(x)\, d\left(x - [x] - \frac{1}{2}\right) =$$

$$= \int_{m}^{n} f(x)\, dx - \left[f(x)\left(x - [x] - \frac{1}{2}\right)\right]_{m-0}^{n+0} + \int_{m}^{n}\left(x - [x] - \frac{1}{2}\right) f'(x)\, dx =$$

$$= \int_{m}^{n} f(x)\, dx + \frac{1}{2}f(m) + \frac{1}{2}f(n) + \int_{m}^{n}\left(x - [x] - \frac{1}{2}\right) f'(x)\, dx.$$

Example 2. We prove a relation between the Riemann zeta-function and $\pi(x)$. Starting from (2.7) and (A9.18), we obtain

$$\ln \varsigma(s) = -\sum_{p} \ln(1 - p^{-s}) = -\int_{2-0}^{\infty} \ln(1 - x^{-s})\, d\pi(x) =$$

$$= \left[-\pi(x)\ln(1 - x^{-s})\right]_{2-0}^{\infty} + \int_{2-0}^{\infty} \pi(x)\, d\ln(1 - x^{-s}) =$$

$$= \int_{2}^{\infty} \frac{\pi(x)\, dx}{x(x^s - 1)} \quad \text{for} \quad s > 1, \tag{A9.31}$$

since the integrated term vanishes at $x = 2 - 0$ and at $x = \infty$.

Further examples can be found in the deductions given on p. 168 and on p. 218.

TABLES

Table 1. The Primes Below 12553 (2–3571)

2	233	547	877	1229	1597	1993	2371	2749	3187
3	239	557	881	1231	1601	1997	2377	2753	3191
5	241	563	883	1237	1607	1999	2381	2767	3203
7	251	569	887	1249	1609	2003	2383	2777	3209
11	257	571	907	1259	1613	2011	2389	2789	3217
13	263	577	911	1277	1619	2017	2393	2791	3221
17	269	587	919	1279	1621	2027	2399	2797	3229
19	271	593	929	1283	1627	2029	2411	2801	3251
23	277	599	937	1289	1637	2039	2417	2803	3253
29	281	601	941	1291	1657	2053	2423	2819	3257
31	283	607	947	1297	1663	2063	2437	2833	3259
37	293	613	953	1301	1667	2069	2441	2837	3271
41	307	617	967	1303	1669	2081	2447	2843	3299
43	311	619	971	1307	1693	2083	2459	2851	3301
47	313	631	977	1319	1697	2087	2467	2857	3307
53	317	641	983	1321	1699	2089	2473	2861	3313
59	331	643	991	1327	1709	2099	2477	2879	3319
61	337	647	997	1361	1721	2111	2503	2887	3323
67	347	653	1009	1367	1723	2113	2521	2897	3329
71	349	659	1013	1373	1733	2129	2531	2903	3331
73	353	661	1019	1381	1741	2131	2539	2909	3343
79	359	673	1021	1399	1747	2137	2543	2917	3347
83	367	677	1031	1409	1753	2141	2549	2927	3359
89	373	683	1033	1423	1759	2143	2551	2939	3361
97	379	691	1039	1427	1777	2153	2557	2953	3371
101	383	701	1049	1429	1783	2161	2579	2957	3373
103	389	709	1051	1433	1787	2179	2591	2963	3389
107	397	719	1061	1439	1789	2203	2593	2969	3391
109	401	727	1063	1447	1801	2207	2609	2971	3407
113	409	733	1069	1451	1811	2213	2617	2999	3413
127	419	739	1087	1453	1823	2221	2621	3001	3433
131	421	743	1091	1459	1831	2237	2633	3011	3449
137	431	751	1093	1471	1847	2239	2647	3019	3457
139	433	757	1097	1481	1861	2243	2657	3023	3461
149	439	761	1103	1483	1867	2251	2659	3037	3463
151	443	769	1109	1487	1871	2267	2663	3041	3467
157	449	773	1117	1489	1873	2269	2671	3049	3469
163	457	787	1123	1493	1877	2273	2677	3061	3491
167	461	797	1129	1499	1879	2281	2683	3067	3499
173	463	809	1151	1511	1889	2287	2687	3079	3511
179	467	811	1153	1523	1901	2293	2689	3083	3517
181	479	821	1163	1531	1907	2297	2693	3089	3527
191	487	823	1171	1543	1913	2309	2699	3109	3529
193	491	827	1181	1549	1931	2311	2707	3119	3533
197	499	829	1187	1553	1933	2333	2711	3121	3539
199	503	839	1193	1559	1949	2339	2713	3137	3541
211	509	853	1201	1567	1951	2341	2719	3163	3547
223	521	857	1213	1571	1973	2347	2729	3167	3557
227	523	859	1217	1579	1979	2351	2731	3169	3559
229	541	863	1223	1583	1987	2357	2741	3181	3571

Table 1. The Primes Below 12553 (3581–7919)

3581	4001	4421	4861	5281	5701	6143	6577	7001	7507
3583	4003	4423	4871	5297	5711	6151	6581	7013	7517
3593	4007	4441	4877	5303	5717	6163	6599	7019	7523
3607	4013	4447	4889	5309	5737	6173	6607	7027	7529
3613	4019	4451	4903	5323	5741	6197	6619	7039	7537
3617	4021	4457	4909	5333	5743	6199	6637	7043	7541
3623	4027	4463	4919	5347	5749	6203	6653	7057	7547
3631	4049	4481	4931	5351	5779	6211	6659	7069	7549
3637	4051	4483	4933	5381	5783	6217	6661	7079	7559
3643	4057	4493	4937	5387	5791	6221	6673	7103	7561
3659	4073	4507	4943	5393	5801	6229	6679	7109	7573
3671	4079	4513	4951	5399	5807	6247	6689	7121	7577
3673	4091	4517	4957	5407	5813	6257	6691	7127	7583
3677	4093	4519	4967	5413	5821	6263	6701	7129	7589
3691	4099	4523	4969	5417	5827	6269	6703	7151	7591
3697	4111	4547	4973	5419	5839	6271	6709	7159	7603
3701	4127	4549	4987	5431	5843	6277	6719	7177	7607
3709	4129	4561	4993	5437	5849	6287	6733	7187	7621
3719	4133	4567	4999	5441	5851	6299	6737	7193	7639
3727	4139	4583	5003	5443	5857	6301	6761	7207	7643
3733	4153	4591	5009	5449	5861	6311	6763	7211	7649
3739	4157	4597	5011	5471	5867	6317	6779	7213	7669
3761	4159	4603	5021	5477	5869	6323	6781	7219	7673
3767	4177	4621	5023	5479	5879	6329	6791	7229	7681
3769	4201	4637	5039	5483	5881	6337	6793	7237	7687
3779	4211	4639	5051	5501	5897	6343	6803	7243	7691
3793	4217	4643	5059	5503	5903	6353	6823	7247	7699
3797	4219	4649	5077	5507	5923	6359	6827	7253	7703
3803	4229	4651	5081	5519	5927	6361	6829	7283	7717
3821	4231	4657	5087	5521	5939	6367	6833	7297	7723
3823	4241	4663	5099	5527	5953	6373	6841	7307	7727
3833	4243	4673	5101	5531	5981	6379	6857	7309	7741
3847	4253	4679	5107	5557	5987	6389	6863	7321	7753
3851	4259	4691	5113	5563	6007	6397	6869	7331	7757
3853	4261	4703	5119	5573	6011	6421	6871	7333	7759
3863	4271	4721	5147	5573	6029	6427	6883	7349	7789
3877	4273	4723	5153	5581	6037	6449	6899	7351	7793
3881	4283	4729	5167	5591	6043	6451	6907	7369	7817
3889	4289	4733	5171	5623	6047	6469	6911	7393	7823
3907	4297	4751	5179	5639	6053	6473	6917	7411	7829
3911	4327	4759	5189	5641	6067	6481	6947	7417	7841
3917	4337	4783	5197	5647	6073	6491	6949	7433	7853
3919	4339	4787	5209	5651	6079	6521	6959	7451	7867
3923	4349	4789	5227	5653	6089	6529	6961	7457	7873
3929	4357	4793	5231	5657	6091	6547	6967	7459	7877
3931	4363	4799	5233	5659	6101	6551	6971	7477	7879
3943	4373	4801	5237	5669	6113	6553	6977	7481	7883
3947	4391	4813	5261	5683	6121	6563	6983	7487	7901
3967	4397	4817	5273	5689	6131	6569	6991	7489	7907
3989	4409	4831	5279	5693	6133	6571	6997	7499	7919

Table 1. The Primes Below 12553 (7927–12553)

7927	8389	8837	9293	9739	10181	10663	11159	11677	12113
7933	8419	8839	9311	9743	10193	10667	11161	11681	12119
7937	8423	8849	9319	9749	10211	10687	11171	11689	12143
7949	8429	8861	9323	9767	10223	10691	11173	11699	12149
7951	8431	8863	9337	9769	10243	10709	11177	11701	12157
7963	8443	8867	9341	9781	10247	10711	11197	11717	12161
7993	8447	8887	9343	9787	10253	10723	11213	11719	12163
8009	8461	8893	9349	9791	10259	10729	11239	11731	12197
8011	8467	8923	9371	9803	10267	10733	11243	11743	12203
8017	8501	8929	9377	9811	10271	10739	11251	11777	12211
8039	8513	8933	9391	9817	10273	10753	11257	11779	12227
8053	8521	8941	9397	9829	10289	10771	11261	11783	12239
8059	8527	8951	9403	9833	10301	10781	11273	11789	12241
8069	8537	8963	9413	9839	10303	10789	11279	11801	12251
8081	8539	8969	9419	9851	10313	10799	11287	11807	12253
8087	8543	8971	9421	9857	10321	10831	11299	11813	12263
8089	8563	8999	9431	9859	10331	10837	11311	11821	12269
8093	8573	9001	9433	9871	10333	10847	11317	11827	12277
8101	8581	9007	9437	9883	10337	10853	11321	11831	12281
8111	8597	9011	9439	9887	10343	10859	11329	11833	12289
8117	8599	9013	9461	9901	10357	10861	11351	11839	12301
8123	8609	9029	9463	9907	10369	10867	11353	11863	12323
8147	8623	9041	9467	9923	10391	10883	11369	11867	12329
8161	8627	9043	9473	9929	10399	10889	11383	11887	12343
8167	8629	9049	9479	9931	10427	10891	11393	11897	12347
8171	8641	9059	9491	9941	10429	10903	11399	11903	12373
8179	8647	9067	9497	9949	10433	10909	11411	11909	12377
8191	8663	9091	9511	9967	10453	10937	11423	11923	12379
8209	8669	9103	9521	9973	10457	10939	11437	11927	12391
8219	8677	9109	9533	10007	10459	10949	11443	11933	12401
8221	8681	9127	9539	10009	10463	10957	11447	11939	12409
8231	8689	9133	9547	10037	10477	10973	11467	11941	12413
8233	8693	9137	9551	10039	10487	10979	11471	11953	12421
8237	8699	9151	9587	10061	10499	10987	11483	11959	12433
8243	8707	9157	9601	10067	10501	10993	11489	11969	12437
8263	8713	9161	9613	10069	10513	11003	11491	11971	12451
8269	8719	9173	9619	10079	10529	11027	11497	11981	12457
8273	8731	9181	9623	10091	10531	11047	11503	11987	12473
8287	8737	9187	9629	10093	10559	11057	11519	12007	12479
8291	8741	9199	9631	10099	10567	11059	11527	12011	12487
8293	8747	9203	9643	10103	10589	11069	11549	12037	12491
8297	8753	9209	9649	10111	10597	11071	11551	12041	12497
8311	8761	9221	9661	10133	10601	11083	11579	12043	12503
8317	8779	9227	9677	10139	10607	11087	11587	12049	12511
8329	8783	9239	9679	10141	10613	11093	11593	12071	12517
8353	8803	9241	9689	10151	10627	11113	11597	12073	12527
8363	8807	9257	9697	10159	10631	11117	11617	12097	12539
8369	8819	9277	9719	10163	10639	11119	11621	12101	12541
8377	8821	9281	9721	10169	10651	11131	11633	12107	12547
8387	8831	9283	9733	10177	10657	11149	11657	12109	12553

Table 2. The Primes Between 10^n and $10^n + 1000$								
10^5+	10^5+	10^5+	10^6+	10^6+	10^7+	10^7+	10^8+	10^8+
3	501	999	3	537	19	747	7	717
19	511		33	541	79	759	37	721
43	517		37	547	103	763	39	793
49	519		39	577	121	769	49	799
57	523		81	579	139	789	73	801
69	537		99	589	141	799	81	837
103	547		117	609	169	813	123	841
109	549		121	619	189	819	127	853
129	559		133	621	223	831	193	891
151	591		151	639	229	849	213	921
153	609		159	651	247	867	217	937
169	613		171	667	253	871	223	939
183	621		183	669	261	873	231	963
189	649		187	679	271	877	237	969
193	669		193	691	303	891	259	
207	673		199	697	339	931	267	
213	693		211	721	349	943	279	
237	699		213	723	357	961	357	
267	703		231	763	363	967	379	
271	733		249	777	379	987	393	
279	741		253	793	439	993	399	
291	747		273	829	451		421	
297	769		289	847	453		429	
313	787		291	849	457		463	
333	799		303	859	481		469	
343	801		313	861	511		471	
357	811		333	889	537		493	
361	823		357	907	583		541	
363	829		367	919	591		543	
379	847		381	921	609		561	
391	853		393	931	643		567	
393	907		397	969	651		577	
403	913		403	973	657		609	
411	927		409	981	667		627	
417	931		423	999	687		643	
447	937		427		691		651	
459	943		429		721		661	
469	957		453		723		669	
483	981		457		733		673	
493	987		507		741		687	

Table 2. The Primes Between 10^n and $10^n + 1000$									
10^9+	10^9+	$10^{10}+$	$10^{10}+$	$10^{11}+$	$10^{11}+$	$10^{12}+$	$10^{13}+$	$10^{14}+$	$10^{15}+$
7	829	19	963	3	801	39	37	31	37
9	861	33	991	19	817	61	51	67	91
21	871	61	993	57	819	63	99	97	159
33	891	69	999	63	861	91	129	99	187
87	901	97		69	901	121	183	133	223
93	919	103		73	943	163	259	139	241
97	931	121		91	951	169	267	169	249
103	933	141		103		177	273	183	259
123	993	147		129		189	279	261	273
181		207		171		193	283	357	279
207		259		183		211	313	367	297
223		277		193		271	343	403	357
241		279		211		303	391	423	399
271		319		223		331	411	469	403
289		343		237		333	433	487	487
297		391		253		339	453	493	513
321		403		283		459	591	541	613
349		469		319		471	609	601	711
363		501		363		537	643	643	741
403		537		367		543	649	657	783
409		583		379		547	657	709	811
411		589		393		561	687	721	843
427		597		403		609	691	753	873
433		601		411		661	717	777	921
439		631		417		669	729	807	
447		643		427		721	751	841	
453		649		447		751	759	843	
459		667		487		787	777	861	
483		679		519		789	853	963	
513		711		567		799	883	993	
531		723		579		841	943		
579		741		621		903	957		
607		753		631		921	987		
613		793		637		931	993		
637		799		669		933			
663		807		699		949			
711		877		703		997			
753		883		721					
787		889		739					
801		949		747					

x	$\pi(x)$	li $-\pi$	$R-\pi$	x	$\pi(x)$	li $-\pi$	$R-\pi$
10	4	2	1	10^5	9,592	38	—5
20	8	2	0	$2\cdot10^5$	17,984	52	—2
30	10	3	0	$3\cdot10^5$	25,997	90	26
40	12	4	1	$4\cdot10^5$	33,860	63	—8
50	15	3	0	$5\cdot10^5$	41,538	68	—8
60	17	4	0	$6\cdot10^5$	49,098	75	—7
70	19	4	0	$7\cdot10^5$	56,543	102	14
80	22	4	0	$8\cdot10^5$	63,951	86	—8
90	24	4	0	$9\cdot10^5$	71,274	88	—8
100	25	5	1	10^6	78,498	130	29
200	46	4	—1	$2\cdot10^6$	148,933	122	—9
300	62	6	0	$3\cdot10^6$	216,816	155	0
400	78	7	0	$4\cdot10^6$	283,146	206	33
500	95	7	—1	$5\cdot10^6$	348,513	125	—64
600	109	9	1	$6\cdot10^6$	412,849	228	24
700	125	8	0	$7\cdot10^6$	476,648	179	—38
800	139	9	1	$8\cdot10^6$	539,777	223	—6
900	154	9	0	$9\cdot10^6$	602,489	187	—53
10^3	168	10	0	10^7	664,579	339	88
$2\cdot10^3$	303	12	0	$2\cdot10^7$	1,270,607	298	—36
$3\cdot10^3$	430	13	0	$3\cdot10^7$	1,857,859	354	—41
$4\cdot10^3$	550	15	1	$4\cdot10^7$	2,433,654	362	—84
$5\cdot10^3$	669	15	0	$5\cdot10^7$	3,001,134	423	—67
$6\cdot10^3$	783	17	1	$6\cdot10^7$	3,562,115	568	39
$7\cdot10^3$	900	14	—3	$7\cdot10^7$	4,118,064	521	—44
$8\cdot10^3$	1,007	19	2	$8\cdot10^7$	4,669,382	709	111
$9\cdot10^3$	1,117	20	1	$9\cdot10^7$	5,216,954	856	228
10^4	1,229	17	—2	10^8	5,761,455	754	97
$2\cdot10^4$	2,262	27	2	$2\cdot10^8$	11,078,937	1,038	153
$3\cdot10^4$	3,245	32	4	$3\cdot10^8$	16,252,325	1,084	30
$4\cdot10^4$	4,203	30	—1	$4\cdot10^8$	21,336,326	1,052	—141
$5\cdot10^4$	5,133	34	0	$5\cdot10^8$	26,355,867	965	—350
$6\cdot10^4$	6,057	26	—9	$6\cdot10^8$	31,324,703	1,342	—81
$7\cdot10^4$	6,935	50	13	$7\cdot10^8$	36,252,931	1,311	—212
$8\cdot10^4$	7,837	39	0	$8\cdot10^8$	41,146,179	1,683	69
$9\cdot10^4$	8,713	44	3	$9\cdot10^8$	46,009,215	2,434	734

Table 3. Accuracy of the Approximations li x and $R(x)$

Table 3. Accuracy of the Approximations li x and $R(x)$			
x	$\pi(x)$	li $x - \pi(x)$	$R(x) - \pi(x)$
10^9	50,847,534	1,701	−79
$2 \cdot 10^9$	98,222,287	3,015	602
$3 \cdot 10^9$	144,449,537	3,192	305
$4 \cdot 10^9$	189,961,812	2,779	−500
$5 \cdot 10^9$	234,954,223	4,558	937
$6 \cdot 10^9$	279,545,368	3,499	−428
$7 \cdot 10^9$	323,804,352	4,068	−138
$8 \cdot 10^9$	367,783,654	5,242	778
$9 \cdot 10^9$	411,523,195	5,059	354
10^{10}	455,052,511	3,104	−1,828
$2 \cdot 10^{10}$	882,206,716	8,163	1,437
$3 \cdot 10^{10}$	1,300,005,926	10,206	2,134
$4 \cdot 10^{10}$	1,711,955,433	8,699	−490
$5 \cdot 10^{10}$	2,119,654,578	11,961	1,799
$6 \cdot 10^{10}$	2,524,038,155	10,163	−872
$7 \cdot 10^{10}$	2,925,699,539	10,280	−1,551
$8 \cdot 10^{10}$	3,325,059,246	12,346	−222
$9 \cdot 10^{10}$	3,722,428,991	15,271	2,013
10^{11}	4,118,054,813	11,588	−2,318
$2 \cdot 10^{11}$	8,007,105,059	13,345	−5,707
$3 \cdot 10^{11}$	11,818,439,135	23,767	851
$4 \cdot 10^{11}$	15,581,005,657	35,814	9,683
$5 \cdot 10^{11}$	19,308,136,142	35,586	6,652
$6 \cdot 10^{11}$	23,007,501,786	30,192	−1,258
$7 \cdot 10^{11}$	26,684,074,310	39,428	5,680
$8 \cdot 10^{11}$	30,341,383,527	34,506	−1,370
$9 \cdot 10^{11}$	33,981,987,586	45,948	8,083
10^{12}	37,607,912,018	38,263	−1,476
$2 \cdot 10^{12}$	73,301,896,139	48,195	−6,432
$3 \cdot 10^{12}$	108,340,298,703	80,927	15,096
$4 \cdot 10^{12}$	142,966,208,126	61,848	−13,314
$5 \cdot 10^{12}$	177,291,661,649	72,126	−11,182
$6 \cdot 10^{12}$	211,381,427,039	99,289	8,669
$7 \cdot 10^{12}$	245,277,688,804	110,790	13,484
$8 \cdot 10^{12}$	279,010,070,811	79,478	−24,020
$9 \cdot 10^{12}$	312,600,354,108	127,831	18,542

Table 3. Accuracy of the Approximations li x and $R(x)$			
x	$\pi(x)$	li $x - \pi(x)$	$R(x) - \pi(x)$
10^{13}	$346,065,536,839$	$108,971$	$-5,773$
$2 \cdot 10^{13}$	$675,895,909,271$	$170,356$	$12,194$
$3 \cdot 10^{13}$	$1,000,121,668,853$	$157,353$	$-33,533$
$4 \cdot 10^{13}$	$1,320,811,971,702$	$221,646$	$3,483$
$5 \cdot 10^{13}$	$1,638,923,764,567$	$253,033$	$11,037$
$6 \cdot 10^{13}$	$1,955,010,428,258$	$323,908$	$60,505$
$7 \cdot 10^{13}$	$2,269,432,871,304$	$197,552$	$-85,431$
$8 \cdot 10^{13}$	$2,582,444,113,487$	$327,644$	$26,520$
$9 \cdot 10^{13}$	$2,894,232,250,783$	$348,266$	$30,170$
10^{14}	$3,204,941,750,802$	$314,890$	$-19,200$
$2 \cdot 10^{14}$	$6,270,424,651,315$	$531,925$	$70,408$
$3 \cdot 10^{14}$	$9,287,441,600,280$	$434,926$	$-122,759$
$4 \cdot 10^{14}$	$12,273,824,155,491$	$567,492$	$-70,423$
$5 \cdot 10^{14}$	$15,237,833,654,620$	$812,601$	$104,541$
$6 \cdot 10^{14}$	$18,184,255,291,570$	$530,687$	$-240,406$
$7 \cdot 10^{14}$	$21,116,208,911,023$	$874,392$	$45,623$
$8 \cdot 10^{14}$	$24,035,890,368,161$	$1,084,477$	$202,253$
$9 \cdot 10^{14}$	$26,944,926,466,221$	$1,179,734$	$247,489$
10^{15}	$29,844,570,422,669$	$1,052,619$	$73,218$
10^{16}	$279,238,341,033,925$	$3,214,632$	$327,052$
$2 \cdot 10^{16}$	$547,863,431,950,008$	$3,776,488$	$-225,875$
$4 \cdot 10^{16}$	$1,075,292,778,753,150$	$5,538,861$	$-10,980$

See references [2], [8], [9], [10] and [12] to Chapter 1.

Prime Factors of Fermat Numbers

F_0 through F_4 are Primes.

F_5 through F_8 are completely factored, and their factorizations are

$F_5 = 641 \cdot 6700417$ (Found by Euler in 1732)

$F_6 = 274177 \cdot 67280421310721$ (Landry and Le Lasseur, 1880)

$F_7 = 59649589127497217 \cdot 5704689200685129054721$
(Morrison and Brillhart, 1974)

$F_8 = 1238926361552897 \cdot 93\,4616397153\,5797776916\,3558199606\,8965840512\text{-}$
$\text{-}3754163818\,8580280321$ (Brent and Pollard, 1980)

Table 4. Prime Factors $p = k \cdot 2^n + 1$ of Fermat Numbers $F_m = 2^{2^m} + 1$			
m	k	n	Factor found by
9	37	16	Western in 1903
10	11131	12	Selfridge in 1953
	395937	14	Brillhart in 1962
11	39	13	Cunningham in 1899
	119	13	Cunningham in 1899
12	7	14	Pervouchine and Lucas in 1877
	397	16	Western in 1903
	973	16	Western in 1903
	11613415	14	Hallyburton and Brillhart in 1974
13	41365885	16	Hallyburton and Brillhart in 1974
14		composite	Selfridge and Hurwitz in 1963
15	579	21	Kraïtchik in 1925
16	1575	19	Selfridge in 1953
17	59251857	19	Gostin in 1980
18	13	20	Western in 1903
19	33629	21	Riesel in 1962
	308385	21	Wrathall in 1963
21	534689	23	Wrathall in 1963
23	5	25	Pervouchine in 1878
25	48413	29	Wrathall in 1963
26	143165	29	Wrathall in 1963
27	141015	30	Wrathall in 1963
29	1120049	31	Gostin and McLaughlin in 1980

m	k	n	Factor found by
30	149041	32	Wrathall in 1963
	127589	33	Wrathall in 1963
32	1479	34	Wrathall in 1963
36	5	39	Seelhoff in 1886
	3759613	38	Gostin and McLaughlin in 1981
38	3	41	Cullen, Cunningham and Western in 1903
	2653	40	Wrathall in 1963
39	21	41	Robinson in 1956
42	43485	45	Wrathall in 1963
52	4119	54	Wrathall in 1963
	21626655	54	Keller in 1982
55	29	57	Robinson in 1956
58	95	61	Robinson in 1957
62	697	64	Shippee in 1977
63	9	67	Robinson in 1956
66	7551	69	Shippee in 1977
71	683	73	Shippee in 1977
73	5	75	Morehead in 1905
75	3447431	77	Gostin in 1983
77	425	79	Robinson and Selfridge in 1957
81	271	84	Robinson and Selfridge in 1957
91	1421	93	Shippee in 1977
93	92341	96	Baillie in 1979
99	16233	104	Gostin and McLaughlin in 1979
117	7	120	Robinson in 1956
125	5	127	Robinson in 1956
144	17	147	Robinson in 1956
147	3125	149	Gostin and McLaughlin in 1979
150	1575	157	Robinson in 1956
	5439	154	Gostin and McLaughlin in 1980
201	4845	204	Gostin and McLaughlin in 1980
205	232905	207	Keller in 1984
207	3	209	Robinson in 1956
215	32111	217	Suyama in 1980
226	15	229	Robinson in 1956
228	29	231	Robinson in 1956

Table 4. Prime Factors $p = k \cdot 2^n + 1$ of Fermat Numbers $F_m = 2^{2^m} + 1$

Table 4. Prime Factors $p = k \cdot 2^n + 1$ **of Fermat Numbers** $F_m = 2^{2^m} + 1$

m	k	n	Factor found by
250	403	252	Robinson and Selfridge in 1957
255	629	257	Baillie in 1979
267	177	271	Robinson and Selfridge in 1957
268	21	276	Robinson in 1956
275	22347	279	Keller in 1984
284	7	290	Robinson in 1956
287	5915	289	Suyama in 1980
298	247	302	Baillie in 1979
316	7	320	Robinson in 1956
329	1211	333	Suyama in 1980
334	27609	341	Keller in 1984
398	120845	401	Keller in 1984
416	8619	418	Suyama in 1981
	38039	419	Keller in 1984
452	27	455	Robinson in 1956
544	225	547	Baillie in 1979
556	127	558	Matthew and Williams in 1976
637	11969	643	Keller in 1984
692	717	695	Atkin and Rickert in 1979
744	17	747	Matthew and Williams in 1976
931	1985	933	Keller in 1980
1551	291	1553	Atkin and Rickert in 1979
1945	5	1947	Robinson in 1957
2023	29	2027	Atkin and Rickert, Williams in 1979
2089	431	2099	Suyama in 1981
2456	85	2458	Atkin and Rickert in 1979
3310	5	3313	Atkin and Rickert, Williams in 1979
4724	29	4727	Cormack and Williams in 1979
6537	17	6539	Cormack and Williams in 1979
6835	19	6838	Keller in 1978
9428	9	9431	Keller in 1984
9448	19	9450	Keller in 1980
23471	5	23473	Keller in 1984

Bibliography

1. Raphael M. Robinson, "A Report on Primes of the Form $k \cdot 2^n + 1$ and on Factors of Fermat Numbers," *Proc. Am. Math. Soc.* **9** (1958) pp. 673–681.

2. C. P. Wrathall, "New Factors of Fermat Numbers," *Math. Comp.* **18** (1964) pp. 324–325.

3. Gary B. Gostin, "A Factor of F_{17}," *Math. Comp.* **35** (1980) pp. 975–976.

4. G. V. Cormack and H. C. Williams, "Some Very Large Primes of the Form $k \cdot 2^m + 1$," *Math. Comp.* **35** (1980) pp. 1419–1421.

5. Gary B. Gostin and Philip B. McLaughlin, Jr, "Six New Factors of Fermat Numbers," *Math. Comp.* **38** (1982) pp. 645–649.

6. Wilfrid Keller, "Factors of Fermat Numbers and Large Primes of the Form $k \cdot 2^n + 1$," *Math. Comp.* **41** (1983) pp. 661–673.

Table 5. Primes of the Form $h \cdot 2^n + 1$	
h	n, search limit within parentheses
1	0, 1, 2, 4, 8, 16 (1048575)
3	1, 2, 5, 6, 8, 12, 18, 30, 36, 41, 66, 189, 201, 209, 276, 353, 408, 438, 534, 2208, 2816, 3168, 3189, 3912 (14000)
5	1, 3, 7, 13, 15, 25, 39, 55, 75, 85, 127, 1947, 3313, 4687, 5947, 13165 (18000) 23473
7	2, 4, 6, 14, 20, 26, 50, 52, 92, 120, 174, 180, 190, 290, 320, 390, 432, 616, 830, 1804, 2256, 6614 (12000)
9	1, 2, 3, 6, 7, 11, 14, 17, 33, 42, 43, 63, 65, 67, 81, 134, 162, 206, 211, 366, 663, 782, 1305, 1411, 1494, 2297, 2826, 3230, 3354, 3417, 3690, 4842, 5802, 6953, 7967 (8500) 9431
11	1, 3, 5, 7, 19, 21, 43, 81, 125, 127, 209, 211, 3225, 4543, 10179 (12000)
13	2, 8, 10, 20, 28, 82, 188, 308, 316, 1000 (17000)
15	1, 2, 4, 9, 10, 12, 27, 37, 38, 44, 48, 78, 112, 168, 229, 297, 339, 517, 522, 654, 900, 1518, 2808, 2875, 3128, 3888, 4410, 6804, 7050, 7392 (8500)
17	3, 15, 27, 51, 147, 243, 267, 347, 471, 747, 2163, 3087, 5355, 6539, 7311, (18000)
19	6, 10, 46, 366, 1246, 2038, 4386, 4438, 6838, 7498, 7998, 9450, 11890 (12000)
21	1, 4, 5, 7, 9, 12, 16, 17, 41, 124, 128, 129, 187, 209, 276, 313, 397, 899, 1532, 1613, 1969, 2245, 2733 (4000)
23	1, 9, 13, 29, 41, 49, 69, 73, 341, 381, 389, 649, 1961, 3929 (4000)
25	2, 4, 6, 10, 20, 22, 52, 64, 78, 184, 232, 268, 340, 448, 554, 664, 740, 748, 1280, 1328, 1640, 3314, 3904, 3938 (4000)
27	2, 4, 7, 16, 19, 20, 22, 26, 40, 44, 46, 47, 50, 56, 59, 64, 92, 175, 215, 275, 407, 455, 1076, 1090, 3080, 3322, 6419, 7639 (8000)
29	1, 3, 5, 11, 27, 43, 57, 75, 77, 93, 103, 143, 185, 231, 245, 391, 1053, 1175, 2027, 3627, 4727, 5443, 7927 (8000)
31	8, 60, 68, 140, 216, 416, 1808, 1944 (4000)
33	1, 6, 13, 18, 21, 22, 25, 28, 66, 93, 118, 289, 412, 453, 525, 726, 828, 1420, 1630, 3076, 3118 (4000)
35	1, 3, 7, 9, 13, 15, 31, 45, 47, 49, 55, 147, 245, 327, 355, 663, 1423, 1443, 2493, 3627 (4000)
37	2, 4, 8, 10, 12, 16, 22, 26, 68, 82, 84, 106, 110, 166, 236, 254, 286, 290, 712, 1240, 1706, 1804, 1904, 2240, 2632, 3104 (4000)
39	1, 2, 3, 5, 7, 10, 11, 13, 14, 18, 21, 22, 31, 42, 67, 70, 71, 73, 251, 370, 375, 389, 407, 518, 818, 865, 1057, 1602, 2211, 3049 (4000)
41	1, 11, 19, 215, 289, 379, 1991 (4000)
43	2, 6, 12, 18, 26, 32, 94, 98, 104, 144, 158, 252, 778, 1076, 2974, 3022, 3528, (4000)
45	2, 9, 12, 14, 23, 24, 29, 60, 189, 200, 333, 372, 443, 464, 801, 1374, 6146, 6284, 6359, 6923 (7000)
47	583, 1483, 6115 (15000)
49	2, 6, 10, 30, 42, 54, 66, 118, 390, 594, 1202, 2334 (4000)

381

Table 5. Primes of the Form $h \cdot 2^n + 1$

h	n, search limit within parentheses
51	1, 3, 7, 9, 13, 17, 25, 43, 53, 83, 89, 92, 119, 175, 187, 257, 263, 267, 321, 333, 695, 825, 1485, 1917, 2660, 2967, 3447, 3659 (4000)
53	1, 5, 17, 21, 61, 85, 93, 105, 133, 485, 857, 1665, 2133, 2765 (4000)
55	4, 8, 16, 22, 32, 94, 220, 244, 262, 286, 344, 356, 392, 1996, 2744 (4000)
57	2, 3, 7, 8, 10, 16, 18, 19, 40, 48, 55, 90, 96, 98, 190, 398, 456, 502, 719, 1312, 1399, 1828 (4000)
59	5, 11, 27, 35, 291, 1085, 2685 (4000)
61	4, 12, 48, 88, 168, 3328 (4000)
63	1, 4, 5, 9, 10, 14, 17, 18, 21, 25, 37, 38, 44, 60, 65, 94, 133, 153, 228, 280, 314, 326, 334, 340, 410, 429, 626, 693, 741, 768, 1150, 1290, 1441, 2424 2478, 3024, 3293 (4000)
65	1, 3, 5, 11, 17, 21, 29, 47, 85, 93, 129, 151, 205, 239, 257, 271, 307, 351, 397, 479, 553, 1317, 1631, 1737, 1859, 1917, 1999, 2353, 3477 (4000)
67	2, 6, 14, 20, 44, 66, 74, 102, 134, 214, 236, 238, 342, 354, 382, 454, 470, 598, 726, 870, 1148, 1366, 1692, 1782, 1870, 3602 (4000)
69	1, 2, 10, 14, 19, 26, 50, 55, 145, 515, 842, 1450, 2159, 2290, 2306, 2335, 3379 (4000)
71	3, 5, 9, 19, 23, 27, 57, 59, 65, 119, 299, 417, 705, 2255 (4000)
73	2, 6, 14, 24, 30, 32, 42, 44, 60, 110, 212, 230, 1892, 1974, 2210, 3596 (4000)
75	1, 3, 4, 6, 7, 10, 12, 34, 43, 51, 57, 60, 63, 67, 87, 102, 163, 222, 247, 312, 397, 430, 675, 831, 984, 1018, 1054, 1615, 2017, 2157 (4000)
77	3, 7, 19, 23, 95, 287, 483, 559, 655, 667, 1639 (4000)
79	2, 10, 46, 206, 538, 970, 1330, 1766, 2162 (4000)
81	1, 4, 5, 7, 12, 15, 16, 21, 25, 27, 32, 35, 36, 39, 48, 89, 104, 121, 125, 148, 152, 267, 271, 277, 296, 324, 344, 396, 421, 436, 447, 539, 577, 592, 711, 809, 852, 1384, 1972, 2624, 2829, 3497, 3945, 3995 (4000)
83	1, 5, 157, 181, 233, 373, 2425, 2773, 3253 (4000)
85	4, 6, 10, 30, 34, 36, 38, 74, 88, 94, 148, 200, 624, 1300, 2458, 2556, 3638, 3834 (4000)
87	2, 6, 8, 18, 26, 56, 78, 86, 104, 134, 207, 518, 602, 1268, 1302, 2324, 2372 (4000)
89	1, 7, 9, 21, 37, 61, 589, 711, 1537, 1921, 3217 (4000)
91	8, 168, 260, 696, 5028, 5536 (6000)
93	2, 4, 6, 10, 12, 30, 42, 44, 52, 70, 76, 108, 122, 164, 170, 226, 298, 398, 686, 1020, 1110, 1478, 1646, 2032, 2066, 2800, 2816 (4000)
95	1, 3, 5, 7, 13, 17, 21, 53, 57, 61, 83, 89, 111, 167, 175, 237, 533, 621, 661, 753, 993, 1039, 1849, 1987, 3437 (4000)
97	2, 4, 14, 20, 40, 266, 400, 652, 722, 2026, 2732, 3880 (4000)
99	1, 2, 5, 6, 10, 11, 22, 31, 33, 34, 41, 42, 53, 58, 65, 82, 126, 143, 162, 170, 186, 189, 206, 211, 270, 319, 369, 410, 433, 631, 894, 1617, 2025 (4000)

Table 5. Primes of the Form $h \cdot 2^n + 1$	
h	n, search limit within parentheses
101	3, 9, 17, 21, 27, 39, 45, 47, 71, 95, 117, 123, 143, 173, 387, 389, 513, 633, 827, 971, 1103, 3767, 3831 (4000)
103	16, 18, 30, 40, 58, 138, 250, 616, 622, 736, 3670 (4000)
105	1, 2, 5, 7, 8, 12, 14, 23, 27, 33, 38, 49, 61, 62, 85, 93, 94, 107, 155, 182, 215, 273, 382, 392, 413, 434, 490, 1631, 3063, 3331, 3461, 3619 (4000)
107	3, 7, 23, 27, 291, 303, 311, 479, 567, 3087 (4000)
109	6, 14, 58, 62, 318, 1574, 2034 (4000)
111	1, 4, 28, 32, 44, 47, 71, 128, 137, 193, 676, 2344 (4000) 10883
113	1, 5, 13, 33, 145, 365, 409, 509, 553, 673, 733, 961, 1045, 1541, 2473, 3461 (4000)
115	2, 12, 20, 26, 42, 114, 228, 396, 456, 482, 1298, 3048 (4000)
117	3, 4, 6, 10, 16, 30, 36, 91, 94, 156, 382, 454, 643, 867, 1416, 2656, 2851 (4000)
119	1, 3, 7, 13, 21, 23, 45, 63, 553, 1115, 2471, 2773 (4000)
121	8, 12, 44, 84, 96, 228, 264, 320, 732, 788, 1808 (4000)
123	6, 8, 17, 21, 29, 32, 46, 57, 69, 128, 141, 268, 333, 476, 742, 832, 1173, 1677 (4000)
125	1, 5, 7, 17, 25, 35, 67, 281, 331, 491, 581, 941, 1205, 1279, 1411, 1631, 1895, 2735, 3475 (4000)
127	2, 12, 18, 24, 54, 72, 114, 180, 214, 504, 558, 964, 1098, 1420, 2764 (4000)
129	3, 5, 21, 27, 59, 75, 111, 287, 414, 786, 966, 1071, 2433, 2817, 3165 (4000)
131	1, 3, 9, 13, 19, 21, 25, 51, 55, 97, 153, 165, 199, 261, 285, 361, 373, 465, 475, 529, 765, 2065, 3553 (4000)
133	4, 6, 10, 16, 30, 124, 174, 192, 336, 600, 720, 1092, 1138, 1588, 1652, 1812, 3012, 3308 (4000)
135	1, 2, 4, 6, 10, 15, 18, 20, 30, 31, 35, 38, 39, 51, 85, 90, 106, 108, 202, 238, 253, 282, 330, 361, 452, 459, 646, 895, 922, 1201, 1402, 1441, 1462, 1523, 1611, 1770, 1923, 2053, 2099, 2242, 2796 (4000)
137	3, 27, 39, 83, 203, 395, 467, 875, 1979 (4000)
139	2, 14, 914, 12614 (13000)
141	1, 3, 5, 7, 8, 12, 15, 20, 31, 33, 37, 41, 61, 65, 91, 93, 103, 117, 133, 137, 141, 160, 291, 303, 343, 488, 535, 555, 556, 640, 756, 897, 917, 1745, 1805, 2053, 2372, 3375 (4000)
143	53, 77, 293, 333, 393, 809, 825 (4000)
145	6, 16, 28, 70, 76, 250, 276, 312, 562, 636, 1366, 1552, 1968 (4000)
147	8, 11, 15, 18, 19, 26, 44, 60, 84, 90, 91, 134, 155, 179, 258, 275, 475, 620, 824, 888, 1731, 2194, 2328, 2568, 2915, 3554 (4000)
149	3, 7, 9, 15, 17, 27, 33, 35, 57, 125, 127, 137, 191, 513, 819, 827, 921, 931, 1047, 1147, 1599, 1815, 2499, 2995 (4000)

Table 6. Primes of the Form $h \cdot 2^n - 1$

h	n, search limit within parentheses
1	2, 3, 5, 7, 13, 17, 19, 31, 61, 89, 107, 127, 521, 607, 1279, 2203, 2281, 3217, 4253, 4423, 9689, 9941, 11213, 19937, 21701, 23209, 44497, 86243 (100000) 132049
3	1, 2, 3, 4, 6, 7, 11, 18, 34, 38, 43, 55, 64, 76, 94, 103, 143, 206, 216, 306, 324, 391, 458, 470, 827, 1274, 3276, 4204, 5134, 7559 (10000) 12676
5	2, 4, 8, 10, 12, 14, 18, 32, 48, 54, 72, 148, 184, 248, 270, 274, 420 (1000)
7	1, 5, 9, 17, 21, 29, 45, 177 (1000)
9	1, 3, 7, 13, 15, 21, 43, 63, 99, 109, 159, 211, 309, 343, 415, 469, 781, 871, 939, 1551, 3115, 3349, 5589, 5815, 5893, 7939, 8007 (10000) 11547
11	2, 26, 50, 54, 126, 134, 246, 354, 362, 950 (1000)
13	3, 7, 23, 287, 291, 795 (1000)
15	1, 2, 4, 5, 10, 14, 17, 31, 41, 73, 80, 82, 116, 125, 145, 157, 172, 202, 224, 266, 289, 293, 463 (1000)
17	2, 4, 6, 16, 20, 36, 54, 60, 96, 124, 150, 252, 356, 460, 612, 654, 664, 698, 702, 972 (1000)
19	1, 3, 5, 21, 41, 49, 89, 133, 141, 165, 189, 293, 305, 395, 651, 665, 771, 801, 923, 953 (1000)
21	1, 2, 3, 7, 10, 13, 18, 27, 37, 51, 74, 157, 271, 458, 530, 891 (1000)
23	4, 6, 12, 46, 72, 244, 264, 544, 888 (1000)
25	3, 9, 11, 17, 23, 35, 39, 75, 105, 107, 155, 215, 335, 635, 651, 687 (1000)
27	1, 2, 4, 5, 8, 10, 14, 28, 37, 38, 70, 121, 122, 160, 170, 253, 329, 362, 454, 485, 500, 574, 892, 962 (1000)
29	4, 16, 76, 148, 184 (1000)
31	1, 5, 7, 11, 13, 23, 33, 35, 37, 47, 115, 205, 235, 271, 409, 739, 837, 887 (1000)
33	2, 3, 6, 8, 10, 22, 35, 42, 43, 46, 56, 91, 102, 106, 142, 190, 208, 266, 330, 360, 382, 462, 503, 815 (1000)
35	2, 6, 10, 20, 44, 114, 146, 156, 174, 260, 306, 380, 654, 686, 702, 814, 906 (1000)
37	1 (1000)
39	3, 24, 105, 153, 188, 605, 795, 813, 839 (1000)
41	2, 10, 14, 18, 50, 114, 122, 294, 362, 554, 582, 638, 758 (1000)
43	7, 31, 67, 251, 767 (1000)
45	1, 2, 3, 4, 5, 6, 8, 9, 14, 15, 16, 22, 28, 29, 36, 37, 54, 59, 85, 93, 117, 119, 161, 189, 193, 256, 308, 322, 327, 411, 466, 577, 591, 902, 928, 946 (1000)
47	4, 14, 70, 78 (1000)
49	1, 5, 7, 9, 13, 15, 29, 33, 39, 55, 81, 95, 205, 279, 581, 807, 813 (1000)

Table 6. Primes of the Form $h \cdot 2^n - 1$	
h	n, search limit within parentheses
51	1, 9, 10, 19, 22, 57, 69, 97, 141, 169, 171, 195, 238, 735, 885 (1000)
53	2, 6, 8, 42, 50, 62, 362, 488, 642, 846 (1000)
55	1, 3, 5, 7, 15, 33, 41, 57, 69, 75, 77, 131, 133, 153, 247, 305, 351, 409, 471 (1000)
57	1, 2, 4, 5, 8, 10, 20, 22, 25, 26, 32, 44, 62, 77, 158, 317, 500, 713 (1000)
59	12, 16, 72, 160, 256, 916 (1000)
61	3, 5, 9, 13, 17, 19, 25, 39, 63, 67, 75, 119, 147, 225, 419, 715, 895 (1000)
63	2, 3, 8, 11, 14, 16, 28, 32, 39, 66, 68, 91, 98, 116, 126, 164, 191, 298, 323, 443, 714, 758, 759 (1000)
65	4, 6, 12, 22, 28, 52, 78, 94, 124, 162, 174, 192, 204, 304, 376, 808, 930, 972 (1000)
67	5, 9, 21, 45, 65, 77, 273, 677 (1000)
69	1, 4, 5, 7, 9, 11, 13, 17, 19, 23, 29, 37, 49, 61, 79, 99, 121, 133, 141, 164, 173, 181, 185, 193, 233, 299, 313, 351, 377, 540, 569, 909 (1000)
71	2, 14, 410 (1000)
73	7, 11, 19, 71, 79, 131 (1000)
75	1, 3, 5, 6, 18, 19, 20, 22, 28, 29, 39, 43, 49, 75, 85, 92, 111, 126, 136, 159, 162, 237, 349, 381, 767, 969 (1000)
77	2, 4, 14, 26, 58, 60, 64, 100, 122, 212, 566, 638 (1000)
79	1, 3, 7, 15, 43, 57, 61, 75, 145, 217, 247 (1000)
81	3, 5, 11, 17, 21, 27, 81, 101, 107, 327, 383, 387, 941 (1000)
83	2, 4, 8, 10, 14, 18, 22, 24, 26, 28, 36, 42, 58, 64, 78, 158, 198, 206, 424, 550, 676, 904 (1000)
85	5, 11, 71, 113, 115, 355, 473, 563, 883 (1000)
87	1, 2, 8, 9, 10, 12, 22, 29, 32, 50, 57, 69, 81, 122, 138, 200, 296, 514, 656, 682, 778, 881 (1000)
89	4, 8, 12, 24, 48, 52, 64, 84, 96 (1000)
91	1, 3, 9, 13, 15, 17, 19, 23, 47, 57, 67, 73, 77, 81, 83, 191, 301, 321, 435, 867, 869, 917 (1000)
93	3, 4, 7, 10, 15, 18, 19, 24, 27, 39, 60, 84, 111, 171, 192, 222, 639, 954 (1000)
95	2, 6, 26, 32, 66, 128, 170, 288, 320, 470 (1000)
97	1, 9, 45, 177, 585 (1000)
99	1, 4, 5, 7, 8, 11, 19, 25, 28, 35, 65, 79, 212, 271, 361, 461 (1000)

\quad	Table 6. Primes of the Form $h \cdot 2^n - 1$
h	n, search limit within parentheses
101	10, 18, 54, 70 (1000)
103	3, 7, 11, 19, 63, 75, 95, 127, 155, 163, 171, 283, 563 (1000)
105	2, 3, 5, 6, 8, 9, 25, 32, 65, 113, 119, 155, 177, 299, 335, 426, 462, 617, 896 (1000)
107	10, 12, 18, 24, 28, 40, 90, 132, 214, 238, 322, 532, 858, 940 (1000)
109	9, 149, 177, 419, 617 (1000)
113	8, 14, 74, 80, 274, 334, 590, 608, 614, 650 (1000)
115	1, 3, 11, 13, 19, 21, 31, 49, 59, 69, 73, 115, 129, 397, 623, 769 (1000)
119	12, 16, 52, 160, 192, 216, 376, 436 (1000)
121	1, 3, 21, 27, 37, 43, 91, 117, 141, 163, 373, 421 (1000)
125	2, 4, 44, 182, 496, 904 (1000)
127	25, 113 (1000)
131	2, 14, 34, 38, 42, 78, 90, 178, 778, 974 (1000)
133	3, 11, 15, 19, 31, 59, 75, 103, 163, 235, 375, 615, 767 (1000)
137	2, 18, 38, 62 (1000)
139	1, 5, 7, 9, 15, 19, 21, 35, 37, 39, 41, 49, 69, 111, 115, 141, 159, 181, 201, 217, 487, 567, 677, 765, 811, 841, 917 (1000)
143	2, 4, 6, 8, 12, 18, 26, 32, 34, 36, 42, 60, 78, 82, 84, 88, 154, 174, 208, 256, 366, 448, 478, 746 (1000)
145	5, 13, 15, 31, 77, 151, 181, 245, 445, 447, 883 (1000)
149	4, 16, 48, 60, 240, 256, 304 (1000)
151	5, 221, 641 (1000)
155	2, 8, 14, 16, 44, 46, 82, 172, 196, 254, 556, 806 (1000)
157	1, 5, 33, 121, 125, 305, 445, 473, 513 (1000)
161	2, 6, 18, 22, 34, 54, 98, 122, 146, 222, 306, 422, 654, 682, 862 (1000)
163	3, 31, 63, 303 (1000)
167	4, 6, 8, 10, 16, 32, 38, 42, 52, 456, 576, 668 (1000)
169	1, 5, 11, 17, 67, 137, 157, 203, 209, 227, 263, 917 (1000)
173	2, 4, 6, 16, 32, 50, 76, 80, 96, 104, 162, 212, 230, 260, 480, 612 (1000)
175	1, 3, 9, 21, 23, 41, 47, 57, 69, 83, 193, 249, 291, 421, 433, 997 (1000)
179	8, 68, 108 (1000)
181	3, 5, 7, 9, 11, 17, 23, 31, 35, 43, 47, 83, 85, 99, 101, 195, 267, 281, 363, 391, 519, 623, 653, 673, 701 (1000)
185	2, 6, 10, 18, 26, 40, 46, 78, 230, 542 (1000)
187	1, 17, 21, 53, 253 (1000)

Table 6. Primes of the Form $h \cdot 2^n - 1$	
h	n, search limit within parentheses
191	226 (1000)
193	3, 15, 27, 63, 97, 135, 543 (1000)
197	2, 16, 20, 22, 40, 82, 112, 178, 230, 302, 304, 328, 374, 442, 472, 500, 580, 694 (1000)
199	1, 5, 7, 15, 19, 23, 25, 27, 43, 65, 99, 125, 141, 165, 201, 221, 331, 369, 389, 445, 461, 463, 467, 513, 583, 835 (1000)

Bibliography

1. Raphael M. Robinson, "A Report on Primes of the Form $k \cdot 2^n + 1$, and on Factors of Fermat numbers," *Proc. Am. Math. Soc.* **9** (1958) pp. 673–681.

2. G. Matthew and H. C. Williams, "Some New Primes of the Form $k \cdot 2^n + 1$," *Math. Comp.* **31** (1977) pp. 797–798.

3. R. Baillie, "New Primes of the Form $k \cdot 2^n + 1$," *Math. Comp.* **33** (1979) pp. 1333–1336.

4. G. V. Cormack and H. C. Williams, "Some Very Large Primes of the Form $k \cdot 2^m + 1$," *Math. Comp.* **34** (1980) pp. 1419–1421.

5. Wilfrid Keller, "Factors of Fermat Numbers and Large Primes of the Form $k \cdot 2^n + 1$," *Math. Comp.* **41** (1983) pp. 661–673.

6. Hans Riesel, "A Note on Prime Numbers of the Forms $N = (6a+1)2^{2n-1} - 1$ and $M = (6a - 1)2^{2n} - 1$", *Ark. för Mat.* **3** (1955) pp. 245–253.

7. H. C. Williams and C. R. Zarnke, A Report on Prime Numbers of the Forms $M = (6a+1)2^{2m-1} - 1$ and $M' = (6a-1)2^{2m} - 1$," *Math. Comp.* **22** (1968) pp. 420–422.

8. Hans Riesel, "Lucasian Criteria For the Primality of $N = h \cdot 2^n - 1$, *Math. Comp.* **23** (1969) pp. 869–875.

9. Ingemar Jönsson, "On Certain Primes of Mersenne-Type," *Nordisk Tidskrift för Informationsbehandling (BIT)* **12** (1972) pp. 117–118.

10. W. Borho, "Some Large Primes and Amicable Numbers," *Math. Comp.* **36** (1981) pp. 303–304.

Table 7. Factors of Mersenne numbers $M_n = 2^n - 1$	
3	7
5	31
7	127
9	$7 \cdot 73$
11	$23 \cdot 89$
13	8191
15	$7 \cdot 31 \cdot 151$
17	131071
19	524287
21	$7^2 \cdot 127 \cdot 337$
23	$47 \cdot 178481$
25	$31 \cdot 601 \cdot 1801$
27	$7 \cdot 73 \cdot 262657$
29	$233 \cdot 1103 \cdot 2089$
31	2147483647
33	$7 \cdot 23 \cdot 89 \cdot 599479$
35	$31 \cdot 71 \cdot 127 \cdot 122921$
37	$223 \cdot 616318177$
39	$7 \cdot 79 \cdot 8191 \cdot 121369$
41	$13367 \cdot 164511353$
43	$431 \cdot 9719 \cdot 2099863$
45	$7 \cdot 31 \cdot 73 \cdot 151 \cdot 631 \cdot 23311$
47	$2351 \cdot 4513 \cdot 13264529$
49	$127 \cdot 4432676798593$
51	$7 \cdot 103 \cdot 2143 \cdot 11119 \cdot 131071$
53	$6361 \cdot 69431 \cdot 20394401$
55	$23 \cdot 31 \cdot 89 \cdot 881 \cdot 3191 \cdot 201961$
57	$7 \cdot 32377 \cdot 524287 \cdot 1212847$
59	$179951 \cdot 3203431780337$
61	2305843009213693951
63	$7^2 \cdot 73 \cdot 127 \cdot 337 \cdot 92737 \cdot 649657$
65	$31 \cdot 8191 \cdot 145295143558111$
67	$193707721 \cdot 761838257287$
69	$7 \cdot 47 \cdot 178481 \cdot 10052678938039$
71	$228479 \cdot 48544121 \cdot 212885833$
73	$439 \cdot 2298041 \cdot 9361973132609$
75	$7 \cdot 31 \cdot 151 \cdot 601 \cdot 1801 \cdot 100801 \cdot 10567201$
77	$23 \cdot 89 \cdot 127 \cdot 581283643249112959$
79	$2687 \cdot 202029703 \cdot 1113491139767$

	Table 7. Factors of Mersenne numbers $M_n = 2^n - 1$
81	$7 \cdot 73 \cdot 2593 \cdot 71119 \cdot 262657 \cdot 97685839$
83	$167 \cdot 57912614113275649087721$
85	$31 \cdot 131071 \cdot 9520972806333758431$
87	$7 \cdot 233 \cdot 1103 \cdot 2089 \cdot 4177 \cdot 9857737155463$
89	$618970019642690137449562111$
91	$127 \cdot 911 \cdot 8191 \cdot 112901153 \cdot 23140471537$
93	$7 \cdot 2147483647 \cdot 658812288653553079$
95	$31 \cdot 191 \cdot 524287 \cdot 420778751 \cdot 30327152671$
97	$11447 \cdot 13842607235828485645766393$
99	$7 \cdot 23 \cdot 73 \cdot 89 \cdot 199 \cdot 153649 \cdot 599479 \cdot 33057806959$
101	$7432339208719 \cdot 341117531003194129$
103	$2550183799 \cdot 3976656429941438590393$
105	$M_{35} \cdot 7^2 \cdot 151 \cdot 337 \cdot 29191 \cdot 106681 \cdot 152041$
107	$162259276829213363391578010288127$
109	$745988807 \cdot 870035986098720987332873$
111	$M_{37} \cdot 7 \cdot 321679 \cdot 26295457 \cdot 319020217$
113	$3391 \cdot 23279 \cdot 65993 \cdot 1868569 \cdot 1066818132868207$
115	$31 \cdot 47 \cdot 14951 \cdot 178481 \cdot 4036961 \cdot 2646507710984041$
117	$M_{39} \cdot 73 \cdot 937 \cdot 6553 \cdot 86113 \cdot 7830118297$
119	$127 \cdot 239 \cdot 20231 \cdot 131071 \cdot 62983048367 \cdot 131105292137$
121	$23 \cdot 89 \cdot 727 \cdot 1786393878363164227858270210279$
123	$M_{41} \cdot 7 \cdot 3887047 \cdot 177722253954175633$
125	$31 \cdot 601 \cdot 1801 \cdot 269089806001 \cdot 4710883168879506001$
127	$170141183460469231731687303715884105727$
129	$M_{43} \cdot 7 \cdot 1105303606504929475345 9639$
131	$263 \cdot 10350794431055162386718619237468234569$
133	$127 \cdot 524287 \cdot 163537220852725398851434325720959$
135	$M_{45} \cdot 271 \cdot 262657 \cdot 348031 \cdot 49971617830801$
137	$32032215596496435569 \cdot 5439042183600204290159$
139	$5625767248687 \cdot 123876132205208335762278423601$
141	$M_{47} \cdot 7 \cdot 4375578271 \cdot 646675035253258729$
143	$23 \cdot 89 \cdot 8191 \cdot 724153 \cdot 158822951431 \cdot 5782172113400990737$
145	$31 \cdot 233 \cdot 1103 \cdot 2089 \cdot 2679895157783862814690027494144991$
147	$M_{49} \cdot 7^3 \cdot 337 \cdot 27416723625287255350 68727$
149	$86656268566282183151 \cdot 8235109336690846723986161$
151	$18121 \cdot 55871 \cdot 165799 \cdot 2332951 \cdot 728908838338825366 4437433$
153	$M_{51} \cdot 73 \cdot 919 \cdot 7558248842417934708343 8319$
155	$M_{31} \cdot 31^2 \cdot 311 \cdot 11471 \cdot 73471 \cdot 4649919401 \cdot 18158209813151$
157	$852133201 \cdot 60726444167 \cdot 1654058017289 \cdot 2134387368610417$
159	$M_{53} \cdot 7 \cdot 6679 \cdot 13960201 \cdot 540701761 \cdot 229890275929$

	Table 7. Factors of Mersenne numbers $M_n = 2^n - 1$
161	$47 \cdot 127 \cdot 1289 \cdot 178481 \cdot 3188767 \cdot 45076044553 \cdot 14808607715315782481$
163	$150287 \cdot 704161 \cdot 110211473 \cdot 27669118297 \cdot 36230454570129675721$
165	$M_{55} \cdot 7 \cdot 151 \cdot 599479 \cdot 2048568835297380486760231$
167	$2349023 \cdot 79638304766856507377778616296087448490695649$
169	$4057 \cdot 8191 \cdot 6740339310641 \cdot 3340762283952395329506327023033$
171	$M_{57} \cdot 73 \cdot 93507247 \cdot 30426456347925413012037847$
173	$730753 \cdot 1505447 \cdot 70084436712553223 \cdot 15528574328857227767 9887$
175	$M_{35} \cdot 601 \cdot 1801 \cdot 39551 \cdot 60816001 \cdot 535347624791488552837151$
177	$M_{59} \cdot 7 \cdot 184081 \cdot 27989941729 \cdot 9213624084535989031$
179	$359 \cdot 1433 \cdot 1489459109360039866456940197095433721664951999121$
181	$43441 \cdot 1164193 \cdot 7648337 \cdot 7923871097285295625344647665764672671$
183	$M_{61} \cdot 7 \cdot 367 \cdot 55633 \cdot 3720170862530514630397335 2041$
185	$M_{37} \cdot 31 \cdot 1587855697992791 \cdot 7248808599285760001152755641$
187	$23 \cdot 89 \cdot 131071 \cdot 707983 \cdot 103267081674384386099885005627 8950666491537$
189	$M_{63} \cdot 262657 \cdot 1560007 \cdot 2076174855442583929 70753527$
191	$383 \cdot 7068569257 \cdot 39940132241 \cdot 332584516519201 \cdot 87274497124602996457$
193	$13821503 \cdot 6165444023324834061 6559 \cdot 14732265321145317331353282383$
195	$M_{65} \cdot 7 \cdot 79 \cdot 151 \cdot 121369 \cdot 13430419684509926257 2814573351$
197	$7487 \cdot 26828803997912886929710867041891989490486893845712448833$
199	$164504919713 \cdot 4884164093883941177660049098586324302977543600799$
201	$M_{67} \cdot 7 \cdot 1609 \cdot 22111 \cdot 874494233974258579426788 33145441$
203	$M_{29} \cdot 127 \cdot 136417 \cdot 121793911 \cdot 113480555808832720110908560 53175361113$
205	$M_{41} \cdot 31 \cdot 2940521 \cdot 70171342151 \cdot 36557250655087971816740789 59681$
207	$M_{69} \cdot 73 \cdot 79903 \cdot 634569679 \cdot 2232578641663 \cdot 42166482463639$
209	$23 \cdot 89 \cdot 524287 \cdot 94803416684681 \cdot 1512348937147247 \cdot$ $\cdot 534695054132396023231 9657$
211	$15193 \cdot 6027295643383884916 1 \cdot$ $\cdot 359387570449582375738819989 4268773153439$
213	$M_{71} \cdot 7 \cdot 66457 \cdot 2849881972114740679 \cdot 4205268574191396793$
215	$M_{43} \cdot 31 \cdot 1721 \cdot 731516431 \cdot 514851898711 \cdot 29792728974047764444862191$
217	$M_{31} \cdot 127 \cdot 5209 \cdot 62497 \cdot 6268703933840364033151 \cdot$ $\cdot 3784288044314244840 82633$
219	$M_{73} \cdot 7 \cdot 3943 \cdot 671165898617413417 \cdot 4815314615204347717321$
221	$1327 \cdot 8191 \cdot 131071 \cdot$ $\cdot 236545439841839977260508620921436345855283986624706923 3$
223	$18287 \cdot 196687 \cdot 1466449 \cdot 2916841 \cdot 1469495262398780123809 \cdot$ $\cdot 5962425999871161284150 63$
225	$M_{75} \cdot 73 \cdot 631 \cdot 23311 \cdot 115201 \cdot 617401 \cdot 1348206751 \cdot 13861369826299351$

	Table 7. Factors of Mersenne numbers $M_n = 2^n - 1$
227	26986333437777017 · · 79921777382059796264915069508677209535456601216886631
229	1504073 · 20492753 · 59833457464970183 · · 4677951201875837235342800003487432365 93
231	M_{77} · 7^2 · 337 · 463 · 599479 · 49823976511782561513383022047620 57
233	1399 · 135607 · 622577 · · 1168681298790776002703448563247662600850665328534921784 31
235	M_{47} · 31 · 2391314881 · 72296287361 · 7320230039515800584547353714697475 1
237	M_{79} · 7 · 1423 · 49297 · 23728823512345609279 · 31357373417090093431
239	479 · 1913 · 5737 · 176383 · 134000609 · · 71100087178244581231050142792537540968637680628 79
241	22000409 · ·16061947437235228941273750872021683922580565632899087995333234043 9
243	M_{81} · 487 · 16753783618801 · 192971705688577 · 3712990163251158343
245	M_{49} · 31 · 71 · 1471 · 122921 · · 2523599020345710168562142988517085297385258216 31
247	8191 · 15809 · 524287 · 6459570124697 · 402004106269663 · · 12828161176172650604534969562121 69
249	M_{83} · 7 · 1621324657 · 8241594690167137359552274418432855740327
251	503 · 54217 · 178230287214063289511 · 6167688219869525750136 7 · · 120703961782498930399696 81
253	23^2 · 47 · 89 · 178481 · 4103188409 · 1999577363284353667695 77 · · 44667711762797798403039426178 361
255	M_{85} · 7 · 103 · 151 · 2143 · 11119 · 106591 · 949111 · · 57024515776397755458386431 51
257	535006138814359 · 1155685395246619182673033 · · 37455059850181093658177663009631318139 3

A more extensive table can be found in

John Brillhart, D. H. Lehmer, J. L. Selfridge, Bryant Tuckerman and S. S. Wagstaff, Jr., *Factorisations of $b^n \pm 1$, $b = 2, 3, 5, 6, 7, 10, 11, 12$ up to High Powers*, American Mathematical Society, Providence R. I., 1983.

Table 8. Factors of $N_n = 2^n + 1$	
1	3
2	5
3	3^2
4	17
5	$3 \cdot 11$
6	$5 \cdot 13$
7	$3 \cdot 43$
8	257
9	$3^3 \cdot 19$
10	$5^2 \cdot 41$
11	$3 \cdot 683$
12	$17 \cdot 241$
13	$3 \cdot 2731$
14	$5 \cdot 29 \cdot 113$
15	$3^2 \cdot 11 \cdot 331$
16	65537
17	$3 \cdot 43691$
18	$5 \cdot 13 \cdot 37 \cdot 109$
19	$3 \cdot 174763$
20	$17 \cdot 61681$
21	$3^2 \cdot 43 \cdot 5419$
22	$5 \cdot 397 \cdot 2113$
23	$3 \cdot 2796203$
24	$97 \cdot 257 \cdot 673$
25	$3 \cdot 11 \cdot 251 \cdot 4051$
26	$5 \cdot 53 \cdot 157 \cdot 1613$
27	$3^4 \cdot 19 \cdot 87211$
28	$17 \cdot 15790321$
29	$3 \cdot 59 \cdot 3033169$
30	$5^2 \cdot 13 \cdot 41 \cdot 61 \cdot 1321$
31	$3 \cdot 715827883$
32	$641 \cdot 6700417$
33	$3^2 \cdot 67 \cdot 683 \cdot 20857$
34	$5 \cdot 137 \cdot 953 \cdot 26317$
35	$3 \cdot 11 \cdot 43 \cdot 281 \cdot 86171$
36	$17 \cdot 241 \cdot 433 \cdot 38737$
37	$3 \cdot 1777 \cdot 27581083$
38	$5 \cdot 229 \cdot 457 \cdot 525313$
39	$3^2 \cdot 2731 \cdot 22366891$
40	$257 \cdot 4278255361$

392

Table 8. Factors of $N_n = 2^n + 1$	
41	$3 \cdot 83 \cdot 8831418697$
42	$5 \cdot 13 \cdot 29 \cdot 113 \cdot 1429 \cdot 14449$
43	$3 \cdot 2932031007403$
44	$17 \cdot 353 \cdot 2931542417$
45	$3^3 \cdot 11 \cdot 19 \cdot 331 \cdot 18837001$
46	$5 \cdot 277 \cdot 1013 \cdot 1657 \cdot 30269$
47	$3 \cdot 283 \cdot 165768537521$
48	$193 \cdot 65537 \cdot 22253377$
49	$3 \cdot 43 \cdot 4363953127297$
50	$5^3 \cdot 41 \cdot 101 \cdot 8101 \cdot 268501$
51	$3^2 \cdot 307 \cdot 2857 \cdot 6529 \cdot 43691$
52	$17 \cdot 858001 \cdot 308761441$
53	$3 \cdot 107 \cdot 28059810762433$
54	$5 \cdot 13 \cdot 37 \cdot 109 \cdot 246241 \cdot 279073$
55	$3 \cdot 11^2 \cdot 683 \cdot 2971 \cdot 48912491$
56	$257 \cdot 5153 \cdot 54410972897$
57	$3^2 \cdot 571 \cdot 174763 \cdot 160465489$
58	$5 \cdot 107367629 \cdot 536903681$
59	$3 \cdot 2833 \cdot 37171 \cdot 1824726041$
60	$17 \cdot 241 \cdot 61681 \cdot 4562284561$
61	$3 \cdot 768614336404564651$
62	$5 \cdot 5581 \cdot 8681 \cdot 49477 \cdot 384773$
63	$3^3 \cdot 19 \cdot 43 \cdot 5419 \cdot 77158673929$
64	$274177 \cdot 67280421310721$
65	$3 \cdot 11 \cdot 131 \cdot 2731 \cdot 409891 \cdot 7623851$
66	$5 \cdot 13 \cdot 397 \cdot 2113 \cdot 312709 \cdot 4327489$
67	$3 \cdot 7327657 \cdot 6713103182899$
68	$17^2 \cdot 354689 \cdot 2879347902817$
69	$3^2 \cdot 139 \cdot 2796203 \cdot 168749965921$
70	$5^2 \cdot 29 \cdot 41 \cdot 113 \cdot 7416361 \cdot 47392381$
71	$3 \cdot 56409643 \cdot 13952598148481$
72	$97 \cdot 257 \cdot 577 \cdot 673 \cdot 487824887233$
73	$3 \cdot 1753 \cdot 1795918038741070627$
74	$5 \cdot 149 \cdot 593 \cdot 184481113 \cdot 231769777$
75	$3^2 \cdot 11 \cdot 251 \cdot 331 \cdot 4051 \cdot 1133836730401$
76	$17 \cdot 1217 \cdot 148961 \cdot 24517014940753$
77	$3 \cdot 43 \cdot 617 \cdot 683 \cdot 78233 \cdot 35532364099$
78	$5 \cdot 13^2 \cdot 53 \cdot 157 \cdot 313 \cdot 1249 \cdot 1613 \cdot 3121 \cdot 21841$
79	$3 \cdot 201487636602438195784363$
80	$65537 \cdot 414721 \cdot 44479210368001$

	Table 8. Factors of $N_n = 2^n + 1$
81	$3^5 \cdot 19 \cdot 163 \cdot 87211 \cdot 135433 \cdot 272010961$
82	$5 \cdot 10169 \cdot 181549 \cdot 12112549 \cdot 43249589$
83	$3 \cdot 499 \cdot 1163 \cdot 2657 \cdot 155377 \cdot 13455809771$
84	$17 \cdot 241 \cdot 3361 \cdot 15790321 \cdot 88959882481$
85	$3 \cdot 11 \cdot 43691 \cdot 26831423036065352611$
86	$5 \cdot 173 \cdot 101653 \cdot 500177 \cdot 1759217765581$
87	$3^2 \cdot 59 \cdot 3033169 \cdot 96076791871613611$
88	$257 \cdot 229153 \cdot 119782433 \cdot 43872038849$
89	$3 \cdot 179 \cdot 62020897 \cdot 18584774046020617$
90	$5^2 \cdot 13 \cdot 37 \cdot 41 \cdot 61 \cdot 109 \cdot 181 \cdot 1321 \cdot 54001 \cdot 29247661$
91	$3 \cdot 43 \cdot 2731 \cdot 224771 \cdot 1210483 \cdot 25829691707$
92	$17 \cdot 291280009243618888211558641$
93	$3^2 \cdot 529510939 \cdot 715827883 \cdot 2903110321$
94	$5 \cdot 3761 \cdot 7484047069 \cdot 140737471578113$
95	$3 \cdot 11 \cdot 2281 \cdot 174763 \cdot 3011347479614249131$
96	$641 \cdot 6700417 \cdot 18446744069414584321$
97	$3 \cdot 971 \cdot 1553 \cdot 31817 \cdot 1100876018364883721$
98	$5 \cdot 29 \cdot 113 \cdot 197 \cdot 19707683773 \cdot 4981857697937$
99	$3^3 \cdot 19 \cdot 67 \cdot 683 \cdot 5347 \cdot 20857 \cdot 242099935645987$
100	$17 \cdot 401 \cdot 61681 \cdot 340801 \cdot 2787601 \cdot 3173389601$
101	$3 \cdot 845100400152152934331135470251$
102	$5 \cdot 13 \cdot 137 \cdot 409 \cdot 953 \cdot 3061 \cdot 13669 \cdot 26317 \cdot 1326700741$
103	$3 \cdot 415141630193 \cdot 8142767081771726171$
104	$257 \cdot 78919881726271091143763623681$
105	$3^2 \cdot 11 \cdot 43 \cdot 211 \cdot 281 \cdot 331 \cdot 5419 \cdot 86171 \cdot 664441 \cdot 1564921$
106	$5 \cdot 15358129 \cdot 586477649 \cdot 1801439824104653$
107	$3 \cdot 643 \cdot 84115747449047881488635567801$
108	$17 \cdot 241 \cdot 433 \cdot 38737 \cdot 33975937 \cdot 138991501037953$
109	$3 \cdot 104124649 \cdot 2077756847362348863128179$
110	$5^2 \cdot 41 \cdot 397 \cdot 2113 \cdot 415878438361 \cdot 3630105520141$
111	$3^2 \cdot 1777 \cdot 3331 \cdot 17539 \cdot 25781083 \cdot 107775231312019$
112	$449 \cdot 2689 \cdot 65537 \cdot 183076097 \cdot 358429848460993$
113	$3 \cdot 227 \cdot 48817 \cdot 636190001 \cdot 491003369344660409$
114	$5 \cdot 13 \cdot 229 \cdot 457 \cdot 131101 \cdot 160969 \cdot 525313 \cdot 275415303169$
115	$3 \cdot 11 \cdot 691 \cdot 2796203 \cdot 1884103651 \cdot 345767385170491$
116	$17 \cdot 59393 \cdot 82280195167144119832390568177$
117	$3^3 \cdot 19 \cdot 2731 \cdot 22366891 \cdot 5302306226370307681801$
118	$5 \cdot 1181 \cdot 3541 \cdot 157649 \cdot 174877 \cdot 5521693 \cdot 104399276341$
119	$3 \cdot 43 \cdot 43691 \cdot 823679683 \cdot 143162553165560959297$
120	$97 \cdot 257 \cdot 673 \cdot 394783681 \cdot 4278255361 \cdot 46908728641$

	Table 8. Factors of $N_n = 2^n + 1$
121	$3 \cdot 683 \cdot 117371 \cdot 1105418458279780045573606110 7$
122	$5 \cdot 733 \cdot 1709 \cdot 3456749 \cdot 368140581013 \cdot 667055378149$
123	$3^2 \cdot 83 \cdot 739 \cdot 165313 \cdot 8831418697 \cdot 13194317913029593$
124	$17 \cdot 290657 \cdot 3770202641 \cdot 1141629180401976895873$
125	$3 \cdot 11 \cdot 251 \cdot 4051 \cdot 229668251 \cdot 5519485418336288303251$
126	$5 \cdot 13 \cdot 29 \cdot 37 \cdot 109 \cdot 113 \cdot 1429 \cdot 14449 \cdot 40388473189 \cdot 118750098349$
127	$3 \cdot 5671372782015641057722910123862 8035243$
128	$59649589127497217 \cdot 5704689200685129054721$
129	$3^2 \cdot 1033 \cdot 1591582393 \cdot 2932031007403 \cdot 15686603697451$
130	$5^2 \cdot 41 \cdot 53 \cdot 157 \cdot 521 \cdot 1613 \cdot 51481 \cdot 34110701 \cdot 108140989558681$
131	$3 \cdot 1049 \cdot 4744297 \cdot 1823311286812077817843918136 11$
132	$17 \cdot 241 \cdot 353 \cdot 7393 \cdot 1761345169 \cdot 2931542417 \cdot 98618273953$
133	$3 \cdot 43 \cdot 4523 \cdot 174763 \cdot 106788290443848295284382097033$
134	$5 \cdot 269 \cdot 15152453 \cdot 42875177 \cdot 2559066073 \cdot 9739278030221$
135	$3^4 \cdot 11 \cdot 19 \cdot 331 \cdot 811 \cdot 15121 \cdot 87211 \cdot 18837001 \cdot 385838642647891$
136	$257 \cdot 383521 \cdot 2368179743873 \cdot 373200722470799764577$
137	$3 \cdot 1097 \cdot 15619 \cdot 32127963626435681 \cdot 105498212027592977$
138	$5 \cdot 13 \cdot 277 \cdot 1013 \cdot 1657 \cdot 30269 \cdot 5415624023749 \cdot 70334392823809$
139	$3 \cdot 4506937 \cdot 5154263952466179530007417425036569 9$
140	$17 \cdot 61681 \cdot 15790321 \cdot 8417984207765786201186788968 1$
141	$3^2 \cdot 283 \cdot 1681003 \cdot 35273039401 \cdot 111349165273 \cdot 165768537521$
142	$5 \cdot 569 \cdot 148587949 \cdot 4999465853 \cdot 5585522857 \cdot 472287102421$
143	$3 \cdot 683 \cdot 2003 \cdot 2731 \cdot 6156182033 \cdot 10425285443 \cdot 15500487753323$
144	$193 \cdot 1153 \cdot 6337 \cdot 65537 \cdot 22253377 \cdot 38941695937 \cdot 278452876033$
145	$3 \cdot 11 \cdot 59 \cdot 3033169 \cdot 7553921 \cdot 999802854724715300883845411$
146	$5 \cdot 293 \cdot 9929 \cdot 649301712182209 \cdot 944473296560185147392 1$
147	$3^2 \cdot 43 \cdot 5419 \cdot 748819 \cdot 4363953127297 \cdot 26032885845392093851$
148	$17 \cdot 20988936657440586486151264256610222593863921$
149	$3 \cdot 1193 \cdot 650833 \cdot 38369587 \cdot 79845595735042598563591 24657$
150	$N_{50} \cdot 13 \cdot 61 \cdot 1201 \cdot 1321 \cdot 63901 \cdot 13334701 \cdot 1182468601$
151	$3 \cdot 18717738334417 \cdot 5083405082410077967730646062149 9$
152	$257 \cdot 27361 \cdot 69394460463940481 \cdot 11699557817717358904481$
153	$3^3 \cdot 19 \cdot 307 \cdot 2857 \cdot 6529 \cdot 43691 \cdot 123931 \cdot 26159806891 \cdot 27439122228481$
154	$5 \cdot 29 \cdot 113 \cdot 397 \cdot 2113 \cdot 8317 \cdot 869467061 \cdot 3019242689 \cdot 76096559910757$
155	$3 \cdot 11 \cdot 11161 \cdot 715827883 \cdot 5947603221397891 \cdot 29126056043168521$
156	$17 \cdot 241 \cdot 858001 \cdot 308761441 \cdot 84159375948762099254554456081$
157	$\cdot 3 \cdot 15073 \cdot 2350291 \cdot 17751783757817897 \cdot 96833299198971305921$
158	$5 \cdot 317 \cdot 38136461186650731796 9 \cdot 6044629098062150 75725313$
159	$3^2 \cdot 107 \cdot 6043 \cdot 28059810762433 \cdot 44751303665181020844276 98737$
160	$641 \cdot 3602561 \cdot 6700417 \cdot 944556849534845630559918385580 81$

Table 8. Factors of $N_n = 2^n + 1$

161	$3 \cdot 43 \cdot 2796203 \cdot 810346749275979232714980036156441026 5219$
162	$N_{54} \cdot 3618757 \cdot 106979941 \cdot 168410989 \cdot 4977454861$
163	$3 \cdot 11281292593 \cdot 1023398150341859 \cdot 3375705470503904 15041769$
164	$17 \cdot 13121 \cdot 8562191377 \cdot 1224386412246561215510639205 6552353$
165	$N_{33} \cdot 11^2 \cdot 331 \cdot 2971 \cdot 48912491 \cdot 415365721 \cdot 2252127523412251$
166	$5 \cdot 997 \cdot 13063537 \cdot 46202197673 \cdot 209957719973 \cdot 148067197374074653$
167	$3 \cdot 623574031927851911766905528625614088386531218 33643$
168	$N_{56} \cdot 97 \cdot 673 \cdot 2017 \cdot 25629623713 \cdot 1538595959564161$
169	$3 \cdot 2731 \cdot 4929910764223610387 \cdot 1852623864601108673 2742614043$
170	$N_{34} \cdot 5 \cdot 41 \cdot 1021 \cdot 4421 \cdot 550801 \cdot 23650061 \cdot 7226904352843746841$
171	$3^3 \cdot 19^2 \cdot 571 \cdot 174763 \cdot 160465489 \cdot 191774583879402681163 49766612211$
172	$17 \cdot 3855260977 \cdot 64082150767423457 \cdot 14253432751031 26327372769$
173	$3 \cdot 347 \cdot 4153 \cdot 35374479827 \cdot 47635010587 \cdot 1643464 2477281892216 23609$
174	$N_{58} \cdot 13 \cdot 349 \cdot 29581 \cdot 27920807689 \cdot 22170214192500421$
175	$N_{35} \cdot 251 \cdot 1051 \cdot 4051 \cdot 110251 \cdot 347833278451 \cdot 34010032331525251$
176	$65537 \cdot 5304641 \cdot 275509565477848842604777623828011 666349761$
177	$N_{59} \cdot 3 \cdot 13099 \cdot 4453762543897 \cdot 1898685496465999273$
178	$5 \cdot 1069 \cdot 5790177919949999561061 49 \cdot 12379400392854 5064364330189$
179	$3 \cdot 58745093521 \cdot 434786819066587937349595056277570 7707143803$
180	$N_{60} \cdot 433 \cdot 38737 \cdot 168692292721 \cdot 469775495062434961$
181	$3 \cdot 1811 \cdot 31675363 \cdot 178101636301126245793428117339780 85990447907$
182	$N_{26} \cdot 29 \cdot 113 \cdot 1093^2 \cdot 4733 \cdot 8861085190774909 \cdot 556338525912325157$
183	$3^2 \cdot 768614336404564651 \cdot 177230399437988782976979507 7302561451$
184	$257 \cdot 43717618369 \cdot 549675408461419937 \cdot 39702995674 72902879791777$
185	$N_{37} \cdot 11 \cdot 1481 \cdot 28136651 \cdot 778429365397887608540618330873281$
186	$N_{62} \cdot 13 \cdot 373 \cdot 951088215727633 \cdot 4611545283086450689$
187	$3 \cdot 683 \cdot 43691 \cdot 21911658253768880847501577164245790 62015865776131$
188	$17 \cdot 1108107457 \cdot 23592342593 \cdot 4501946625921233 \cdot$ $\cdot 181352306852476069537$
189	$N_{63} \cdot 3 \cdot 379 \cdot 87211 \cdot 119827 \cdot 127391413339 \cdot 56202143607667$
190	$N_{38} \cdot 5 \cdot 41 \cdot 761 \cdot 54721 \cdot 276696631250953741 \cdot 2416923620660807201$
191	$3 \cdot 104618362256444679397263157053461106935039257407 7339085483$
192	$N_{64} \cdot 769 \cdot 442499826945303593556473164314770689$
193	$3 \cdot 6563 \cdot 35679139 \cdot 1871670769 \cdot 7455099975844049 \cdot$ $\cdot 1280761337388845898643$
194	$5 \cdot 389 \cdot 3881 \cdot 4657 \cdot 5821 \cdot 3555339061 \cdot 4959325597 \cdot 394563864677 \cdot$ $\cdot 17637260034881$
195	$N_{65} \cdot 3 \cdot 331 \cdot 107251 \cdot 22366891 \cdot 571403921126076957182161$

Table 8. Factors of $N_n = 2^n + 1$

196	$17 \cdot 7057 \cdot 273617 \cdot 1007441 \cdot 15790321 \cdot 375327457 \cdot 1405628248417 \cdot$ $\cdot\ 364565561997841$
197	$3 \cdot 197002597249 \cdot 1348959352853811313 \cdot$ $\cdot\ 2519515738672530122591440010843$
198	$N_{66} \cdot 37 \cdot 109 \cdot 42373 \cdot 235621 \cdot 8463901912489 \cdot 15975607282273$
199	$3 \cdot 26782300737649837925699368205686043375370049896379880588 3563$
200	$N_{40} \cdot 1601 \cdot 25601 \cdot 82471201 \cdot 4323632031270028855065431726184 01$
201	$N_{67} \cdot 3 \cdot 2011 \cdot 9649 \cdot 6324667 \cdot 5915154911853267687444 8563$
202	$5 \cdot 809 \cdot 9491060093 \cdot 5218735279937 \cdot 600503817460697 \cdot$ $\cdot\ 53425037363873248657$
203	$N_{29} \cdot 43 \cdot 596834617 \cdot 3692022713 \cdot 25271581461556596241868896585573 1$
204	$N_{68} \cdot 241 \cdot 8161 \cdot 40932193 \cdot 1467129352609 \cdot 737539985835313$
205	$N_{41} \cdot 11 \cdot 2125820563389437533390243893834597846757304863651$
206	$5 \cdot 41201 \cdot 17325013 \cdot 520379897 \cdot 473000157711296729 \cdot$ $\cdot\ 117070097457656623005977$
207	$N_{69} \cdot 3 \cdot 19 \cdot 6113142872404227834840443898241613032969$
208	$65537 \cdot 928513 \cdot 18558466369 \cdot 23877647873 \cdot 21316654212673 \cdot$ $\cdot\ 715668470267111297$
209	$3 \cdot 419 \cdot 683 \cdot 174763 \cdot 3410623284654639440707 \cdot$ $\cdot\ 160779201878039402409551431 7003$
210	$N_{70} \cdot 13 \cdot 61 \cdot 421 \cdot 1321 \cdot 1429 \cdot 14449 \cdot 146919792181$ $\cdot\ 1041815865690181$
211	$3 \cdot 4643 \cdot 9878177 \cdot 5344743097 \cdot 199061567251 \cdot$ $\cdot\ 22481127512575175864234185190299$
212	$17 \cdot 1692645313 \cdot \text{composite number}$
213	$N_{71} \cdot 3 \cdot 5113 \cdot 17467 \cdot 102241 \cdot 20352554576630130693322627192 9$
214	$5 \cdot 857 \cdot 843589 \cdot 8174912477117 \cdot 23528569104401 \cdot$ $\cdot\ 3786680906166005726421925339 7$
215	$N_{43} \cdot 11 \cdot 9084611 \cdot$ $\cdot\ 5990460837870566137743018260871169892413072 1$
216	$N_{72} \cdot 209924353 \cdot 4261383649 \cdot 2492906081826536045170819 3$
217	$N_{31} \cdot 43 \cdot 16233337 \cdot$ $\cdot\ 1405086085901642802259342330988668427458089059 47$
218	$5 \cdot 5669 \cdot 666184021 \cdot 74323515777853 \cdot 1746518852140345553 \cdot$ $\cdot\ 171857646012809566969$
219	$N_{73} \cdot 3 \cdot 9070197542196643 \cdot 3278244690156222434135906137$
220	$N_{44} \cdot 61681 \cdot 109121 \cdot 148721 \cdot 3404676001 \cdot 11035465708081 \cdot$ $\cdot\ 2546717317681681$

A more extensive table can be found in the reference given after Table 7.

Table 9. Factors of $P_n = (10^n - 1)/9$

3	$3 \cdot 37$
5	$41 \cdot 271$
7	$239 \cdot 4649$
9	$3^2 \cdot 37 \cdot 333667$
11	$21649 \cdot 513239$
13	$53 \cdot 79 \cdot 265371653$
15	$3 \cdot 31 \cdot 37 \cdot 41 \cdot 271 \cdot 2906161$
17	$2071723 \cdot 5363222357$
19	1111111111111111111
21	$3 \cdot 37 \cdot 43 \cdot 239 \cdot 1933 \cdot 4649 \cdot 10838689$
23	11111111111111111111111
25	$41 \cdot 271 \cdot 21401 \cdot 25601 \cdot 182521213001$
27	$3^3 \cdot 37 \cdot 757 \cdot 333667 \cdot 440334654777631$
29	$3191 \cdot 16763 \cdot 43037 \cdot 62003 \cdot 77843839397$
31	$2791 \cdot 6943319 \cdot 57336415063790604359$
33	$3 \cdot 37 \cdot 67 \cdot 21649 \cdot 513239 \cdot 1344628210313298373$
35	$41 \cdot 71 \cdot 239 \cdot 271 \cdot 4649 \cdot 123551 \cdot 102598800232111471$
37	$2028119 \cdot 247629013 \cdot 2212394296770203368013$
39	$3 \cdot 37 \cdot 53 \cdot 79 \cdot 265371653 \cdot 900900900900990990990991$
41	$83 \cdot 1231 \cdot 538987 \cdot 201763709900322803748657942361$
43	$173 \cdot 1527791 \cdot 1963506722254397 \cdot 2140992015395526641$
45	$3^2 \cdot 31 \cdot 37 \cdot 41 \cdot 271 \cdot 238681 \cdot 333667 \cdot 2906161 \cdot 4185502830133110721$
47	$35121409 \cdot 316362908763458525001406154038726382279$
49	$239 \cdot 4649 \cdot 505885997 \cdot 1976730144598190963568023014679333$
51	$3 \cdot 37 \cdot 613 \cdot 210631 \cdot 2071723 \cdot 52986961 \cdot 5363222357 \cdot$ $\cdot 13168164561429877$
53	$107 \cdot 1659431 \cdot 1325815267337711173 \cdot 4719885879949142566200071$
55	$41 \cdot 271 \cdot 1321 \cdot 21649 \cdot 62921 \cdot 513239 \cdot 83251631 \cdot$ $\cdot 1300635692678058358830121$
57	$3 \cdot 37 \cdot 21319 \cdot 10749631 \cdot 1111111111111111111 \cdot$ $\cdot 3931123022305129377976519$
59	$2559647034361 \cdot 4340876285657460212144534289928559826755746751$
61	$733 \cdot 4637 \cdot 329401 \cdot 974293 \cdot 1360682471 \cdot 106007173861643 \cdot$ $\cdot 7061709990156159479$
63	$P_{21} \cdot 3 \cdot 10837 \cdot 23311 \cdot 45613 \cdot 333667 \cdot 45121231 \cdot 1921436048294281$
65	$P_{13} \cdot 41 \cdot 271 \cdot 162503518711 \cdot 5538396997364024056286510640780600481$
67	$493121 \cdot 79863595778924342083 \cdot$ $\cdot 28213380943176667001263153660999177245677$
69	$P_{23} \cdot 3 \cdot 37 \cdot 277 \cdot 203864078068831 \cdot 15953520863292246443348978893$

Table 9. Factors of $P_n = (10^n - 1)/9$

71	241573142393627673576957439049 · · 459948113478868463102217288952230 34301839
73	12171337159 · 1855193842151350117 · · 49207341634646326934001739482502131487446637
75	P_{25} · 3 · 31 · 37 · 151 · 4201 · 2906161 · 157639855537391917091641709400631 51
77	239 · 4649 · 5237 · 21649 · 42043 · 513239 · 29920507 · · 13661466857600232937149644755591 5740910181043
79	317 · 6163 · 10271 · 307627 · composite number
81	P_{27} · 3 · 163 · 9397 · 2462401 · 676421558270641 · · 13065489780800777842504 6117
83	3367147378267 · composite number
85	P_{17} · 41 · 271 · 262533041 · 8119594779271 · · 42221001194055301701793311902914887 89678081
87	P_{29} · 3 · 37 · 4003 · 72559 · · 3101702516580297590451577932373394983427 63245483
89	497867 · 103733951 · 104984505733 · composite number
91	53 · 79 · 239 · 547 · 4649 · 14197 · 17837 · 4262077 · 265371653 · · 43442141653 · 316877365766624209 · · 1107421864705300542913 18013
93	P_{31} · 3 · 37 · 900900900900900900900900900900990990990990990990990- -990990991
95	P_{19} · 41 · 191 · 271 · 59281 · 63841 · composite number
97	12004721 · composite number
99	P_{33} · 3 · 199 · 397 · 34849 · 333667 · · 3628537243429904693247662354742 68869786311886053883

Bibliography

1. John Brillhart, D. H. Lehmer, J. L. Selfridge, Bryant Tuckerman and S. S. Wagstaff, Jr., *Factorisations of* $b^n \pm 1$, $b = 2, 3, 5, 6, 7, 10, 11, 12$ *up to High Powers,* American Mathematical Society, Providence R. I., 1983.

399

\multicolumn{2}{c}{**Table 10. Factors of** $Q_n = 10^n + 1$}	

1	11
2	101
3	$7 \cdot 11 \cdot 13$
4	$73 \cdot 137$
5	$11 \cdot 9091$
6	$101 \cdot 9901$
7	$11 \cdot 909091$
8	$17 \cdot 5882353$
9	$7 \cdot 11 \cdot 13 \cdot 19 \cdot 52579$
10	$101 \cdot 3541 \cdot 27961$
11	$11^2 \cdot 23 \cdot 4093 \cdot 8779$
12	$73 \cdot 137 \cdot 99990001$
13	$11 \cdot 859 \cdot 1058313049$
14	$29 \cdot 101 \cdot 281 \cdot 121499449$
15	$7 \cdot 11 \cdot 13 \cdot 211 \cdot 241 \cdot 2161 \cdot 9091$
16	$353 \cdot 449 \cdot 641 \cdot 1409 \cdot 69857$
17	$11 \cdot 103 \cdot 4013 \cdot 21993833369$
18	$101 \cdot 9901 \cdot 999999000001$
19	$11 \cdot 909090909090909091$
20	$73 \cdot 137 \cdot 1676321 \cdot 5964848081$
21	$7^2 \cdot 11 \cdot 13 \cdot 127 \cdot 2689 \cdot 459691 \cdot 909091$
22	$89 \cdot 101 \cdot 1052788969 \cdot 1056689261$
23	$11 \cdot 47 \cdot 139 \cdot 2531 \cdot 549797184491917$
24	$17 \cdot 5882353 \cdot 9999999900000001$
25	$11 \cdot 251 \cdot 5051 \cdot 9091 \cdot 78875943472201$
26	$101 \cdot 521 \cdot 1900381976777332243781$
27	$7 \cdot 11 \cdot 13 \cdot 19 \cdot 52579 \cdot 70541929 \cdot 14175966169$
28	$73 \cdot 137 \cdot 7841 \cdot 127522001020150503761$
29	$11 \cdot 59 \cdot 154083204930662557781201849$
30	$61 \cdot 101 \cdot 3541 \cdot 9901 \cdot 27961 \cdot 4188901 \cdot 39526741$
31	$11 \cdot 909090909090909090909090909091$
32	$19841 \cdot 976193 \cdot 6187457 \cdot 834427406578561$
33	$7 \cdot 11^2 \cdot 13 \cdot 23 \cdot 4093 \cdot 8779 \cdot 599144041 \cdot 183411838171$
34	$101 \cdot 28559389 \cdot 1491383821 \cdot 2324557465671829$
35	$11 \cdot 9091 \cdot 909091 \cdot 4147571 \cdot 265212793249617641$
36	$73 \cdot 137 \cdot 3169 \cdot 98641 \cdot 99990001 \cdot 3199044596370769$
37	$11 \cdot 7253 \cdot 422650073734453 \cdot 296557347313446299$
38	$101 \cdot 722817036322379041 \cdot 1369778187490592461$
39	$7 \cdot 11 \cdot 13^2 \cdot 157 \cdot 859 \cdot 6397 \cdot 216451 \cdot 1058313049 \cdot 388847808493$
40	$17 \cdot 5070721 \cdot 5882353 \cdot 19721061166646717498359681$

	Table 10. Factors of $Q_n = 10^n + 1$
41	$11 \cdot 2670502781396266997 \cdot 3404193829806058997303$
42	$29 \cdot 101 \cdot 281 \cdot 9901 \cdot 226549 \cdot 121499449 \cdot 4458192223320340849$
43	$11 \cdot 57009401 \cdot 2182600451 \cdot 7306116556571817748755241$
44	$73 \cdot 137 \cdot 617 \cdot 16205834846012967584927082656402106953$
45	$7 \cdot 11 \cdot 13 \cdot 19 \cdot 211 \cdot 241 \cdot 2161 \cdot 9091 \cdot 29611 \cdot 52579 \cdot 3762091 \cdot$ $\cdot 8985695684401$
46	$101 \cdot 1289 \cdot 18371524594609 \cdot 41810033000071669867932658901$
47	$11 \cdot 6299 \cdot 4855067598095567 \cdot 2972627050091390067711611927$
48	$97 \cdot 353 \cdot 449 \cdot 641 \cdot 1409 \cdot 69857 \cdot 206209 \cdot 66554101249 \cdot 75118313082913$
49	$11 \cdot 197 \cdot 909091 \cdot 50761416243655329949187817263959 39035533$
50	$101 \cdot 3541 \cdot 27961 \cdot 60101 \cdot 7019801 \cdot 14103673319201 \cdot 1680588011350901$
51	$7 \cdot 11 \cdot 13 \cdot 103 \cdot 4013 \cdot 21993833369 \cdot 291078844423 \cdot$ $\cdot 377526955309799110357$
52	$73 \cdot 137 \cdot 1580801 \cdot 632527440202150745090622412245443923049201$
53	$11 \cdot 90909090909090909090909090909090909090909090909090909091$
54	$101 \cdot 109 \cdot 9901 \cdot 153469 \cdot 999999000001 \cdot$ $\cdot 59779577156334533866654838281$
55	$Q_{11} \cdot 331 \cdot 5171 \cdot 9091 \cdot 20163494891 \cdot 31872784116567457977 6721$
56	$17 \cdot 113 \cdot 5882353 \cdot 73765755896403138401 \cdot$ $\cdot 1199683691448463702 26083377$
57	$7 \cdot 11 \cdot 13 \cdot 1458973 \cdot 909090909090909091 \cdot$ $\cdot 7532018062713284625479 77919407$
58	$101 \cdot 349 \cdot 38861 \cdot 618049 \cdot$ $\cdot 1181180637520183640867963573625 8669583187541$
59	$11 \cdot 1889 \cdot 1090805842068098677837 \cdot$ $\cdot 4411922770996074109644535362851087$
60	$Q_{20} \cdot 99990001 \cdot 10000999999989999899990000000010001$
61	$11 \cdot 81131 \cdot$ $\cdot 11205222530116836855321528257890437575145023592596037161$
62	$101 \cdot 2049349 \cdot$ $\cdot 48312854955451223730555458835903982239730714968 5578249$
63	$Q_{21} \cdot 19 \cdot 52579 \cdot 5274739 \cdot 18977242267323558587448573 2659$
64	$1265011073 \cdot 15343168188889137818369 \cdot$ $515217525265213267447869906815873$
65	$Q_{13} \cdot 131 \cdot 9091 \cdot 8396862596258693901610602298557167100076327481$
66	$Q_{22} \cdot 9901 \cdot 5419170769 \cdot 789390798020221 \cdot 2361000305507449$
67	$11 \cdot 90909090909090909090909090909090909090909090909090909090-$ -90909091
68	$73 \cdot 137 \cdot 152533657 \cdot$ $\cdot 6555274617188258326423007086888436687780 3237222654400793$
69	$Q_{23} \cdot 7 \cdot 13 \cdot 31051 \cdot 143574021480139 \cdot 2464944534764905919 2745899$
70	$Q_{14} \cdot 421 \cdot 3541 \cdot 27961 \cdot 3471301 \cdot 13489841 \cdot 60368344121 \cdot$ $\cdot 848654483879497562821$

	Table 10. Factors of $Q_n = 10^n + 1$
71	$11 \cdot 290249 \cdot 313210694641810683554152093234053895417069794931561$- -89716729115659
72	$Q_{24} \cdot 8929 \cdot 11199462425803561429051394333072012543397 9169$
73	$11 \cdot 293 \cdot$ composite number
74	$101 \cdot 149 \cdot 3109 \cdot 111149 \cdot 708840373781 \cdot 669031686661427842829 \cdot$ $\cdot 405481405140627747580718 40361$
75	$Q_{25} \cdot 7 \cdot 13 \cdot 211 \cdot 241 \cdot 2161 \cdot$ $\cdot 1000009999999989999989999900000000 0100001$
76	$73 \cdot 137 \cdot 457 \cdot 1403417 \cdot$ composite number
77	$11^2 \cdot 23 \cdot 463 \cdot 4093 \cdot 8779 \cdot 24179 \cdot 590437 \cdot 909091 \cdot 7444361 \cdot$ $\cdot 4539402627853030477 \cdot 492463016031572620 7887$
78	$Q_{26} \cdot 3121 \cdot 9901 \cdot 53397071018461 \cdot 6060517860310398033985611921721$
79	$11 \cdot 1423 \cdot 9615060929 \cdot 66443174541490579097997510158021076958 3929$- -38976011506949065646573
80	$Q_{16} \cdot 1634881 \cdot 18453761 \cdot 947147262401 \cdot$ $\cdot 349954396040122577928041596214187 605761$
81	$P_{27} \cdot 1459 \cdot 2458921051 \cdot 456502382570032651 \cdot$ $\cdot 6106003860898583499 39139$
82	$101 \cdot 68389 \cdot 144774599701851189374007660603168623753834536241353$- $-15606455731040065067 49609$
83	$11 \cdot 167 \cdot 997 \cdot 3565183 \cdot 2097307081 \cdot$ composite number
84	$Q_{28} \cdot 99990001 \cdot 11189053009 \cdot 603812429055411913 \cdot$ $\cdot 14802942340075 0506553$
85	$Q_{17} \cdot 9091 \cdot 87211 \cdot 787223761 \cdot$ $\cdot 160220794821014452066741918303580917664386555 934641$
86	$101 \cdot 338669 \cdot 292350055629830335522265394854270659844892508 58537$- $-09961673200056984872843366529$
87	$Q_{29} \cdot 7 \cdot 13 \cdot 638453709757 \cdot 135080726389891 \cdot$ $\cdot 12741947328981484717664 04179653$
88	$17 \cdot 5882353 \cdot 10100113 \cdot 990087922778685842425722365680473079 85554$- $-2210271310825918482298267356 0177$
89	$179 \cdot$ composite number
90	$Q_{30} \cdot 181 \cdot 999999000001 \cdot 4999437541453012143121 \cdot$ $\cdot 11050977950029947 98105101$

Table 10. Factors of $Q_n = 10^n + 1$
91	11 · 859 · 909091 · 21705503 · 1058313049 · 50678387411703889101759125- -785290439894389920385627096501794498837
92	73 · 137 · 2393 · 4178437150016715837818641871709193476807772628503- -969494400330129962348520684914375257
93	Q_{31} · 7 · 13 · 373 · 44641 · 3590254957 · · 18381907262281244633158190677786966663091011
94	101 · 45121 · composite number
95	Q_{19} · 9091 · 1812604116731 · 121450506296081 · · 4996731930447843676185843959746621491531100801
96	Q_{32} · 193 · 769 · 1253224535459902849 · · 53763491189967221358575546107279034709697
97	101 · composite number
98	Q_{14} · 9999999999999900000000000000000099999999999999900000000000000- -99999999999999900000000000000001
99	Q_{33} · 19 · 52579 · 7093127053 · · 141122524877886182282233539317796144938305111168717
100	Q_{20} · 401 · 1201 · 1601 · 129694419029057750551385771184564274499075- -700947656757821537291527196801

Bibliography

1. John Brillhart, D. H. Lehmer, J. L. Selfridge, Bryant Tuckerman and S. S. Wagstaff, Jr., *Factorisations of $b^n \pm 1$, $b = 2, 3, 5, 6, 7, 10, 11, 12$ up to High Powers*, American Mathematical Society, Providence R. I., 1983.

Table 11a. Factors of $3^n + 2^n$

1	5
2	13
3	$5 \cdot 7$
4	97
5	$5^2 \cdot 11$
6	$13 \cdot 61$
7	$5 \cdot 463$
8	$17 \cdot 401$
9	$5 \cdot 7 \cdot 577$
10	$13 \cdot 4621$
11	$5 \cdot 35839$
12	$97 \cdot 5521$
13	$5 \cdot 79 \cdot 4057$
14	$13 \cdot 369181$
15	$5^2 \cdot 7 \cdot 11 \cdot 31 \cdot 241$
16	$3041 \cdot 14177$
17	$5 \cdot 3673 \cdot 7039$
18	$13 \cdot 37 \cdot 61 \cdot 73 \cdot 181$
19	$5 \cdot 419 \cdot 555029$
20	$41 \cdot 97 \cdot 281 \cdot 3121$
21	$5 \cdot 7^2 \cdot 463 \cdot 92233$
22	$13 \cdot 2414250301$
23	$5 \cdot 461 \cdot 3083 \cdot 13249$
24	$17 \cdot 401 \cdot 41432641$
25	$5^3 \cdot 11 \cdot 1201 \cdot 513101$
26	$13^2 \cdot 1093 \cdot 13761229$
27	$5 \cdot 7 \cdot 577 \cdot 377604937$
28	$97 \cdot 78569 \cdot 3001769$
29	$5 \cdot 59 \cdot 1567 \cdot 148466603$
30	$13 \cdot 61 \cdot 2341 \cdot 4621 \cdot 24001$
31	$5 \cdot 1489 \cdot 3659 \cdot 22674269$
32	$1153 \cdot 1607133116929$
33	$5 \cdot 7 \cdot 199 \cdot 35839 \cdot 22270249$
34	$13 \cdot 1282861452271981$
35	$5^2 \cdot 11 \cdot 463 \cdot 392943879511$
36	$97 \cdot 2593 \cdot 3889 \cdot 5521 \cdot 27793$
37	$5 \cdot 149 \cdot 604408111852283$
38	$13 \cdot 103911691734684541$
39	$5 \cdot 7 \cdot 79 \cdot 4057 \cdot 357241 \cdot 1011271$
40	$17 \cdot 401 \cdot 1783433557073281$

	Table 11a. Factors of $3^n + 2^n$
41	$5 \cdot 83 \cdot 11071 \cdot 35507 \cdot 223574641$
42	$13 \cdot 61 \cdot 43261 \cdot 369181 \cdot 8639401$
43	$5 \cdot 173 \cdot 947 \cdot 5333 \cdot 14449 \cdot 5200421$
44	$89 \cdot 97 \cdot 617 \cdot 184879309516177$
45	$5^2 \cdot 7 \cdot 11 \cdot 31 \cdot 241 \cdot 577 \cdot 2791 \cdot 127559071$
46	$13 \cdot 53269 \cdot 12798522145252969$
47	$5 \cdot 6299 \cdot 75577 \cdot 11170373961701$
48	$193 \cdot 3041 \cdot 14177 \cdot 38977 \cdot 245953921$
49	$5 \cdot 463 \cdot 224617 \cdot 3440291 \cdot 133768139$
50	$13 \cdot 4621 \cdot 11701 \cdot 9802501 \cdot 104189401$
51	$5 \cdot 7 \cdot 103 \cdot 409 \cdot 3673 \cdot 7039 \cdot 57529 \cdot 982057$
52	$97 \cdot 13001 \cdot 62401 \cdot 82104167839721$
53	$5 \cdot 107 \cdot 24574686173 \cdot 1474296171913$
54	$13 \cdot 37 \cdot 61 \cdot 73 \cdot 109 \cdot 181 \cdot 13177 \cdot 43633 \cdot 2393389$
55	$5^2 \cdot 11^2 \cdot 35839 \cdot 55661 \cdot 28909244547481$
56	$17 \cdot 113 \cdot 401 \cdot 673 \cdot 1009493350615396529$
57	$5 \cdot 7 \cdot 419 \cdot 9007 \cdot 555029 \cdot 2439487 \cdot 8778799$
58	$13 \cdot 2437 \cdot 7011853 \cdot 77701093 \cdot 272881417$
59	$5 \cdot 19471 \cdot 145142890373531870546641$
60	$41 \cdot 97 \cdot 281 \cdot 3121 \cdot 5521 \cdot 2201414671278481$
61	$5 \cdot 21107 \cdot 5764013 \cdot 209061986047163521$
62	$13 \cdot 373 \cdot 34686397 \cdot 2268330947579252221$
63	$5 \cdot 7^2 \cdot 127 \cdot 463 \cdot 577 \cdot 7687 \cdot 92233 \cdot 194208941809$
64	$769 \cdot 121899667073 \cdot 36629538145348481$
65	$5^2 \cdot 11 \cdot 79 \cdot 131 \cdot 4057 \cdot 5981 \cdot 149166625778220961$
66	$13 \cdot 61 \cdot 395297101 \cdot 2414250301 \cdot 40834167001$
67	$5 \cdot 8556286591 \cdot 12384629330 \cdot 174978917747$
68	$97 \cdot 137 \cdot 236641 \cdot 93067415281 \cdot 950310165593$
69	$5 \cdot 7 \cdot 461 \cdot 3083 \cdot 13249 \cdot 1266021208542598810111$
70	$13 \cdot 2521 \cdot 4621 \cdot 13441 \cdot 369181 \cdot 3330924131977261$
71	$5 \cdot 1279 \cdot 5113 \cdot 606341 \cdot 378770204035162384997$
72	$17 \cdot 401 \cdot 3169 \cdot 41432641 \cdot 25169360989905069409$
73	$5 \cdot 439 \cdot 22777 \cdot 3606347 \cdot 374846134168397471243$
74	$13 \cdot 7549 \cdot 21313 \cdot 3094090369 \cdot 31330151333999377$
75	$5^3 \cdot 7 \cdot 11 \cdot 31 \cdot 151 \cdot 241 \cdot 751 \cdot 1201 \cdot 6151 \cdot 513101 \cdot 19679141551$
76	$97 \cdot 143350267555241 \cdot 131233622361283804841$
77	$5 \cdot 463 \cdot 35839 \cdot 135367 \cdot 173713 \cdot 2805980251094900881$
78	$13^2 \cdot 61 \cdot 157 \cdot 313 \cdot 1093 \cdot 1873 \cdot 32917 \cdot 13761229 \cdot 34959235741$
79	$5 \cdot 3793 \cdot 443742053 \cdot 3709120153 \cdot 1578428033020043$
80	$3041 \cdot 14177 \cdot 191408953253761 \cdot 17911723223817601$

Table 11b. Factors of $3^n - 2^n$	
3	19
5	211
7	$29 \cdot 71$
9	$19 \cdot 1009$
11	$23^2 \cdot 331$
13	$53 \cdot 29927$
15	$19 \cdot 211 \cdot 3571$
17	129009091
19	$1559 \cdot 745181$
21	$19 \cdot 29 \cdot 43 \cdot 71 \cdot 6217$
23	$47 \cdot 2002867877$
25	$101 \cdot 211 \cdot 39756701$
27	$19 \cdot 1009 \cdot 397760329$
29	68629840493971
31	617671248800299
33	$19 \cdot 23^2 \cdot 67 \cdot 331 \cdot 24939553$
35	$29 \cdot 71 \cdot 211 \cdot 6091 \cdot 18906721$
37	$8891471 \cdot 50642213021$
39	$19 \cdot 53 \cdot 29927 \cdot 134473349011$
41	$821 \cdot 32309 \cdot 99139 \cdot 13869481$
43	$431 \cdot 1196347 \cdot 636618868367$
45	$19 \cdot 211 \cdot 1009 \cdot 3571 \cdot 204521695561$
47	$1129 \cdot 303168989 \cdot 77681973839$
49	$29 \cdot 71 \cdot 197 \cdot 589955028851094677$
51	$19 \cdot 307 \cdot 3163 \cdot 129009091 \cdot 904841491$
53	$19383245658672820642055731$
55	$23^2 \cdot 211 \cdot 331 \cdot 33000001 \cdot 143083295851$
57	$19^2 \cdot 457 \cdot 1559 \cdot 745181 \cdot 8191816751977$
59	$14130386091162273752461387579$
61	$93941 \cdot 203778431 \cdot 6643288986915481$
63	$19 \cdot 29 \cdot 43 \cdot 71 \cdot 1009 \cdot 3529 \cdot 6217 \cdot 110251 \cdot 278775379$
65	$53 \cdot 211 \cdot 29927 \cdot 2010191 \cdot 47891351 \cdot 319716931$
67	$269 \cdot 17685389279 \cdot 19487546413505693209$
69	$19 \cdot 47 \cdot 139 \cdot 691 \cdot 102811 \cdot 174571 \cdot 270619 \cdot 2002867877$
71	$67049419 \cdot 111998979662707645844109121$
73	$293 \cdot 645845309 \cdot 3571539558007960529941 63$
75	$19 \cdot 101 \cdot 211 \cdot 601 \cdot 1051 \cdot 3571 \cdot 39756701 \cdot 16751715789451$
77	$23^2 \cdot 29 \cdot 71 \cdot 331 \cdot 5237 \cdot 117699514703 \cdot 24634250588641$
79	$317 \cdot 146309 \cdot 157531818764909 \cdot 6743424997709927$

Table 12a. Factors of $4^n + 3^n$	
1	7
2	5^2
3	$7 \cdot 13$
4	337
5	$7 \cdot 181$
6	$5^2 \cdot 193$
7	$7^2 \cdot 379$
8	$17 \cdot 4241$
9	$7 \cdot 13 \cdot 19 \cdot 163$
10	$5^3 \cdot 8861$
11	$7 \cdot 727 \cdot 859$
12	$337 \cdot 51361$
13	$7 \cdot 313 \cdot 31357$
14	$5^2 \cdot 29 \cdot 376853$
15	$7 \cdot 13 \cdot 31 \cdot 181 \cdot 2131$
16	4338014017
17	$7 \cdot 919 \cdot 2690659$
18	$5^2 \cdot 73 \cdot 193 \cdot 196201$
19	$7 \cdot 39434309773$
20	$41 \cdot 241 \cdot 337 \cdot 331241$
21	$7^2 \cdot 13 \cdot 43 \cdot 379 \cdot 631 \cdot 673$
22	$5^2 \cdot 1277 \cdot 3169 \cdot 174197$
23	$7 \cdot 139 \cdot 72418178167$
24	$17 \cdot 97 \cdot 1249 \cdot 4241 \cdot 32257$
25	$7 \cdot 181 \cdot 186601 \cdot 4765801$
26	$5^2 \cdot 53 \cdot 3400861504301$
27	$7 \cdot 13 \cdot 19 \cdot 109 \cdot 163 \cdot 3727 \cdot 157411$
28	$337 \cdot 10990561 \cdot 19461121$
29	$7 \cdot 4003 \cdot 10288676582887$
30	$5^3 \cdot 193 \cdot 2281 \cdot 8861 \cdot 2364841$
31	$7 \cdot 110460751 \cdot 5965019443$
32	$11969 \cdot 1541364950613953$
33	$7 \cdot 13 \cdot 67 \cdot 727 \cdot 859 \cdot 1321 \cdot 2179 \cdot 6733$
34	$5^2 \cdot 409 \cdot 28866951820151833$
35	$7^2 \cdot 181 \cdot 379 \cdot 421 \cdot 8191 \cdot 101855671$
36	$337 \cdot 51361 \cdot 659521 \cdot 413696161$
37	$7 \cdot 2698559459340638836021$
38	$5^2 \cdot 2123213 \cdot 36249341 \cdot 39269353$
39	$7 \cdot 13^2 \cdot 79 \cdot 313 \cdot 547 \cdot 937 \cdot 31357 \cdot 642877$
40	$17 \cdot 4241 \cdot 16768214728492007041$

Table 12a. Factors of $4^n + 3^n$

41	$7 \cdot 64453 \cdot 10718197205615817217$
42	$5^2 \cdot 29 \cdot 193 \cdot 12601 \cdot 376853 \cdot 29110508281$
43	$7 \cdot 1291 \cdot 22447 \cdot 381415976737303369$
44	$89 \cdot 337 \cdot 2729 \cdot 3257 \cdot 1160913331730153$
45	$7 \cdot 13 \cdot 19 \cdot 31 \cdot 163 \cdot 181 \cdot 541 \cdot 2131 \cdot 9901 \cdot 68583241$
46	$5^2 \cdot 1980707608031856300839039713$
47	$7 \cdot 20023 \cdot 4336597 \cdot 32586965663296063$
48	$4338014017 \cdot 18263712835003367041$
49	$7^3 \cdot 379 \cdot 74163546829 \cdot 32871239562379$
50	$5^4 \cdot 101 \cdot 701 \cdot 8861 \cdot 101501 \cdot 132001 \cdot 241296001$

Table 12b. Factors of $4^n - 3^n$

3	37
5	$11 \cdot 71$
7	14197
9	$37 \cdot 6553$
11	$23 \cdot 174659$
13	$131 \cdot 500111$
15	$11 \cdot 37 \cdot 61 \cdot 71 \cdot 601$
17	17050729021
19	$25309 \cdot 10814953$
21	$37 \cdot 14197 \cdot 8352709$
23	$47 \cdot 1933 \cdot 17389 \cdot 44483$
25	$11 \cdot 71 \cdot 1440528320401$
27	$37 \cdot 2053 \cdot 6553 \cdot 36174709$
29	$59 \cdot 349 \cdot 47387 \cdot 295324633$
31	$311 \cdot 21577 \cdot 687147718331$
33	$23 \cdot 37 \cdot 174659 \cdot 496393931749$
35	$11 \cdot 71 \cdot 14197 \cdot 106471574188981$
37	$38629 \cdot 3554591 \cdot 137564452439$
39	$37 \cdot 131 \cdot 157 \cdot 16693 \cdot 500111 \cdot 47572981$
41	$83 \cdot 97007 \cdot 2305103 \cdot 260546273807$
43	$431 \cdot 179514905332642395926027$
45	$11 \cdot 37 \cdot 61 \cdot 71 \cdot 601 \cdot 6553 \cdot 461521 \cdot 386375761$
47	$408431 \cdot 4849537385691028702738$7
49	$8821 \cdot 14197 \cdot 39887 \cdot 63444425561697419$
51	$37 \cdot 17050729021 \cdot 8037383607237646741$
53	$107 \cdot 75822073861085059827821596586$3
55	$11^2 \cdot 23 \cdot 71 \cdot 174659 \cdot 942179591 \cdot 39921168784201$
57	$37 \cdot 9349 \cdot 25309 \cdot 10814953 \cdot 219357741909100321$
59	$3323069848158428764872172603052750$77

Table 13a. Factors of $5^n + 2^n$	
1	7
2	29
3	$7 \cdot 19$
4	641
5	$7 \cdot 11 \cdot 41$
6	$29 \cdot 541$
7	$7^2 \cdot 1597$
8	$17 \cdot 22993$
9	$7 \cdot 19 \cdot 37 \cdot 397$
10	$29 \cdot 61 \cdot 5521$
11	$7 \cdot 23 \cdot 303293$
12	$641 \cdot 380881$
13	$7 \cdot 53 \cdot 131 \cdot 25117$
14	$29 \cdot 210466621$
15	$7 \cdot 11 \cdot 19 \cdot 41 \cdot 508771$
16	$97 \cdot 1573071713$
17	$7 \cdot 108991369171$
18	$29 \cdot 181 \cdot 541 \cdot 1343341$
19	$7 \cdot 2724783836059$
20	$641 \cdot 3761 \cdot 39558401$
21	$7^2 \cdot 19 \cdot 127 \cdot 1597 \cdot 2525293$
22	$29 \cdot 89 \cdot 192149 \cdot 4807441$
23	$7 \cdot 47 \cdot 168499 \cdot 215038823$
24	$17 \cdot 22993 \cdot 129937 \cdot 1173553$
25	$7 \cdot 11 \cdot 41 \cdot 251 \cdot 1601 \cdot 3851 \cdot 61001$
26	$29 \cdot 3121 \cdot 16463734208221$
27	$7 \cdot 19 \cdot 37 \cdot 397 \cdot 3813697527769$
28	$641 \cdot 58116853330557841$
29	$7 \cdot 59 \cdot 12703 \cdot 108287 \cdot 327866809$
30	$29 \cdot 61 \cdot 241 \cdot 541 \cdot 5521 \cdot 7321 \cdot 99901$
31	$7 \cdot 1029263 \cdot 2452721 \cdot 263510293$
32	$193 \cdot 102593 \cdot 1175885676400129$
33	$7 \cdot 19 \cdot 23 \cdot 331 \cdot 463 \cdot 1321 \cdot 303293 \cdot 619807$
34	$29 \cdot 3469 \cdot 4293929 \cdot 1347482056121$
35	$7^2 \cdot 11 \cdot 41 \cdot 1597 \cdot 3011 \cdot 768041 \cdot 35659681$
36	$73 \cdot 641 \cdot 380881 \cdot 816488284824217$
37	$7 \cdot 29432983 \cdot 353148886170294037$
38	$29 \cdot 125447545072128127601 35261$
39	$7 \cdot 19 \cdot 53 \cdot 131 \cdot 1093 \cdot 25117 \cdot 71753557059367$
40	$17 \cdot 22993 \cdot 756641 \cdot 30751460163770081$

Table 13b. Factors of $(5^n - 2^n)/3$

3	$3 \cdot 13$
5	1031
7	25999
9	$3^2 \cdot 13 \cdot 5563$
11	$1409 \cdot 11551$
13	406898311
15	$3 \cdot 13 \cdot 31 \cdot 1031 \cdot 8161$
17	$82009 \cdot 3101039$
19	6357828601279
21	$3 \cdot 13 \cdot 43 \cdot 1009 \cdot 3613 \cdot 25999$
23	$7591 \cdot 197479 \cdot 2650751$
25	$151 \cdot 1031 \cdot 95801 \cdot 6660751$
27	$3^3 \cdot 13 \cdot 5563 \cdot 1271899175923$
29	$1277 \cdot 1207387 \cdot 40269058729$
31	$373 \cdot 4161405605965366483$
33	$3 \cdot 13 \cdot 67 \cdot 1409 \cdot 11551 \cdot 61843 \cdot 14754631$
35	$71 \cdot 1031 \cdot 25999 \cdot 3143561 \cdot 162156121$
37	$242531920472780405314165 51$
39	$3 \cdot 13^2 \cdot 79 \cdot 157 \cdot 15679 \cdot 15113671 \cdot 406898311$

Table 14b. Factors of $(5^n - 3^n)/2$

3	7^2
5	$11 \cdot 131$
7	$43 \cdot 883$
9	$7^2 \cdot 109 \cdot 181$
11	$67 \cdot 363067$
13	609554401
15	$7^2 \cdot 11 \cdot 61 \cdot 131 \cdot 3541$
17	$2347 \cdot 162507523$
19	9536162033329
21	$7^3 \cdot 43 \cdot 883 \cdot 18306583$
23	5960417405949649
25	$11 \cdot 131 \cdot 103408180634401$
27	$7^2 \cdot 109 \cdot 181 \cdot 3853528045489$
29	$59 \cdot 349 \cdot 4522957755638831$
31	2328306127701998147089
33	$7^2 \cdot 67 \cdot 363067 \cdot 48834018937441$
35	$11 \cdot 43 \cdot 71 \cdot 131 \cdot 491 \cdot 883 \cdot 79801 \cdot 9560461$
37	$1259 \cdot 6661 \cdot 4338054444662614919$
39	$7^2 \cdot 9127 \cdot 609554401 \cdot 3336288286423$

	Table 14a. Factors of $(5^n + 3^n)/2$
1	2^2
2	17
3	$2^2 \cdot 19$
4	353
5	$2^2 \cdot 421$
6	$13 \cdot 17 \cdot 37$
7	$2^2 \cdot 10039$
8	198593
9	$2^2 \cdot 19 \cdot 12979$
10	$17 \cdot 101 \cdot 2861$
11	$2^2 \cdot 23 \cdot 266333$
12	$353 \cdot 346561$
13	$2^2 \cdot 53 \cdot 2882777$
14	$17 \cdot 29 \cdot 6195029$
15	$2^2 \cdot 19 \cdot 31 \cdot 421 \cdot 15391$
16	$97 \cdot 786757409$
17	$2^2 \cdot 95383574161$
18	$13 \cdot 17 \cdot 37 \cdot 613 \cdot 380557$
19	$2^2 \cdot 2384331073699$
20	$41 \cdot 353 \cdot 6961 \cdot 473321$
21	$2^2 \cdot 19 \cdot 10039 \cdot 312496801$
22	$17 \cdot 70124034472801$
23	$2^2 \cdot 47 \cdot 139 \cdot 228092436443$
24	$198593 \cdot 150068046721$
25	$2^2 \cdot 421 \cdot 751 \cdot 117825508651$
26	$17 \cdot 157 \cdot 7963801 \cdot 35052733$
27	$2^2 \cdot 19 \cdot 379 \cdot 919 \cdot 12979 \cdot 10843039$
28	$113 \cdot 281 \cdot 353 \cdot 1661770035577$
29	$2^2 \cdot 233 \cdot 99927351691777397$
30	$13 \cdot 17 \cdot 37 \cdot 101 \cdot 241 \cdot 2861 \cdot 21841 \cdot 37441$
31	$2^2 \cdot 102859 \cdot 353897 \cdot 15990462533$
32	$143797249 \cdot 80957968182017$
33	$2^2 \cdot 19 \cdot 23 \cdot 331 \cdot 266333 \cdot 377733728131$
34	$17^2 \cdot 137 \cdot 7350751721413580329$
35	$2^2 \cdot 421 \cdot 2731 \cdot 4621 \cdot 10039 \cdot 6820715791$
36	$73 \cdot 353 \cdot 346561 \cdot 814728505027897$
37	$2^2 \cdot 2202019 \cdot 49051493 \cdot 84202867883$
38	$17 \cdot 1901 \cdot 5628583749420166055621$
39	$2^2 \cdot 19 \cdot 53 \cdot 79 \cdot 2882777 \cdot 991454008116079$
40	$8161 \cdot 198593 \cdot 1176881 \cdot 2384132086001$

Table 15a. Factors of $5^n + 4^n$	
1	3^2
2	41
3	$3^3 \cdot 7$
4	881
5	$3^2 \cdot 461$
6	$13 \cdot 37 \cdot 41$
7	$3^2 \cdot 10501$
8	$17 \cdot 26833$
9	$3^4 \cdot 7 \cdot 3907$
10	$41 \cdot 263761$
11	$3^2 \cdot 23 \cdot 256147$
12	$73 \cdot 881 \cdot 4057$
13	$3^2 \cdot 1483 \cdot 96487$
14	$41 \cdot 1289 \cdot 120569$
15	$3^3 \cdot 7 \cdot 461 \cdot 362581$
16	$4801 \cdot 32677121$
17	$3^2 \cdot 103 \cdot 841552667$
18	$13 \cdot 37 \cdot 41 \cdot 196917841$
19	$3^2 \cdot 2149818248341$
20	$881 \cdot 193841 \cdot 564881$
21	$3^3 \cdot 7^2 \cdot 43 \cdot 127 \cdot 6343 \cdot 10501$
22	$41 \cdot 89 \cdot 129449 \cdot 5084641$
23	$3^2 \cdot 47 \cdot 829 \cdot 33029 \cdot 1035323$
24	$17 \cdot 26833 \cdot 131282857921$
25	$3^2 \cdot 101 \cdot 461 \cdot 95401 \cdot 7482901$
26	$41 \cdot 53 \cdot 687813952605677$
27	$3^5 \cdot 7 \cdot 109 \cdot 3907 \cdot 23869 \cdot 431947$
28	$881 \cdot 42366584084741281$
29	$3^2 \cdot 20728082811027490541$
30	$13 \cdot 37 \cdot 41 \cdot 263761 \cdot 179265994321$
31	$3^2 \cdot 1489 \cdot 39929 \cdot 1878043 \cdot 4638407$
32	$641 \cdot 5953 \cdot 59925889 \cdot 101900353$
33	$3^3 \cdot 7 \cdot 23 \cdot 67^2 \cdot 17623 \cdot 256147 \cdot 1322443$
34	$41 \cdot 613 \cdot 23171597383513843357$
35	$3^2 \cdot 461 \cdot 5741 \cdot 10501 \cdot 102481 \cdot 113585081$
36	$73 \cdot 881 \cdot 937 \cdot 1009 \cdot 4057 \cdot 55441 \cdot 1064377$
37	$3^2 \cdot 149 \cdot 3109 \cdot 24421 \cdot 714808900594241$
38	$41 \cdot 5724920477 \cdot 1550233221940373$
39	$3^3 \cdot 7 \cdot 1483 \cdot 96487 \cdot 67271415191054101$
40	$17 \cdot 1601 \cdot 26833 \cdot 12455134257765816641$

	Table 15b. Factors of $5^n - 4^n$
3	61
5	$11 \cdot 191$
7	$29 \cdot 2129$
9	$19 \cdot 61 \cdot 1459$
11	$6359 \cdot 7019$
13	$79 \cdot 14602459$
15	$11 \cdot 31 \cdot 61 \cdot 191 \cdot 7411$
17	$1259 \cdot 2381 \cdot 248779$
19	$830339 \cdot 22639679$
21	$29 \cdot 61 \cdot 211 \cdot 2129 \cdot 594511$
23	$139 \cdot 599 \cdot 23599 \cdot 6031199$
25	$11 \cdot 191 \cdot 251 \cdot 147451 \cdot 3818201$
27	$19 \cdot 61 \cdot 1459 \cdot 46549 \cdot 94425589$
29	$59 \cdot 349 \cdot 6091 \cdot 1482830563601$
31	$42719 \cdot 108897707976754259$
33	$61 \cdot 6359 \cdot 7019 \cdot 42730786509181$
35	$11 \cdot 29 \cdot 71 \cdot 191 \cdot 421 \cdot 2129 \cdot 32971 \cdot 22756301$
37	$5215151 \cdot 2588307251 \cdot 5388832441$
39	$61 \cdot 79 \cdot 3121 \cdot 14602459 \cdot 8280971940541$

	Table 16b. Factors of $6^n - 5^n$
3	$7 \cdot 13$
5	4651
7	$29 \cdot 6959$
9	$7 \cdot 13 \cdot 19 \cdot 37 \cdot 127$
11	313968931
13	11839990891
15	$7 \cdot 13 \cdot 4651 \cdot 1038811$
17	$137 \cdot 409 \cdot 443 \cdot 651169$
19	$191 \cdot 3090503945981$
21	$7^2 \cdot 13 \cdot 29 \cdot 3739 \cdot 6959 \cdot 44647$
23	777809294098524691
25	$4651 \cdot 6048648636003601$
27	$7 \cdot 13 \cdot 19 \cdot 37 \cdot 127 \cdot 1783 \cdot 70138897327$
29	$8212573801 \cdot 4463812400371$
31	$311 \cdot 1489 \cdot 146197 \cdot 4391027 \cdot 4446331$
33	$7 \cdot 13 \cdot 10997449 \cdot 151604179 \cdot 313968931$
35	$29 \cdot 71 \cdot 4651 \cdot 6959 \cdot 7377791 \cdot 3490454731$
37	$149 \cdot 25679 \cdot 61051 \cdot 5006767 \cdot 52853144173$
39	$7 \cdot 13^2 \cdot 859 \cdot 11839990891 \cdot 185018332369573$

Table 16a. Factors of $6^n + 5^n$

n	Factors
1	11
2	61
3	$11 \cdot 31$
4	$17 \cdot 113$
5	$11 \cdot 991$
6	$61 \cdot 1021$
7	$11 \cdot 43 \cdot 757$
8	2070241
9	$11 \cdot 31 \cdot 35281$
10	$61 \cdot 181 \cdot 6361$
11	$11^2 \cdot 23 \cdot 353 \cdot 419$
12	$17 \cdot 113 \cdot 1009 \cdot 1249$
13	$11 \cdot 79 \cdot 157 \cdot 104677$
14	$61 \cdot 1384716061$
15	$11 \cdot 31 \cdot 991 \cdot 1481671$
16	2973697798081
17	$11 \cdot 1608145354351$
18	$61 \cdot 109 \cdot 1021 \cdot 2269 \cdot 6841$
19	$11 \cdot 1158887 \cdot 49297553$
20	$17 \cdot 41 \cdot 113 \cdot 241 \cdot 197642201$
21	$11 \cdot 31 \cdot 43 \cdot 757 \cdot 12517 \cdot 161323$
22	$61 \cdot 30713 \cdot 71527296677$
23	$11 \cdot 47 \cdot 68771 \cdot 22547040163$
24	$433 \cdot 2070241 \cdot 5352419857$
25	$11 \cdot 101 \cdot 151 \cdot 991 \cdot 12601 \cdot 13713251$
26	$53 \cdot 61 \cdot 53223583142271257$
27	$11 \cdot 31 \cdot 163 \cdot 35281 \cdot 525715668307$
28	$17 \cdot 113 \cdot 53089 \cdot 60580061533489$
29	$11 \cdot 59 \cdot 233 \cdot 10267 \cdot 23852396714239$
30	$61 \cdot 181 \cdot 1021 \cdot 6361 \cdot 7561 \cdot 409472221$
31	$11 \cdot 58901 \cdot 84011 \cdot 547399 \cdot 44674039$
32	$193 \cdot 577 \cdot 645313 \cdot 9104321 \cdot 12199937$
33	$11^2 \cdot 23 \cdot 31 \cdot 67 \cdot 353 \cdot 419 \cdot 55989980564893$
34	$61 \cdot 470645699290500173623566$
35	$11 \cdot 43 \cdot 631 \cdot 757 \cdot 991 \cdot 7690744212894121$
36	$17 \cdot 73 \cdot 113 \cdot 1009 \cdot 1249 \cdot 58445821686260377$
37	$11 \cdot 38917859 \cdot 144732124061802361589$
38	$61 \cdot 59988636488701 \cdot 101571993910081$
39	$11 \cdot 31 \cdot 79 \cdot 157 \cdot 104677 \cdot 5036406912869773351$
40	$2801 \cdot 21601 \cdot 2070241 \cdot 106791706920474161$

414

	Table 17a. Factors of $7^n + 2^n$
1	3^2
2	53
3	$3^3 \cdot 13$
4	2417
5	$3^2 \cdot 1871$
6	$53 \cdot 2221$
7	$3^2 \cdot 71 \cdot 1289$
8	$17 \cdot 339121$
9	$3^4 \cdot 13 \cdot 19 \cdot 2017$
10	$53 \cdot 5329741$
11	$3^2 \cdot 23 \cdot 9552313$
12	$337 \cdot 2417 \cdot 16993$
13	$3^2 \cdot 131 \cdot 157 \cdot 523433$
14	$29 \cdot 53 \cdot 113 \cdot 617 \cdot 6329$
15	$3^3 \cdot 13 \cdot 1871 \cdot 7229191$
16	$97 \cdot 449 \cdot 8513 \cdot 89633$
17	$3^2 \cdot 1021 \cdot 25316194811$
18	$37 \cdot 53 \cdot 2221 \cdot 8101 \cdot 46153$
19	$3^2 \cdot 29147 \cdot 43453662797$
20	$41 \cdot 641 \cdot 2417 \cdot 1256152201$
21	$3^3 \cdot 13 \cdot 71 \cdot 673 \cdot 1289 \cdot 3361 \cdot 7687$
22	$53 \cdot 11617 \cdot 6350194410253$
23	$3^2 \cdot 3040971926676589439$
24	$17 \cdot 13729 \cdot 339121 \cdot 2420529889$
25	$3^2 \cdot 151 \cdot 1871 \cdot 963301 \cdot 547514651$
26	$53 \cdot 10973 \cdot 127817 \cdot 126287152681$
27	$3^5 \cdot 13 \cdot 19 \cdot 2017 \cdot 5347 \cdot 123661 \cdot 820909$
28	$2417 \cdot 1355548993 \cdot 140395520497$
29	$3^2 \cdot 59 \cdot 349 \cdot 3307 \cdot 5253987285355843$
30	$53 \cdot 2221 \cdot 5329741 \cdot 35926149051781$
31	$3^2 \cdot 13604858587 \cdot 1288554224343277$
32	$876097 \cdot 1260622595721616043201$
33	$3^3 \cdot 13 \cdot 23 \cdot 463 \cdot 9552313 \cdot 11142253 \cdot 19432909$
34	$53 \cdot 2381 \cdot 428842772879257262178881$
35	$3^2 \cdot 71 \cdot 631 \cdot 1289 \cdot 1871 \cdot 3889560108609443241$

Table 17b. Factors of $(7^n - 2^n)/5$	
3	67
5	$5 \cdot 11 \cdot 61$
7	164683
9	$67 \cdot 163 \cdot 739$
11	$199 \cdot 1987261$
13	$1301 \cdot 14894543$
15	$5 \cdot 11 \cdot 31 \cdot 61 \cdot 67 \cdot 136261$
17	$137 \cdot 443 \cdot 766606297$
19	2279779036969771
21	$43 \cdot 67 \cdot 164683 \cdot 235448977$
23	$47 \cdot 116462754638606501$
25	$5^2 \cdot 11 \cdot 61 \cdot 101 \cdot 158305897173001$
27	$67 \cdot 163 \cdot 379 \cdot 739 \cdot 1783 \cdot 2409792661$
29	$20942288579 \cdot 30750275870441$
31	$21559938839 \cdot 146359767727821$
33	$67 \cdot 199 \cdot 331 \cdot 4423 \cdot 1987261 \cdot 39860027527$
35	$5 \cdot 11 \cdot 61 \cdot 211 \cdot 164683 \cdot 649886695996040161$

Table 18b. Factors of $(7^n - 3^n)/4$	
3	79
5	$41 \cdot 101$
7	205339
9	$79 \cdot 109 \cdot 1171$
11	$89 \cdot 5553791$
13	$56993 \cdot 424997$
15	$41 \cdot 79 \cdot 101 \cdot 751 \cdot 4831$
17	$647 \cdot 89888093063$
19	2849723505777919
21	$43 \cdot 79 \cdot 127 \cdot 379 \cdot 4159 \cdot 205339$
23	$47 \cdot 8741 \cdot 16654666834177$
25	$41 \cdot 101 \cdot 80962848274370701$
27	$79 \cdot 109 \cdot 1171 \cdot 3641653 \cdot 447381523$
29	$59 \cdot 2437 \cdot 439409 \cdot 12741092841523$
31	$18043 \cdot 2186102394754588111993$
33	$67 \cdot 79 \cdot 89 \cdot 331 \cdot 146323 \cdot 5553791 \cdot 15252931$
35	$41 \cdot 101 \cdot 211 \cdot 70351 \cdot 205339 \cdot 659611 \cdot 11375071$

416

Table 18a. Factors of $(7^n + 3^n)/2$

1	5
2	29
3	$5 \cdot 37$
4	$17 \cdot 73$
5	$5^2 \cdot 11 \cdot 31$
6	$13 \cdot 29 \cdot 157$
7	$5 \cdot 71 \cdot 1163$
8	$113 \cdot 25537$
9	$5 \cdot 19 \cdot 37 \cdot 5743$
10	$29 \cdot 4871281$
11	$5 \cdot 23 \cdot 8597843$
12	$17 \cdot 73 \cdot 5576881$
13	$5 \cdot 9689060473$
14	$29 \cdot 197 \cdot 2017 \cdot 29429$
15	$5^2 \cdot 11 \cdot 31 \cdot 37 \cdot 691 \cdot 10891$
16	$97 \cdot 171303987713$
17	$5 \cdot 9419 \cdot 2469801923$
18	$13 \cdot 29 \cdot 157 \cdot 13756052521$
19	$5 \cdot 191 \cdot 20521 \cdot 290824451$
20	$17 \cdot 73 \cdot 32148376222561$
21	$5 \cdot 37 \cdot 71 \cdot 1163 \cdot 3529 \cdot 5180449$
22	$29 \cdot 661 \cdot 101982917209141$
23	$5 \cdot 139 \cdot 81663709 \cdot 241107667$
24	$113 \cdot 193 \cdot 25537 \cdot 54721 \cdot 3143137$
25	$5^3 \cdot 11 \cdot 31 \cdot 2414101 \cdot 6516301601$
26	$29 \cdot 53 \cdot 521 \cdot 5861482247985157$
27	$5 \cdot 19 \cdot 37 \cdot 271 \cdot 5743 \cdot 6005976772267$
28	$17 \cdot 73 \cdot 1128289 \cdot 164256660577969$
29	$5 \cdot 1103 \cdot 63743 \cdot 23598983 \cdot 194062607$
30	$13 \cdot 29 \cdot 61 \cdot 157 \cdot 693961 \cdot 923341 \cdot 4871281$
31	$5 \cdot 157775382035463480011326669$
32	$257 \cdot 4673 \cdot 45980996645427023296$1
33	$5 \cdot 23 \cdot 37 \cdot 199 \cdot 294757 \cdot 8597843 \cdot 1801353907$
34	$29 \cdot 84389 \cdot 949213 \cdot 11648120341225553$
35	$5^2 \cdot 11 \cdot 31 \cdot 71 \cdot 1163 \cdot 47111 \cdot 5711453309691431$

Table 19a. Factors of $7^n + 4^n$	
1	11
2	$5 \cdot 13$
3	$11 \cdot 37$
4	2657
5	$11 \cdot 1621$
6	$5 \cdot 13 \cdot 1873$
7	$11 \cdot 29 \cdot 2633$
8	$17 \cdot 193 \cdot 1777$
9	$11 \cdot 37 \cdot 99793$
10	$5^2 \cdot 13 \cdot 872381$
11	$11^2 \cdot 23 \cdot 67 \cdot 10627$
12	$1993 \cdot 2617 \cdot 2657$
13	$11 \cdot 79 \cdot 111572059$
14	$5 \cdot 13 \cdot 449 \cdot 23247953$
15	$11 \cdot 37 \cdot 1621 \cdot 1831 \cdot 3931$
16	$563777 \cdot 58954561$
17	$11 \cdot 1667 \cdot 3469 \cdot 3657347$
18	$5 \cdot 13 \cdot 1873 \cdot 4861 \cdot 2751733$
19	$11 \cdot 1036288187570917$
20	$41 \cdot 401 \cdot 2657 \cdot 18121 \cdot 100801$
21	$11 \cdot 29 \cdot 37 \cdot 43 \cdot 379 \cdot 1009 \cdot 1093 \cdot 2633$
22	$5 \cdot 13 \cdot 617 \cdot 174241 \cdot 559512713$
23	$11 \cdot 2531 \cdot 5107 \cdot 389621 \cdot 494041$
24	$17 \cdot 193 \cdot 1777 \cdot 32859423538561$
25	$11 \cdot 1621 \cdot 75210013210917601$
26	$5 \cdot 13^2 \cdot 157 \cdot 7229 \cdot 1589849 \cdot 6156853$
27	$11 \cdot 37 \cdot 109 \cdot 99793 \cdot 532981 \cdot 27849313$
28	$2657 \cdot 3361 \cdot 51509237566325281$
29	$11 \cdot 28711 \cdot 10195351303566120931$
30	$5^2 \cdot 13 \cdot 61 \cdot 1873 \cdot 241441 \cdot 872381 \cdot 2881861$
31	$11 \cdot 154877 \cdot 92610374476949641001$
32	$577 \cdot 648449 \cdot 2951791308975956929$
33	$11^2 \cdot 23 \cdot 37 \cdot 67 \cdot 991 \cdot 10627 \cdot 106404808991779$
34	$5 \cdot 13 \cdot 137 \cdot 2857 \cdot 212710632348967318 0673$
35	$11 \cdot 29 \cdot 71 \cdot 701 \cdot 1621 \cdot 2633 \cdot 12669161 \cdot 441247871$

	Table 19b. Factors of $(7^n - 4^n)/3$
3	$3 \cdot 31$
5	5261
7	$113 \cdot 2381$
9	$3^2 \cdot 19 \cdot 31 \cdot 2521$
11	657710813
13	$53 \cdot 608942777$
15	$3 \cdot 31 \cdot 421 \cdot 5261 \cdot 7681$
17	$103 \cdot 752793961547$
19	$4789 \cdot 72277 \cdot 10977061$
21	$3 \cdot 31 \cdot 113 \cdot 2381 \cdot 11173 \cdot 665953$
23	$47 \cdot 139 \cdot 1979 \cdot 470167 \cdot 1500797$
25	$5261 \cdot 15901 \cdot 67751 \cdot 78871651$
27	$3^3 \cdot 19 \cdot 31 \cdot 1459 \cdot 2521 \cdot 374471275609$
29	$59 \cdot 523 \cdot 151903 \cdot 228982248320251$
31	$81391701437 \cdot 646156690987649$
33	$3 \cdot 31 \cdot 4214893 \cdot 657710813 \cdot 9995616313$
35	$113 \cdot 2381 \cdot 5261 \cdot 89208041273469370981$

	Table 20b. Factors of $(7^n - 5^n)/2$
3	109
5	6841
7	372709
9	$19 \cdot 73 \cdot 109 \cdot 127$
11	$23 \cdot 419 \cdot 100057$
13	$131 \cdot 365146211$
15	$31 \cdot 109 \cdot 6841 \cdot 102031$
17	$1531 \cdot 75724224211$
19	$3877 \cdot 1467606615817$
21	$43 \cdot 109 \cdot 3067 \cdot 52081 \cdot 372709$
23	$139 \cdot 540961 \cdot 181909325071$
25	$2351 \cdot 6841 \cdot 41682356332151$
27	$19 \cdot 73 \cdot 109 \cdot 127 \cdot 1459 \cdot 3187 \cdot 367979653$
29	$59 \cdot 27285758400832683314699$
31	$683 \cdot 2357 \cdot 4093 \cdot 2206829 \cdot 5425074587$
33	$23 \cdot 67 \cdot 109 \cdot 419 \cdot 463 \cdot 100057 \cdot 136951 \cdot 8656891$
35	$6841 \cdot 14561 \cdot 372709 \cdot 49339501 \cdot 103400501$

Table 20a. Factors of $(7^n + 5^n)/2$

1	$2 \cdot 3$
2	37
3	$2 \cdot 3^2 \cdot 13$
4	$17 \cdot 89$
5	$2 \cdot 3 \cdot 11 \cdot 151$
6	$37 \cdot 1801$
7	$2 \cdot 3 \cdot 29 \cdot 2591$
8	$97 \cdot 31729$
9	$2 \cdot 3^3 \cdot 13 \cdot 30133$
10	$37 \cdot 61 \cdot 101 \cdot 641$
11	$2 \cdot 3 \cdot 168846239$
12	$17 \cdot 89 \cdot 4654801$
13	$2 \cdot 3 \cdot 79 \cdot 313 \cdot 330643$
14	$37 \cdot 84533 \cdot 109397$
15	$2 \cdot 3^2 \cdot 11 \cdot 13 \cdot 151 \cdot 211 \cdot 29131$
16	16692759230113
17	$2 \cdot 3 \cdot 103 \cdot 6869 \cdot 27490123$
18	$37 \cdot 1801 \cdot 12247162201$
19	$2 \cdot 3 \cdot 5860133 \cdot 162367883$
20	$17 \cdot 41 \cdot 89 \cdot 643912383161$
21	$2 \cdot 3^2 \cdot 13 \cdot 29 \cdot 2591 \cdot 15897110041$
22	$37 \cdot 270073 \cdot 195753141937$
23	$2 \cdot 3 \cdot 47 \cdot 1910059 \cdot 25416641143$
24	$97 \cdot 31729 \cdot 2471809 \cdot 12595489$
25	$2 \cdot 3 \cdot 11 \cdot 151 \cdot 1183151 \cdot 56879588851$
26	$37 \cdot 53 \cdot 267306469 \cdot 8955728393$
27	$2 \cdot 3^4 \cdot 13 \cdot 23599 \cdot 30133 \cdot 21941786467$
28	$17 \cdot 89 \cdot 18381329 \cdot 8270551286369$
29	$2 \cdot 3 \cdot 10093 \cdot 26586842533836180377$
30	$37 \cdot 61 \cdot 101 \cdot 181 \cdot 641 \cdot 1801 \cdot 236606079421$
31	$2 \cdot 3 \cdot 131483365539765736673701739$
32	$2689 \cdot 3329 \cdot 4018753 \cdot 15350429395841$
33	$2 \cdot 3^2 \cdot 13 \cdot 331 \cdot 2311 \cdot 2707 \cdot 47248543 \cdot 168846239$
34	$37 \cdot 1608133 \cdot 454762188593360145997$
35	$2 \cdot 3 \cdot 11 \cdot 29 \cdot 71 \cdot 151 \cdot 281 \cdot 2591 \cdot 41776001 \cdot 303477791$

	Table 21a. Factors of $7^n + 6^n$
1	13
2	$5 \cdot 17$
3	$13 \cdot 43$
4	3697
5	$13 \cdot 31 \cdot 61$
6	$5 \cdot 17 \cdot 1933$
7	$13 \cdot 29 \cdot 2927$
8	$353 \cdot 21089$
9	$13 \cdot 43 \cdot 90217$
10	$5^2 \cdot 17 \cdot 41 \cdot 19681$
11	$13 \cdot 23 \cdot 419 \cdot 18679$
12	$3697 \cdot 4332721$
13	$13^2 \cdot 650589967$
14	$5 \cdot 17 \cdot 337 \cdot 26412541$
15	$13 \cdot 31 \cdot 43 \cdot 61 \cdot 181 \cdot 27271$
16	36054040477057
17	$13 \cdot 1259 \cdot 15247581929$
18	$5 \cdot 17 \cdot 37 \cdot 109 \cdot 1933 \cdot 2610721$
19	$13 \cdot 571 \cdot 881411 \cdot 1835363$
20	$601 \cdot 3697 \cdot 9721 \cdot 3863521$
21	$13 \cdot 29 \cdot 43 \cdot 211 \cdot 2927 \cdot 57979489$
22	$5 \cdot 17 \cdot 287321 \cdot 365069 \cdot 453289$
23	$13 \cdot 599 \cdot 3616088039441957$
24	$353 \cdot 21089 \cdot 26371388480641$
25	$13 \cdot 31 \cdot 61 \cdot 701 \cdot 312701 \cdot 254143801$
26	$5 \cdot 17 \cdot 1769762333 \cdot 63538355921$
27	$13 \cdot 43 \cdot 19009 \cdot 90217 \cdot 69614507377$
28	$3697 \cdot 1487215241 \cdot 84777660521$
29	$13 \cdot 59 \cdot 1103 \cdot 133691 \cdot 28794633922133$
30	$5^2 \cdot 17 \cdot 41 \cdot 541 \cdot 1933 \cdot 19681 \cdot 63464533801$
31	$13 \cdot 24449081 \cdot 268169221 \cdot 1866639023$
32	$257 \cdot 4328351499431389288675841$
33	$13 \cdot 23 \cdot 43 \cdot 67 \cdot 419 \cdot 727 \cdot 18679 \cdot 1587058994071$
34	$5 \cdot 17^2 \cdot 613 \cdot 61418366576438054494937$
35	$13 \cdot 29 \cdot 31 \cdot 61 \cdot 71 \cdot 2927 \cdot 2568525080137867901$

	Table 21b. Factors of $7^n - 6^n$
3	127
5	$11 \cdot 821$
7	543607
9	$19 \cdot 127 \cdot 12547$
11	$89 \cdot 18140783$
13	$53 \cdot 79 \cdot 20021093$
15	$11 \cdot 127 \cdot 821 \cdot 3729391$
17	$1021 \cdot 290701 \cdot 726751$
19	$422029 \cdot 25565862643$
21	$127 \cdot 4999 \cdot 543607 \cdot 1554841$
23	$47 \cdot 565511002489942841$
25	$11 \cdot 151 \cdot 821 \cdot 6551 \cdot 146934809201$
27	$19 \cdot 127 \cdot 919 \cdot 1999 \cdot 12421 \cdot 12547 \cdot 93637$
29	318306010252639083385431
31	$2729 \cdot 614483 \cdot 93295181256835141$
33	$89 \cdot 127 \cdot 397 \cdot 81973 \cdot 18140783 \cdot 1151419039$
35	$11 \cdot 821 \cdot 543607 \cdot 76813086046161451351$

	Table 22b. Factors of $(8^n - 3^n)/5$
3	97
5	$5 \cdot 1301$
7	418993
9	$19 \cdot 97 \cdot 14563$
11	$23 \cdot 67 \cdot 1114829$
13	$8087 \cdot 13595999$
15	$5 \cdot 97 \cdot 571 \cdot 1301 \cdot 19531$
17	$307 \cdot 1466970478531$
19	28823037382718881
21	$43 \cdot 97 \cdot 211 \cdot 418993 \cdot 5002537$
23	$47 \cdot 2511897064955584991$
25	$5^2 \cdot 1301 \cdot 2251 \cdot 103201704493751$
27	$19 \cdot 97 \cdot 14563 \cdot 536059 \cdot 33610182283$
29	$450893 \cdot 68638237857143004653$
31	$1980704062856484905159341969$
33	$23 \cdot 67 \cdot 97 \cdot 1114829 \cdot 760706165024355001$
35	$5 \cdot 71 \cdot 1301 \cdot 418993 \cdot 246923251 \cdot 169787274581$

	Table 22a. Factors of $8^n + 3^n$
1	11
2	73
3	$7^2 \cdot 11$
4	4177
5	$11 \cdot 3001$
6	$13 \cdot 73 \cdot 277$
7	$11 \cdot 29 \cdot 6581$
8	$17 \cdot 113 \cdot 8737$
9	$7^2 \cdot 11 \cdot 271 \cdot 919$
10	$61 \cdot 73 \cdot 241141$
11	$11^2 \cdot 5171 \cdot 13729$
12	$577 \cdot 4177 \cdot 28513$
13	$11 \cdot 53 \cdot 942980117$
14	$73 \cdot 757 \cdot 1009 \cdot 78877$
15	$7^2 \cdot 11 \cdot 31 \cdot 3001 \cdot 701671$
16	$193 \cdot 599009 \cdot 2434721$
17	$11 \cdot 103 \cdot 250037 \cdot 7948691$
18	$13 \cdot 37 \cdot 73 \cdot 277 \cdot 12421 \cdot 149113$
19	$11 \cdot 49552609 \cdot 264393361$
20	$41 \cdot 4177 \cdot 18041 \cdot 373156321$
21	$7^3 \cdot 11 \cdot 29 \cdot 6581 \cdot 9871 \cdot 1297633$
22	$73 \cdot 1010780497619442001$
23	$11 \cdot 5800241477 \cdot 9251900237$
24	$17 \cdot 113 \cdot 8737 \cdot 18913 \cdot 14876801377$
25	$11 \cdot 101 \cdot 151 \cdot 3001 \cdot 16301 \cdot 4603396951$
26	$73 \cdot 4140156916523276157601$
27	$7^2 \cdot 11 \cdot 271 \cdot 919 \cdot 1567 \cdot 7993 \cdot 33967 \cdot 42337$
28	$4177 \cdot 4630790786175950104801$
29	$11 \cdot 59 \cdot 23843221095645788095 4939$
30	$13 \cdot 61 \cdot 73 \cdot 241 \cdot 277 \cdot 4441 \cdot 15061 \cdot 19861 \cdot 241141$
31	$11 \cdot 11719 \cdot 76825670157115949022871$
32	$34032769 \cdot 2327996364746758943233$
33	$7^2 \cdot 11^2 \cdot 2311 \cdot 5171 \cdot 13729 \cdot 11615473 \cdot 56096767$
34	$73 \cdot 9181 \cdot 7565658083195841147767221$
35	$11 \cdot 29 \cdot 3001 \cdot 6581 \cdot 666230671 \cdot 9664432411831$

	Table 23a. Factors of $8^n + 5^n$
1	13
2	89
3	$7^2 \cdot 13$
4	4721
5	$11 \cdot 13 \cdot 251$
6	$89 \cdot 3121$
7	$13 \cdot 167329$
8	$17 \cdot 1009873$
9	$7^2 \cdot 13 \cdot 19 \cdot 11251$
10	$89 \cdot 181 \cdot 67261$
11	$13 \cdot 23 \cdot 28892183$
12	$2857 \cdot 4721 \cdot 5113$
13	$13^2 \cdot 25169 \cdot 129533$
14	$29 \cdot 89 \cdot 1706373509$
15	$7^2 \cdot 11 \cdot 13 \cdot 251 \cdot 1051 \cdot 19051$
16	$641 \cdot 29569 \cdot 14858689$
17	$13 \cdot 4463317 \cdot 38821813$
18	$89 \cdot 3121 \cdot 214381 \cdot 302581$
19	$13 \cdot 11087250889398769$
20	$401 \cdot 4721 \cdot 570961 \cdot 1066721$
21	$7^3 \cdot 13 \cdot 127 \cdot 3571 \cdot 27259 \cdot 167329$
22	$89 \cdot 829093937984598001$
23	$13 \cdot 47 \cdot 691 \cdot 241823 \cdot 5781780619$
24	$17 \cdot 1249 \cdot 3217 \cdot 3793 \cdot 18049 \cdot 1009873$
25	$11 \cdot 13 \cdot 251 \cdot 11801 \cdot 76001 \cdot 1173559801$
26	$89 \cdot 6449 \cdot 9049 \cdot 58191403405721$
27	$7^2 \cdot 13 \cdot 19 \cdot 11251 \cdot 31989493 \cdot 555059413$
28	$4721 \cdot 244553 \cdot 16753805796341657$
29	$13 \cdot 59 \cdot 375302268371 \cdot 537568227529$
30	$61 \cdot 89 \cdot 181 \cdot 3121 \cdot 67261 \cdot 1119121 \cdot 5362501$
31	$13 \cdot 11161 \cdot 976243321 \cdot 69917399183089$
32	$769 \cdot 1217 \cdot 84656984224706453739457$
33	$7^2 \cdot 13 \cdot 23 \cdot 122299 \cdot 789493 \cdot 15507823 \cdot 28892183$
34	$89 \cdot 409 \cdot 3061 \cdot 45507498488038589473189$
35	$11 \cdot 13 \cdot 211 \cdot 251 \cdot 7001 \cdot 167329 \cdot 11011211 \cdot 415232161$

Table 23b. Factors of $(8^n - 5^n)/3$	
3	$3 \cdot 43$
5	$41 \cdot 241$
7	$71 \cdot 9479$
9	$3^2 \cdot 37 \cdot 43 \cdot 3079$
11	$67 \cdot 199 \cdot 213533$
13	$53 \cdot 443 \cdot 599 \cdot 13001$
15	$3 \cdot 31 \cdot 41 \cdot 43 \cdot 241 \cdot 296551$
17	$193597 \cdot 3875812253$
19	48032038196509249
21	$3 \cdot 43 \cdot 71 \cdot 9479 \cdot 52711 \cdot 671791$
23	$46490959 \cdot 4232248607231$
25	$41 \cdot 241 \cdot 1274453794816087601$
27	$3^3 \cdot 37 \cdot 43 \cdot 163 \cdot 919 \cdot 3079 \cdot 11719 \cdot 3471121$
29	$70877 \cdot 727750509785297587213$
31	$311^2 \cdot 34130870089122903753409$
33	$3 \cdot 43 \cdot 67 \cdot 199 \cdot 213533 \cdot 16411693 \cdot 35051956237$
35	$41 \cdot 71 \cdot 241 \cdot 9479 \cdot 56911 \cdot 1513751 \cdot 23602381601$

Table 24b. Factors of $8^n - 7^n$	
3	13^2
5	$11 \cdot 1451$
7	1273609
9	$13^2 \cdot 199 \cdot 2791$
11	6612607849
13	$157 \cdot 3329 \cdot 866477$
15	$11 \cdot 13^2 \cdot 31 \cdot 181 \cdot 1451 \cdot 2011$
17	20191169299698041
19	$8403853 \cdot 15792314893$
21	$13^2 \cdot 43 \cdot 883 \cdot 1060249 \cdot 1273609$
23	$47 \cdot 5452933 \cdot 2196464096219$
25	$11 \cdot 401 \cdot 1451 \cdot 47051 \cdot 120998384851$
27	$13^2 \cdot 199 \cdot 271 \cdot 2791 \cdot 92468566625599$
29	$1515225991548593546355552921$
31	$9745744932248196392577951049$
33	$13^2 \cdot 67 \cdot 727 \cdot 142231 \cdot 80868019 \cdot 6612607849$
35	$11 \cdot 1451 \cdot 1273609 \cdot 1976872083389140097161$

	Table 24a. Factors of $8^n + 7^n$
1	$3 \cdot 5$
2	113
3	$3^2 \cdot 5 \cdot 19$
4	$73 \cdot 89$
5	$3 \cdot 5^2 \cdot 661$
6	$113 \cdot 3361$
7	$3 \cdot 5 \cdot 194713$
8	$17 \cdot 1326001$
9	$3^3 \cdot 5 \cdot 19 \cdot 68059$
10	$113 \cdot 12001921$
11	$3 \cdot 5 \cdot 23 \cdot 30629743$
12	$73 \cdot 89 \cdot 12707521$
13	$3 \cdot 5 \cdot 43109654953$
14	$29 \cdot 113 \cdot 281 \cdot 5512669$
15	$3^2 \cdot 5^2 \cdot 19 \cdot 61 \cdot 661 \cdot 231661$
16	314707907280257
17	$3 \cdot 5 \cdot 137 \cdot 1327 \cdot 5101 \cdot 178603$
18	$37 \cdot 113 \cdot 541 \cdot 3361 \cdot 2583793$
19	$3 \cdot 5 \cdot 886541 \cdot 11694445661$
20	$41 \cdot 73 \cdot 89 \cdot 7121 \cdot 649867081$
21	$3^2 \cdot 5 \cdot 19 \cdot 48679 \cdot 194713 \cdot 1207039$
22	$113 \cdot 804497 \cdot 854673513233$
23	$3 \cdot 5 \cdot 20747 \cdot 1984751394414571$
24	$17 \cdot 35617 \cdot 1326001 \cdot 6120408673$
25	$3 \cdot 5^3 \cdot 101 \cdot 661 \cdot 453851 \cdot 3442955851$
26	$53 \cdot 113 \cdot 1770497 \cdot 29388290965021$
27	$3^4 \cdot 5 \cdot 19 \cdot 487 \cdot 68059 \cdot 303859 \cdot 32046463$
28	$73 \cdot 89 \cdot 337 \cdot 673 \cdot 14897 \cdot 902131700513$
29	$3 \cdot 5 \cdot 59 \cdot 2440308821 \cdot 73141808154431$
30	$113 \cdot 3361 \cdot 25741 \cdot 12001921 \cdot 10742679301$
31	$3 \cdot 5 \cdot 6707530464211925337205357 69$
32	$33857 \cdot 1347511751681 \cdot 1760802861761$
33	$3^2 \cdot 5 \cdot 19 \cdot 23 \cdot 4093 \cdot 12509641 \cdot 20802277 \cdot 30629743$
34	$113 \cdot 4535149873407849307 6599756481$
35	$3 \cdot 5^2 \cdot 71 \cdot 661 \cdot 194713 \cdot 398163641 \cdot 150040605751$

426

Table 25a. Factors of $9^n + 2^n$

1	11
2	$5 \cdot 17$
3	$11 \cdot 67$
4	6577
5	$11 \cdot 41 \cdot 131$
6	$5 \cdot 13^2 \cdot 17 \cdot 37$
7	$11 \cdot 434827$
8	$449 \cdot 95873$
9	$11 \cdot 19 \cdot 67 \cdot 73 \cdot 379$
10	$5^2 \cdot 17 \cdot 2441 \cdot 3361$
11	$11^2 \cdot 259347617$
12	$6577 \cdot 42942001$
13	$11 \cdot 1483 \cdot 155818417$
14	$5 \cdot 17 \cdot 29^2 \cdot 337 \cdot 949621$
15	$11 \cdot 41 \cdot 67 \cdot 131 \cdot 52013371$
16	$1601 \cdot 20353 \cdot 56867009$
17	$11 \cdot 11716571 \cdot 129398561$
18	$5 \cdot 13^2 \cdot 17 \cdot 37 \cdot 397 \cdot 1153 \cdot 616933$
19	$11 \cdot 761483 \cdot 161270444129$
20	$241 \cdot 6577 \cdot 702881 \cdot 10912481$
21	$11 \cdot 43 \cdot 67 \cdot 211 \cdot 1723 \cdot 21841 \cdot 434827$
22	$5 \cdot 17 \cdot 89 \cdot 3433 \cdot 20549 \cdot 1845277897$
23	$11 \cdot 139 \cdot 5796558613245591553$
24	$97 \cdot 449 \cdot 95873 \cdot 102337 \cdot 186669409$
25	$11 \cdot 41 \cdot 131 \cdot 401 \cdot 2801 \cdot 10818260054401$
26	$5 \cdot 17 \cdot 53 \cdot 313 \cdot 1301 \cdot 6761 \cdot 23557 \cdot 22113521$
27	$11 \cdot 19 \cdot 67 \cdot 73 \cdot 163 \cdot 379 \cdot 186841 \cdot 4928387059$
28	$113 \cdot 6577 \cdot 704180474767069120577$
29	$11 \cdot 59 \cdot 7257517253075878020737249$
30	$5^2 \cdot 13^2 \cdot 17 \cdot 37 \cdot 61 \cdot 181 \cdot 541 \cdot 2441 \cdot 3361 \cdot 8821 \cdot 36901$
31	$11 \cdot 1117057467289 \cdot 31049141130244483$
32	$1248533862155521 \cdot 2750172762126337$
33	$11^2 \cdot 67 \cdot 259347617 \cdot 14698072207689632059$
34	$5 \cdot 17^2 \cdot 409 \cdot 240109 \cdot 240335597 \cdot 8155068743293$
35	$11 \cdot 41 \cdot 131 \cdot 434827 \cdot 4547500721 \cdot 21426474239171$

	Table 25b. Factors of $(9^n - 2^n)/7$
3	103
5	8431
7	$7 \cdot 97609$
9	$103 \cdot 127 \cdot 4231$
11	$23 \cdot 194913401$
13	363123688591
15	$31 \cdot 103 \cdot 8431 \cdot 1092601$
17	$137 \cdot 17390179040183$
19	$1103 \cdot 174958129474481$
21	$7 \cdot 103 \cdot 97609 \cdot 222110725159$
23	$47 \cdot 566767 \cdot 47531034164447$
25	$8431 \cdot 420551 \cdot 5303951 \cdot 5453401$
27	$103 \cdot 127 \cdot 271 \cdot 4231 \cdot 2809567 \cdot 197131969$
29	6728755281780349763400676111
31	$394073 \cdot 746047 \cdot 185385978753609833$
33	$23 \cdot 103 \cdot 2750353 \cdot 194913401 \cdot 3476238566263$
35	$7 \cdot 71 \cdot 8431 \cdot 97609 \cdot 303297961 \cdot 2882670197681$

	Table 26b. Factors of $(9^n - 5^n)/4$
3	151
5	$11 \cdot 31 \cdot 41$
7	$29 \cdot 40559$
9	$19 \cdot 151 \cdot 33589$
11	7833057871
13	$79 \cdot 8040016219$
15	$11 \cdot 31 \cdot 41 \cdot 151 \cdot 541 \cdot 45061$
17	4169104690053361
19	$571 \cdot 14479 \cdot 40847631499$
21	$29 \cdot 151 \cdot 40559 \cdot 154016989561$
23	$205391 \cdot 45304159 \cdot 238120979$
25	$11 \cdot 31 \cdot 41 \cdot 9001 \cdot 70951 \cdot 83701 \cdot 240151$
27	$19 \cdot 109 \cdot 151 \cdot 33589 \cdot 3337309 \cdot 414702991$
29	$59 \cdot 1995817165670224538909299$
31	$4626074353079 \cdot 20617936002606749$
33	$151 \cdot 7833057871 \cdot 6531824931593529481$
35	$11 \cdot 29 \cdot 31 \cdot 41 \cdot 71 \cdot 281 \cdot 701 \cdot 40559 \cdot 49554331 \cdot 54908561$

Table 26a. Factors of $(9^n + 5^n)/2$

1	7
2	53
3	$7 \cdot 61$
4	3593
5	$7 \cdot 4441$
6	$13 \cdot 53 \cdot 397$
7	$7^2 \cdot 49603$
8	$17 \cdot 1277569$
9	$7 \cdot 61 \cdot 455941$
10	$53 \cdot 32986321$
11	$7 \cdot 23 \cdot 67 \cdot 89 \cdot 16369$
12	$3593 \cdot 39336721$
13	$7 \cdot 181649037961$
14	$53 \cdot 197 \cdot 1095819173$
15	$7 \cdot 61 \cdot 4441 \cdot 54295441$
16	$673 \cdot 970561 \cdot 1418561$
17	$7 \cdot 1191281759937121$
18	$13 \cdot 37 \cdot 53 \cdot 397 \cdot 2521 \cdot 2941453$
19	$7 \cdot 229 \cdot 647 \cdot 651247415327$
20	$3593 \cdot 49201 \cdot 34386854641$
21	$7^2 \cdot 43 \cdot 61 \cdot 421 \cdot 49603 \cdot 20383567$
22	$53 \cdot 881 \cdot 10545191852840921$
23	$7 \cdot 47 \cdot 13469528937053884763$
24	$17 \cdot 337 \cdot 1277569 \cdot 5449132496593$
25	$7 \cdot 101 \cdot 4441 \cdot 114322768664733701$
26	$53 \cdot 521 \cdot 35933 \cdot 3622321 \cdot 898837837$
27	$7 \cdot 61 \cdot 455941 \cdot 149341769351686621$
28	$617 \cdot 3593 \cdot 113177 \cdot 1042940999359729$
29	$7 \cdot 202406023 \cdot 1662192519802761487$
30	$13 \cdot 53 \cdot 397 \cdot 1321 \cdot 15361 \cdot 32986321 \cdot 115765561$
31	$7 \cdot 1489 \cdot 2692500039629 \cdot 6797345516201$
32	$1217 \cdot 4807681 \cdot 293429712112056751489$
33	$7 \cdot 23 \cdot 61 \cdot 67 \cdot 89 \cdot 16369 \cdot 885721 \cdot 18198400004401$
34	$53 \cdot 56237 \cdot 466570538274120409533357413$
35	$7^2 \cdot 4441 \cdot 49603 \cdot 13967748481 \cdot 8301310004401$

Table 27a. Factors of $(9^n + 7^n)/2$

1	2^3
2	$5 \cdot 13$
3	$2^3 \cdot 67$
4	4481
5	$2^3 \cdot 11 \cdot 431$
6	$5 \cdot 13 \cdot 4993$
7	$2^3 \cdot 29 \cdot 43 \cdot 281$
8	$17^2 \cdot 84449$
9	$2^3 \cdot 67 \cdot 399043$
10	$5^2 \cdot 13 \cdot 401 \cdot 14461$
11	$2^3 \cdot 23 \cdot 617 \cdot 146917$
12	$73 \cdot 4481 \cdot 452857$
13	$2^3 \cdot 79 \cdot 1171 \cdot 1782769$
14	$5 \cdot 13 \cdot 181192427137$
15	$2^3 \cdot 11 \cdot 67 \cdot 151 \cdot 431 \cdot 274471$
16	$449 \cdot 2100504587329$
17	$2^3 \cdot 239 \cdot 8093 \cdot 546400843$
18	$5 \cdot 13 \cdot 4993 \cdot 28513 \cdot 8197921$
19	$2^3 \cdot 287851 \cdot 295780328377$
20	$41 \cdot 4481 \cdot 33304460909081$
21	$2^3 \cdot 29 \cdot 43 \cdot 67 \cdot 281 \cdot 297151 \cdot 985279$
22	$5 \cdot 13 \cdot 661 \cdot 14081 \cdot 817104789821$
23	$2^3 \cdot 277 \cdot 28751 \cdot 2303359 \cdot 30290219$
24	$17^2 \cdot 84449 \cdot 1638097339153921$
25	$2^3 \cdot 11 \cdot 431 \cdot 1801 \cdot 5264651305119901$
26	$5 \cdot 13^2 \cdot 1301 \cdot 1873 \cdot 1571209177192097$
27	$2^3 \cdot 67 \cdot 163 \cdot 379 \cdot 541 \cdot 3943 \cdot 399043 \cdot 1032697$
28	$113^2 \cdot 4481 \cdot 10193 \cdot 449063710848593$
29	$2^3 \cdot 233 \cdot 99761 \cdot 12673391882462976037$
30	$5^2 \cdot 13 \cdot 61 \cdot 401 \cdot 4993 \cdot 8581 \cdot 14461 \cdot 4305472081$
31	$2^3 \cdot 35055421 \cdot 680490686196188952907$
32	$5547814913 \cdot 309562260261976083457$
33	$2^3 \cdot 23 \cdot 67 \cdot 617 \cdot 8779 \cdot 146917 \cdot 151141 \cdot 10423276711$
34	$5 \cdot 13 \cdot 1804381 \cdot 1185927713628157345794037$
35	$2^3 \cdot 11 \cdot 29 \cdot 43 \cdot 71 \cdot 281 \cdot 431 \cdot 1668031 \cdot 11618881 \cdot 68448521$

	Table 27b. Factors of $(9^n - 7^n)/2$
3	193
5	21121
7	1979713
9	$19 \cdot 37 \cdot 193 \cdot 1279$
11	$2663 \cdot 5520791$
13	$53 \cdot 859 \cdot 26851943$
15	$31 \cdot 193 \cdot 21121 \cdot 795871$
17	$137 \cdot 60016610166713$
19	$457 \cdot 2129 \cdot 688344052841$
21	$193 \cdot 32089 \cdot 1979713 \cdot 4439401$
23	$47 \cdot 1381 \cdot 68063301125552099$
25	$21121 \cdot 395201 \cdot 42922814046401$
27	$19 \cdot 37 \cdot 109 \cdot 193 \cdot 1279 \cdot 26964037 \cdot 56941921$
29	$59 \cdot 5743 \cdot 11369 \cdot 610931745533942197$
31	$185753 \cdot 1135531 \cdot 904010195547430331$
33	$193 \cdot 2663 \cdot 2971 \cdot 427417 \cdot 5520791 \cdot 4287261067$
35	$21121 \cdot 30941 \cdot 1979713 \cdot 967255143583330421$

	Table 28b. Factors of $9^n - 8^n$
3	$7 \cdot 31$
5	$41 \cdot 641$
7	2685817
9	$7 \cdot 31 \cdot 1166833$
11	$23 \cdot 990918479$
13	$6553 \cdot 303999697$
15	$7 \cdot 31 \cdot 41 \cdot 641 \cdot 3271 \cdot 9151$
17	$103 \cdot 140052251320207$
19	$191 \cdot 15391 \cdot 26183 \cdot 15678079$
21	$7^2 \cdot 31 \cdot 2685817 \cdot 24559217887$
23	$47 \cdot 93703 \cdot 1878420820626737$
25	$41 \cdot 151 \cdot 641 \cdot 171382356359199751$
27	$7 \cdot 31 \cdot 1166833 \cdot 2185921 \cdot 100693360657$
29	$4555386192335572300559213161$
31	$371616904162662789429456905017$
33	$7 \cdot 23 \cdot 31 \cdot 1783 \cdot 990918479 \cdot 3432621786269743$
35	$41 \cdot 71 \cdot 641 \cdot 2685817 \cdot 491377853869613295551$

	Table 28a. Factors of $9^n + 8^n$
1	17
2	$5 \cdot 29$
3	$17 \cdot 73$
4	10657
5	$11 \cdot 17 \cdot 491$
6	$5 \cdot 13 \cdot 29 \cdot 421$
7	$17 \cdot 404713$
8	$577 \cdot 103681$
9	$17 \cdot 19 \cdot 73 \cdot 22123$
10	$5^2 \cdot 29 \cdot 61 \cdot 101 \cdot 1021$
11	$17 \cdot 67^2 \cdot 523777$
12	$1489 \cdot 10657 \cdot 22129$
13	$17 \cdot 181860096601$
14	$5 \cdot 29 \cdot 281 \cdot 1709 \cdot 391693$
15	$11 \cdot 17 \cdot 73 \cdot 491 \cdot 691 \cdot 52051$
16	$193 \cdot 1217 \cdot 16001 \cdot 567937$
17	$17^2 \cdot 137 \cdot 307 \cdot 1557293467$
18	$5 \cdot 13 \cdot 29 \cdot 37 \cdot 421 \cdot 2557 \cdot 2239057$
19	$17 \cdot 87939229749932233$
20	$10657 \cdot 1082801 \cdot 1153489361$
21	$17 \cdot 43 \cdot 73 \cdot 404713 \cdot 5493541243$
22	$5 \cdot 29 \cdot 89 \cdot 2861 \cdot 19801 \cdot 29173 \cdot 49633$
23	$17 \cdot 252449 \cdot 442291 \cdot 4980233747$
24	$97 \cdot 577 \cdot 2689 \cdot 103681 \cdot 5414541217$
25	$11 \cdot 17 \cdot 491 \cdot 202001 \cdot 40743613162001$
26	$5 \cdot 29 \cdot 53 \cdot 17975058401 \cdot 48960441661$
27	$17 \cdot 19 \cdot 73 \cdot 22123 \cdot 116110335992252113$
28	$113 \cdot 449 \cdot 7393 \cdot 10657 \cdot 135759965703601$
29	$17 \cdot 59 \cdot 4850320241432619510751739$
30	$5^2 \cdot 13 \cdot 29 \cdot 61 \cdot 101 \cdot 421 \cdot 1021 \cdot 1201 \cdot 781861 \cdot 1861501$
31	$17 \cdot 3907 \cdot 58932526052971118780044579$
32	$35129119828067768222513930939617$
33	$17 \cdot 67^2 \cdot 73 \cdot 1123 \cdot 55441 \cdot 263737 \cdot 523777 \cdot 658219$
34	$5 \cdot 29 \cdot 613 \cdot 2789 \cdot 556583333 \cdot 2052506375529397$
35	$11 \cdot 17 \cdot 491 \cdot 206641 \cdot 404713 \cdot 1045801 \cdot 316762351921$

Table 29a. Factors of $10^n + 3^n$

1	13
2	109
3	$13 \cdot 79$
4	$17 \cdot 593$
5	$11 \cdot 13 \cdot 701$
6	$109 \cdot 9181$
7	$13 \cdot 29 \cdot 43 \cdot 617$
8	100006561
9	$13 \cdot 37 \cdot 79 \cdot 26317$
10	$109 \cdot 91743661$
11	$13 \cdot 23 \cdot 199 \cdot 1680647$
12	$17 \cdot 73 \cdot 593 \cdot 1358857$
13	$13^2 \cdot 131 \cdot 5903 \cdot 76519$
14	$109 \cdot 197 \cdot 11117 \cdot 418909$
15	$11 \cdot 13 \cdot 31 \cdot 79 \cdot 271 \cdot 701 \cdot 15031$
16	$9697 \cdot 1031246781793$
17	$13 \cdot 137 \cdot 64091 \cdot 876070453$
18	$109 \cdot 1549 \cdot 9181 \cdot 645107509$
19	$13 \cdot 769230769320173959$
20	$17 \cdot 41 \cdot 593 \cdot 241942703137481$
21	$13 \cdot 29 \cdot 43 \cdot 79 \cdot 617 \cdot 1265546009911$
22	$109 \cdot 353 \cdot 83689 \cdot 3105492024053$
23	$13 \cdot 47 \cdot 163666121113083704057$
24	$1777 \cdot 182353 \cdot 30858241 \cdot 100006561$
25	$11 \cdot 13 \cdot 701 \cdot 275586251 \cdot 361983185651$
26	$53 \cdot 109 \cdot 677 \cdot 946570613 \cdot 27011950577$
27	$13 \cdot 37 \cdot 79 \cdot 26317 \cdot 4037797 \cdot 247654926037$
28	$17 \cdot 113 \cdot 593 \cdot 511897 \cdot 2516473 \cdot 6814642577$
29	$13 \cdot 59 \cdot 233 \cdot 559562645836014955041253$
30	$61^2 \cdot 109 \cdot 1381 \cdot 9181 \cdot 91743661 \cdot 2119600921$
31	$13 \cdot 769230769230769278282568944919$
32	$769 \cdot 1518977 \cdot 85609598896830600941057$
33	$13 \cdot 23 \cdot 67 \cdot 79 \cdot 199 \cdot 1680647 \cdot 111165319 \cdot 16995268027$
34	$109 \cdot 3805862635595633 \cdot 24105735821360477$
35	$11 \cdot 13 \cdot 29 \cdot 43 \cdot 617 \cdot 701 \cdot 1471 \cdot 5881 \cdot 149875441321363801$

	Table 29b. Factors of $(10^n - 3^n)/7$
3	139
5	14251
7	$7 \cdot 211 \cdot 967$
9	$19 \cdot 139 \cdot 54091$
11	$1321 \cdot 3191 \cdot 3389$
13	$157 \cdot 9099179623$
15	$139 \cdot 14251 \cdot 72117691$
17	$103 \cdot 138696255021997$
19	$2281 \cdot 626291726613499$
21	$7 \cdot 139 \cdot 211 \cdot 967 \cdot 719581832971$
23	$53591 \cdot 5561263 \cdot 47933226683$
25	$101 \cdot 14251 \cdot 65851 \cdot 15072068930951$
27	$19 \cdot 139 \cdot 757 \cdot 12853 \cdot 54091 \cdot 102779893009$
29	$187921 \cdot 237859 \cdot 265470467 \cdot 1203901187$
31	$37511 \cdot 380840667689858857033657189$
33	$139 \cdot 1321 \cdot 3169 \cdot 3191 \cdot 3389 \cdot 14020579 \cdot 1619189881$
35	$7 \cdot 71 \cdot 211 \cdot 967 \cdot 14251 \cdot 215412961 \cdot 45890149374581$

	Table 30b. Factors of $(10^n - 7^n)/3$
3	$3 \cdot 73$
5	$11 \cdot 2521$
7	$197 \cdot 15527$
9	$3^2 \cdot 37 \cdot 73 \cdot 13159$
11	$23 \cdot 8317 \cdot 170809$
13	$27067 \cdot 121957993$
15	$3 \cdot 11 \cdot 31 \cdot 73 \cdot 2521 \cdot 1762141$
17	$54503 \cdot 374953 \cdot 1627309$
19	$229 \cdot 14539448478623911$
21	$3 \cdot 73 \cdot 197 \cdot 15527 \cdot 497322616771$
23	$920667599 \cdot 36195702394381$
25	$11 \cdot 101 \cdot 2521 \cdot 1189963375319721901$
27	$3^3 \cdot 37 \cdot 73 \cdot 13159 \cdot 347327340199303483$
29	$57943 \cdot 575259479685462073763717$
31	$3333280741539321718064461652419$
33	$3 \cdot 23 \cdot 73 \cdot 199 \cdot 8317 \cdot 170809 \cdot 15603787 \cdot 15001783567$
35	$11 \cdot 197 \cdot 211 \cdot 281 \cdot 2521 \cdot 15527 \cdot 13912291 \cdot 476398474091$

	Table 30a. Factors of $10^n + 7^n$
1	17
2	149
3	$17 \cdot 79$
4	12401
5	$17 \cdot 6871$
6	$13 \cdot 149 \cdot 577$
7	$17 \cdot 631 \cdot 1009$
8	$353 \cdot 299617$
9	$17 \cdot 19 \cdot 79 \cdot 40771$
10	$149 \cdot 69009901$
11	$17 \cdot 5998666279$
12	$97^2 \cdot 8689 \cdot 12401$
13	$17 \cdot 53 \cdot 11206314107$
14	$29 \cdot 149 \cdot 23299750769$
15	$17 \cdot 61 \cdot 79 \cdot 151 \cdot 6871 \cdot 11821$
16	$7297 \cdot 1374980530433$
17	$17^2 \cdot 239 \cdot 90917 \cdot 15961301$
18	$13 \cdot 109 \cdot 149 \cdot 577 \cdot 8221947589$
19	$17 \cdot 588905817363845479$
20	$41^3 \cdot 12401 \cdot 117094937081$
21	$17 \cdot 43 \cdot 79 \cdot 631 \cdot 1009 \cdot 27213068317$
22	$149 \cdot 1013 \cdot 40056677 \cdot 1654623301$
23	$17 \cdot 47 \cdot 277 \cdot 461 \cdot 980373065211481$
24	$353 \cdot 17137 \cdot 299617 \cdot 551832457873$
25	$17 \cdot 6871 \cdot 85622788605303312001$
26	$149 \cdot 937 \cdot 2393 \cdot 33073 \cdot 9051042800957$
27	$17 \cdot 19 \cdot 79 \cdot 163 \cdot 919 \cdot 40771 \cdot 547453 \cdot 11721889$
28	$12401 \cdot 80642367442436454640401$
29	$17 \cdot 59 \cdot 1887592267139 \cdot 52820786204071$
30	$13 \cdot 149 \cdot 577 \cdot 28081 \cdot 69009901 \cdot 461721800821$
31	$17 \cdot 964257667 \cdot 610049155069329241837$
32	$193 \cdot 1772609 \cdot 7930832257 \cdot 36856642754689$
33	$17 \cdot 67 \cdot 79 \cdot 7789 \cdot 8087773 \cdot 29409469 \cdot 5998666279$
34	$149 \cdot 1429 \cdot 3469 \cdot 38372533757 \cdot 352824777556193$
35	$17 \cdot 71 \cdot 631 \cdot 1009 \cdot 6871 \cdot 18938862939319597778561$

	Table 31a. Factors of $10^n + 9^n$
1	19
2	181
3	$7 \cdot 13 \cdot 19$
4	16561
5	$11 \cdot 19 \cdot 761$
6	$181 \cdot 8461$
7	$19 \cdot 778051$
8	$17 \cdot 1249 \cdot 6737$
9	$7 \cdot 13 \cdot 19 \cdot 802441$
10	$41 \cdot 181 \cdot 241 \cdot 7541$
11	$19 \cdot 23 \cdot 89 \cdot 3378013$
12	$73 \cdot 16561 \cdot 1060777$
13	$19 \cdot 53 \cdot 12454683047$
14	$29 \cdot 181 \cdot 23409562289$
15	$7 \cdot 11 \cdot 13 \cdot 19 \cdot 761 \cdot 5521 \cdot 15091$
16	$97 \cdot 2689 \cdot 45442946977$
17	$19 \cdot 103 \cdot 1361 \cdot 43806340997$
18	$181 \cdot 8461 \cdot 750988536481$
19	$19^2 \cdot 4219 \cdot 7487 \cdot 12161 \cdot 81853$
20	$9521 \cdot 16561 \cdot 329201 \cdot 2160721$
21	$7^2 \cdot 13 \cdot 19 \cdot 60397 \cdot 778051 \cdot 1950649$
22	$181 \cdot 3301 \cdot 3490609 \cdot 5267028889$
23	$19 \cdot 47 \cdot 1289 \cdot 9293 \cdot 648371 \cdot 15696259$
24	$17 \cdot 1249 \cdot 6737 \cdot 23484481 \cdot 321418561$
25	$11 \cdot 19 \cdot 401 \cdot 761 \cdot 3438451 \cdot 48873277451$
26	$181 \cdot 707981 \cdot 830788923528534881$
27	$7 \cdot 13 \cdot 19 \cdot 802441 \cdot 63556543 \cdot 11999931247$
28	$16561 \cdot 944609 \cdot 543579737 \cdot 1237519417$
29	$19 \cdot 59 \cdot 29059 \cdot 3214418477147441944451$
30	$41 \cdot 61 \cdot 181 \cdot 241 \cdot 2221 \cdot 7541 \cdot 8461 \cdot 30781 \cdot 2190481$
31	$19 \cdot 373 \cdot 44579 \cdot 32860049085650464449533$
32	$9601 \cdot 55861633561729 \cdot 192855438828289$
33	$7 \cdot 13 \cdot 19 \cdot 23 \cdot 89 \cdot 331 \cdot 5743 \cdot 3378013 \cdot 45360403163527$
34	$181 \cdot 409 \cdot 91121 \cdot 16834421 \cdot 90509776889642449$
35	$11 \cdot 19 \cdot 281 \cdot 761 \cdot 28211 \cdot 778051 \cdot 4061051 \cdot 25729715411$

	Table 31b. Factors of $10^n - 9^n$
3	271
5	$31 \cdot 1321$
7	5217031
9	$37 \cdot 199 \cdot 271 \cdot 307$
11	68618940391
13	$2081 \cdot 3583918391$
15	$31 \cdot 151 \cdot 271 \cdot 631 \cdot 751 \cdot 1321$
17	$239 \cdot 613 \cdot 216751 \cdot 2623883$
19	8649148282327007911
21	$43 \cdot 271 \cdot 5217031 \cdot 14649144637$
23	$277 \cdot 329014663827969310123$
25	$31 \cdot 1321 \cdot 259001 \cdot 875145743369801$
27	$37 \cdot 163 \cdot 199 \cdot 271 \cdot 307 \cdot 757 \cdot 12460512714031$
29	95289871302753755165078396311
31	$311 \cdot 3092758705959659705374116081$
33	$67^2 \cdot 271 \cdot 397 \cdot 136951 \cdot 213524917 \cdot 68618940391$
35	$31 \cdot 71 \cdot 1321 \cdot 2801 \cdot 8681 \cdot 5217031 \cdot 264339106286921$

Table 32. Quadratic Residues: $(a/p) = +1$, if $p \equiv l \pmod{k}$

a	k	l
−1	4	1
2	8	±1
−2	8	1, 3
3	12	±1
−3	6	1
5	10	±1
−5	20	1, 3, 7, 9
6	24	±(1, 5)
−6	24	1, 5, 7, 11
7	28	±(1, 3, 9)
−7	14	1, 9, 11
10	40	±(1, 3, 9, 13)
−10	40	1, 7, 9, 11, 13, 19, 23, 37
11	44	±(1, 5, 7, 9, 19)
−11	22	1, 3, 5, 9, 15
13	26	±(1, 3, 9)
−13	52	1, 7, 9, 11, 15, 17, 19, 25, 29, 31, 47, 49
14	56	±(1, 5, 9, 11, 13, 25)
−14	56	1, 3, 5, 9, 13, 15, 19, 23, 25, 27, 39, 45
15	60	±(1, 7, 11, 17)
−15	30	1, 17, 19, 23
17	34	±(1, 9, 13, 15)
−17	68	1, 3, 7, 9, 11, 13, 21, 23, 25, 27, 31, 33, 39, 49, 53, 63
19	76	±(1, 3, 5, 9, 15, 17, 25, 27, 31)
−19	38	1, 5, 7, 9, 11, 17, 23, 25, 35
21	42	±(1, 5, 17)
−21	84	1, 5, 11, 17, 19, 23, 25, 31, 37, 41, 55, 71
22	88	±(1, 3, 7, 9, 13, 21, 25, 27, 29, 39)
−22	88	1, 9, 13, 15, 19, 21, 23, 25, 29, 31, 35, 43, 47, 49, 51, 61, 71, 81, 83, 85
23	92	±(1, 7, 9, 11, 13, 15, 19, 25, 29, 41, 43)
−23	46	1, 3, 9, 13, 25, 27, 29, 31, 35, 39, 41
26	104	±(1, 5, 9, 11, 17, 19, 21, 23, 25, 37, 45, 49)
−26	104	1, 3, 5, 7, 9, 15, 17, 21, 25, 27, 31, 35, 37, 43, 45, 47, 49, 51, 63, 71, 75, 81, 85, 93
29	58	±(1, 5, 7, 9, 13, 23, 25)
−29	116	1, 3, 5, 9, 11, 13, 15, 19, 25, 27, 31, 33, 39, 43, 45, 47, 49, 53, 55, 57, 65, 75, 79, 81, 93, 95, 99, 109

		Table 32. Quadratic Residues: $(a/p) = +1$, if $p \equiv l \pmod{k}$
a	k	l
30	120	$\pm(1, 7, 13, 17, 19, 29, 37, 49)$
-30	120	1, 11, 13, 17, 23, 29, 31, 37, 43, 47, 49, 59, 67, 79, 101, 113
31	124	$\pm(1, 3, 5, 9, 11, 15, 23, 25, 27, 33, 41, 43, 45, 49, 55)$
-31	62	1, 5, 7, 9, 19, 25, 33, 35, 39, 41, 45, 47, 49, 51, 59
33	66	$\pm(1, 17, 25, 29, 31)$
-33	132	1, 7, 17, 19, 23, 25, 29, 37, 41, 43, 47, 49, 59, 65, 71, 79, 97, 101, 119, 127
34	136	$\pm(1, 3, 5, 9, 11, 15, 25, 27, 29, 33, 37, 45, 47, 49, 55, 61)$
-34	136	1, 5, 7, 9, 19, 23, 25, 29, 31, 33, 35, 37, 39, 43, 45, 49, 59, 61, 63, 67, 71, 79, 81, 83, 89, 95, 109, 115, 121, 123, 125, 133
35	140	$\pm(1, 9, 13, 17, 19, 23, 29, 31, 33, 43, 59, 67)$
-35	70	1, 3, 9, 11, 13, 17, 27, 29, 33, 39, 47, 51
37	74	$\pm(1, 3, 7, 9, 11, 21, 25, 27, 33)$
-37	148	1, 9, 15, 19, 21, 23, 25, 31, 33, 35, 39, 41, 43, 49, 51, 53, 55, 59, 65, 73, 77, 79, 81, 85, 87, 91, 101, 103, 119, 121, 131, 135, 137, 141, 143, 145
38	152	$\pm(1, 9, 11, 13, 15, 17, 21, 25, 29, 31, 35, 37, 43, 49, 53, 69, 71, 73)$
-38	152	1, 3, 7, 9, 13, 17, 21, 23, 25, 27, 29, 37, 39, 47, 49, 51, 53, 55, 59, 63, 67, 69, 73, 75, 81, 87, 91, 107, 109, 111, 117, 119, 121, 137, 141, 147
39	156	$\pm(1, 5, 7, 19, 23, 25, 31, 35, 41, 49, 61, 67)$
-39	78	1, 5, 11, 25, 41, 43, 47, 49, 55, 59, 61, 71
41	82	$\pm(1, 5, 9, 21, 23, 25, 31, 33, 37, 39)$
-41	164	1, 3, 5, 7, 9, 11, 15, 19, 21, 25, 27, 33, 35, 37, 45, 47, 49, 55, 57, 61, 63, 67, 71, 73, 75, 77, 79, 81, 95, 99, 105, 111, 113, 121, 125, 133, 135, 141, 147, 151
42	168	$\pm(1, 11, 13, 17, 19, 25, 29, 41, 47, 53, 61, 79)$
-42	168	1, 13, 17, 23, 25, 29, 31, 41, 43, 53, 55, 59, 61, 67, 71, 83, 89, 95, 103, 121, 131, 149, 157, 163
43	172	$\pm(1, 3, 7, 9, 13, 17, 19, 21, 25, 27, 39, 41, 49, 51, 53, 55, 57, 63, 71, 75, 81)$
-43	86	1, 9, 11, 13, 15, 17, 21, 23, 25, 31, 35, 41, 47, 49, 53, 57, 59, 67, 79, 81, 83
46	184	$\pm(1, 3, 5, 7, 9, 15, 21, 25, 27, 35, 37, 41, 45, 49, 53, 59, 61, 63, 73, 75, 79, 81)$
-46	184	1, 5, 9, 11, 19, 21, 25, 31, 37, 39, 41, 43, 45, 47, 49, 51, 53, 55, 61, 67, 71, 73, 81, 83, 87, 91, 95, 99, 105, 107, 109, 119, 121, 125, 127, 149, 151, 155, 157, 167, 169, 171, 177, 181

Table 32. Quadratic Residues: $(a/p) = +1$, if $p \equiv l \pmod{k}$

a	k	l
47	188	$\pm$(1, 9, 11, 15, 17, 19, 21, 23, 25, 31, 35, 37, 39, 43, 49, 53, 61, 65, 67, 81, 87, 89, 91)
—47	94	1, 3, 7, 9, 17, 21, 25, 27, 37, 49, 51, 53, 55, 59, 61, 63, 65, 71, 75, 79, 81, 83, 89
51	204	$\pm$(1, 5, 7, 13, 25, 29, 31, 35, 41, 47, 49, 59, 65, 79, 83, 91)
—51	102	1, 5, 11, 13, 19, 23, 25, 29, 41, 43, 49, 55, 65, 67, 71, 95
53	106	$\pm$(1, 7, 9, 11, 13, 15, 17, 25, 29, 37, 43, 47, 49)
—53	212	1, 3, 9, 13, 17, 19, 23, 25, 27, 29, 31, 35, 37, 39, 49, 51, 55, 57, 67, 69, 71, 75, 77, 79, 81, 83, 87, 89, 93, 97, 103, 105, 111, 113, 117, 121, 127, 139, 147, 149, 151, 153, 165, 167, 169, 171, 179, 191, 197, 201, 205, 207
55	220	$\pm$(1, 3, 9, 13, 17, 19, 23, 27, 39, 47, 49, 51, 57, 67, 69, 73, 79, 81, 89, 103)
—55	110	1, 7, 9, 13, 17, 31, 43, 49, 57, 59, 63, 69, 71, 73, 81, 83, 87, 89, 91, 107
57	114	$\pm$(1, 7, 25, 29, 41, 43, 49, 53, 55)
—57	228	1, 11, 23, 25, 29, 31, 35, 41, 47, 49, 53, 61, 65, 67, 73, 79, 83, 85, 89, 91, 103, 113, 119, 121, 127, 131, 151, 157, 169, 173, 185, 191, 211, 215, 221, 223
58	232	$\pm$(1, 3, 7, 9, 11, 19, 21, 23, 25, 27, 33, 37, 43, 49, 57, 61, 63, 65, 69, 71, 75, 77, 81, 85, 99, 101, 103, 111)
—58	232	1, 9, 15, 21, 25, 31, 33, 35, 37, 39, 47, 49, 51, 55, 57, 59, 61, 65, 67, 69, 77, 79, 81, 83, 85, 91, 95, 101, 107, 115, 119, 121, 123, 127, 129, 133, 135, 139, 143, 157, 159, 161, 169, 179, 187, 189, 191, 205, 209, 213, 215, 219, 221, 225, 227, 229
59	236	$\pm$(1, 5, 9, 11, 17, 21, 23, 25, 29, 31, 39, 41, 43, 45, 47, 49, 53, 55, 57, 67, 81, 83, 85, 91, 99, 103, 105, 111, 115)
—59	118	1, 3, 5, 7, 9, 15, 17, 19, 21, 25, 27, 29, 35, 41, 45, 49, 51, 53, 57, 63, 71, 75, 79, 81, 85, 87, 95, 105, 107
61	122	$\pm$(1, 3, 5, 9, 13, 15, 19, 25, 27, 39, 41, 45, 47, 49, 57)
—61	244	1, 5, 7, 9, 11, 13, 23, 25, 31, 35, 41, 43, 45, 49, 51, 55, 57, 59, 63, 65, 67, 71, 73, 77, 79, 81, 87, 91, 97, 99, 109, 111, 113, 115, 117, 121, 125, 137, 139, 141, 143, 149, 151, 155, 159, 161, 169, 175, 191, 197, 205, 207, 211, 215, 217, 223, 225, 227, 229, 241

a	k	l

Table 32. Quadratic Residues: $(a/p) = +1$, if $p \equiv l \pmod{k}$

a	k	l
62	248	$\pm$(1, 9, 13, 15, 19, 21, 23, 25, 29, 33, 35, 37, 41, 49, 51, 53, 55, 59, 61, 67, 77, 79, 81, 85, 97, 107, 113, 117, 119, 121)
−62	248	1, 3, 7, 9, 11, 13, 21, 25, 27, 29, 33, 37, 39, 41, 43, 47, 49, 53, 61, 63, 71, 75, 77, 81, 83, 85, 87, 91, 95, 97, 99, 103, 111, 113, 115, 117, 121, 123, 129, 139, 141, 143, 147, 159, 169, 175, 179, 181, 183, 189, 191, 193, 197, 203, 213, 225, 229, 231, 233, 243
65	130	$\pm$(1, 7, 9, 29, 33, 37, 47, 49, 51, 57, 61, 63)
−65	260	1, 3, 9, 11, 19, 23, 27, 29, 31, 33, 37, 43, 49, 57, 59, 61, 69, 71, 73, 81, 87, 93, 97, 99, 101, 103, 107, 111, 119, 121, 127, 129, 137, 147, 151, 171, 177, 181, 183, 193, 197, 207, 209, 213, 219, 239, 243, 253
66	264	$\pm$(1, 5, 13, 17, 19, 25, 31, 41, 43, 49, 53, 59, 61, 65, 85, 95, 97, 103, 109, 125)
−66	264	1, 5, 7, 13, 17, 23, 25, 35, 41, 47, 49, 53, 61, 65, 67, 71, 79, 83, 85, 91, 97, 107, 109, 115, 119, 125, 127, 131, 151, 161, 163, 169, 175, 191, 205, 221, 227, 233, 235, 245
67	268	$\pm$(1, 3, 7, 9, 11, 17, 21, 25, 27, 29, 31, 33, 37, 43, 49, 51, 63, 65, 73, 75, 77, 79, 81, 87, 89, 93, 95, 99, 111, 115, 119, 121, 129)
−67	134	1, 9, 15, 17, 19, 21, 23, 25, 29, 33, 35, 37, 39, 47, 49, 55, 59, 65, 71, 73, 77, 81, 83, 89, 91, 93, 103, 107, 121, 123, 127, 129, 131
69	138	$\pm$(1, 5, 11, 13, 17, 25, 31, 49, 53, 55, 65)
−69	276	1, 5, 7, 13, 17, 19, 25, 35, 43, 47, 49, 53, 59, 65, 67, 71, 73, 79, 85, 89, 91, 95, 103, 113, 119, 121, 125, 131, 133, 137, 149, 167, 169, 175, 179, 193, 199, 215, 221, 235, 239, 245, 247, 265
70	280	$\pm$(1, 3, 9, 11, 17, 23, 27, 31, 33, 37, 51, 53, 61, 69, 73, 81, 83, 93, 97, 99, 101, 111, 121, 127)
−70	280	1, 9, 17, 19, 33, 37, 39, 43, 47, 53, 59, 61, 67, 69, 71, 73, 79, 81, 87, 93, 97, 101, 103, 107, 121, 123, 131, 139, 143, 151, 153, 163, 167, 169, 171, 181, 191, 197, 223, 229, 239, 249, 251, 253, 257, 267, 269, 277
71	284	$\pm$(1, 5, 7, 9, 11, 23, 25, 29, 31, 35, 37, 39, 45, 47, 49, 51, 55, 57, 59, 63, 67, 73, 77, 81, 89, 99, 101, 109, 115, 121, 123, 125, 127, 129, 139)
−71	142	1, 3, 5, 9, 15, 19, 25, 27, 29, 37, 43, 45, 49, 57, 73, 75, 77, 79, 81, 83, 87, 89, 91, 95, 101, 103, 107, 109, 111, 119, 121, 125, 129, 131, 135

		Table 32. Quadratic Residues: $(a/p) = +1$, if $p \equiv l \pmod{k}$
a	k	l
73	146	$\pm(1, 3, 9, 19, 23, 25, 27, 35, 37, 41, 49, 55, 57, 61, 65, 67, 69, 71)$
−73	292	1, 7, 9, 11, 15, 25, 31, 37, 39, 41, 43, 47, 49, 51, 57, 59, 61, 63, 65, 69, 77, 81, 83, 85, 87, 89, 95, 97, 99, 103, 105, 107, 109, 115, 121, 131, 135, 137, 139, 145, 149, 151, 159, 163, 165, 167, 169, 173, 175, 179, 181, 191, 199, 201, 213, 217, 221, 225, 237, 239, 247, 257, 259, 263, 265, 269, 271, 273, 275, 279, 287, 289
74	296	$\pm(1, 5, 7, 9, 13, 19, 25, 29, 33, 35, 41, 43, 45, 47, 49, 51, 59, 61,$ 63, 65, 69, 71, 73, 81, 91, 93, 95, 109, 117, 121, 125, 127, 131, 133, 137, 145)
−74	296	1, 3, 5, 9, 11, 13, 15, 23, 25, 27, 29, 31, 33, 39, 41, 45, 49, 55, 61, 65, 67, 69, 73, 75, 79, 81, 83, 87, 93, 99, 103, 107, 109, 115, 117, 119, 121, 123, 125, 133, 135, 137, 139, 143, 145, 147, 155, 165, 167, 169, 183, 191, 195, 199, 201, 205, 207, 211, 219, 225, 233, 237, 239, 243, 245, 249, 253, 261, 275, 277, 279, 289
77	154	$\pm(1, 9, 13, 15, 17, 19, 23, 25, 37, 41, 53, 61, 67, 71, 73)$
−77	308	1, 3, 9, 13, 17, 25, 27, 31, 37, 39, 41, 43, 47, 51, 53, 59, 61, 73, 75, 79, 81, 93, 95, 101, 103, 107, 111, 113, 115, 117, 123, 127, 129, 137, 141, 145, 151, 153, 159, 169, 173, 177, 183, 199, 211, 219, 221, 223, 225, 237, 239, 241, 243, 251, 263, 279, 285, 289, 293, 303
78	312	$\pm(1, 7, 11, 23, 25, 29, 31, 37, 41, 43, 49, 53, 59, 77, 83, 85, 89,$ 95, 101, 109, 121, 137, 139, 151)
−78	312	1, 19, 25, 29, 35, 37, 41, 47, 49, 53, 55, 67, 71, 77, 79, 85, 89, 101, 103, 107, 109, 115, 119, 121, 127, 131, 137, 155, 161, 163, 167, 173, 179, 187, 199, 215, 217, 229, 239, 251, 253, 269, 281, 289, 295, 301, 305, 307
79	316	$\pm(1, 3, 5, 7, 9, 13, 15, 21, 25, 27, 35, 39, 43, 45, 47, 49, 59, 63,$ 65, 71, 73, 75, 81, 89, 91, 97, 101, 103, 105, 107, 117, 121, 125, 127, 129, 135, 139, 141, 147)
−79	158	1, 5, 9, 11, 13, 19, 21, 23, 25, 31, 45, 49, 51, 55, 65, 67, 73, 81, 83, 87, 89, 95, 97, 99, 101, 105, 111, 115, 117, 119, 121, 123, 125, 129, 131, 141, 143, 151, 155
82	328	$\pm(1, 3, 9, 11, 13, 19, 23, 25, 27, 29, 31, 33, 35, 39, 49, 53, 57, 67,$ 69, 73, 75, 81, 85, 87, 93, 99, 101, 103, 105, 109, 113, 117, 119, 121, 127, 143, 147, 149, 157, 159)
−82	328	1, 7, 9, 13, 15, 25, 29, 33, 43, 47, 49, 51, 53, 55, 57, 59, 63, 69, 71, 73, 79, 81, 83, 85, 91, 93, 95, 101, 105, 107, 109, 111, 113, 115, 117, 121, 131, 135, 139, 149, 151, 155, 157, 163, 167, 169, 175, 181, 183, 185, 187, 191, 195, 199, 201, 203, 209, 225, 229, 231, 239, 241, 251, 253, 261, 263, 267, 283, 289, 291, 293, 297, 301, 305, 307, 309, 311, 317, 323, 325

a	k	l
Table 32. Quadratic Residues:		$(a/p) = +1$, if $p \equiv l \pmod{k}$
83	332	$\pm$(1, 9, 15, 17, 19, 21, 25, 29, 33, 35, 37, 39, 41, 43, 47, 49, 55, 61, 65, 67, 69, 71, 77, 79, 81, 91, 93, 103, 107, 109, 113, 115, 121, 135, 139, 143, 153, 155, 159, 161, 163)
—83	166	1, 3, 7, 9, 11, 17, 21, 23, 25, 27, 29, 31, 33, 37, 41, 49, 51, 59, 61, 63, 65, 69, 75, 77, 81, 87, 93, 95, 99, 109, 111, 113, 119, 121, 123, 127, 131, 147, 151, 153, 161
85	170	$\pm$(1, 3, 7, 9, 19, 21, 23, 27, 37, 49, 57, 59, 63, 69, 73, 81)
—85	340	1, 9, 11, 21, 31, 37, 39, 43, 47, 49, 57, 67, 69, 71, 73, 79, 81, 83, 87, 89, 91, 97, 99, 101, 103, 113, 121, 123, 127, 131, 133, 139, 149, 159, 161, 169, 173, 177, 183, 189, 193, 197, 199, 203, 211, 223, 229, 231, 233, 247, 263, 277, 279, 281, 287, 299, 307, 311, 313, 317, 321, 327, 333, 337
86	344	$\pm$(1, 5, 7, 9, 11, 17, 25, 29, 35, 37, 39, 41, 45, 49, 55, 57, 59, 61, 63, 67, 69, 71, 77, 81, 83, 85, 93, 97, 99, 107, 119, 121, 125, 139, 141, 145, 149, 151, 153, 157, 159, 169)
—86	344	1, 3, 5, 9, 15, 17, 19, 23, 25, 27, 29, 31, 37, 41, 45, 47, 49, 51, 57, 61, 69, 75, 77, 79, 81, 85, 87, 91, 93, 95, 97, 103, 111, 115, 121, 123, 125, 127, 131, 135, 141, 143, 145, 147, 149, 153, 155, 157, 163, 167, 169, 171, 179, 183, 185, 193, 205, 207, 211, 225, 227, 231, 235, 237, 239, 243, 245, 255, 261, 271, 273, 277, 279, 281, 285, 289, 291, 305, 309, 311, 323, 331, 333, 337
87	348	$\pm$(1, 13, 17, 19, 23, 25, 31, 35, 41, 43, 49, 55, 59, 71, 77, 79, 83, 89, 101, 107, 109, 113, 121, 127, 137, 163, 167, 169)
—87	174	1, 7, 11, 13, 17, 25, 41, 47, 49, 67, 77, 89, 91, 95, 101, 103, 109, 113, 115, 119, 121, 131, 137, 139, 143, 151, 155, 169
89	178	$\pm$(1, 5, 9, 11, 17, 21, 25, 39, 45, 47, 49, 53, 55, 57, 67, 69, 71, 73, 79, 81, 85, 87)
—89	356	1, 3, 5, 7, 9, 15, 17, 19, 21, 23, 25, 27, 31, 35, 43, 45, 49, 51, 53, 57, 59, 63, 69, 73, 75, 81, 83, 85, 93, 95, 97, 103, 105, 109, 115, 119, 121, 125, 127, 129, 133, 135, 143, 147, 151, 153, 155, 157, 159, 161, 163, 169, 171, 173, 175, 177, 189, 191, 207, 211, 215, 217, 219, 225, 233, 239, 243, 245, 249, 255, 257, 265, 269, 277, 279, 285, 289, 291, 295, 301, 309, 315, 317, 319, 323, 327, 343, 345
91	364	$\pm$(1, 3, 5, 9, 11, 15, 25, 27, 29, 33, 41, 45, 53, 55, 67, 71, 73, 75, 81, 87, 89, 97, 99, 103, 113, 121, 123, 125, 131, 135, 139, 145, 151, 159, 163, 165)
—91	182	1, 5, 9, 19, 23, 25, 29, 31, 33, 41, 43, 45, 47, 51, 53, 59, 73, 79, 81, 83, 89, 95, 97, 107, 111, 113, 115, 121, 125, 127, 145, 155, 165, 167, 171, 179

a	k	l
		Table 32. Quadratic Residues: $(a/p) = +1$, if $p \equiv l \pmod{k}$
93	186	$\pm(1, 7, 11, 17, 19, 23, 25, 29, 49, 53, 65, 67, 77, 83, 89)$
—93	372	1, 17, 25, 29, 35, 43, 47, 49, 53, 55, 59, 65, 71, 77, 79, 89, 91, 95, 97, 107, 109, 115, 121, 127, 131, 133, 137, 139, 143, 151, 157, 161, 169, 185, 191, 193, 197, 199, 205, 209, 223, 227, 247, 253, 259, 269, 271, 287, 289, 299, 305, 311, 331, 335, 349, 353, 359, 361, 365, 367
94	376	$\pm(1, 3, 5, 9, 13, 15, 17, 23, 25, 27, 29, 31, 39, 45, 49, 51, 59, 65, 69, 75, 77, 81, 83, 85, 87, 89, 93, 97, 109, 115, 117, 121, 125, 127, 131, 133, 135, 145, 147, 151, 153, 155, 167, 169, 177, 181)$
—94	376	1, 5, 7, 9, 11, 13, 17, 19, 25, 29, 35, 43, 45, 49, 55, 63, 65, 67, 69, 71, 77, 79, 81, 85, 89, 91, 93, 95, 97, 99, 103, 107, 109, 111, 117, 119, 121, 123, 125, 133, 139, 143, 145, 153, 159, 163, 169, 171, 175, 177, 179, 181, 183, 187, 191, 203, 209, 211, 215, 219, 221, 225, 227, 229, 239, 241, 245, 247, 249, 261, 263, 271, 275, 289, 293, 301, 303, 315, 317, 319, 323, 325, 335, 337, 339, 343, 345, 349, 353, 355, 361, 373
95	380	$\pm(1, 7, 9, 13, 23, 31, 33, 37, 43, 47, 49, 51, 53, 59, 61, 63, 71, 79, 81, 83, 87, 91, 97, 101, 113, 117, 121, 123, 149, 151, 161, 163, 169, 173, 179, 187)$
—95	190	1, 3, 9, 11, 13, 27, 33, 37, 39, 49, 53, 61, 67, 81, 97, 99, 101, 103, 107, 111, 113, 117, 119, 121, 127, 131, 139, 143, 147, 149, 159, 161, 167, 169, 173, 183
97	194	$\pm(1, 3, 9, 11, 25, 27, 31, 33, 35, 43, 47, 49, 53, 61, 65, 73, 75, 79, 81, 85, 89, 91, 93, 95)$
—97	388	1, 7, 9, 15, 19, 23, 25, 33, 39, 49, 51, 53, 55, 59, 61, 63, 65, 67, 71, 73, 81, 83, 85, 87, 89, 93, 101, 105, 107, 109, 111, 113, 121, 123, 127, 129, 131, 133, 135, 139, 141, 143, 145, 155, 161, 169, 171, 175, 179, 185, 187, 193, 197, 199, 205, 207, 211, 215, 221, 223, 225, 229, 231, 235, 237, 239, 241, 251, 263, 269, 271, 273, 285, 289, 293, 297, 309, 311, 313, 319, 331, 341, 343, 345, 347, 351, 353, 357, 359, 361, 367, 371, 375, 377, 383, 385
101	202	$\pm(1, 5, 9, 13, 17, 19, 21, 23, 25, 31, 33, 37, 43, 45, 47, 49, 65, 71, 77, 79, 81, 85, 87, 95, 97)$
—101	404	1, 3, 5, 7, 9, 11, 13, 15, 17, 21, 25, 27, 33, 35, 37, 39, 45, 49, 51, 55, 59, 63, 65, 67, 75, 77, 81, 83, 85, 91, 97, 99, 103, 105, 111, 117, 119, 121, 125, 127, 135, 137, 139, 143, 147, 151, 153, 157, 163, 165, 167, 169, 175, 177, 181, 185, 187, 189, 191, 193, 195, 197, 199, 201, 221, 225, 231, 233, 243, 245, 249, 255, 259, 263, 271, 273, 275, 281, 289, 291, 295, 297, 309, 311, 315, 317, 325, 331, 333, 335, 343, 347, 351, 357, 361, 363, 373, 375, 381, 385

Gauss' Formulas for Cyclotomic Polynomials. Let n be an odd, square-free integer > 3 and $\Phi_n(z)$ the nth cyclotomic polynomial, as defined on p. 320. Then $\Phi_n(z)$ is of degree $\varphi(n)$ and can be written in the form

$$4\Phi_n(z) = A_n^2(z) - (-1)^{(n-1)/2} nz^2 B_n^2(z),$$

where $A_n(z)$ and $B_n(z)$ have integer coefficients and are of degree $\varphi(n)/2$ and $\varphi(n)/2 - 2$, respectively. For example,

$$4\Phi_7(z) = 4(z^6 + z^5 + z^4 + z^3 + z^2 + z + 1) =$$
$$= (2z^3 + z^2 - z - 2)^2 + 7z^2(z+1)^2,$$

$$4\Phi_{21}(z) = 4(z^{12} - z^{11} + z^9 - z^8 + z^6 - z^4 + z^3 - z + 1) =$$
$$= (2z^6 - z^5 + 5z^4 - 7z^3 + 5z^2 - z + 2)^2 -$$
$$- 21z^2(z^4 - z^3 + z^2 - z + 1)^2.$$

$A_n(z)$ and $B_n(z)$ exhibit certain symmetric properties. Thus $A_n(z)$ is symmetric if its degree is even, otherwise it is anti-symmetric. $B_n(z)$ is symmetric in most cases. However, if n is composite and of the form $4k + 3$, then B_n is anti-symmetric. As a result of these properties only half of the coefficients of A_n and B_n need to be displayed in Table 33 below. The above examples are described by

$$7 \quad A{:}\ 2\ 1\ |\ -1\ -2 \qquad B{:}\ 1\ |\ 1$$
$$21 \quad A{:}\ 2\ -1\ 5\ -7^*\ 5\ldots \qquad B{:}\ 1\ -1\ 1^*\ -1\ 1$$

where a vertical bar is placed at the point of symmetry in the case of an even number of terms, and an asterisk indicates the symmetric term when the number of terms is odd. The terms following the points of symmetry serve as reminders of the change of sign for subsequent terms in the corresponding anti-symmetric cases.

Table 33. $4\Phi_n(z) = A_n^2(z) - (-1)^{(n-1)/2} nz^2 B_n^2(z)$	
n	Coefficients of $A_n(z)$ and $B_n(z)$
5	$A{:}\ 2\ 1^*\ 2$ $\qquad\qquad\qquad$ $B{:}\ 1$
7	$A{:}\ 2\ 1\ |\ -1\ -2$ $\qquad\qquad$ $B{:}\ 1\ |\ 1$
11	$A{:}\ 2\ 1\ -2\ |\ 2\ldots$ $\qquad\quad$ $B{:}\ 1\ 0\ |\ 0\ 1$
13	$A{:}\ 2\ 1\ 4\ -1^*\ 4\ldots$ $\qquad$ $B{:}\ 1\ 0\ 1^*\ 0\ 1$
15	$A{:}\ 2\ -1\ -4^*\ -1\ldots$ $\quad$ $B{:}\ 1\ 0^*\ -1$
17	$A{:}\ 2\ 1\ 5\ 7\ 4^*\ 7\ldots$ $\qquad$ $B{:}\ 1\ 1\ 1\ 2^*\ 1\ldots$
19	$A{:}\ 2\ 1\ -4\ 3\ 5\ |\ -5\ldots$ $\ $ $B{:}\ 1\ 0\ -1\ 1\ |\ 1\ldots$
21	$A{:}\ 2\ -1\ 5\ -7^*\ 5\ldots$ $\quad$ $B{:}\ 1\ -1\ 1^*\ -1\ 1$

n	Coefficients of $A_n(z)$ and $B_n(z)$

Table 33. $4\Phi_n(z) = A_n^2(z) - (-1)^{(n-1)/2} n z^2 B_n^2(z)$

n	Coefficients of $A_n(z)$ and $B_n(z)$
23	A: 2 1 -5 -8 -7 -4 | 4 ... B: 1 1 0 -1 -2 | -2 ...
29	A: 2 1 8 -31 -2 3 9* 3 ... B: 1 0 1 -1 0 1 1* 1 ...
31	A: 2 1 -7 -11 2 8 -3 -5 | 5 ... B: 1 1 -1 -2 0 1 -1 | -1 ...
33	A: 2 -1 8 5 2 14* 2 ... B: 1 0 1 2 0* 2 ...
35	A: 2 -1 -9 13 -5 -13 24* -13 ... B: 1 -1 -1 3 -3 0* 3 ...
37	A: 2 1 10 -4 15 -5 17 -8 11 -4* 11 ... B: 1 0 2 -1 3 -1 2 -1 2* -1 ...
39	A: 2 -1 -10 -4 11 5 -4* 5 ... B: 1 0 -2 -1 1 0* -1 ...
41	A: 2 1 11 16 14 29 30 22 36 34 20* 34 ... B: 1 1 2 4 3 4 6 4 4 6* 4 ...
43	A: 2 1 -10 6 16 -20 -4 27 -15 -7 17 | -17 ... B: 1 0 -2 2 2 -4 1 3 -3 1 | 1 ...
47	A: 2 1 -11 -17 -9 6 29 37 20 -2 -16 -11 | 11 ... B: 1 1 -1 -3 -4 -3 1 4 4 2 -1 | -1 ...
51	A: 2 -1 -13 20 5 -46 38 26 -64* 26 ... B: 1 -1 -2 5 -2 -6 8 0* -8 ...
53	A: 2 1 14 -6 8 -14 -4 19 -12 24 -9 -11 27 -25* 27 ... B: 1 0 2 -2 0 0 -1 4 -2 1 1 -3 5* -3 ...
55	A: 2 -1 -14 -7 30 22 -26 -29 13 15 -8* 15 ... B: 1 0 -3 -2 4 4 -3 -3 1 0* -1 ...
57	A: 2 -1 14 8 11 35 11 29 38 8* 38 ... B: 1 0 2 3 1 5 3 2 6* 2 ...
59	A: 2 1 -14 8 7 -35 22 12 -33 18 23 -29 14 18 -29 | 29 ... B: 1 0 -2 3 -1 -4 4 -1 -4 3 1 -4 4 -1 | -1 ...
61	A: 2 1 16 -7 32 -20 63 -33 72 -54 89 -62 88 -89 95 -81* 95 ... B: 1 0 3 -2 6 -3 9 -6 10 -7 12 -10 11 -11 13* -11 ...

Table 33. $4\Phi_n(z) = A_n^2(z) - (-1)^{(n-1)/2} n z^2 B_n^2(z)$	
n	Coefficients of $A_n(z)$ and $B_n(z)$
65	A: 2 $-$ 1 16 8 5 37 $-$ 6 6 43 $-$ 30 22 44 $-$ 34* 44 ...
	B: 1 0 2 3 $-$ 1 4 2 $-$ 3 6 0 $-$ 2 8* $-$ 2 ...
67	A: 2 1 $-$ 16 9 33 $-$ 44 $-$ 18 79 $-$ 39 $-$ 48 75 $-$ 35 $-$ 14 69 $-$ 89 10
	106 $\mid$ $-$ 106 ...
	B: 1 0 $-$ 3 3 4 $-$ 8 19 $-$ 8 $-$ 17 $-$ 8 5 5 $-$ 14 8 $\mid$ 8 ...
69	A: 2 $-$ 1 17 $-$ 25 35 $-$ 46 59 $-$ 40 29 $-$ 34 20 $-$ 7* 20 ...
	B: 1 $-$ 1 3 $-$ 5 6 $-$ 5 6 $-$ 5 2 $-$ 2 3* $-$ 2 ...
71	A: 2 1 $-$ 17 $-$ 26 $-$ 5 31 58 64 60 33 $-$ 29 $-$ 89 $-$ 106 $-$ 91 $-$ 70
	$-$44 $-$ 16 $-$ 3 $\mid$ 3 ...
	B: 1 1 $-$ 2 $-$ 5 $-$ 5 $-$ 2 2 6 11 13 9 2 $-$ 3 $-$ 6 $-$ 8 $-$ 8 $-$ 7 $\mid$ $-$ 7 ...
73	A: 2 1 19 28 61 106 158 251 322 442 544 652 783 868 983 1050 1113 1164
	1156* 1164 ...
	B: 1 1 4 7 20 27 40 50 64 78 89 104 113 124 131 134 138* 134 ...
77	A: 2 $-$ 1 19 $-$ 29 30 $-$ 56 35 23 $-$ 50 103 $-$ 141 94 $-$ 18 $-$ 57 145
	$-$189* 145 ...
	B: 1 $-$ 1 3 $-$ 6 4 $-$ 3 18 $-$ 13 13 $-$ 12 4 9 $-$ 17 19* $-$ 17 ...
79	A: 2 1 $-$ 19 $-$ 29 24 69 4 $-$ 64 19 75 $-$ 67 $-$ 125 81 166 $-$ 60 $-$ 137
	87 110 $-$ 152 $-$ 148 $\mid$ 148 ...
	B: 1 1 $-$ 3 $-$ 6 1 8 0 $-$ 7 5 11 $-$ 9 $-$ 17 8 18 $-$ 7 $-$ 13 14 14
	$-$18 $\mid$ $-$ 18 ...
83	A: 2 1 $-$ 20 11 20 $-$ 65 58 42 $-$ 137 115 47 $-$ 223 184 62 $-$ 290 273
	44 $-$ 343 319 7 $-$ 348 $\mid$ 348 ...
	B: 1 0 $-$ 3 4 $-$ 1 $-$ 8 12 $-$ 3 $-$ 13 21 $-$ 7 $-$ 20 30 $-$ 12 $-$ 23 40
	$-$17 $-$ 23 43 $-$ 22 $\mid$ $-$ 22 ...
85	A: 2 $-$ 1 21 $-$ 32 65 $-$ 98 159 $-$ 189 223 $-$ 250 287 $-$ 301 331 $-$ 377
	425 $-$ 453 459* $-$ 453 ...
	B: 1 $-$ 1 4 $-$ 7 12 $-$ 15 21 $-$ 25 28 $-$ 29 33 $-$ 37 41 $-$ 45 49 $-$ 51*
	49 ...
87	A: 2 $-$ 1 $-$ 22 $-$ 10 47 44 $-$ 7 $-$ 25 $-$ 52 $-$ 64 38 101 17 $-$ 46
	$-$46* $-$ 46 ...
	B: 1 0 $-$ 4 $-$ 3 4 5 3 2 $-$ 6 $-$ 10 1 9 4 0* $-$ 4 ...

	Table 33. $4\Phi_n(z) = A_n^2(z) - (-1)^{(n-1)/2}nz^2B_n^2(z)$
n	Coefficients of $A_n(z)$ and $B_n(z)$

n	Coefficients of $A_n(z)$ and $B_n(z)$
89	A: 2 1 23 34 52 118 123 146 237 209 203 303 257 230 368 332 266 410 364 242 365 333 198* 333 ...
	B: 1 1 4 8 8 15 20 17 25 29 21 29 36 26 34 44 30 34 44 27 27 40* 27 ...
91	A: 2 $-$1 $-$23 34 44 $-$133 14 233 $-$176 $-$226 342 100 $-$385 22 327 $-$51 $-$272 21 254* 21 ...
	B: 1 $-$1 $-$4 8 3 $-$20 9 26 $-$28 $-$18 40 3 $-$385 31 $-$4 $-$27 0* 27 ...
93	A: 2 $-$1 23 $-$34 59 $-$91 104 $-$85 41 11 $-$52 65 $-$34 $-$13 74 $-$109* 74 ...
	B: 1 $-$1 4 $-$7 9 $-$10 9 $-$5 $-$1 6 $-$7 4 3 $-$8 9* $-$8 ...
95	A: 2 $-$1 $-$24 $-$12 40 67 49 $-$24 $-$87 $-$95 $-$83 $-$16 61 88 100 62 4 $-$69 $-$122* $-$69 ...
	B: 1 0 $-$4 $-$4 1 7 10 5 $-$1 $-$7 $-$12 $-$9 $-$4 2 8 10 9 0* $-$9 ...
97	A: 2 1 25 37 93 171 278 466 643 936 1219 1556 1960 2295 2750 3111 3520 3930 4225 4632 4846 5114 5302 5347 5468* 5347 ...
	B: 1 1 5 9 17 30 44 69 92 125 160 194 239 274 319 358 394 437 462 498 518 534 551 546* 551 ...
101	A: 2 1 26 $-$12 40 $-$65 33 9 40 49 $-$110 $-$4 13 63 86 $-$166 $-$3 6 90 77 $-$185 19 53 76 66 $-$206* 66 ...
	B: 1 0 4 $-$4 3 $-$5 5 6 $-$3 0 $-$10 5 11 $-$4 0 $-$15 8 12 $-$5 $-$1 $-$13 11 14 $-$10 0* $-$10 ...
103	A: 2 1 $-$25 $-$38 44 122 26 $\quad$ 123 $-$96 $-$22 10 150 221 $-$45 $-$334 $-$213 93 269 293 82 $-$296 $-$441 $-$129 297 426 195 | $-$195
	B: 1 1 $-$4 $-$8 2 14 7 $-$6 $-$8 $-$10 $-$10 13 31 8 $-$25 $-$29 $-$11 13 36 30 $-$11 $-$43 $-$33 6 37 | 37 ...
105	A: 2 1 27 39 87 164 226 332 433 507 587 637 638* 637 ...
	B: 1 1 5 9 14 24 32 41 51 57 61 64* 61 ...
107	A: 2 1 $-$26 14 39 $-$104 83 100 $-$254 200 83 $-$412 372 12 $-$447 563 $-$120 $-$406 598 $-$285 $-$318 515 $-$253 $-$194 403 $-$78 $-$237 | 237 ...
	B: 1 0 $-$4 5 0 $-$13 18 $-$2 $-$22 35 $-$14 $-$28 48 $-$31 $-$21 58 $-$42 $-$6 53 $-$48 $-$1 37 $-$38 $-$3 32 $-$15 | $-$15 ...

Table 33. $4\Phi_n(z) = A_n^2(z) - (-1)^{(n-1)/2}nz^2B_n^2(z)$	
n	Coefficients of $A_n(z)$ and $B_n(z)$
109	A: 2 1 28 $-$ 13 84 $-$ 77 211 $-$ 170 372 $-$ 339 599 $-$ 543 883 $-$ 874 1236 $-$ 1225 1603 $-$ 1634 1966 $-$ 2004 2308 $-$ 2408 2631 $-$2689 2842 $-$ 2909 2947 $-$ 2931* 2947 ...
	B: 1 0 5 $-$ 4 13 $-$ 11 28 $-$ 24 45 $-$ 41 71 $-$ 67 100 $-$ 100 137 $-$ 137 170 $-$ 174 206 $-$ 212 236 $-$ 245 265 $-$ 270 277 $-$ 281 285* $-$281 ...
111	A: 2 $-$1 $-$ 28 $-$ 13 74 77 $-$ 49 $-$ 76 17 $-$ 4 $-$ 67 2 41 $-$ 37 $-$ 46 5 14 47 86* 47 ...
	B: 1 0 $-$ 5 $-$ 4 7 9 $-$ 2 $-$ 3 4 $-$ 1 $-$ 6 1 3 $-$ 4 $-$ 5 $-$ 4 $-$ 5 0* 5 ...
113	A: 2 1 29 43 80 185 179 255 345 194 286 273 78 301 262 193 473 376 353 537 436 431 583 504 437 539 370 292 424* 292 ...
	B: 1 1 5 10 11 23 25 21 34 20 15 30 11 22 39 23 40 47 33 46 51 41 50 53 35 40 38 22* 38 ...
115	A: 2 $-$1 $-$ 29 43 70 $-$ 203 19 391 $-$ 357 $-$ 290 692 $-$ 236 $-$ 499 698 $-$ 280 $-$ 398 869 $-$ 569 $-$ 462 1165 $-$ 648 $-$ 601 1246* $-$ 601 ...
	B: 1 $-$1 $-$ 5 10 5 $-$ 29 15 37 $-$ 52 $-$ 6 64 $-$ 49 $-$ 14 62 $-$ 62 5 75 $-$92 7 96 $-$ 101 0* 101 ...
119	A: 2 $-$1 $-$ 30 $-$ 15 65 105 49 $-$ 54 $-$ 127 $-$ 170 $-$ 169 $-$ 61 112 203 156 83 26 7 68 139 150 45 $-$ 134 $-$ 282 $-$ 336* $-$ 282 ...
	B: 1 0 $-$ 5 $-$ 5 3 11 12 7 0 $-$ 11 $-$ 21 $-$ 20 7 4 7 6 1 $-$ 2 3 14 25 26 16 0* $-$ 16 ...
123	A: 2 $-$1 $-$ 31 47 59 $-$ 226 158 293 $-$ 682 329 701 $-$ 1354 614 1142 $-$2137 1004 1469 $-$ 2794 1370 1559 $-$ 3046* 1559 ...
	B: 1 $-$1 $-$ 5 11 0 $-$ 28 37 8 $-$ 75 79 18 $-$ 136 136 21 $-$ 196 195 14 $-$234 233 0* $-$ 233 ...
127	A: 2 1 $-$ 31 $-$ 47 70 190 26 $-$ 248 $-$ 169 45 11 74 399 242 $-$ 447 $-$521 39 191 34 374 498 $-$ 294 $-$ 778 $-$ 231 258 136 319 513 $-$79 $-$ 636 $-$ 346 16 $\mid$ $-$ 16 ...
	B: 1 1 $-$ 5 $-$ 10 4 22 10 $-$ 15 $-$ 13 $-$ 5 $-$ 18 $-$ 3 44 39 $-$ 23 $-$ 39 $-$7 $-$ 12 $-$ 16 38 66 $-$ 2 $-$ 53 $-$ 26 $-$ 10 $-$ 15 25 63 16 $-$32 $-$ 24 $\mid$ $-$ 24 ...

	Table 33. $4\Phi_n(z) = A_n^2(z) - (-1)^{(n-1)/2}nz^2B_n^2(z)$
n	Coefficients of $A_n(z)$ and $B_n(z)$
129	A: 2 − 1 32 17 74 152 143 299 425 458 653 866 872 1091 1331 1277 1502 1658 1592 1754 1802 1712* 1802 …
	B: 1 0 5 6 8 21 23 31 50 57 64 90 97 101 129 130 134 154 150 152 162* 152 …
131	A: 2 1 − 32 17 64 − 152 79 195 − 333 144 294 − 422 65 390 − 438 22 357 − 374 124 169 − 297 206 112 − 460 412 16 − 468 414 37 − 386 306 94 − 411 | 411 …
	B: 1 0 − 5 6 2 − 19 19 5 − 32 28 12 − 39 24 20 − 42 23 10 − 28 23 −1 − 22 34 − 12 − 28 48 − 24 − 22 37 − 12 − 26 38 −17 | − 17 …
133	A: 2 − 1 33 − 50 135 − 238 420 − 600 783 − 982 1098 − 1139 1078 −987 940 − 827 817 − 792 870 − 909 850 − 776 576 − 393 139 100 − 216 311* − 216 …
	B: 1 − 1 6 − 11 22 − 34 52 − 71 84 − 94 97 − 96 89 − 76 73 − 69 74 − 74 75 − 76 66 − 54 31 − 11 − 6 22 − 23* 22 …
137	A: 2 1 35 52 114 267 279 445 597 404 603 532 213 600 422 344 847 536 582 783 486 631 763 940 1046 1184 1372 895 994 906 384 970 805 716 1326* 716 …
	B: 1 1 6 12 15 33 37 37 58 37 34 53 18 40 60 31 66 62 39 60 53 61 76 98 106 90 111 72 52 84 40 73 100 56* 100 …
139	A: 2 1 − 34 18 120 − 170 − 179 499 − 10 − 828 520 787 − 1047 −285 1140 − 288 − 696 461 120 − 88 81 − 467 286 727 −988 − 390 1494 − 387 − 1435 1051 869 − 1165 − 328 730 290 | − 290 …
	B: 1 0 − 6 6 14 − 27 − 11 58 − 20 − 74 69 49 − 99 2 82 − 38 −33 28 − 4 17 − 3 − 56 53 54 − 110 − 2 131 − 65 − 101 101 47 − 84 − 20 40 | 40 …
141	A: 2 − 1 35 − 52 125 − 226 338 − 382 503 − 535 542 − 595 749 −838 1019 − 1204 1295 − 1372 1466 − 1456 1451 − 1564 1547 − 1504* 1547 …
	B: 1 − 1 6 − 11 18 − 26 38 − 39 43 − 48 53 − 57 74 − 87 98 −111 118 − 118 124 − 129 124 − 129 134* − 129 …

Table 33. $4\Phi_n(z) = A_n^2(z) - (-1)^{(n-1)/2} n z^2 B_n^2(z)$	
n	Coefficients of $A_n(z)$ and $B_n(z)$
143	A: 2 -1 -36 -18 96 153 79 -132 -429 -429 0 508 747 361 -387 -703 -351 402 841 287 -704 -1144 -581 648 1307 724 -474 -1186 -662 529 1104* 529 ... B: 1 0 -6 -6 5 17 24 9 -29 -52 -38 9 59 55 1 -49 -52 11 77 66 -12 -93 -96 -5 86 94 12 -80 -83 0* 83 ...
145	A: 2 -1 36 18 120 172 309 546 728 1215 1532 2184 2826 3513 4495 5277 6414 7356 8428 9530 10442 11489 12291 13098 13745 14252 14669 14871 14968* 14871 ... B: 1 0 6 6 16 27 40 70 89 138 176 232 301 361 451 522 616 702 784 876 947 1025 1086 1141 1186 1215 1237 1242* 1237 ...
149	A: 2 1 38 -18 96 -152 109 -101 194 117 -90 -113 -40 -27 554 -279 51 -323 -45 462 -4 -160 -48 -481 631 -70 -40 38 -564 499 71 -120 273 -510 317 255 -388 475* -388 ... B: 1 0 6 -6 8 -15 13 4 9 2 -17 -10 27 -4 33 -39 -11 11 6 26 -12 -42 27 -19 40 -4 -42 24 -21 29 18 -43 28 -11 3 36 -49* 36 ...

Bibliography

1. C. F. Gauss *Untersuchungen über höhere Arithmetik,* Chelsea, New York, 1965, p. 425–428 (articles 356–357).

2. Maurice Kraïtchik, *Recherches sur la Théorie des Nombres,* Gauthiers-Villars, Paris, 1924, pp. 93–129.

3. Maurice Kraïtchik, *Recherches sur la Théorie des Nombres,* Tome II, *Factorization,* Gauthiers-Villars, Paris, 1929, pp. 1–5.

Lucas' Formulas for Cyclotomic Polynomials. Let n be a square-free integer ≥ 3 and $\Phi_n(z)$ the nth cyclotomic polynomial, as defined on p. 320. Then $\Phi_n(z)$ is of degree $\varphi(n)$ and can, for n odd, be written in the form

$$\Phi_n(z) = U_n^2(z) - (-1)^{(n-1)/2} n z V_n^2(z),$$

where $U_n(z)$ and $V_n(z)$ have integer coefficients and are of degree $\varphi(n)/2$ and $\varphi(n)/2 - 1$, respectively. Replacing z by $(-1)^{(n-1)/2} z$ we always get a minus sign in the right-hand-side:

$$\Phi_n\big((-1)^{(n-1)/2} z\big) = C_n^2(z) - n z D_n^2(z).$$

For n of the form $4k+2$ we get a minus sign in the right-hand-side by considering $\Phi_{n/2}(-z^2)$:

$$\Phi_{n/2}(-z^2) = C_n^2(z) - n z D_n^2(z).$$

In this case the degrees of $C_n(z)$ and $D_n(z)$ are $\varphi(n)$ and $\varphi(n) - 1$, respectively. For example,

$$\Phi_7(-z) = z^6 - z^5 + z^4 - z^3 + z^2 - z + 1 =$$
$$= (z^3 + 3z^2 + 3z + 1)^2 - 7z(z^2 + z + 1)^2,$$

$$\Phi_{21}(z) = z^{12} - z^{11} + z^9 - z^8 + z^6 - z^4 + z^3 - z + 1 =$$
$$= (z^6 + 10z^5 + 13z^4 + 7z^3 + 13z^2 + 10z + 2)^2 -$$
$$- 21z(z^5 + 3z^4 + 2z^3 + 2z^2 + 3z + 1)^2.$$

$$\Phi_{15}(-z^2) = \frac{(z^{30}+1)(z^2+1)}{(z^{10}+1)(z^6+1)} = z^{16} + z^{14} - z^{10} - z^8 - z^6 + z^2 + 1 =$$
$$= (z^8 + 15z^7 + 38z^6 + 45z^5 + 43z^4 + 45z^3 + 38z^2 + 15z + 1)^2 -$$
$$- 30z(z^7 + 5z^6 + 8z^5 + 8z^4 + 8z^3 + 8z^2 + 5z + 1)^2.$$

$C_n(z)$ and $D_n(z)$ are symmetric polynomials. Thus only half of the coefficients of C_n and D_n need to be displayed in Table 34 below. The above examples are described by

$$\begin{array}{lll} 7 & C: 1\ 3 \mid 3\ 1 & D: 1\ 1^*\ 1 \\ 21 & C: 1\ 10\ 13\ 7^*\ 13\ \ldots & D: 1\ 3\ 2 \mid 2\ 3\ 1 \\ 30 & C: 1\ 15\ 38\ 45\ 43^*\ 45\ \ldots & D: 1\ 5\ 8\ 8 \mid 8\ 8\ 5\ 1 \end{array}$$

where a vertical bar is placed at the point of symmetry in the case of an *even* number of terms, and an asterisk indicates the symmetric term when the number of terms is *odd*.

Table 34. $\Phi_n((-1)^{(n-1)/2}z)$ or $\Phi_{n/2}(-z^2) = C_n^2(z) - nzD_n^2(z)$

n	Coefficients of $U_n(z)$ and $V_n(z)$	
2	C: 1 | 1	D: 1
3	C: 1 | 1	D: 1
5	C: 1 3* 1	D: 1 | 1
6	C: 1 3* 1	D: 1 | 1
7	C: 1 3 | 3 1	D: 1 1* 1
10	C: 1 5 7* 5 1	D: 1 2 | 2 1
11	C: 1 5 −1 | −1 5 1	D: 1 1 −1* 1 1
13	C: 1 7 15 19* 15 7 1	D: 1 3 5 | 5 3 1
14	C: 1 7 3 −7* 3 7 1	D: 1 2 −1 | −1 2 1
15	C: 1 8 13* 8 1	D: 1 3 | 3 1
17	C: 1 9 11 −5 −15* −5 ...	D: 1 3 1 −3 | −3 ...
19	C: 1 9 17 27 31 | 31 ...	D: 1 3 5 7 7* 7 ...
21	C: 1 10 13 7* 13 ...	D: 1 3 2 | 2 3 1
22	C: 1 11 27 33 21 11* 21 ...	D: 1 4 7 6 3 | 3 ...
23	C: 1 11 9 −19 −15 25 | 25 ...	D: 1 3 −1 −5 1 7* 1 ...
26	C: 1 13 19 −13 −11 13 7* 13 ...	D: 1 4 1 −4 1 2 | 2 ...
29	C: 1 15 33 13 15 57 45 19* 45 ...	D: 1 5 5 1 7 1 15 | 5 ...
30	C: 1 15 38 45 43* 45 ...	D: 1 5 8 8 | 8 8 5 1
31	C: 1 15 43 83 125 151 169 173 | 173 ...	D: 1 5 11 19 25 29 31 31* 31 ...
33	C: 1 16 37 19 −32 −59* −32 ...	D: 1 5 6 −1 −9 | −9 ...
34	C: 1 17 59 119 181 221 243 255 257* 255 ...	D: 1 6 15 26 35 40 43 44 | 44 ...
35	C: 1 18 48 11 −55 −11 47* −11 ...	D: 1 6 7 −5 −8 5 | 5 ...
37	C: 1 19 79 183 285 349 397 477 579 627* 579 ...	D: 1 7 21 39 53 61 71 87 101 | 101 ...
38	C: 1 19 47 −19 −135 −57 179 209 −83 −285* −83 ...	D: 1 6 5 −14 −21 10 39 14 −37 | −37 ...
39	C: 1 20 73 119 142 173 193* 173 ...	D: 1 7 16 21 25 30 | 30 ...
41	C: 1 21 67 49 7 35 15 11 −23 −65 −31* −65 ...	D: 1 7 11 3 3 5 1 1 −9 −7 | −7 ...
42	C: 1 21 74 105 55 −42 −91* −42 ...	D: 1 7 15 14 1 −12 | −12 ...

Table 34. $\Phi_n((-1)^{(n-1)/2}z)$ or $\Phi_{n/2}(-z^2) = C_n^2(z) - nzD_n^2(z)$	
n	Coefficients of $U_n(z)$ and $V_n(z)$
43	C: 1 21 81 169 223 225 213 223 229 197 159 | 159 ...
	D: 1 7 19 31 35 33 33 35 33 27 23* 27 ...
46	C: 1 23 103 253 469 759 1131 1541 1917 2231 2463 2553* 2463 ...
	D: 1 8 25 52 89 138 197 256 307 348 373 | 373 ...
47	C: 1 23 65 $-$15 $-$169 $-$97 179 287 $-$37 $-$375 $-$149 311 | 311 ...
	D: 1 7 7 $-$15 $-$25 5 41 25 $-$37 $-$49 15 57* 15 ...
51	C: 1 26 121 245 334 431 529 548* 529 ...
	D: 1 9 26 41 53 68 77 75 | 75 ...
53	C: 1 27 113 103 $-$155 $-$219 263 513 $-$59 $-$465 75 551 93 $-$357* 93 ...
	D: 1 9 19 $-$1 $-$35 $-$3 67 41 $-$51 $-$39 57 57 $-$31 | $-$31 ...
55	C: 1 28 158 471 950 1419 1637 1472 1024 570 381* 570 ...
	D: 1 10 39 94 162 212 216 171 105 58 | 58 ...
57	C: 1 28 121 175 34 $-$125 $-$23 100 $-$95 $-$281* $-$95 ...
	D: 1 9 22 17 $-$9 $-$14 9 5 $-$30 | $-$30 ...
58	C: 1 29 159 435 729 841 799 899 1233 1421 1103 551 393 725 967* 725 ...
	D: 1 10 37 78 107 108 107 138 181 174 107 52 69 118 | 118 ...
59	C: 1 29 111 55 $-$85 47 11 53 131 $-$245 41 103 $-$111 227 $-$103 | $-$103 ...
	D: 1 9 15 $-$5 $-$5 9 $-$3 21 $-$9 $-$25 25 $-$11 9 19 $-$31* 19 ...
61	C: 1 31 191 637 1541 2979 4881 7029 9125 10953 12397 13511 14379 15053 15511 15667* 15511 ...
	D: 1 11 47 131 281 497 761 1037 1291 1501 1663 1789 1887 1961 2001 | 2001 ...
62	C: 1 31 139 93 $-$391 $-$589 331 1209 249 $-$1333 $-$841 837 913 $-$217 $-$385 $-$31* $-$385 ...
	D: 1 10 21 $-$14 $-$77 $-$32 117 124 $-$85 $-$178 7 146 43 $-$60 $-$19 | $-$19 ...
65	C: 1 32 138 69 $-$290 $-$79 582 133 $-$791 $-$145 921 22 $-$1057* 22 ...
	D: 1 10 19 $-$14 $-$38 37 67 $-$53 $-$86 67 89 $-$91 | $-$91 ...
66	C: 1 33 182 429 697 924 905 693 364 33 $-$73* 33 ...
	D: 1 11 37 69 102 117 100 67 22 $-$6 | $-$6 ...

Table 34. $\Phi_n((-1)^{(n-1)/2}z)$ or $\Phi_{n/2}(-z^2) = C_n^2(z) - nzD_n^2(z)$	
n	Coefficients of $U_n(z)$ and $V_n(z)$
67	C: 1 33 193 565 1055 1429 1599 1803 2225 2637 2617 2195 1869 1875 1865 1469 991 | 991 ...
	D: 1 11 43 99 155 187 205 243 301 329 297 243 225 233 209 147 111* 147 ...
69	C: 1 34 181 367 466 529 409 256 325 397 562 721* 562 ...
	D: 1 11 34 51 61 60 37 33 44 55 81 | 81 ...
70	C: 1 35 228 770 1798 3255 4911 6545 8065 9450 10629 11445 11737* 11445 ...
	D: 1 12 53 146 297 487 686 875 1049 1204 1326 1394 | 1394 ...
71	C: 1 35 169 155 — 109 233 597 39 101 445 163 293 89 — 203 249 — 49 —505 37 | 37 ...
	D: 1 11 25 1 — 5 63 43 — 9 43 37 21 35 — 19 1 29 — 47 — 35 23* —35 ...
73	C: 1 37 265 963 2257 3703 4313 3301 891 — 1823 — 3889 — 5149 —5777 — 5479 — 3633 — 205 3805 6933 8097* 6933 ...
	D: 1 13 63 181 353 489 471 259 — 59 — 347 — 539 — 649 — 677 — 559 —243 213 651 913 | 913 ...
74	C: 741 37 203 259 — 143 37 927 259 — 751 629 1279 — 777 — 639 1369 —33 — 1221 653 333 — 1145* 333 ...
	D: 1 12 33 10 — 25 62 99 — 54 — 35 158 41 — 144 65 128 — 127 — 44 113 — 78 | — 78 ...
77	C: 1 38 202 178 — 601 — 952 749 2129 — 102 — 2759 — 802 2434 1146 —1607 — 505 1253* — 505 ...
	D: 1 12 30 — 15 — 112 — 37 202 165 — 206 — 271 131 273 — 59 —176 82 | 82 ...
78	C: 1 39 254 663 853 468 — 37 39 532 663 173 — 468 — 743* — 468 ...
	D: 1 13 51 93 82 20 — 10 32 77 54 — 19 — 76 | — 76 ...
79	C: 1 39 267 923 2303 4745 8613 14173 21537 30733 41639 53901 66993 80339 93269 105099 115265 123343 128931 131767 | 131767 ...
	D: 1 13 59 169 379 729 1257 1983 2915 4049 5359 6793 8289 9777 11179 12423 13455 14229 14705 14865* 14705 ...

	Table 34. $\Phi_n((-1)^{(n-1)/2}z)$ or $\Phi_{n/2}(-z^2) = C_n^2(z) - nzD_n^2(z)$
n	Coefficients of $U_n(z)$ and $V_n(z)$
82	C: 1 41 307 1107 2445 3485 2903 451 — 1879 — 1271 2507 5699 4033 —1927 — 6161 — 3321 4733 10045 7035 — 1025 — 5279* —1025 ...
	D: 1 14 69 192 341 382 203 — 102 — 227 42 497 618 143 — 520 — 631 54 911 1060 343 — 456 | — 456 ...
83	C: 1 41 239 285 — 571 — 1337 29 2303 1467 — 2257 — 2591 1813 3329 —1525 — 4627 403 5583 1707 — 4945 — 3199 3893 | 3893 ...
	D: 1 13 37 — 3 — 123 — 105 143 269 — 41 — 351 — 67 377 153 — 437 —323 419 523 — 235 — 589 55 577* 55 ...
85	C: 1 42 308 1089 2540 4541 6922 9638 12479 14835 16081 16127 15338 14049 12495 11121 10557* 11121 ...
	D: 1 14 67 188 378 617 894 1201 1493 1694 1761 1715 1599 1441 1274 1161 | 1161 ...
86	C: 1 43 279 473 — 91 — 129 1459 1161 — 723 387 1859 1161 41 — 989 911 2709 — 1147 — 2709 1143 1505 — 1131 — 2021* —1131 ...
	D: 1 14 47 30 — 37 66 189 6 — 61 152 179 68 — 67 — 54 259 136 — 309 —98 223 14 — 197 | — 197 ...
87	C: 1 44 337 1049 1546 413 — 2657 — 5164 — 3593 2381 8284 8519 1879 —6850 — 10811* — 6850 ...
	D: 1 15 70 151 135 — 104 — 453 — 539 — 104 615 1001 630 — 287 —1047 | — 1047 ...
89	C: 1 45 323 729 471 59 373 429 875 683 — 655 285 529 115 807 — 813 —223 739 — 975 121 — 663 — 1213 633* — 1213 ...
	D: 1 15 59 75 19 19 47 59 111 — 11 — 47 79 9 65 17 — 111 79 — 23 —73 25 — 161 — 3 | — 3 ...
91	C: 1 46 398 1712 5027 11578 22484 38604 60518 88474 122293 161111 203238 246235 287132 322848 350578 368130 374137* 368130 ...
	D: 1 16 92 319 820 1722 3129 5117 7730 10974 14798 19068 23566 28004 32059 35418 37815 39062 | 39062 ...

Table 34. $\Phi_n((-1)^{(n-1)/2}z)$ or $\Phi_{n/2}(-z^2) = C_n^2(z) - nzD_n^2(z)$

n	Coefficients of $U_n(z)$ and $V_n(z)$
93	C: 1 46 337 913 1006 1 — 785 190 1555 889 — 956 — 851 1471 2458 409 —1217* 409 ...
	D: 1 15 64 113 63 — 58 — 53 105 158 — 9 — 133 22 245 175 —78 | — 78 ...
94	C: 1 47 399 1645 4577 9823 17375 26461 35645 43475 49159 52969 55921 58891 62139 65189 67549 69231 71159 74589 80033 86527 91867 93953* 91867 ...
	D: 1 16 89 294 711 1376 2249 3212 4105 4806 5285 5620 5917 6240 6856 6573 7059 7228 7491 7950 8589 9230 9635 | 9635 ...
95	C: 1 48 368 861 210 — 1221 262 2862 — 281 — 3800 1716 4758 — 3817 —4699 6320 3294 — 8738 — 1343 9439* — 1343 ...
	D: 1 16 67 78 — 64 — 101 198 215 — 296 — 203 491 118 — 672 88 777 —391 — 798 632 | 632 ...
97	C: 1 49 449 2007 5721 11483 16823 17887 13077 5803 1919 3009 3211 —4915 — 20741 — 33801 — 33645 — 20927 — 6763 — 691 —875 2557 15261 31337 38721* 31337 ...
	D: 1 17 103 361 857 1467 1837 1637 951 321 209 383 43 — 1255 — 2887 —3621 — 2873 — 1343 — 257 — 57 — 21 797 2399 3733 | 3733 ...
101	C: 1 51 417 1105 929 119 471 459 121 479 147 503 1067 1597 1457 1331 1261 3 853 1099 867 767 1289 1351 1245 2523* 1245 ...
	D: 1 17 77 119 45 17 63 17 33 35 19 83 131 167 123 151 53 21 121 89 85 87 157 99 205 | 205 ...
102	C: 1 51 434 1479 2569 2244 551 — 255 946 1887 329 — 1938 — 1409 1479 2486 153 — 1613* 153 ...
	D: 1 17 87 209 262 144 — 11 17 164 140 — 96 — 211 — 2 243 156 —110 | — 110 ...
103	C: 1 51 451 1873 4863 8591 10313 7323 673 — 3697 211 11049 18805 13567 — 3257 — 17867 — 16279 1771 21519 26327 12521 —7659 — 16905 — 7101 14633 31987 | 31987 ...
	D: 1 17 97 313 667 977 933 411 — 223 — 289 517 1585 1781 585 — 1173 —1925 — 851 1245 2605 2097 205 — 1429 — 1407 317 2453 3401* 2453 ...

\multicolumn{2}{c}{**Table 34.** $\Phi_n((-1)^{(n-1)/2}z)$ or $\Phi_{n/2}(-z^2) = C_n^2(z) - nzD_n^2(z)$}	
n	Coefficients of $U_n(z)$ and $V_n(z)$

Table 34. $\Phi_n((-1)^{(n-1)/2}z)$ or $\Phi_{n/2}(-z^2) = C_n^2(z) - nzD_n^2(z)$

n	Coefficients of $U_n(z)$ and $V_n(z)$
105	C: 1 53 486 1857 3981 5542 5363 3421 269 $-$ 3264 $-$ 5849 $-$ 6769 -6821^* $-$ 6769 ...
	D: 1 18 101 282 481 556 447 191 $-$ 149 $-$ 464 $-$ 634 $-$ 666 $\mid$ $-$ 666 ...
106	C: 1 53 503 2279 6897 16377 33287 60579 100757 155025 223243 304167 395325 493165 594147 695943 797485 898297 998243 1097471 1195649 1290709 1378615 1454479 1513561 1551469 1564599* 1551469 ...
	D: 1 18 111 400 1057 2304 4415 7668 12249 18206 25471 33866 43091 52784 62651 72534 82365 92114 101785 111384 120807 129750 137763 144372 149133 151646 $\mid$ 151646 ...
107	C: 1 53 415 849 $-$ 569 $-$ 3051 95 6905 1585 $-$ 11549 $-$ 3329 16577 3289 $-$ 22583 $-$ 919 29411 $-$ 3591 $-$ 36177 9501 41177 -16637 $-$ 43539 24965 43967 $-$ 32915 $-$ 42547 39097 $\mid$ 39097 ...
	D: 1 17 69 47 $-$ 197 $-$ 243 383 623 $-$ 565 $-$ 1077 801 1469 $-$ 1235 -1775 1893 1973 $-$ 2703 $-$ 2033 3509 1865 $-$ 4225 $-$ 1433 4859 871 $-$ 5315 $-$ 281 5497* $-$ 281 ...
109	C: 1 55 559 2773 9307 24541 54987 109351 197803 330591 516469 760821 1063573 1417653 1808601 2215303 2611733 2970087 3264879 3476861 3596301 3625117 3577115 3475919 3351209 3234047 3151519 3121875* 3151519 ...
	D: 1 19 127 505 1483 3579 7525 14239 24717 39889 60457 86699 118285 154181 192669 231463 267939 299495 323945 339867 346871 345721 338247 327053 315109 305265 299741 $\mid$ 299741 ...
110	C: 1 55 468 1210 488 $-$ 1925 $-$ 2169 440 2710 1430 $-$ 2541 $-$ 2090 1417 1870 482 $-$ 1155 $-$ 1066 275 $-$ 90 110 1041* 110 ...
	D: 1 18 83 111 $-$ 68 $-$ 240 $-$ 99 174 254 $-$ 62 $-$ 298 $-$ 18 197 120 -37 $-$ 138 $-$ 28 26 $-$ 24 70 $\mid$ 70 ...
111	C: 1 56 541 2171 5032 8231 10243 9584 6835 3665 1354 1109 2407 4514 6859 7823 7444 6359 5593* 6359 ...
	D: 1 19 112 331 633 902 975 793 496 219 91 156 319 547 718 735 661 552 $\mid$ 552 ...

Table 34. $\Phi_n((-1)^{(n-1)/2}z)$ or $\Phi_{n/2}(-z^2) = C_n^2(z) - nzD_n^2(z)$	
n	Coefficients of $U_n(z)$ and $V_n(z)$
113	C: 1 57 523 1547 831 — 3467 — 3809 6851 12207 — 4875 — 18615 1323 23091 5631 — 18175 — 5045 13175 75 — 10615 8009 15473 —11141 — 23043 7145 25813 — 3403 — 26935 — 645 24281* —645 ...
	D: 1 19 97 155 — 101 — 463 51 1123 561 — 1417 — 1169 1485 1787 —899 — 1497 675 921 — 835 — 333 1543 421 — 2131 — 1053 2095 1457 — 1947 — 1759 1601 | 1601 ...
114	C: 1 57 542 2109 4909 9120 15443 23655 33550 45771 59633 73986 88783 103227 115838 126597 135451 140790 142325* 140790 ...
	D: 1 19 109 315 636 1124 1813 2655 3687 4926 6253 7618 9006 10284 11371 12302 12985 13296 | 13296 ...
115	C: 1 58 618 3141 10270 24539 45722 69092 87049 93640 88321 75788 62713 54621 54485 62069 73632 83487 87629 86415 83281 81088 80433* 81088 ...
	D: 1 20 139 554 1532 3213 5370 7387 8574 8604 7697 6427 5389 4988 5359 6313 7377 8053 8160 7913 7645 7515 | 7515 ...
118	C: 1 59 619 3009 8961 17995 24599 20237 1685 — 21889 — 33465 —22951 2725 25783 33211 25429 11813 1475 — 1897 — 531 41 —3481 — 6977 — 1711 14757 31683 33031 13865 — 12375 —24485* — 12375 ...
	D: 1 20 135 504 1225 2038 2231 1148 — 967 — 2776 — 2849 — 1010 1443 2922 2829 1720 539 — 98 — 137 18 — 117 — 532 — 541 506 2231 3226 2353 24 — 1959 | — 1959 ...
119	C: 1 60 580 1852 1877 — 112 1274 4549 2852 1275 — 144 — 743 5649 3290 — 4563 — 220 2393 2555 3009 — 4651 — 4090 3726 528 —4381 — 4333* — 4381 ...
	D: 1 20 108 203 80 — 10 302 387 155 84 — 117 200 593 — 154 — 321 197 180 334 — 36 — 587 43 321 — 249 — 414 | — 414 ...

Bibliography

1. Maurice Kraïtchik, *Recherches sur la Théorie des Nombres*, Gauthiers-Villars, Paris, 1924, pp. 87–88.

INDEX

460